MEASURES OF INFORMATION AND THEIR APPLICATIONS

MEASURES OF INFORMATION AND THEIR APPLICATIONS

J.N. KAPUR

JOHN WILEY & SONS
NEW YORK • CHICHESTER • BRISBANE • TORONTO • SINGAPORE

First Published in 1994 by
WILEY EASTERN LIMITED
4835/24 Ansari Road, Daryaganj
New Delhi 110 002, India

Distributors:

Australia and New Zealand:
JACARANDA WILEY LIMITED
PO Box 1226, Milton Old 4046, Australia

Canada:
JOHN WILEY & SONS CANADA LIMITED
22 Worcester Road, Rexdale, Ontario, Canada

Europe and Africa:
JOHN WILEY & SONS LIMITED
Baffins Lane, Chichester, West Sussex, England

South East Asia:
JOHN WILEY & SONS (PTE) LIMITED
05-04, Block B, Union Industrial Building
37 Jalan Pemimpin, Singapore 2057

Africa and South Asia:
WILEY EASTERN LIMITED
4835/24 Ansari Road, Daryaganj
New Delhi 110 002, India

North and South America and rest of the World:
JOHN WILEY & SONS INC.
605, Third Avenue, New York, NY 10158, USA

Library of Congress Cataloging-in-Publication Data

ISBN 0-470-22064-3 John Wiley & Sons Inc.
ISBN 81-224-0484-7 Wiley Eastern Limited

Typeset by Printek India and Printed in India at
Baba Barkha Nath Printers, New Delhi, India

PREFACE

My interest in Measures of Information started in 1965 i.e. 17 years after the appearance of Shannon's epoch-making paper in which he gave his measure of entropy $\sum_{i=1}^{n} p_i \ln p_i$ and which laid the foundations of the development of Information Theory and its great applications, and 4 years after the appearance of Renyis' paper on Measures of Information and Entropy in which he gave the measures of entropy and information containing a parameter α. I spent the next 4 years in developing generalised measures of entropy and directed divergence and in studying the possible applications of these generalised measures from the point of view of Communication Theory. However I could not find any serious applications of these generalised measures there and I accordingly shifted my interest to other areas of applied mathematics.

My interest was revived in 1980, when I listened to a lecture on Applications of Maximum Entropy Principle to Marketing at University of Manitoba where I was working as a Visiting Professor. I began an extensive search for literature on the applications of this principle and was fascinated by the diversity of these applications in at least two dozen different fields in journals devoted to Physics, Chemistry, Statistics, Mathematics, Business, Environment, Planning, Geography, Literature, Marketing, Elections, Electrical, Mechanical and Civil Engineering, Computer Science and so on. I found different scientists using different terminologies sometimes not aware of each other's work and very often rediscovering the same results independently. This was happening because there was no standard book on the subject and papers on applications of maximum entropy principle were spread over many dozens of journals.

I therefore, decided to write a comprehensive book and spent the next five years teaching courses on the subject, filling in some gaps in knowledge and writing the book. The book on *Maximum Entropy Models in Science and Engineering* was ready by 1986, though it was published in 1989 by Wiley Eastern, New Delhi and John Wiley, New York. It contained chapters on Maximum Entropy Principle, Maximum Entropy discrete and continuous univariate and multi-variate probability distributions. Applications to Statistical Mechanics, Thermo-Dynamics Contingency Tables, Statistics, and Regional and Urban Planning, Marketing, Elections, Economics, Finance, Insurance Spectral Analysis, Image Reconstruction, Pattern Recognition, Biology, Medicine and Agriculture.

The book essentially dealt with applications of Jaynes' maximum entropy principle based on Shannon's measure of entropy. It also gave some applications of Kullback's principle of minimum discrimination information based on Kullback-Leibler measure of directed divergence. It made a reference to the existence of other measures of entropy and directed divergence, but did not make use of them.

However, soon after finalisation of this book, I realised that the use of only Shannon's and Kullback-Liebler Measures unnecessarily restricted the scope of applications or the principles of maximum entropy and minimum cross-entropy and in order to increase the scope of the applications of these principles, we need the flexibility which can be provided by other parametric measures of entropy and cross-entropy. I also realised that while finding maximum entropy distribution for given moment constraints was important, the inverse problem of finding moments constraints for a given maximum entropy probability distribution is also significant. I discussed all these problems with my friend Prof. H.K. Kesavan of Waterloo University when I visited that university every year as a Visiting Professor from 1981 onwards. Our discussions resulted in a joint monograph on *Generalised Maximum entropy principle with applications* which was published by Sand-Ford Educational Press of the University of Waterloo, Canada in 1990.

As we continued our discussions, we realised that we could have many other entropy optimisation principles, apart from Jayne's and Kullback's principles, since we had to deal with the concepts of entropy, cross-entropy and inter-dependence and since we could maximise or minimise any one of these and since for every principle, we could associate an inverse principle and a dual principle. Accordingly we wrote a joint book on *Entropy Optimisation Principles and their Applications* which was published by Academic Press, San Diago in 1992.

During all these years I had an opportunity to give a very large number of seminars on Maximum Entropy Principle and its applications at about 50 universities all over the world and had a chance to discuss with many distinguished scientists and engineers in a dozen universities when I visited these as a Visiting Professor in India, USA, Canada and Australia. I found that there were a large number of misunderstandings about both the original principle and the generalisations being introduced and these misunderstandings were preventing many scientists and engineers from exploiting the full potential of the powerful principles. I accordingly wrote my four books on *Insight into Entropy Optimisation Principles* in the form of four dialogues to clarify all the issues which arose. The book was published by Mathematical Sciences Trust Society.

I went on publishing the results of my research investigations in the form of research papers in various journals and more than one hundred papers were published till 1989. Later on I found that the impact of these papers was greatly diffused as these were published in different journals all over the world and few persons could have access to all these journals. I therefore decided to publish my later investigation in the form of books and the present book represents four years effort of my research work. Different chapters represent different research papers on related topics by same investigator. The papers can be read independently and they can be read in any order. The aim is to answer all the questions which arose in my mind or minds of my friends, as we pondered over the various measures of information and their applications.

I hope this book will be found useful by all those who are interested in using entropy optimisation principles to solve problems in science and technology. It is also hoped that it will lead to further investigation by other mathematicians, statisticians, scientists, engineers and others and with the co-operation of all, these powerful principles will play their destined role in the development of science and technology.

I am grateful to my students Anju and Sandeep for the help in proof reading and to Shri H.S. Poplai and his colleagues for help at all stages in the course of printing.

J.N. KAPUR

C-766, New Friends Colony,
NEW DELHI - 110 065.

CONTENTS

PART C–APPLICATIONS OF MEASURES OF INFORMATION AND ENTROPY OPTIMIZATION PRINCIPLE

PART D–INFORMATION THEORY AND ENTROPY OPTIMIZATION

Part A

Measures of Information

1

AN UNORTHODOX MEASURE OF ENTROPY

[The properties of the entropy measure $-\ln(\max(p_1, p_2, \ldots, p_n))$ and its use in entropy maximisation problems are discussed.]

1.1 INTRODUCTION

The measure of entropy we consider for probability distribution, $P = (p_1, p_2, \ldots, p_n)$ is

$$S(P) = -\ln p_{\max}, \tag{1}$$

where
$$p_{\max} = \max(p_1, p_2, \ldots, p_n) \tag{2}$$

It is easily seen to have the following properties:

(i) It is a continuous function of $p_1, p_2, \ldots, p_n$ since if $p_1, p_2, \ldots, p_n$ change by small amounts then their maximum value changes by a small amount.

(ii) It is a permutationally symmetric function of $p_1, p_2, \ldots, p_n$ since if $p_1, p_2, \ldots, p_n$ are permuted among themselves, $p_{\max}$ does not change.

(iii) Since maximizing it means minimizing $p_{\max}$, it is maximum subject to $\sum_{i=1}^{n} p_i = 1$, when $p_1 = p_2 = \ldots = p_n = 1/n$ and its maximum value is $\ln n$ which increases with n.

(iv) It is always ≥ 0 and its minimum value zero occurs for each of the n degenerate distributions.

(v) If P and Q are two independent probability distributions, then since

$$\max_{ij}(p_i q_j) = \left(\max_i p_i\right)\left(\max_j q_j\right), \tag{3}$$

we get,

$$S(P*Q) = S(P) + S(Q), \tag{4}$$

so that the measure is additive.

(vi) It is neither sub-additive nor super-additive.
Thus, for

.1	.2	.3
.3	.4	.7
.4	.6	1.0

$S(P*Q) = -\ln .4 = \ln 2.5$
$S(P) + S(Q) = -\ln .7 - \ln .6$
$= \ln 2.4328$

while for

.2	.1	.3	$S(P * Q) = \ln 2.5$
.3	.4	.7	$S(P) + S(Q) = -\ln .7 - \ln .5$
.5	.5	1.0	$= \ln 2.857$

(vii) $S(P)$ is not differentiable everywhere.
Consider the probability distribution $(p, 1-p)$

$$S(P) = -\ln(1-p), \quad 0 \le p \le \frac{1}{2}$$

$$= -\ln p, \quad \frac{1}{2} \le p \le 1 \tag{5}$$

As p increases from 0 to 1/2, $S(P)$ increases from 0 to ln2 and as p increases from 1/2 to 1, $S(P)$ decreases from ln2 to 0.

However,

$$S'\left(\frac{1}{2}-0\right) = 2, \quad S'\left(\frac{1}{2}+0\right) = -2 \tag{6}$$

$\therefore S(P)$ is not differentiable at $p = 1/2$. Its graph is given in Figure 1.1.

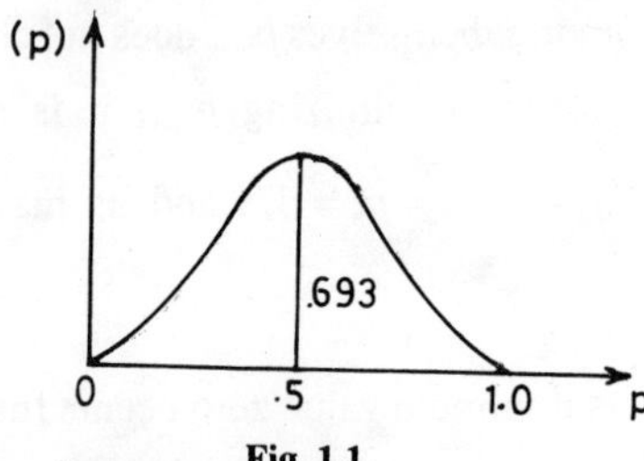

Fig. 1.1

Similarly, if the probability distribution is $(p, q, 1-p-q)$, $S(P)$ is not differentiable when $p = q$ or $p = 1-p-q$ or $q = 1-p-q$ or when $p = q = 1-p-q$. However, for the surface defined by (7) (Figure 1.2), at each point within the six regions the surface function is continuous and differentiable, but for the point on the boundaries, though it is continuous, it is not differentiable.

$S(P) = -\ln p$ in regions I and II

$S(P) = -\ln q$ in regions III and IV

$S(P) = -\ln(1-p-q)$ in regions V and VI $\quad$ (7)

It is obvious that when $S(P)$ is maximized subject to the above constraints, there will be one global maximum.

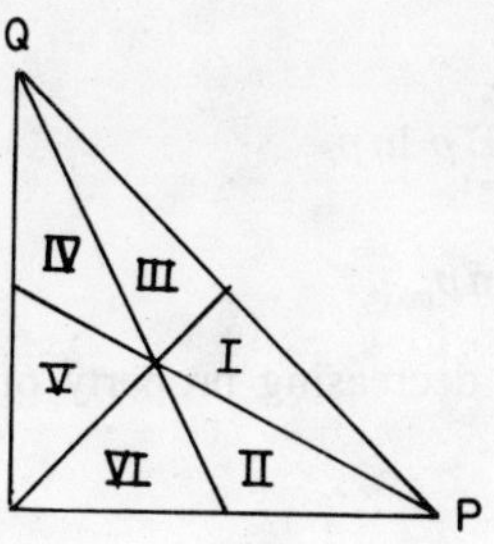

Fig. 1.2

(viii) Since

$$\min_i \left(\frac{\lambda}{p_i} + \frac{1-\lambda}{q_i} \right) \geq \lambda \min_i \frac{1}{p_i} + (1-\lambda) \min_i \frac{1}{q_i}, 0 \leq \lambda \leq 1, \tag{8}$$

$\min_i \frac{1}{p_i}$ is a concave function of $p_1, p_2, \ldots, p_n$ or $\frac{1}{p_{\max}}$ is a concave function of $p_1, p_2, \ldots, p_n$ or $-\ln p_{\max}$ is a concave function of $p_1, p_2, \ldots, p_n$.

1.2 RELATIONSHIP WITH RENYI'S ENTROPY MEASURE

Renyi's [6] entropy is defined by

$$R_\alpha(P) = \frac{1}{1-\alpha} \ln \sum_{i=1}^{n} p_i^\alpha, \ \alpha \neq 1, \ \alpha > 0 \tag{9}$$

It is wellknown that for fixed P, $R_\alpha(P)$ is a monotonic decreasing function of α. Also,

$$\underset{\alpha \to \infty}{Lt} \frac{1}{1-\alpha} \ln \sum_{i=1}^{n} p_i^\alpha = -\underset{\alpha \to \infty}{Lt} \frac{\sum_{i=1}^{n} p_i^\alpha \ln p_i}{\sum_{i=1}^{n} p_i^\alpha}$$

$$= -\underset{\alpha \to \infty}{Lt} \frac{\sum_{i=1}^{n} \left(\frac{p_i}{p_{\max}} \right)^\alpha \ln p_i}{\sum_{i=1}^{n} \left(\frac{p_i}{p_{\max}} \right)^\alpha} \tag{10}$$

As $\alpha \to \infty, \left(\frac{p_i}{p_{\max}} \right)^\alpha \to 0$ except when$p_i = p_{\max}$, so that,

$$\underset{\alpha \to \infty}{Lt} \frac{1}{1-\alpha} \ln \sum_{i=1}^{n} p_i^\alpha = -\ln p_{\max} \tag{11}$$

This proof will hold even if two or more largest probabilities are equal.

When $\alpha = 0, \quad R_0(P) = \ln n$

$$\text{As} \quad \alpha \to 1, \quad R_\alpha(P) \to -\sum_{i=1}^{n} p_i \ln p_i \tag{12}$$

$$\text{As} \quad \alpha \to \infty, \quad R_\alpha(P) \to -\ln p_{\max},$$

so that from the monotonic decreasing property of Renyi's entropy, we get the inequality

$$\ln n \geq -\sum_{i=1}^{n} p_i \ln p_i \geq -\ln p_{\max} \tag{13}$$

and the equality sign will hold only if P coincides with the uniform distribution.

1.3 APPLICATION VIA GENERALISED MAXIMUM ENTROPY PRINCIPLE

Suppose we are given the equations,

$$\sum_{i=1}^{n} p_i = 1, p_1 \geq 0, p_2 \geq 0 \ldots p_n \geq 0 \quad \sum_{i=1}^{n} p_i g_{ri} = a_r,$$

$$r = 1, 2, \ldots m\,; \quad m+1 < n \tag{14}$$

These equations are not sufficient to determine $p_1, p_2, \ldots, p_n$ uniquely, then according to Kapur & Kesavan's [2, 3] generalised maximum entropy principle, we should choose $p_1, p_2, \ldots, p_n$ so as to maximize a general measure of entropy subject to (14).

In our case, we choose the general measure as S (P), so that our problem is,

$$\text{maximize} -\ln p_{\max} \text{ or maximize} \ln (1/p_{\max}) \tag{15}$$

or, maximize $(1/p_{\max})$ or minimize $p_{\max}$

subject to (14).

There may be an infinity of probability distributions consistent with (15). Each will have a largest component. We have to choose that whose largest component is smallest. In some sense, the generalised maximum entropy principle becomes a minimax probability principle.

1.4 AN ILLUSTRATIVE EXAMPLE : MEPD FOR n = 3

We consider the simple case when $n = 3$ and the constraints are

$$p_1 + p_2 + p_3 = 1, p_1 + 2p_2 + 3p_3 = m$$

$$p_1 \geq 0, p_2 \geq 0, p_3 \geq 0 \tag{16}$$

When $m = 1$, the only probability distribution which satisfies the constraints (16) is (1, 0, 0) and this is obviously the MEPD. In this case $p_{\max} = 1$ and max S (P) $= 0$.

For any other value of *m*,

$$p_1 + 2p_2 + 3(1 - p_1 - p_2) = m$$

$$2\,p_1 + p_2 = 3 - m \quad \text{or} \quad p_2 = 3 - m - 2p_1, \tag{17}$$

so that

$$P = (p_1, 3 - m - 2p_1, m - 2 + p_1) \tag{18}$$

For $m = 1.1, P = (p_1, 1.9 - 2p_1, p_1 - .9), p_1 \geq .9, p_2 \leq .95$, MEPD is (.9, .1, 0)
For $m = 1.2, P = (p_1, 1.8 - 2p_1, p_1 - .8), p_1 \geq .8, p_2 \leq .9$, MEPD is (.8, .2, 0) Proceeding in the same way we get the values given in Table 1.1 below.

Table 1.1

MEPD, S_{max}, S_{min} and $S_{max} - S_{min}$ for various values of m.

m	MEPD	S_{max}	Min EPD	S_{min}	$S_{max} - S_{min}$
1.0	(1, 0, 0)	0	(1, 0, 0)	0	0
1.1	(.9, .1, 0)	.1054	(.95, 0, .05)	.0513	.0514
1.2	(.8, .2, 0)	.2231	(.90, 0, .10)	.1054	.177
1.3	(.7, .3, 0)	.3967	(.85, 0, .15)	.1625	.1842
1.4	(.6, .4, 0)	.5106	(.80, 0, .20)	.2231	.2875
1.5	(.5, .5, 0)	.6931	(.75, 0, .25)	.2877	.4054
1.6	(14/30, 14/30, 2/30)	.7421	(.70, 0 0.5)	.3567	.4795
1.7	(13/30, 13/30, 9/30)	.8362	(.30, .70, 0)	.3567	.4795
1.8	(12/30, 12/30, 6/30)	.9163	(.2, .8, 0)	.2231	.6932
1.9	(11/30, 11/30, 8/30)	1.003	(.1, .9, 0)	.1054	.8979
2.0	(10/30, 10/30, 10/30)	1.0986	(0, 1, 0)	0	1.0986
2.1	(8/30, 11/30, 11/30)	1.0033	(0, .9, .1)	.1054	.8975
2.2	(6/30, 12/30, 12/30)	.9163	(0, .8, .2)	(.2231)	.6932
2.3	(4/30, 13/30, 13/30)	.8362	(0, .7, .3)	(.3567)	.4755
2.4	(2/30, 14/30, 14/30)	.7421	(.3, 0, .7)	(.2577)	.4054
2.5	(0, .5, .5)	.6931	(.25, 0, .75)	.2577	.4054
2.6	(0, .4, .6)	.5106	(.20, 0, .80)	.2231	.2875
2.7	(0, .3, .7)	.3967	(.19, 0,.814)	.1625	.1842
2,8	(0, .2, .8)	.2237	(.10, 0, .90)	.1054	.1077
2.9	(0, .1, .9)	.1054	(.05, 0, .95)	.0513	.0513
3.0	(0, 0, 1)	0	(0, 0, 1)	0	0

It will be noted that,

(i) MEPD for the value $4 - m$ is the same as the MEPD for value m, but it is in reverse order since

$$p_1 + 2p_2 - 3p_3 = m \Rightarrow 3p_1 + 2p_2 + p_3 = 4(p_1 + p_2 + p_3) - (p_1 + 2p_2 + 3p_3)$$

$$= 4 - m \tag{19}$$

(ii) The value of S_{max} for m is the value for S_{max} for $4 - m$, so that the curve for S_{max} against m is symmetrical about $m = 2$. This is illustrated in Figure 1.3.

(iii) Either one of the probabilities is zero or two of the probabilities in the MEPD are equal.

1.5. MINIMUM ENTROPY PROBABILITY DISTRIBUTION

For calculating minimum entropy probability distribution and S_{min}, we note that since there are only two constraints, one of the probabilities should be zero and the other two probabilities should be as equal as possible.

If the non-zero components are p_1, p_3 we get

$$p_1(m-1) = p_3(3-m) = (1-p_1)(3-m)$$

or

$$p_1 = \frac{3-m}{2}, \quad p_3 = \frac{m-1}{2} \tag{20}$$

If the non-zero probabilities are p_1 and p_2, we get

$$p_1(m-1) = (1-p_1)(2-m) \quad \text{or} \quad p_1 = 2-m, p_2 = m-1 \tag{21}$$

In the first case, the larger component is $(3 - m)/2$ and in the second case it is $m - 1$ and these two are equal if, $3 - m = 2m - 2$ or $m = 5/3$.
At $m = 5/3$, we can take either (2/3, 0, 2/3) or (1/3, 2/3, 0). These give us the minimum entropy probability distributions. The values of S_{min} are given in Table 1.1
The following additional values may be noted.

m	Min EPD	S_{min}
5/3	(2/3, 0, 1/3) or (1/3, 2/3, 0)	.4055
7/3	(0, 2/3, 1/3) or (1/3, 0, 2/3)	.4055

Just like maximum entropy probability distribution, minimum entropy probability distribution is also symmetric about $m = 2$ and the graph of S_{min} against m is also symmetrical about $m = 2$. However, while S_{max} has one maximum value at $m = 2$, S_{min} has two maxima at $m = 5/3$ and 7/3, and these maximum value are equal. Figure 1.3 gives the graphs of S_{max}, S_{min} and $S_{max} - S_{min}$ against m.

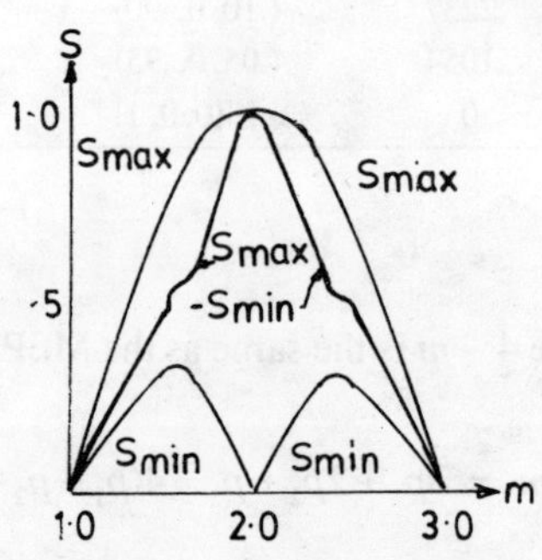

Fig. 1.3

1.6 CONCLUDING REMARKS

(a) On the basis of the above results, we make the following conjectures:

(i) For any value of n, S_{max} is a concave function with its maximum at $(n + 1)/2$ and maximum value of S_{max} is $\ln n$.

(ii) The minimum entropy curve is piece-wise concave, being concave between 1 and 2, 2 and 3, ... , $n - 1$ and n. In each sub-interval S_{min} has a maximum and the $(n - 1)$ maximum values are equal and each is equal to $\ln \frac{n}{n-1}$.

(iii) $S_{max} - S_{min}$ is a piece-wise convex curve.

(iv) All the three curves S_{max}, m; S_{min}, m; $S_{max} - S_{min}$, m are symmetrical about $m = (n + 1)/2$.

(b) The truth of conjectures (i) & (ii) is demonstrated in [3]. The truth of the last conjecture is demonstrated here. The validity of the remaining conjecture (iii) will be discussed elsewhere.

We have to find min max $(p_1, p_2, \ldots, p_n)$ subject to

$$\sum_{i=1}^{n} p_i = 1, \quad \sum_{i=1}^{n} i p_i = m \tag{22}$$

Consider the related problem of finding min max $(p_1, p_2, \ldots, p_n)$ subject to,

$$\sum_{i=1}^{n} p_i = 1, \quad \sum_{i=1}^{n} i p_i = n + 1 - m \tag{23}$$

Suppose, $(q_1, q_2, \ldots, q_n)$ is a p. d. satisfying (22), so that

$$\sum_{i=1}^{n} q_i = 1, \quad \sum_{i=1}^{n} i q_i = m, \tag{24}$$

then $(q_n, q_{n-1}, \ldots, q_1)$ satisfies (24) since the sum of its components is unity, and

$$1q_n + 2q_{n-1} + 3q_{n-2} + \ldots + nq_1$$

$$= (n + 1 - n)q_n + (\overline{n + 1} - \overline{n - 1})q_{n-1} + (\overline{n + 1} - \overline{n - 2})q_{n-2} + \ldots + (n + 1 - 1)q_1$$

$$= (n + 1)(q_n + q_{n-1} + q_{n-2} + \ldots + 1) - (q_1 + 2q_2 + \ldots + nq_n)$$

$$= n + 1 - m \tag{25}$$

Now $\min\max(q_1, q_2, \ldots, q_n) = \min\max(q_n, q_{n-1}, \ldots, q_1)$

Thus whether we find min max $(p_1, p_2, \ldots, p_n)$ subject to (22) or (23), we get the same result. As such S_{max}, m curve is symmetrical about $m = (n + 1)/2$. Of course, if the maximizing probability distribution for constraints (22) is $(p_1, p_2, \ldots, p_n)$, then the maximizing probability distribution for constraints (23) is $(p_n, p_{n-1}, \ldots, p_1)$.

To find minimum entropy, we have to find minimum $- \ln \max (p_1, p_2, \ldots, p_n)$, i. e. we have to maximize max $(p_1, p_2, \ldots, p_n)$ and again by an argument similar to that used in (9) S_{max}, m curve is also symmetrical about $m = (n + 1)/2$.

Since

$$(S_{max})_m = (S_{max})_{n+1-m} \tag{26}$$

$$(S_{min})_m = (S_{min})_{n+1-m} \tag{27}$$

we get,

$$(S_{max} - S_{min})_m = (S_{max} - S_{min})_{n+1-m}, \tag{28}$$

so that the curve of $S_{max} - S_{min}$ against m is also symmetrical about $m = (n + 1)/2$.

(c) An alternative way of deriving our present measure is the following:

The most certain distributions are the degenerate distributions

$$\Delta_1 = (1, 0, ..., 0), \Delta_2 = (0, 1, ..., 0), ..., \Delta_n = (0, 0, 0, ..., 0, 1) \tag{29}$$

The further away the distribution P is from these, the more uncertain it would be. Now,

$$D(\Delta_i\text{:P}) = -\ln p_i, \quad i = 1, 2, ..., n \tag{30}$$

where, D (Q : P) is Kullback-Leibler [5] measure of directed divergence of Q from P.

The average directed divergence of $\Delta_1, \Delta_2, \ldots, \Delta_n$ from P can be measured by

$$-\frac{1}{n}\sum_{i=1}^{n} \ln p_i \quad \text{or} \quad -\sum_{i=1}^{n} p_i \ln p_i \quad \text{or} \; (-\ln p_i)_{\min} \quad \text{or} \quad (-\ln p_i)_{\max} \tag{31}$$

The first and fourth of these are convex functions of $p_1, p_2, \ldots, p_n$, while the second and third are concave functions of $p_1, p_2, \ldots, p_n$.

The second gives us Shannon's [7] measure of entropy and the third gives our new measure of entropy.

(d) The first and fourth of the measures (31) are minimum subject to

$\sum_{i=1}^{n} p_i = 1$, when

$$p_i = p_2 = ... = p_n = 1/n \tag{32}$$

so that,

$$\sum_{i=1}^{n} \ln p_i \tag{33}$$

and
$$-(-\ln p_i)_{\max} = -\left(\ln\frac{1}{p_i}\right)_{\max} = -\ln\frac{1}{(p_i)_{\min}} = \ln \min(p_1, p_2, ..., p_2) \tag{34}$$

are maximum when (32) is satisfied. Though both (33) and (34) are negative, these can be used as measures of entropy for entropy maximization purposes. (33) gives us Burg's [1] measure of entropy and (34) gave us still another unorthodox measure of entropy. The maximum entropy principle here becomes in some sense maximum probability principle. Thus our new approach gives the two most important orthodox measures, viz those due to Burg and Shannon as well as two unorthodox measures, viz. $-\ln p_{max}$ and $\ln p_{min}$.

(e) The maximin probability distribution will in general be different from the minimax probability distribution. One can easily prepare a Table for maximin probability distribution corresponding to Table 1.1 for minimax probability distribution.

REFERENCES

1. J.P. Burg (1972), "The Relationship between Maximum Entropy Spectra and Maximum Likelihood Spectra, in Modern Spectral Analysis, ed. D.G. Childers, pp. 130–131.
2. J.N. Kapur and H.K. Kesavan (1987), "Generalised Maximum Entropy Principle (with Applications)," Sandford Educational Press, University of Waterloo Canada.
3. J.N. Kapur and H.K. Kesavan (1992), "Entropy Optimization Principles with Applications," Academic Press, New York.
4. J.N. Kapur (1992), "The Minimum Entropy Probability Distribution,".Chapter 23 in present book.
5. S. Kullback and R.A. Leibler (1951), On Information and Sufficiency, Ann. Math Stat, **22** , pp. 79-86.
6. A Renyi (1961), "Measures of Entropy and Information," in Proceedings Fourth Berkeley Symp. Prob. Math Stat **1**. pp. 547–561.
7. C.E. Shannon (1948), "A Mathematical Theory of Communication", Bell Systems Tech Journal; vol 27, pp. 379–423, 623–659.

2

A NEW PARAMETRIC MEASURE OF ENTROPY

$$[K_a(P) = \sum_{i=1}^{n} [\ln (1 + ap_i) - p_i \ln (1 + a)] = \sum_{i=1}^{n} \ln (1 + ap_i) - \ln (1 + a)$$

is proposed for the probability distribution, $P = (p_1, p_2, \ldots, p_n)$ and its properties are studied. In particular, it is shown that it is a monotonic increasing function of a. The corresponding measures of directed divergence are also obtained.]

2.1 INTRODUCTION

Let $P = (p_1, p_2, \ldots, p_n)$ be a probability distribution, then Shannon [9] gave the measure

$$S(P) = -\sum_{i=1}^{n} p_i \ln p_i \tag{1}$$

to measure its uncertainty or entropy. It can also be regarded as a measure of equality of $p_1, p_2, \ldots, p_n$ among themselves.

Later Renyi [8], Havrda and Charvat [2] and Kapur [3] gave the measures,

$$\frac{1}{1-\alpha} \ln \sum_{i=1}^{n} p_i^{\alpha},\ \frac{1}{1-\alpha}\left(\sum_{i=1}^{n} p_i^{\alpha} - 1\right),\ \frac{1}{\beta-\alpha} \ln \frac{\sum_{i=1}^{n} p_i^{\alpha}}{\sum_{i=1}^{n} p_i^{\beta}},\quad \alpha >, \beta > 0,\ \alpha \neq \beta \tag{2}$$

and Burg [1] and Kapur [4] gave the measures

$$\sum_{i=1}^{n} \ln p_i, -\sum_{i=1}^{n} p_i \ln p_i + \frac{1}{a} \sum_{i=1}^{n} (1 + ap_i) \ln (1 + ap_i) - \frac{1}{a}(1 + a) \ln (1 + a) \tag{3}$$

Shannon's and Burg's measures do not have any parameter, while the other measures have one or two parameters. When maximized by Lagrange's method, subject to linear constraints on probabilities, Shannon, Burg's and Kapur's second measure always give non-negative probabilities, while in the case of measures (2), non-negativity conditions have to be explicitly imposed. Shannon's measure has been the most successful and most widely used measure. Burg's measure has also been successful, but it is always negative and as such it is hard to interpret it as a measure of uncertainty. However, it can be used for entropy maximization purposes and it has been so used [5]. Moreover, its maximum value decreases with n and this is not a desirable property for a measure of entropy. Moreover, its computation also causes problems when p_i is very small since then $\ln p_i$ is large and negative and its derivative is also large.

In the present discussion, we modify Burg's measure to get a parametric measure of entropy. This does not suffer from the weaknesses of Burg's measure. We also study the properties of the measure and show in particular that it is a monotonic increasing function of the parameter. We also investigate the corresponding measures of directed divergence.

2.2 SOME PROPERTIES OF THE NEW MEASURE OF ENTROPY

The measure is defined by

$$K_a(P) = \sum_{i=1}^{n} \ln(1+ap_i) - \ln(1+a),\ a>0 \tag{4}$$

It has the following properties:

(i) It is a continuous function of $p_1, p_2, \ldots, p_n$, so that it changes by a small amount when $p_1, p_2, \ldots, p_n$ change by small amounts.

(ii) It is a permutationally symmetric function of $p_1, p_2, \ldots, p_n$, i.e. the function does not change when $p_1, p_2, \ldots, p_n$ are permuted among themselves.

(iii) It is maximum subject to $\sum_{i=1}^{n} p_i = 1$ when

$$p_i = p_2 = \ldots = p_n = 1/n \tag{5}$$

(iv) The maximum value is an increasing function of n. In fact the maximum value is given by

$$f(n) = n \ln\frac{n+a}{n} - \ln(1+a), \tag{6}$$

so that
$$f'(n) = \ln\frac{n+a}{n} - \frac{a}{n+a} \tag{7}$$

and
$$f''(n) = -\frac{a^2}{n(n+a)^2} < 0, \tag{8}$$

so that $f(n)$ is a concave function of n and $f'(n)$ is a monotonic decreasing function of n. Now

$$f'(n) = \ln\frac{n+a}{n} - \frac{a}{n+a} = \ln y - 1 + \frac{1}{y}$$

$$= \frac{y\ln y - y + 1}{y};\ y = 1 + \frac{a}{n} > 0 \tag{9}$$

Since $y > 0$ and $y \ln y - y + 1 > 0$ when $y \geq 0$, (10)

we get
$$f'(n) > 0, \tag{11}$$

so that $f(n)$ is always increasing. Now,

$$\underset{n\to\infty}{Lt}\ f(n) = \underset{n\to\infty}{Lt}\ \ln\left(1+\frac{a}{n}\right)^n - \ln(1+a) \tag{12}$$

$$= \ln e^a - \ln(1+a) = a - \ln(1+a) \tag{13}$$

Now if $\phi(a) = a - \ln(1+a),\ \phi'(a) = 1 - \dfrac{1}{1+a} = \dfrac{a}{1+a} > 0$ (14)

Also $\phi(o) = 0$ and $\phi'(a) > 0$, so that $\phi(a) > 0$ when $a > 0$ then the graph of $f(n)$ is as given in Figure 2.1.

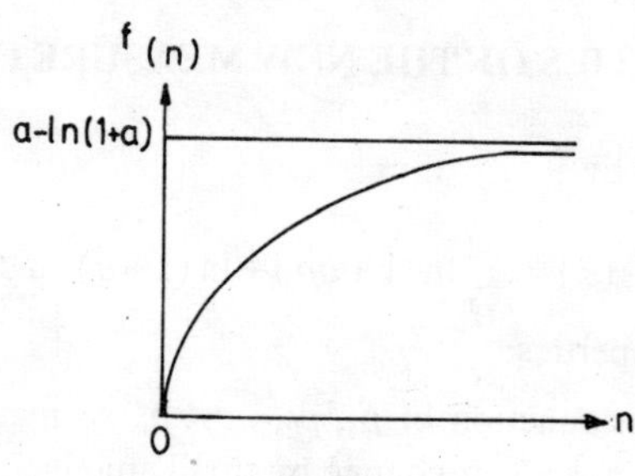

Fig. 2.1

On the other hand for Burg's measure of entropy

$$B(P) = \sum_{i=1}^{n} \ln p_i, \tag{15}$$

the maximum value of $B(P)$ occurs when P is the uniform distribution and so maximum value of $B(P)$ is

$$\psi(n) = n \ln \frac{1}{n} = -n \ln n, \tag{16}$$

so that
$$\psi'(n) = -(1 + \ln n) < 0, \tag{17}$$

so that $\psi(n)$ is a monotonic decreasing function of n.
Thus though the measure $K_a(P)$ is inspired by $B(P)$, it does not suffer from the weaknesses of Burg's measure.

(v) Now,
$$\frac{\partial}{\partial p_i} K_a(P) = \frac{a}{1+ap_i}, \frac{\partial^2}{\partial p_i^2} K_a(P) = -\frac{a^2}{(1+ap_i)^2}$$

$$\frac{\partial^2}{\partial p_i \partial p_j} K_a(P) = 0, \tag{18}$$

so that $K_a(P)$ is a concave function of $p_1, p_2, \ldots, p_n$.

(vi) Since $K_a(P)$ is a concave function and its domain is

$$p_1 \geq 0, p_2 \geq 0, p_3 \geq 0, \ldots, p_n \geq 0; \quad \sum_{i=1}^{n} p_i = 1, \tag{19}$$

its minimum value occurs at each of the degenerate distributions

$$\Delta_i = (0, 0, \ldots, 1, 0, 0, \ldots, 0); \quad i = 1, 2, \ldots, n \tag{20}$$

where in Δ_i, unity occurs in the ith place and for each of these, its value is zero. Thus,

$$K_a(P) \geq 0 \tag{21}$$

and it vanishes only when P coincides with one of the degenerate distributions, i.e. when there is perfect certainty and uncertainty is zero. Thus unlike Burg's measure, it can be used as a measure of uncertainty.

(vii) Let $p_i < p_j$ and let us increase p_i by x and decrease p_j by x so that $p_i + x$ is still $\leq p_j - x$, then

$$K_a(P') - K_a(P) = \ln(1 + a(p_i + x)) + \ln(1 + a(p_j - x))$$
$$- \ln(1 + ap_i) - \ln(1 + ap_j) = h(x), \text{ say} \tag{22}$$

so that

$$h'(x) = \frac{a}{1 + ap_i + ax} - \frac{a}{1 + ap_j - ax} \tag{23}$$

$$h''(x) = -\frac{a^2}{(1 + ap_i + ax)^2} - \frac{a^2}{(1 + ap_j - ax)^2} < 0 \tag{24}$$

so that $h(x)$ is maximum when, $\quad p_i + x = p_j - x$ (25)

Thus as the probabilities become more and more equal, the measure increases, so that the measure can be used as a measure of equality of probabilities.

Thus, $K_a(P)$ satisfies all the important properties satisfied by Shannon's measure of entropy except additivity and recursivity. However, these properties are unimportant for entropy maximization purposes, since whenever a measure is maximized, its monotonic increasing functions are also maximized, but while the original measure may be additive and recursive, its monotonic functions need not be additive and recursive, although these will lead to the same maximizing probability distribution.

2.3. MONOTONIC CHARACTER OF THE NEW MEASURE OF ENTROPY

Now,

$$\frac{d}{da}(K_a(P)) = \sum_{i=1}^{n} \frac{p_i}{1 + ap_i} - \frac{1}{1 + a} = \sum_{i=1}^{n} \frac{p_i}{1 + ap_i} - \sum_{i=1}^{n} \frac{p_i}{1 + a}$$

$$= \sum_{i=1}^{n} p_i \left[\frac{1}{1 + ap_i} - \frac{1}{1 + a}\right] = \sum_{i=1}^{n} \frac{ap_i(1 - p_i)}{(1 + a)(1 + ap_i)} > 0, \tag{26}$$

so that $K_a(P)$, given P, is a monotonic increasing function of a.
For the probability distribution $(p, 1 - p)$

$$K_a(P) = \ln(1 + ap) + \ln(1 + a - ap) - \ln(1 + a) \tag{27}$$

$$= \ln\left(1 + \frac{a^2 p(1 - p)}{1 + a}\right) \tag{28}$$

and

$$\max_p K_a(P) = \ln\left(1 + \frac{a^2}{4(1 + a)}\right) = \ln\frac{(a + 2)^2}{4(1 + a)} \tag{29}$$

Table 2.1 max$_P$ $K_a(P)$ against a

a	0.5	1.0	2.0	5.0	10.0	20.0	∞
$\max_P K_a(P)$	.00408	.1178	.2877	.7138	1.1856	1.7613	∞

The graphs of $K_a(p, 1-p)$ for different values of a are shown in Figure 2.2.

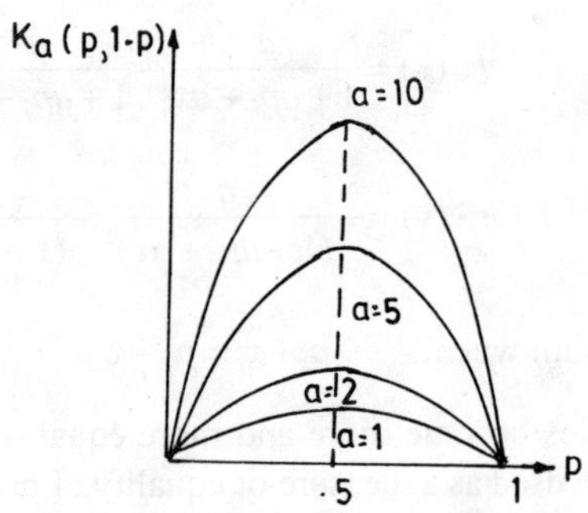

Fig. 2.2

2.4 RELATIONSHIP WITH BURG'S ENTROPY

We can write,

$$K_a(P) = \sum_{i=1}^{n} \ln(1+ap_i) - \ln(1+a) = \sum_{i=1}^{n} \left[\ln a + \ln\left(p_i + \frac{1}{a}\right) \right] - \ln a - \ln\left(1+\frac{1}{a}\right)$$

$$= (n-1)\ln a + \sum_{i=1}^{n} \ln\left(p_i + \frac{1}{a}\right) - \ln\left(1 + \frac{1}{a}\right), \tag{30}$$

so that when a is large, maximising $K_a(P)$ will give the same result as maximizing $B(P)$. Alternatively if we maximize $K_a(P)$ subject to constraints

$$\sum_{i=1}^{n} p_i = 1, \quad \sum_{i=1}^{n} p_i g_{ri} = a_r, r = 1, 2, ..., m, \tag{31}$$

we get,

$$\frac{a}{1+ap_i} = \lambda_0 + \lambda_1 g_{1i} + ... + \lambda_m g_{mi}. \tag{32}$$

Now letting $a \to \infty$, we get

$$\frac{1}{p_i} = \lambda_0' + \lambda_1' g_{1i} + ... + \lambda_m' g_{mi}, \tag{33}$$

where λ's are determined by using constraints (31) so that (33) gives the maximum entropy probability distribution when Burg's entropy $B(P)$ is maximized subject to (31).

Thus as $a \to \infty$, the MEPD for $K_a(P)$ approaches the MEPD for $B(P)$. Of course $K_a(P)$ does not approach $B(P)$ as $a \to \infty$. This is expected as $K_a(P) \geq 0$ and $B(P)$ is always negative. In fact,

$$K_a(P) - B(P) = (n-1)\ln a - \ln\left(1+\frac{1}{a}\right) + \sum_{i=1}^{n} \ln \frac{\left(p_i + \frac{1}{a}\right)}{p_i}, \tag{34}$$

so that

$$K_a(P) - B(P) \to \infty \quad as \quad a \to \infty \tag{35}$$

2.5 CONCAVITY OF S_{max} WHEN MEAN IS PRESCRIBED

Maximizing $K_a(P)$ subject to

$$\sum_{i=1}^{n} p_i = 1, \quad \sum_{i=1}^{n} i p_i = m. \tag{36}$$

we get,

$$p_i + \frac{1}{a} = \frac{1}{\lambda + \mu i}, \tag{37}$$

where,

$$\sum_{i=1}^{n} \frac{1}{\lambda + \mu i} = 1 + \frac{n}{a}, \sum_{i=1}^{n} \frac{i}{\lambda + \mu i} = m + \frac{n(n+1)}{2a} \tag{38}$$

so that

$$\lambda\left(1+\frac{n}{a}\right) + \mu\left(m + \frac{n(n+1)}{2a}\right) = n \tag{39}$$

or,

$$\lambda = \frac{n - \mu\left(m + \frac{n(n+1)}{2a}\right)}{1 + n/a} \tag{40}$$

and,

$$\lambda + \mu i = \frac{n + \mu\left[(1+n/a)i - m - \frac{n(n+1)}{2a}\right]}{1 + n/a} \tag{41}$$

Also,

$$p_i = \frac{a+n}{an + \mu\left[(n+a)i - am - \frac{n(n+1)}{2}\right]} - \frac{1}{a}, \tag{42}$$

where μ is determined from the equation

$$\frac{n}{a} + 1 = \sum_{i=1}^{n} \frac{a+n}{an + \mu\left[(n+a)i - am - \frac{n(n+1)}{2}\right]}$$

or,

$$\frac{1}{a} = \sum_{i=1}^{n} \frac{1}{an + \mu\left[(n+a)i - am - \frac{n(n+1)}{2}\right]} \tag{43}$$

Now,
$$S_{\max} = \sum_{i=1}^{n} \ln(1+ap_i) - \ln(1+a)$$

$$= \sum_{i=1}^{n} \ln\frac{a}{\lambda+\mu i} - \ln(1+a)$$

$$= \ln\frac{a^n}{1+a} - \sum_{i=1}^{n} \ln(\lambda+\mu i)$$

$$= \ln\frac{a^n}{1+a} - \sum_{i=1}^{n} \ln\left(n+\mu\left[\left(1+\frac{n}{a}\right)i - m - \frac{n(n+1)}{2a}\right]\right)$$

$$+n\ln\left(1+\frac{n}{a}\right) \tag{44}$$

$$\frac{dS_{\max}}{dm} = -\sum_{i=1}^{n} \frac{\left[\frac{d\mu}{dm}\left[\left(1+\frac{n}{a}\right)i - m - \frac{n(n+1)}{2a}\right] - \mu\right]}{n+\mu\left[(1+n/a)\,i - m - \frac{n(n+1)}{2a}\right]}$$

$$= -\sum_{i=1}^{n} \frac{p_i + 1/a}{a+n}\, a\left[\frac{d\mu}{dm}\left[\left(1+\frac{n}{a}\right)i - m - \frac{n(n+1)}{2a}\right] - \mu\right]$$

$$= \frac{d\mu}{am}\left[m + \frac{n(n+1)}{2} - m - \frac{n(n+1)}{2}\right]$$

$$+\frac{\mu a}{a+n}\left(1+\frac{n}{a}\right) \tag{45}$$

so that,
$$\frac{dS_{\max}}{dm} = \mu \tag{46}$$

$$\frac{d^2 S_{\max}}{dm^2} = \frac{d\mu}{dm} \tag{47}$$

Now from (41) and (42)

$$\frac{d\mu}{dm} = \frac{\mu\sum_{i=1}^{n}(p_i+1/a)^2}{\sum_{i=1}^{n}(p_i+1/a)^2\left[\left(1+\frac{n}{a}\right)i - m - \frac{n(n+1)}{2a}\right]} \tag{48}$$

$S_{\max}$ will be a concave function of m if $\frac{d\mu}{dm} < 0$, i.e. if either $\mu > 0$, denominater < 0 or $\mu < 0$, denominater > 0.

Case I. If $\mu > 0$, $m < \frac{n+1}{2}$, probabilities are decreasing

$$\left(1+\frac{n}{a}\right)i - m - \frac{n(n+1)}{2a} = (i-m) + (n/a)\left(i - \frac{n+1}{2}\right)$$

$$= (i-m)(1-n/a) + (n/a)\left(m - \frac{n+1}{2}\right) \tag{49}$$

$$\sum_{i=1}^{n}(i-m)(p_i+1/a)^2=\sum_{i=1}^{n}\left(p_i+\frac{1}{a}\right)^2\left[\frac{\sum_{i=1}^{n} i(p_i+1/a)^2}{\sum_{i=1}^{n}(p_i+1/a)^2}-\frac{\sum_{i=1}^{n} ip_i}{\sum_{i=1}^{n} p_i}\right] \quad (50)$$

Both terms within square brackets are weighted means of 1, 2, . . . , n; the weights in the first case are proportional to $(p_i + 1/a)^2$ and the weights in the second case are proportional to p_i. The weights decrease faster in the first case and therefore the expression (50) is negative. Since $m \le \frac{n+1}{2}$, using (49) we find the denominator of (48) is negative since $\mu > 0$. In that case $\frac{d\mu}{dm} < 0$.

Case II. $\mu < 0$, $m > \frac{n+1}{2}$, probabilities are increasing. By using the same arguments, denominator (47) is positive and $\frac{d\mu}{dm}$ is again negative.

Thus we have proved that for all positive values of m, $S_{\max}$ is a concave function of m.

2.6 COMPARISON WITH SHANNON'S MEASURE

Table 2.2 gives the values of p_i's for the case when $a = 1$, $n = 6$. Table 2.3 gives the corresponding values of p_i's for Shannon's measure. Only the values of m upto 3.5 are given since p_i when mean is m is equal to the value of p_{7-i} when mean is $7 - m$.

Table 2.2 Probabilities for Burg's modified measure when a = 1

m	p_1	p_2	p_3	p_4	p_5	p_6
1.0	1.0000	0	0	0	0	0
1.5	.5988	.3024	.0988	0	0	0
2.0	.4175	.2897	.1835	.0933	0	0
2.5	.3193	.2499	.1875	.1310	.0796	.0327
3.0	.2405	.2088	.1790	.1506	.1234	.0975
3.5	.1667	.1667	.1667	.1667	.1667	.1667

Table 2.3 Probabilities for Shannons measure

m	p_1	p_2	p_3	p_4	p_5	p_6
1.0	1.0000	0.0000	0.0000	0.0000	0.0000	0.0000
1.5	0.6637	0.2238	0.0755	0.0255	0.0086	0.0029
2.0	0.4781	0.2548	0.1357	0.0723	0.0385	0.0205
2.5	0.3475	0.2398	0.1654	0.1142	0.0788	0.0544
3.0	0.2468	0.2072	0.1740	0.1461	0.1227	0.1031
3.5	0.1667	0.1667	0.1667	0.1667	0.1667	0.1667

2.7 CORRESPONDING MEASURES OF DIRECTED DIVERGENCE

(a) Consider the measure of directed divergence,

$$D_{a,1}(Q,P)=\sum_{i=1}^{n}(1+aq_i)\ln\frac{1+aq_i}{1+ap_i}$$

It is easily shown that $D_{a,1}(Q,P)\geq 0$, vanishes iff $Q=P$ and is a convex function of both $p_1, p_2, \ldots, p_n$ and $q_1, q_2, \ldots, q_n$. If U is the uniform distribution, we get

$$D_{a,1}(U,P)=\sum_{i=1}^{n}(1+a/n)\ln\frac{1+a/n}{1+ap_i}=(n+a)\ln\frac{n+a}{n}-\frac{n+a}{a}\sum_{i=1}^{n}\ln(1+ap_i),$$

so that $\sum_{i=1}^{n}\ln(1+ap_i)$ + (constant) is a monotonic decreasing function of

$D_{a,1}(U{:}P)$ and can be used as a measure of entropy. Choosing the constant, so that the measure of entropy vanishes for degenerate distributions, we get the measure of entropy $K_a(P)$, so that we can regard $D_{a,1}(Q,P)$ as a measure of directed divergence corresponding to $K_a(P)$.

(b) Next consider the measure of directed divergence

$$D_{a,2}(P{:}Q)=\sum_{i=1}^{n}\left(\frac{1+ap_i}{1+aq_i}-\ln\frac{1+ap_i}{1+aq_i}-1\right)$$

Since $x-\ln x-1\geq 0$ when $x>0$ ad vanishes iff $x=1$, $D_{a,2}(P{:}Q)\geq 0$ and vanishes iff $P=Q$. It is also easily shown that $D_{a,2}(P{:}Q)$ is a convex function of both $p_1, p_2, \ldots, p_n$ and $q_1, q_2, \ldots, q_n$ so that, $D_{a,2}(P:Q)$ is a proper measure of directed divergence. Now,

$$D_{a,2}(P{:}U)=\sum_{i=1}^{n}\left(\frac{1+ap_i}{1+a/n}-\ln\frac{1+ap_i}{1+a/n}-1\right)=\frac{n+a}{1+a/n}-\sum_{i=1}^{n}\ln(1+ap_i)+n\ln(1+a/n)-n.$$

so that again $\sum_{i=1}^{n}\ln(1+ap_i)$ + (constant) is a monotonic decreasing function

of $D_{a,2}(P{:}U)$ and $K_a(P)$ can be used as a measure of entropy, so that, $D_{a,2}(P{:}Q)$ can also be regarded as measure of directed divergence corresponding to $K_a(P)$.

(c) It may be noted that

$$\underset{a\to\infty}{Lt}\sum_{i=1}^{n}\frac{1+aq_i}{1+a}\ln\frac{1+aq_i}{1+ap_i}=\sum_{i=1}^{n}q_i\ln\frac{q_i}{p_i}=D_1(Q{:}P)$$

$$\underset{a\to\infty}{Lt}\sum_{i=1}^{n}\left(\frac{1+ap_i}{1+aq_i}-\ln\frac{1+ap_i}{1+aq_i}-1\right)=\sum_{i=1}^{n}\left(\frac{p_i}{q_i}-\ln\frac{p_i}{q_i}-1\right)=D_2(P{:}Q)$$

where $D_1(Q{:}P)$ is Kullback and Leibler's [7] measure and $D_2(P{:}Q)$ is also measure of directed divergence [6].

REFERENCES

1. J.P. Burg, (1972), "The Relationship between Maximum Entropy Spectra and Maximum Likelihood Spectra" in Modern Spectral Analysis, ed. D.G. Childers, pp. 130–131; M.S.A.
2. J.H. Havrda, and F. Charvat (1967), "Quantification methods of classification Processes: Concept of structural α Entropy," Kybernatica, **3**, pp. 30–35.
3. J.N. Kapur, (1967), "Generalised Entropy of order α and Type β" The Mathematics Seminar **4**, 78–94.
4. J.N. Kapur, (1986), "Four Families of Measures of Entropy," Ind. Jour. Pure and App. Maths, **17**, no. 4, pp. 429–449.
5. J.N. Kapur, (1991), Burg and Shannon's measure of entropy as limiting cases of a family of measures of entropy. Proc. Nat. Aead. Sciences. **61**A Part III, 371–387.
6. J.N. Kapur, (1986), "Matisuta's generalised measure of affinity". Istanbal Uni Fac Sai. **47**A, 73–79.
7. S. Kullback, and R.A. Leibler (1951), " On information and sufficiency. Ann. Math. Stat, **22**, pp. 79–86.
8. A. Renyi, (1961), "On Measures of Information and Entropy" Berkeley Symop Prob. Stat **1**, 547–561.
9. C.E. Shannon; (1948), "A mathematical theory of communication," Bell system Tech. J., **27**, pp. 379–423, 623–659.

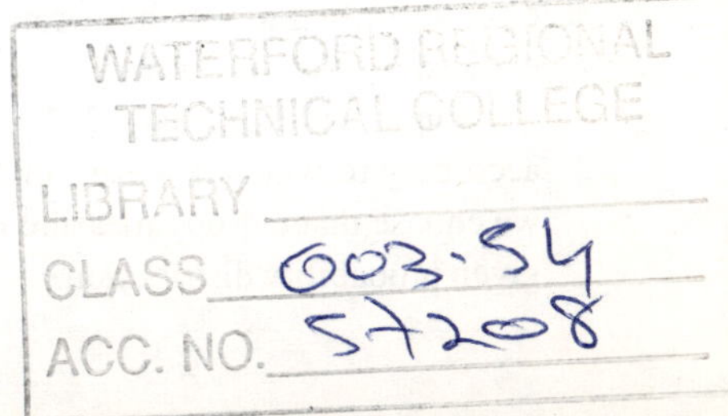

3

ON THE USE OF MAXIMUM-ENTROPY PRINCIPLE FOR GENERATING NEW ENTROPIES

[Two entropies, *viz.* the entropy of a probability distribution and the entropy of a given information about a probability distribution, are considered and compared. A new proposed 'discrete' analogue of Boltzmann-Shannon measure is also considered. The generation of both types of entropies is also discussed. A third type of entropy, *viz.* the entropy of a finite set of positive numbers is also introduced.]

3.1 INTRODUCTION

Let $P = (p_1, p_2, \ldots, p_n)$ be a probability distribution, then there is some uncertainty about the outcomes. Shannon [19] gave the measure

$$S(P) = -\sum_{i=1}^{n} p_i \ln p_i \tag{1}$$

to measure this uncertainty and called it the Entropy of the Probability Distribution P. Later Renyi [18], Havrda and Charvat [4], Kapur [6,7,8,9], Sharma-Taneja [20], Sharma-Mittal [21] and others proposed other functions of $p_1, p_2, \ldots, p_n$ to measure this uncertainty and called these functions as Generalised Measures of Entropy or simply Generalised Entropies.

If we have some information about $p_1, p_2, \ldots, p_n$ in the form of moment constraints,

$$\sum_{i=1}^{n} p_i = 1, \quad \sum_{i=1}^{n} p_i g_r(x_i) = a_r, r = 1, 2, \ldots, m; \quad m \ll n, \tag{2}$$

there may be an infinity of probability distributions consistent with this information. According to the principle of maximum entropy due to Jaynes [5], out of these infinity of distributions, the distribution which best represents our knowledge is the one which maximizes (1) subject to (2).

Kapur and Kesavan [12–14] considered a generalised maximum-entropy principle according to which we can choose the distribution which maximizes any generalised measure of entropy subject to (2).

Kapur and Kesavan [12–14] also gave an Inverse Maximum Entropy Principle according to which if we are given a probability distribution P and constraints (2), we choose that entropy measure $H(P)$ whose maximization subject to (2) yeilds the given probability distribution.

Kapur [10] used this principle to generate a number of new measures of entropy by using probability distributions arising in population distribution, innovation diffusion processes, spread of epidemics, chemical kinetics, earthquake frequency studies, income distribution etc., by regarding these as generalised maximum entropy probability distributions and choosing the measures of entropy by maximizing which we get the given distributions as generalised maximum entropy distributions, subject to some very natural and plausible constraints.

This represented one method of using the Maximum Entropy Principles (generalised and inverse) to generate new measures of entropy.

Recently, Forte and Hughes [3] gave a different use of Jaynes maximum entropy principle to define new entropies. They did not generalise Jaynes maximum entropy principle and did not generate new measures of entropy. They confined themselves to defining a new type of entropy. In fact, they introduced a new concept of the entropy of a given Information and used Jaynes principle to get a measure for it.

Given a certain information (2), we obtain the maximum entropy probability distribution by maximizing (1) subject to (2), then the Shannon entropy of this distribution will be greater than the entropy of any other distribution consistent with (2). Let the entropy of this distribution be denoted by $S_{\max}$, then Forte and Hughes [2] defined $S_{\max}$ as the entropy of the information (2).

$S_{\max}$ will obviously be a function of $a_1, a_2, \ldots, a_m$. Kapur [11] has proved that $S_{\max}$ will always be a concave function of $a_1, a_2, \ldots, a_m$.

$S_{\max}$ represents the maximum uncertainty that remains after the given information (2) has all been used, since according to the maximum entropy principle, we should not use any other information not given to us. Thus $S_{\max}$ represents the uncertainty associated with the given information. It is not the uncertainty due to the given information; it is the uncertainty inspite of the given information.

Lazo and Rathie [17] derived and tabulated entropies for various continuous univariate probability distributions. Ali Ahmed and Gokhale [1] extended their tables to continuous multi-variate probability distributions. Since as shown by Kapur [11], most of these distributions are maximum entropy distributions, subject to some simple moment constraint, these tables can be interpreted as giving $S_{\max}$ for specified moment values.

3.2 CROSS-ENTROPY OF A GIVEN INFORMATION WITH RESPECT TO A GIVEN PROBABILITY DISTRIBUTION

Given *a priori* distribution, $Q = (q_1, q_2, \ldots, q_n)$ and given moment constraints (2), there will be a certain distribution P which will minimize Kullback-Leibler [15] measure

$$D(P:Q) = \sum_{i=1}^{n} p_i \ln \frac{p_i}{q_i} \tag{3}$$

subject to (2). The minimum value of $D(P:Q)$ which will arise for this distribution will be defined as the cross-entropy of (2) with reference to the given *a priori* distribution Q.

Thus given that the *a priori* probability distribution is a Poisson distribution with parameter m_0 and given that

$$\sum_{i=0}^{\infty} p_i = 1, \quad \sum_{i=0}^{\infty} i p_i = m, \tag{4}$$

the distribution which minimizes (3) subject to (4) is Poisson distribution with parameter m and the minimum cross-entropy is

$$\sum_{i=0}^{\infty} e^{-m} \frac{m^i}{i!} \ln \frac{e^{-m} m^i / i!}{e^{-m_0} m_0^i / i!}$$

$$= \sum_{i=0}^{\infty} e^{-m} \frac{m^i}{i!} \left[(m_0 - m) + i \ln \frac{m}{m_0} \right]$$

$$= m_o - m + m \ln \frac{m}{m_0} \tag{5}$$

This gives the cross-entropy of the information (4) relative to the *a priori* Poisson distribution with parameter m_0. This cross-entropy is a convex function of m with its minimum value at m_0. In fact, Kapur [11] has proved that this minimum cross-entropy will always be a convex function of the specified moment values.

3.3 THE 'DISCRETE' ANALOGUE OF BOLTZMANN-SHANNON ENTROPY

By maximizing Boltzmann-Shannon entropy measure

$$-\int_a^b f(x) \ln f(x) dx \tag{6}$$

subject to

$$\int_{x_{i-1}}^{x_i} f(x)\, dx = p_i, \; i = 1, 2, \ldots, n \tag{7}$$

where,

$$a = x_0 < x_1 < x_2 \ldots\ldots < x_{n-1} < x_n = b, \tag{8}$$

we get,

$$f(x) = \frac{p_i}{x_i - x_{i-1}}, \; x_{i-1} \le x < x_i, \; i = 1, 2, \ldots, n-1 \tag{9}$$

$$f(x) = \frac{p_n}{x_n - x_{n-1}}, \; x_{n-1} \le x \le x_n, \tag{10}$$

so that

$$S_{\max} = -\sum_{i=1}^{n} \int_{x_{i-n}}^{xi} \frac{p_i}{x_i - x_{i-1}} \ln \frac{p_i}{x_i - x_{i-1}}\, dx$$

$$= -\sum_{i=1}^{n} p_i \ln \frac{p_i}{x_i - x_{i-1}} \qquad (11)$$

Forte and Hughes call (11) as the correct discrete analogue of (6). This is in contrast to the measure (1), whose connection with (6) cannot be rigorously estabilished [11].

We make the following remarks on this interpretation of the result obtained.

(*i*) For (11), the random variate is not discrete, while for (1) it is discrete. For (11) it is a continuous random variate, though its density function is only piecewise continuous and not continuous over $[a, b]$. Its distribution function is however continuous. For (1), the variate is discrete and can take values $x_1, x_2, \ldots, x_n$. As such, while (11) may be called a discrete analogue, since it is a sum and not an integral, it must be clearly understood that it does not give entropy for a discrete variate

(*ii*) The measures (1) and (11) measure conceptually different entities. While (1) is the entropy of the probability distribution $(p_1, p_2, \ldots, p_n)$, (11) is the entropy of the maximum entropy probability distribution obtained by maximizing (6) subject to (7).

(*iii*) While (1) is always ≥ 0, (11) can be positive or negative. Thus if the intervals are equal, (11) gives

$$-\sum_{i=1}^{n} p_i \ln \frac{p_i}{\frac{b-a}{n}} = -\sum_{i=1}^{n} p_i \ln p_i + \ln \frac{b-a}{n} \qquad (12)$$

The first term on the RHS > 0, but the second term can be < 0.

(*iv*) While we can use (1) as a measure of uncertainty, it is difficult to use (11) for this purpose, since it can be negative. However (11) can be used for comparing uncertainties, since differences in uncertainties can be positive or negative.

(*v*) While (1) is maximum subject to $\sum_{i=1}^{n} p_i = 1$,

when all the probabilities are equal, (11) is maximum subject to $\sum_{i=1}^{n} p_i = 1$

when

$$\frac{p_1}{x_1 - x_0} = \frac{p_2}{x_2 - x_1} = \ldots = \frac{p_n}{x_n - x_{n-1}} = \frac{1}{x_n - x_0} = \frac{1}{b-a} \qquad (13)$$

i.e., when the probabilities are proportional to the lengths of intervals. Thus while (1) is maximum for the discrete variate uniform distribution, (11) is maximum for the continous variate uniform distribution.

(*vi*) Thus, if $x_1 - x_{i-1}$ represents the ith income range and mean income is prescribed, and if we want to maximize equality in the number of persons in different intervals, we use (1) and if we want the number of persons to be as nearly proportional as possible to the lengths of income ranges, we use (11).

(*vii*) If we put

$$q_i = \frac{x_i - x_{i-1}}{b-a} \tag{14}$$

then,

$$-\sum_{i=1}^{n} p_i \ln \frac{p_i}{x_i - x_{i-1}}$$

$$= \ln(b-a) - \sum_{i=1}^{n} p_i \ln \frac{p_i}{q_i} \tag{15}$$

Now $\sum_{i=1}^{n} p_i \ln \frac{p_i}{x_i - x_{i-1}}$ is a convex function of $p_1, p_2, \ldots, p_n$ and it gets its minimum value when (13) is satisfied, i.e. when

$$p_i = \frac{x_i - x_{i-1}}{b-a} = q_i, \tag{16}$$

so that the maximum value of the LHS of (15) is ln (*b–a*). Thus we get

$$-\sum_{i=1}^{n} p_i \ln \frac{p_i}{x_i - x_{i-1}} \le \ln(b-a), \tag{17}$$

which is otherwise obvious since ln $(b - a)$ is the maximum values of (6), when there are no constraints except the natural constraint

$$\int_a^b f(x)dx = 1 \tag{18}$$

and the LHS is the maximum value of (6) in the presence of constraint (7).

The measure (11) has already been successfully used by Forte and Sahoo, [3] to define optimal threshold for the gray levels in image processing.

3.4 FERMI-DIRAC AND BOSE-EINSTEIN DISTRIBUTIONS

Let p_{ij} be the probability of there being *j* particles in the ith energy level with energy ε_i, when $j = 0$ or 1, then maximizing the entropy

$$-\sum_{i=1}^{n} \sum_{j=0}^{1} p_{ij} \ln p_{ij} \tag{19}$$

subject to

$$\sum_{j=0}^{1} p_{ij} = 1, i = 1,2,...,n; \sum_{i=1}^{n} \sum_{j=0}^{1} jp_{ij} = N; \sum_{i=1}^{n} \varepsilon_i \sum_{j=0}^{1} jp_{ij} = N\hat{\varepsilon}, \tag{20}$$

we get

$$p_{ij} = a_i e^{-(\lambda+\mu\varepsilon_i)j}, a_i^{-1} = 1 + e^{-(\lambda+\mu\varepsilon_i)} \tag{21}$$

$$\overline{n}_i = \sum_{j=0}^{1} jp_{ij} = p_{i1} = a_{i1} e^{-(\lambda+\mu\varepsilon_i)} = \frac{1}{e^{\lambda+\mu\varepsilon_i}+1}, \tag{22}$$

which gives us the Fermi-Dirac distribution. In this case

$$S_{\max} = -\sum_{i=1}^{n} \overline{n}_i \ln \overline{n}_i - \sum_{i=1}^{n} (1-\overline{n}_i) \ln(1-\overline{n}_i), \tag{23}$$

where $\overline{n}_i$ is a function of λ and μ, and λ and μ are functions of N and $N\hat{\varepsilon}$.

This $S_{\max}$ is the Fermi-Dirac entropy corresponding to the information (20).

$S_{\max}$ is essentially a function of N and $\hat{\varepsilon}$,

but it can also be regarded as a function of $\overline{n}_1, \overline{n}_2, ..., \overline{n}_n$.

Now $(\overline{n}_1, \overline{n}_2, ..., \overline{n}_n)$ does not give a probability distribution and $S_{\max}$ is not the entropy of a probability distribution.

Since,

$$\frac{\overline{n}_1}{N} + \frac{\overline{n}_2}{N} + ... + \frac{\overline{n}_n}{N} = 1, \tag{24}$$

we can also regard $S_{\max}$ in some sense as the entropy of a probability distribution.

However, it is better to define a third type of entropy, *viz.* the entropy of a distribution of a set of a positive numbers with known sum N. In this case $S_{\max}$ can be used as a measure of equality of the numbers $\overline{n}_1, \overline{n}_2, ..., \overline{n}_n$. The more equal these numbers are, the greater is the measure.

Thus the concept of entropy of an information set leads to the concept of an entropy measure of a set of positive numbers with known sum.

In the same way, in the case when j can take the values 0, 1, 2, . . .

we get

$$\overline{n}_i = \frac{1}{e^{\lambda+\mu\varepsilon_1}-1} \tag{25}$$

which gives Bose-Einstein distribution and the measure

$$S_{\max} = -\sum_{i=1}^{n} \overline{n}_i \ln \overline{n}_i + \sum_{i=1}^{n} (1+\overline{n}_i) \ln(1+\overline{n}_i), \tag{26}$$

which can also be regarded as a measure of equality of $\overline{n}_1, \overline{n}_2, ..., \overline{n}_n$.

Similarly corresponding to the intermediate statistics distribution

$$\overline{n}_i = \frac{1}{e^{(\lambda+\mu\varepsilon_i)}+a}, i = 1,2, ..., n \tag{27}$$

we get the measure

$$S_{\max} = -\sum_{i=1}^{n} \bar{n}_i \ln \bar{n}_i + \frac{1}{a} \sum_{i=1}^{n} (1 + a\bar{n}_i) \ln(1 + a\bar{n}_i) \tag{28}$$

3.5 SUMMARY AND CONCLUDING REMARKS

(i) We have discussed three different interpretations of entropy,

(*a*) entropy of a probability distribution $(p_1, p_2, \ldots, p_n)$. This is a function of $p_1, p_2, \ldots, p_n$ which measures in some sense the uncertainty associated with the probability distribution.

(*b*) entropy of a certain information given in the form of moments of a probability distribution. This is the maximum uncertainty that remains after the given information has been used and it is a concave function of the moment values.

(*c*) entropy of a set of positive numbers whose sum is known. This is a measure of equality of the numbers.
We can also define cross-entropy of each type.

(*ii*) Kapur and Kesavan [14] and Kapur [18] gave methods of generating entropies of the first kind. Forte and Hughes [1] defined entropies of the second kind. We have shown that the second definition can also lead us to entropies of the third kind.

(*iii*) While Shannon's measure (1) refers to a discrete variate distribution, its discrete analogue of Forte and Hughes refers to a continuous-variate distribution.

(iv) If in addition to (7), we are also given

$$\int_{x_{i-1}}^{x_i} x f(x) dx = \lambda_i x_{i-1} + (1 - \lambda_i) x_i, \quad 0 < \lambda_i < 1 \tag{29}$$

and we maximize (6) subject to (7) and (29), the maximum value will be the sum of n terms where the ith term will depend on p_i, x_i, x_{i-1} and λ_i. This will also be the entropy of the information given by (7) and (29) and it will be a sum, but this will not be a discrete analogue of a entropy measure.

(v) Theil and Fiebig [22] and Lind [16] have also obtained discrete analogues of (1) by using the maximum entropy principle. The former exploited the continuity of the distribution function and the latter used the knowledge of the form of the density function, so that only the parameter had to be estimated in terms of the density function.

REFERENCES

1. Ali Ahmed and D.V. Gokhale (1989), "Entropy Expressions and their estimations for multivariate distributions" IEEE Trans Inf. Theory, 35 (3), 688–692.
2. B. Forte and W.Hughes (1988), "The Maximum Entropy Principle: A tool to define New Entropies", Reports of Mathematical Physics, 26 (2), 227–238.
3. B. Forte and P.K. Sahoo (1986), "Minimal loss of information and optimal threshold for digital images", Manuscript.

4. J. Havrda and F. Charvat (1967). ''Quantification methods of classification processes : Concept of Structural α Entropy. ''Kybernetica. vol. 3. 30–35.
5. E.T. Jaynes. (1957). ''Information theory and statistical mechanics''. Physical Reviews. vol. 106. 620–630.
6. J.N. Kapur (1968). ''On information of order α and type β''. Proc. Ind. Acad. Sci. vol. 48A 65–75.
7. J.N. Kapur (1972). ''Measures of uncertainty. mathematical programming and physics. Jour. Ind. Soc. Agr. Stat.. vol 24. 47–66.
8. J.N. Kapur (1983). ''Comparative assessment of various measures of entropy''. Jour Inf and Opt. Sci., vol. 4 no. 1. 207–232.
9. J.N. Kapur (1986). ''Four families of measures of entropy''. Ind. Jour. Pure and App. Math.. Vol. 17. no. 4. 429–449.
10. J.N. Kapur (1992). ''Generalised maximum entropy principle in Population Dynamics. Innovation Diffusion and Chemical Kinetics''. Jour. of Math & Phy. Sciences. 26. 183–213.
11. J.N. Kapur (1990). ''Maximum-Entropy Models in Science and Engineering. Wiley Eastern and John Wiley & Sons.
12. J.N. Kapur and H.K. Kesavan (1987). Generalised Maximum Entropy Principle with Application. Sandford Educational Press. University of Waterloo. Canada.
13. H.K. Kesavan and J.N. Kapur (1982). ''Generalised Maximum Entropy Principle''. IEEE Trans. Sys. Man. Cyb. 19 (9) 1042–1050
14. J.N. Kapur & H.K. Kesavan, ''On the family of solutions of generalised maximum entropy and minimum cross-entropy problems'', Int Jour Gen. Systems, 16, 199–219.
15. S. Kullback and R.A. Leibler, (1951). ''On information and sufficiency'', Ann Math. Stat.
16. N. Lind (1989). ''Principle of Minimum Information''. IRR. Report No. 11. University of Waterloo.
17. C.C.V. Lazo, and P.N. Rathie (1978), ''On the entropy of continuous probability distributions'', IEEE Trans. Inf. Theory IT 29 (11).
18. A. Renyi (1961). ''On measures of entropy and information''. Proc. 4th Berkeley Symp. Math. Stat. Prob., Vol. 1, 547–561.
19. C.E. Shannon (1948). ''A mathematical theory of communication''. Bell System. Tech. J. vol. 27. 379–423. 623–659.
20. B.D. Sharma & I.J. Taneja (1974). ''Entropy of type (α, β) and other generalised measures of information''. Metrika. vol. 22. 205–214.
21. B.D. Sharma & D.P. Mittal (1975). ''New non-additive measures of entropy for discrete probability distributions'', J. Math. Sci.. vol. 10 28–40.
22. H. Theil & D.G. Fiebig (1984), ''Exploiting continuity : Maximum Entropy Estimation of Continuous Distribution,'' Ballinger. Cambridge.

4

CHARACTERISATION THEOREMS FOR SHANNON AND HAVRDA-CHARVAT MEASURES OF ENTROPY

[Shannon and Havrda-Charvat measures of entropy have been characterised in terms of sum, concavity and recursivity properties.]

4.1 INTRODUCTION

There have been more than a dozen characterisations of measures of entropy due to Shannon and Havrda-Charvat. The characterisation theorems have been given by Shannon (1948), Tverberg (1958), Khinchin (1953), Fadeev (1956), Chaundy and Mcleod (1960), Aczel, Forte and Ng (1974), Daroczy (1966), Burge (1967), Forte and Daroczy (1968), Havrda-Charvat (1967) and others.

The statements of these theorems and proofs of some of them are given in Aczel and Daroczy (1975) and Mathai and Rathie (1975) where full references to other original papers are given.

None of these papers uses the concavity property of these measures. This property is very important when these measures are used in the maximum entropy principle [Jaynes (1957) Kapur (1990), Kapur and Kesavan (1987, 1992)] where we maximize a measure of entropy subject to some linear constraints on probabilities. The importance arises due to the fact that the stationary value of a concave function, when it is obtained by Lagrange's method, will give the globally maximum value. Thus, when we get the stationary value, we have not to worry whether it is maximum or minimum and we have not to worry whether it is the largest maximum value. The problem of checking these can be a complicated problem, since entropy is a function of n variables. However, because of the concavity of Shannon and Havrda-Charvat's measures, these problems are automatically taken care of and the problem of entropy maximization becomes a relatively simple matter.

It is therefore necessary that concavity property of these measures should be considered as a property of primary importance and should be emphasised in characterisation theorems.

The object of the present paper is to characterise both these measures in terms of mainly concavity and recursivity properties.

4.2 DERIVATION OF SHANNON MEASURE

Let $P = (p_1, p_2, \ldots, p_n)$ be a probability distribution so that

$$p_i \geq 0, p_2 \geq 0, \ldots, p_n \geq 0, p_1 + p_2 + \ldots + p_n = 1 \tag{1}$$

Let D be the domain in n-dimensional Euclidean space determined by (1) and let

$$H_n(p_1, p_2, ..., p_n) = \phi(p_1) + \phi(p_2) + ... + \phi(p_n), \quad (2)$$

where $\phi(x)$ is a continuous, differentiable and concave function in (0, 1) for which

$$\phi(0) = \phi(1) = 0 \quad (3)$$

We now study the properties of this function $H_n(p_1, p_2, \ldots, p_n)$.

(*i*) Since the sum of any number of continuous, differentiable and concave functions is itself a continuous, differentiable and concave function $H_n(p_1, p_2, \ldots, p_n)$ is a continuous, differentiable and concave function in the domain D,

(*ii*) $H_n(p_1, p_2, \ldots, p_n)$ is permutationally symmetric, i.e. it does not change when $p_1, p_2, \ldots, p_n$ are permuted among themselves,

(*iii*) $H_{n+1}(p_1, p_2, ..., p_n, 01 = \phi(p_1) + \phi(p_2) + ... + \phi(p_n) + \phi(0)$

$$= \phi(p_1) + \phi(p_2) + ... + \phi(p_n)$$

$$= H_n(p_1, p_2, ..., p_n) \quad (4)$$

Thus our function satisfies the expansionability property.

(*iv*) If we maximize $H_n(p_1, p_2, \ldots, p_n)$ subject to $\sum_{i=1}^{n} p_i = 1$ by using Lagrange's method, we get

$$\frac{\phi'(p_i)}{1} = \frac{\phi'(p_2)}{1} = ... = \frac{\phi'(p_n)}{1} \quad (5)$$

Since $\phi(x)$ is a concave function, $\phi'(x)$ is a monotonic decreasing function and to every value of $\phi'(x)$ there is a unique value of x so that (5) gives

$$\frac{p_1}{1} = \frac{p_2}{1} = ... = \frac{p_n}{1} = \frac{\sum_{i=1}^{n} p_i}{n} = \frac{1}{n} \quad (6)$$

In view of the concavity of $H_n(p_1, p_2, \ldots, p_n)$, this probability distribution will give the globally maximum value of this function so that

$$H_n(p_1, p_2, ..., p_n) \leq H_n\left(\frac{1}{n}, \frac{1}{n}, ..., \frac{1}{n}\right) \quad (7)$$

or

$$\sum_{i=1}^{n} \phi(p_i) \leq n\phi\left(\frac{1}{n}\right) \quad (8)$$

Thus the function is globally maximum for the uniform distribution and its maximum value is $n\ \phi\left(\frac{1}{n}\right)$.

(*v*) Since $\phi(x)$ is a concave function,

$$\phi(\lambda x_1 + (1-\lambda) x_2) \geq \lambda\phi(x_1) + (1-\lambda)\phi(x_2), 0 \leq \lambda \leq 1 \quad (9)$$

Putting

$$x_1 = \frac{1}{n}, \quad x_2 = 0; \quad \lambda = \frac{n}{n+1}, \quad (1-\lambda) = \frac{1}{n+1} \tag{10}$$

(*a*) gives

$$\phi\left(\frac{1}{n+1}\right) \geq \frac{n}{n+1}\phi\left(\frac{1}{n}\right) + \frac{1}{n+1}\phi(0) \tag{11}$$

and since $\phi(0) = 0$, we get

$$(n+1)\,\phi\left(\frac{1}{n+1}\right) \geq n\,\phi\left(\frac{1}{n}\right) \tag{12}$$

but $n\,\phi(1/n)$ is the maximum value of $H_n(p_1, p_2, \ldots, p_n)$ and $(n+1)$ $\phi(1/(n+1))$ is the maximum value of $H_{n+1}(p_1, p_2, \ldots, p_{n+1})$. Thus the maximum value of $H_n(p_1, p_2, \ldots, p_n)$ is a monotonic increasing function of n.

(*vi*) Since $H_n(p_1, p_2, \ldots, p_n)$ is a concave function, its minimum value over the domain D will occur at the corners of D viz. at

$$\Delta_1 = (1, 0, \ldots, 0), \Delta_2 = (0, 1, 0, \ldots, 0), \ldots, \Delta_n = (0, 0, \ldots, 1) \tag{13}$$

and the minimum value at each of the degenerate probability distributions is

$$\phi(1) + \phi(0) + \ldots + \phi(0) = 0 \tag{14}$$

Thus the minimum value of this function is zero and it occurs for the n degenerate distributions. This also implies that

$$H_n(p_1, p_2, \ldots, p_n) \geq 0 \tag{15}$$

(*vii*) Now we assume the recursivity property or the branching property holds so that

$$H_n(p_1, p_2, \ldots, p_n) = H_{n-1}(p_1 + p_2, p_3, \ldots, p_n) + (p_1 + p_2)\,H_2\left(\frac{p_1}{p_1 + p_2}, \frac{p_2}{p_1 + p_2}\right) \tag{16}$$

This gives

$$\phi(p_1) + \phi(p_2) = \phi(p_1 + p_2) + (p_1 + p_2)\left[\phi\left(\frac{p_1}{p_1 + p_2}\right) + \phi\left(\frac{p_2}{p_1 + p_2}\right)\right] \tag{17}$$

This is a functional equation whose only continuous differentiable solution is [see Appendix]

$$\phi(x) = Ax \ln x \tag{18}$$

If it is to be a concave function, A has to be negative, so that

$$\phi(x) = -kx \ln x, \tag{19}$$

where k is an arbitrary positive constant, so that

$$H_n(p_1, p_2, \ldots, p_n) = \sum_{i=1}^{n} \phi(p_i) = -k \sum_{i=1}^{n} p_i \ln p_i \tag{20}$$

4.3 DERIVATION OF HAVRDA-CHARVAT MEASURE

We proceed in the same manner as in section 2, but instead of (17) we take the equation

$$\phi(p_1) + \phi(p_2) = \phi(p_1 + p_2) + (p_1 + p_2)^\alpha \left[\phi\left(\frac{p_1}{p_1 + p_2}\right) + \phi\left(\frac{p_2}{p_1 + p_2}\right)\right] \tag{21}$$

This has a functional equation of which the only continuous differentiable solution is [see Appendix]

$$\phi(x) = A\,(x^\alpha - x) \tag{22}$$

To make it concave for all values of $\alpha \neq 1$, we take $A = \dfrac{1}{\alpha(1-\alpha)}$, so that

$$\phi(x) = \frac{x^\alpha - x}{\alpha(1-\alpha)}, \tag{23}$$

so that our measure of entropy is

$$H_n(p_1, p_2, \ldots, p_n) = \frac{1}{\alpha(1-\alpha)} \sum_{i=1}^{n} (p_i^\alpha - p_i) = \frac{\sum_{i=1}^{n} p_i^\alpha - 1}{\alpha(1-\alpha)}, \alpha \neq 1 \tag{24}$$

which is due to Havrda and Charvat [1967].

4.4 CHARACTERISATION THEOREMS

Theorem 1. The only function $H_n(p_1, p_2, \ldots, p_n)$ which is of the form $\sum_{i=1}^{n} \phi(p_i)$ where $\phi(x)$ is a continuous differentiable concave function defined on [0, 1] for which $\phi(0) = \phi(1) = 0$ and which satisfies the recursivity relation (16) is an arbitrary positive multiple of Shannon measure.

Theorem 2. The only function $H_n(p_1, \ldots, p_n)$ which is of the form $\sum_{i=1}^{n} \phi(p_i)$ where $\phi(x)$ is continuous, differentiable, concave function defined in [0, 1] for which $\phi(0) = \phi(1)$ and which satisfies the recursivity relation,

$$H_n(p_1, p_2, \ldots, p_n) = H_{n-1}(p_1 + p_2, p_3, \ldots, p_n) + (p_1 + p_2)^\alpha H_2\left(\frac{p_1}{p_1 + p_2}, \frac{p_2}{p_1 + p_2}\right)$$

is a positive multiple of Havrda-Charvat measure (24).

4.5 MEASURES WHICH CAN BE CHARACTERISED IN TERMS OF SUM-FUNCTION PROPERTY CONCAVITY AND RECURSIVITY

We now consider the more general recursivity relation

$$\phi(p_1)+\phi(p_2)=\phi(p_1+p_1)+g(p_1+p_2)\left[\phi\left(\frac{p_1}{p_1+p_2}\right)+\phi\left(\frac{p_2}{p_1+p_2}\right)\right] \tag{25}$$

This gives

$$\phi(p_1)+\phi(p_2)+\phi(p_3)=[\phi(p_1+p_2)+\phi(p_3)]$$
$$+g(p_1+p_2)\left[\phi\left(\frac{p_1}{p_1+p_2}\right)+\phi\left(\frac{p_2}{p_1+p_2}\right)\right] \tag{26}$$

Using (25) with the first term on the R.H.S. of (26)

$$\phi(p_1)+\phi(p_2)+\phi(p_3)=\phi(p_1+p_2+p_3)+g(p_1+p_2+p_3)\left[\phi\left(\frac{p_1+p_2}{p_1+p_2+p_3}\right)+\phi\left(\frac{p_3}{p_1+p_2+p_3}\right)\right]$$
$$+g(p_1+p_2)\left[\phi\left(\frac{p_1}{p_1+p_2}\right)+\phi\left(\frac{p_2}{p_1+p_2}\right)\right] \tag{27}$$

Using (25) again

$$\phi\left(\frac{p_1}{p_1+p_2+p_3}\right)+\phi\left(\frac{p_2}{p_1+p_2+p_3}\right)$$
$$=\phi\left(\frac{p_1+p_2}{p_1+p_2+p_3}\right)+g\left(\frac{p_1+p_2}{p_1+p_2+p_3}\right)\left[\phi\left(\frac{p_1}{p_1+p_2}\right)+\phi\left(\frac{p_2}{p_1+p_2}\right)\right] \tag{28}$$

From (27) and (28)

$$\phi(p_1)+\phi(p_2)+\phi(p_3)=\phi(p_1+p_2+p_3)+g(p_1+p_2+p_3)$$
$$\left[\phi\left(\frac{p_1}{p_1+p_2+p_3}\right)+\phi\left(\frac{p_2}{p_1+p_2+p_3}\right)r+\phi\left(\frac{p_3}{p_1+p_2+p_3}\right)\right]$$
$$-g\left(\frac{p_1+p_2}{p_1+p_2+p_3}\right)\left[\phi\left(\frac{p_1}{p_1+p_2}\right)+\phi\left(\frac{p_2}{p_1+p_2}\right)\right]$$
$$+g(p_1+p_2)\left[\phi\left(\frac{p_1}{p_1+p_2}\right)+\phi\left(\frac{p_2}{p_1+p_2}\right)\right] \tag{29}$$

This recursivity property (25) can be extended to three probabilities if

$$g\left(\frac{p_1+p_2}{p_1+p_2+p_3}\right)=\frac{g(p_1+p_2)}{g(p_1+p_2+p_3)} \tag{30}$$

which has the solution,

$$g(x)=x^{\alpha}, \tag{31}$$

so that (25) becomes

$$\phi(p_1)+\phi(p_2)=\phi(p_1+p_2)+(p_1+p_2)^{\alpha}\left[\phi\left(\frac{p_1}{p_1+p_2}\right)+\phi\left(\frac{p_2}{p_1+p_2}\right)\right], \tag{32}$$

which is the same as (21).

Thus, the only measures which can be characterised in terms of sum-property, concavity and recursivity properties are Shannon's and Havrda-Charvat measures.

APPENDIX

Consider the functional equation,

$$\phi(x)+\phi(y)=\phi(x+y)+(x+y)^{\alpha}\left[\phi\left(\frac{x}{x+y}\right)+\phi\left(\frac{y}{x+y}\right)\right] \tag{A1}$$

Putting $x=y$, we get

$$2\phi(x)=\phi(2x)+(2x)^{\alpha}\,[2\phi(1/2)] \tag{A2}$$

Putting

$$\phi(x)=x^{\alpha}\,\Psi(x) \tag{A3}$$

we get,

$$2x^{\alpha}\,\Psi(x)=(2x)^{\alpha}\,\Psi(2x)+(2x)^{\alpha}\,2\psi\left(\frac{1}{2}\right)\left(\frac{1}{2}\right)^{\alpha}$$

or,

$$\psi(x)=2^{\alpha-1}\psi(2x)+\psi\left(\frac{1}{2}\right) \tag{A4}$$

so that

$$\psi'(x)=2^{\alpha}\psi'(2x) \tag{A5}$$

of which the solution is

$$\psi'(x)=\frac{A}{x^{\alpha}} \tag{A6}$$

Integrating

$$\psi(x)=\frac{Ax^{-\alpha+1}}{-\alpha+1}+B \tag{A7}$$

$$\phi(x)=\frac{Ax}{1-\alpha}+Bx^{\alpha} \tag{A8}$$

Substituting in (A2)

$$\frac{2Ax}{1-\alpha}+2Bx^{\alpha}=\frac{2Ax}{1-\alpha}+B2^{\alpha}x^{\alpha}+2^{\alpha+1}x^{\alpha}\quad\left[\frac{1}{2}\frac{A}{1-\alpha}+B\left(\frac{1}{2}\right)^{\alpha}\right]$$

or,
$$2B = B2^{\alpha} + \frac{2^{\alpha}}{1-\alpha}A + 2B$$

or,
$$B = -\frac{A}{1-\alpha}$$

so that,
$$\phi(x) = A\frac{x^{\alpha} - x}{\alpha - 1},$$

which automatically satisfies (2)

Since, it is to be a concave function, we choose $A = -\frac{k}{\alpha}$, so that

$$\phi(x) = k\frac{x^{\alpha} - x}{\alpha(1-\alpha)}, k > 0$$

As $\alpha \to 1$, we get

$$\phi(x) = -kx \ln x$$

REFERENCES

1. J. Aczel and Z. Daroczy (1975), On Measures of Information and their Characterizations, Academic Press, New York.
2. J.H. Havrda and F. Charvat (1967), ''Quantification Methods of Classification Processes: Concept of Structural α Entropy'', Kybernetica, Vol. 3, 30–35.
3. E.T. Jaynes (1957), ''Information Theory and Statistical Mechanics'', Physical Reviews, Vol. 106, pp. 620–630.
4. J.N. Kapur (1990), Maximum Entropy Models in Science & Engineering, John Wiley & Sons, New York.
5. J.N. Kapur and H.K. Kesavan (1987), ''Generalised Maximum Entropy Principle with Applications'', Sand Ford Educational Press, Waterloo: Canada.
6. J.N. Kapur and H.K. Kesavan (1992), ''Entropy Optimization Principles and their Applications'', Academic Press, New York.
7. A.M. Mathai and P.N. Rathie (1975), Basic Concepts of Information Theory and Statistics, John Wiley, New York.
8. C.E. Shannon (1948), ''A Mathematical Theory of Communication'', Bell System Tech. J., Vol. 27, pp. 379–423, 623–659.

5

ON RECURSIVITY PROPERTY OF MEASURES OF ENTROPY

[A strong recursivity property is defined and the recursivity and strong recursivity properties of a measure of entropy are examined. It is shown that the only measures which satisfy the sum function property and the strong recursivity property are the Havrda-Charvat and Shannon's measures of entropy.]

5.1 INTRODUCTION

A measure of entropy H_n $(p_1, p_2, \ldots, p_n)$ will be said to have the recursivity property with respect to recursive function g (p_1, p_2) if

$$H_n(p_1, p_2, ..., p_n) = H_{n-1}(p_1 + p_2, p_3, ..., p_n) + g\,(p_1, p_2)H_2\left(\frac{p_1}{p_1 + p_2}, \frac{p_2}{p_1 + p_2}\right) \quad (1)$$

holds for all $n \geq 3$. Here g (p_1, p_2) is a known continuous function. The property is called recursivity property because if we know H_2 (p_1, p_2), we can find H_3 (p_1, p_2, p_3), knowing which we can find H_4 (p_1, p_2, p_3, p_4), knowing which we can find H_5 $(p_1, p_2, \ldots, p_5)$ and continuing in this way we can find H_n $(p_1, p_2, \ldots, p_n)$ for all values of n.

If we consider entropy measures of the sum form, i.e. if we consider entropy measures of the form

$$H_n(p_1, p_2, ..., p_n) = \phi(p_1) + \phi(p_2) + ... + \phi(p_n) \quad (2)$$

we do not really need the recursivity property to find H_n $(p_1, p_2, \ldots, p_n)$ since knowing the function ϕ (p), we can write down the expression for H_n $(p_1, p_2, \ldots, p_n)$, at once.

However (1) and (2) give

$$\phi(p_1) + \phi(p_2) = \phi(p_1 + p_2) + g\,(p_1, p_2 + p_2)\left[\phi\left(\frac{p_1}{p_1 + p_2}\right) + \phi\left(\frac{p_2}{p_1 + p_2}\right)\right] \quad (3)$$

Two problems now arise:

(*i*) Knowing ϕ (p), determine g (p_1, p_2)

(*ii*) Knowing g (p_1, p_2), determine ϕ (p)

The first problem is simple. Thus when

$$(i)\ \phi(p) = p \ln p, \qquad g(p_1, p_2) = (p_1 + p_2) \tag{4}$$

$$(ii)\ \phi(p) = p^{\alpha} - p, \qquad g(p_1, p_2) = (p_1 + p_2)^{\alpha} \tag{5}$$

$$(iii)\ \phi(p) = \ln p, \qquad g(p_1, p_2) = \frac{\ln p_1 + \ln p_2 - \ln(p_1 + p_2)}{\ln p_1 + \ln p_2 - 2\ln(p_1 + p_2)} \tag{6}$$

In the same way for every concave function $\phi(p)$, we have a function which may be simple or complicated, but which will ensure that the simple recursivity property (3) holds. We shall later define a strong recursivity property which will be satisfied by (4) and (5) and not by (6).

5.2 THE SOLUTION OF THE SECOND PROBLEM

The second problem given above is however more difficult. Knowing $g(p_1, p_2)$, (3) is a functional equation to determine $\phi(p)$. It may or may not have a solution and even when it exists, it may not be easy to find it.

However, when $g(p_1, p_2) = p_1 + p_2$, we get $\phi(p) = p \ln p$ and we get Shannon's [4] measure of entropy.

Similarly, when $g(p_1, p_2) = (p_1 + p_2)^{\alpha}$, we get $\phi(p) = (p^{\alpha} - p)/\alpha(\alpha - 1)$ so that we get Havrda-Charvat [2] measure of entropy.

We now consider a more general form

$$g(p_1, p_2) = (p_1^k + p_2^k)^{\alpha} \tag{7}$$

and try to find $\phi(p)$. In this case (3) gives

$$\phi(p_1) + \phi(p_2) = \phi(p_1 + p_2) + (p_1^k + p_2^k)^{\alpha}\left(\phi\left(\frac{p_1}{p_1 + p_2}\right) + \phi\left(\frac{p_2}{p_1 + p_2}\right)\right) \tag{8}$$

Putting $p_1 = p_2 = x$, we get

$$2\phi(x) = \phi(2x) + (2x^k)^{\alpha}[2\phi(1/2)] \tag{9}$$

Let

$$\phi(x) = x^{k\alpha}\,\psi(x) \tag{10}$$

then we get

$$2x^{k\alpha}\,\psi(x) = (2x)^{k\alpha}\,\psi(2x) + 2^{\alpha+1-k\alpha}\,x^{k\alpha}\,\psi(1/2) \tag{11}$$

or

$$2\,\psi(x) = 2^{k\alpha}\,\psi(2x) + 2^{\alpha+1-k\alpha}\,\psi(1/2) \tag{12}$$

Differentiating with respect to x

$$2\psi'(x) = 2^{k\alpha+1}\psi'(2x), \tag{13}$$

which has the solution

$$\psi'(x) = \frac{A}{x^{k\alpha}} \tag{14}$$

Integrating,

$$\psi(x) = A'x^{-k\alpha+1} + B \tag{15}$$

Substituting in (12)

$$A'2^{\alpha} = B(2 - 2^{k\alpha} - 2^{\alpha+1-k\alpha}) \tag{16}$$

From (15) and (16)

$$\phi(x) = B\left[\frac{2 - 2^{k\alpha} - 2^{\alpha+1-k\alpha}}{2^{\alpha}}x + x^{k\alpha}\right] \tag{17}$$

Since we want $\phi(0) = 0$, $\phi(1) = 0$, we get

$$(1 - 2^{k\alpha-1})(1 - 2^{\alpha(1-k)}) = 0 \tag{18}$$

There are now three possibilities.

(*i*) $k\alpha = 1$, This gives $\quad \phi(x) = B(x - x) = 0 \quad$ (19)

which gives a trivial solution.

(*ii*) $\alpha = 0$, $\quad \phi(x) = B(1 - x)$

$$\sum_{i=1}^{n} \phi(p_i) = B(n-1), \tag{20}$$

which is also a trivial solution.

(*iii*) $k = 1$ $\quad \phi(x) = B(x^{\alpha} - x) \quad$ (21)

which gives Havrda-Charvat measure of entropy

Putting $B = [\alpha(1-\alpha)]^{-1}$ and letting $\alpha \to 1$ we get Shannon's measure of entropy.

Thus by generalising $g(p_1, p_2)$ from $(p_1 + p_2)^{\alpha}$ to $(p_1^k + p_2^k)^{\alpha}$, we do not get any new solution.

5.3 DISCUSSION OF ANOTHER POSSIBILITY

In the above discussion, we had $\phi(0) = 0$, but we also wanted $\phi(1) = 0$ so that the minimum value of entropy measure is zero. However, in entropy maximization problems, this condition is not necessary. Here we can have even negative entropy so long as the maximizing probabilities are non-negative. As such we use (17) itself to get the measure

$$\sum_{i=1}^{n} \phi(p_i) = B\left(\sum_{i=1}^{n} p_i^{k\alpha} + \frac{2 - 2^{k\alpha} + 2^{\alpha+1-k\alpha}}{2^{\alpha}} \sum_{i=1}^{n} p_i\right) \tag{22}$$

$$= B\left(\sum_{i=1}^{n} p_i^{\beta} + \frac{2 - 2^{\beta} - 2^{\alpha+1-\beta}}{2^{\alpha}}\right) \tag{23}$$

$$= \frac{c}{\beta(1-\beta)}\left[\sum_{i=1}^{n} p_i^{\beta} - 1\right]$$

$$+ \frac{c}{\beta(1-\beta)}\left[\frac{2^{\alpha} + 2 - 2^{\beta} - 2^{\alpha+1-\beta}}{2^{\alpha}}\right] \tag{24}$$

This differs from Havrda-Charvat measure only by a constant

$$K = \frac{c}{\beta 2^{\alpha}}\left[\frac{2^{\beta}(2^{\alpha-\beta}-1) - 2(2^{\alpha-\beta}-1)}{1-\beta}\right]$$

$$= \frac{c}{\beta 2^{\alpha-1}} \frac{(2^{\beta-1}-1)(2^{\alpha-\beta}-1)}{1-\beta} \tag{25}$$

which is negative if $\alpha > \beta > 1, \quad \beta < \alpha < 1$

and is positive if $\alpha < \beta < 1, \quad \beta > \alpha > 1 \quad (26)$

However, essentially this leads to Havrda-Charvat measure and to Shannon measure when $\beta \to 1$. Thus this extension also does not lead to any new measure.

5.4 A STRONG RECURSIVITY PROPERTY

From (2),

$$H_n(p_1, p_2, \ldots, p_n) = H_{n-1}(p_1 + p_2, p_3, \ldots, p_n)$$

$$+ g(p_1, p_2) H_2\left(\frac{p_1}{p_1+p_2}, \frac{p_2}{p_1+p_2}\right) \tag{27}$$

$$= H_{n-2}(p_1 + p_2 + p_3, p_4, \ldots, p_n)$$

$$+ g(p_1+p_2, p_3)\, H_2\left(\frac{p_1+p_2}{p_1+p_2+p_3}, \frac{p_3}{p_1+p_2+p_3}\right)$$

$$+ g(p_1, p_2)\, H_2\left(\frac{p_1}{p_1+p_2}, \frac{p_2}{p_1+p_2}\right) \tag{28}$$

$$= H_{n-3}(p_1 + p_2 + p_3 + p_4, p_5, \ldots, p_n)$$

$$+ g(p_1+p_2+p_3, p_4)\, H_2\left(\frac{p_1+p_2+p_3}{p_1+p_2+p_3+p_4}, \frac{p_4}{p_1+p_2+p_3+p_4}\right)$$

$$+ g(p_1+p_2, p_3)\, H_2\left(\frac{p_1+p_2}{p_1+p_2+p_3}, \frac{p_3}{p_1+p_2+p_3}\right)$$

$$+ g(p_1, p_2)\, H_2\left(\frac{p_1}{p_1+p_2}, \frac{p_2}{p_1+p_2}\right) \tag{29}$$

In general,

$$H_n(p_1, p_2, ..., p_n) = H_{n-k}(p_1 + p_2 + ... + p_k, p_{k+1}, ..., p_n)$$

$$+ \sum_{i=1}^{k} g(\sum_{j=1}^{i} p_j, p_{i+1}) H_2 \left(\frac{\sum_{j=1}^{i} p_j}{\sum_{j=1}^{i+1} p_j}, \frac{p_{i+1}}{\sum_{j=1}^{i+1} p_j} \right) \tag{30}$$

To calculate H_n, we have first to calculate k entropies of type H_2 and one entropy of type H_{n-k}.

A stronger type of recursivity enables us to express H_n in term of two entropies, one of type H_{n-k+1} and the other of type H_k. In particular we would like to express H_n in terms of H_{n-2} and H_3. From (27)

$$H_3\left(\frac{p_1}{p_1+p_2+p_3}, \frac{p_2}{p_1+p_2+p_3}, \frac{p_3}{p_1+p_2+p_3}\right) = H_2\left(\frac{p_1+p_2}{p_1+p_2+p_3}, \frac{p_3}{p_1+p_2+p_3}\right)$$

$$+g\left(\frac{p_1+p_2}{p_1+p_2+p_3}, \frac{p_3}{p_1+p_2+p_3}\right) H_2\left(\frac{p_1}{p_1+p_2}, \frac{p_2}{p_1+p_2}\right) \tag{31}$$

From (28) and (31)

$$H_n(p_1, p_2, ..., p_n) = H_{n-2}(p_1+p_2+p_3, p_4, ..., p_n) + g(p_1+p_2, p_3)$$

$$\left[H_3\left(\frac{p_1}{p_1+p_2+p_3}, \frac{p_2}{p_1+p_2+p_3}, \frac{p_3}{p_1+p_2+p_3}\right) - g\left(\frac{p_1}{p_1+p_2+p_3}, \frac{p_2}{p_1+p_2+p_3}\right) H_2\left(\frac{p_1}{p_1+p_2}, \frac{p_2}{p_1+p_2}\right)\right]$$

$$+ g(p_1, p_2) H_2\left(\frac{p_1}{p_1+p_2}, \frac{p_2}{p_1+p_2}\right) \tag{32}$$

This will be expressed in terms of H_{n-2} and H_3 only if

$$g(p_1+p_2, p_3)\, g\left(\frac{p_1}{p_1+p_2+p_3}, \frac{p_2}{p_1+p_2+p_3}\right) = g(p_1, p_2) \tag{33}$$

This will be satisfied if $g\ (p_1+p_2, p_3) = f(p_1+p_2+p_3)$

$$\text{or} \quad g\ (p_1, p_2) = f(p_1+p_2) \tag{34}$$

Equations (33) and (34) give

$$f(p_1+p_2+p_3)\ f\left(\frac{p_1+p_2}{p_1+p_2+p_3}\right) = f(p_1+p_2)$$

$$\text{or} \quad f\left(\frac{p_1+p_2}{p_1+p_2+p_3}\right) = \frac{f(p_1+p_2)}{f(p_1+p_2+p_3)} \tag{35}$$

The functional equation (35) has the solution

$$f(x) = x^{\alpha} \tag{36}$$

so that our equations (27), (28), (29) give

$$H_n(p_1, p_2, \ldots, p_n) = H_{n-1}(p_1 + p_2, p_3, \ldots, p_n) + (p_1 + p_2)^\alpha H_2\left(\frac{p_1}{p_1 + p_2}, \frac{p_2}{p_1 + p_2}\right)$$

$$= H_{n-2}(p_1 + p_2 + p_3, p_4, \ldots, p_n)$$

$$+ (p_1 + p_2 + p_3)^\alpha H_3\left(\frac{p_1}{p_1 + p_2 + p_3}, \frac{p_2}{p_1 + p_2 + p_3}, \frac{p_3}{p_1 + p_2 + p_3}\right)$$

$$= H_{n-k+1}(p_1 + p_2 + \ldots + p_k, p_{k+1}, \ldots, p_n)$$

$$+ (p_1 + p_2 + \ldots + p_k)^\alpha H_k\left(\frac{p_1}{\sum_{j=1}^{k} p_j}, \ldots, \frac{p_k}{\sum_{j=1}^{k} p_j}\right) \tag{37}$$

Thus Havrda-Charvat [2] and Shannon [7] measures are the only measures of entropy for which the strong recursivity condition applies, i.e. for which H_n can be expressed in terms of H_{n-k+1} and H_k for all possible value of n and k ($< n$).

5.5 CONCLUDING REMARKS

Shannon's measure is additive and recursive.
Havrda and Charvat's measure is non-additive and recursive.
Renyi's measure is additive but is not recursive.
The maximum entropy probability distributions given by Havrda-Charvat and Renyi's measures are identical. This shows that neither additivity nor recursivity is essential for a measure to be used in maximum entropy principle.

Recursivity is a nice property but, this is not an essential property. Further discussion of these properties and their implications will be found in the references [1, 3–6, 7].

REFERENCES

1. J. Aczel and Z. Daroczy (1975), Measures of Information and their Applications, Academic Press, New York.
2. J.H. Havrda and F. Charvat (1967), Quantification of Methods of Classifications Processes: Concept of structural α entropy, Kybernetica **3**, 30–35.
3. J.N. Kapur (1989), "Maximum Entropy Models in Science and Engineering", Wiley Eastern, New Delhi and John Wiley, New York.
4. J.N. Kapur and H.K. Kesavan (1987), "Generalised Maximum Entropy Principle (with Applications)" SandFord Educational Press, Waterloo University, Canada.
5. J.N. Kapur and H.K. Kesavan (1992), "Entropy Optimization Principles and their Applications Academic press, New York.
6. A Mathai and P.N. Rathie (1975), "Basic Concepts of Information Theory and Statistics", John Wiley, New York.
7. C.E. Shannon (1948), "Mathematical Theory of Communication" Bell System Tech. Jour. **27**, 379–423.

6
ARE ADDITIVITY AND RECURSIVITY IDENTICAL?

[It is shown that the additivity and recursivity properties which are usually treated as two separate properties of measures of entropy are essentially identical under some general conditions.]

6.1 INTRODUCTION

Shannon in his seminal paper [4] on information theory took the recursivity property or the branching principle as one of the three basic assumptions which he used for characterising his measure of entropy

$$S(P) = -\sum_{i=1}^{n} p_i \ln p_i \tag{1}$$

for the probability distribution

$$P = (p_1, p_2, \ldots, p_n) \tag{2}$$

and obtained the additivity property as a property of this measure, so that in a sense, he deduced the additivity property from the recursivity property and the other properties which he assumed for his measure of entropy.

In almost all the books on the subject (Aczel & Daroczy [1], Kapur [2], Mathai and Rathie [3]), these are mentioned as two separate properties of Shannon's measure of entropy.

It is sometimes felt that the properties are separate since the recursivity property belongs to one distribution P, while the additivity property belongs to at least two probability distributions P and Q.

We show here that each property can be deduced from the other, so that the two properties are essentially the same. We may mention here that what we have called additivity property is sometimes called strong additivity property in the literature, where its special case when the distributions are independent, is called the additivity property.

6.2 THE BRANCHING PRINCIPLE

According to this principle

$$\begin{aligned} &H_{m_1+m_2+\ldots+m_n}(p_{11}, p_{12}, \ldots, p_{1m_1};\ p_{21}, p_{22}, \ldots, p_{2m_2}, \ldots, p_{n1}, p_{n2}, \ldots, p_{nm_n}) \\ &\qquad = H_n(p_1, p_2, \ldots, p_n) \\ &\qquad + \sum_{i=1}^{n} p_i H_{m_i}\left(\frac{p_{i1}}{p_i}, \frac{p_{i2}}{p_i}, \ldots, \frac{p_{im_i}}{p_i}\right), \end{aligned} \tag{1}$$

where,

$$p_i = \sum_{j=1}^{m_i} p_{ij}, \quad i = 1, 2, ..., n$$

Graphically this is represented in Figure 6.1.

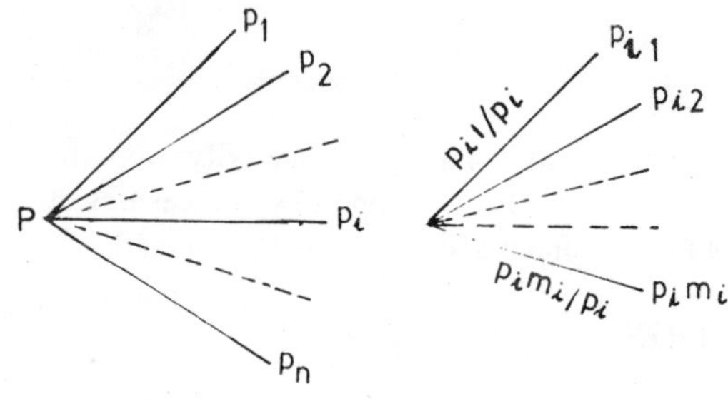

Fig. 6.1

This probability distribution has $m_1 + m_2 + \ldots + m_n$ sub outcomes which are grouped into n outcomes where the ith outcome has m_i sub outcomes ($i = 1, 2, \ldots, n$).

According to the branching principle, the entropy of the probability distribution of $m_1 + m_2 + \ldots + m_n$ sub outcomes

= entropy of the probability distribution of n outcomes

$+ \sum_{i=1}^{n}$ (probability of ith outcome) $\times$

(entropy of the probability distribution obtained by dividing the probabilities of the sub outcomes of the ith outcome by the probability of the ith outcome) (3)

$H_n(p_1, p_2, \ldots, p_n)$ is called the entropy between the n outcomes.

$H_{m_i}\left(\frac{p_{i1}}{p_i}, \frac{p_{i2}}{p_i}, \ldots, \frac{p_{im_i}}{p_i}\right)$ is called the entropy within the ith outcome

$\sum_{i=1}^{n} p_i H_{m_i}\left(\frac{p_{i1}}{p_i}, \frac{p_{i2}}{p_i}, \ldots, \frac{p_{im_i}}{p_i}\right)$ is called the entropy within outcomes.

Thus the branching principle can be considered as the following *decomposition principle*:

Total Entropy = Entropy between outcomes + entropy within outcomes (4)

Sometimes outcomes are called classes or groups and as such we may say

Total Entropy = Entropy between classes + entropy within classes. (5)

or,

Total Entropy = Entropy between groups + entropy within groups (6)

Each entropy may also be regarded as a measure of equality. Thus, $H_{m_i}\left(\frac{p_{i1}}{p_i}, \frac{p_{i2}}{p_i}, \ldots, \frac{p_{im_i}}{p_i}\right)$ will be maximum when

$$\frac{p_{i1}}{p_i} = \frac{p_{i2}}{p_i} = \ldots, \frac{p_{im_i}}{p_i} = \frac{1}{m_i} \tag{7}$$

i.e. when there is maximum equality within the ith class:

Also, $H_n\,(p_1, p_2, \ldots, p_n)$ will be maximum when $p_1 = p_2 = \ldots = p_n = \frac{1}{n}$ i.e. when there is maximum equality between the classes. Thus (3) can be written as

$$\begin{aligned}\text{Overall measure of equality} &= \text{measure of equality between classes}\\ &\quad + \sum_{i=1}^{n} p_i\ (\text{measure of equality within the } i^{\text{th}} \text{ class})\end{aligned} \tag{8}$$

We thus arrive at the overall measure of equality in a number of steps

(*i*) Find the measure of equality within the ith class,

(*ii*) Multiply this measure by p_i and sum up as i varies from 1 to n,

(*iii*) Find the entropy between the n classes,

(*iv*) Find the sum of entropies between and within classes.

If we like, we may further decompose each sub outcome into a number of sub outcomes (Figure 6.2)

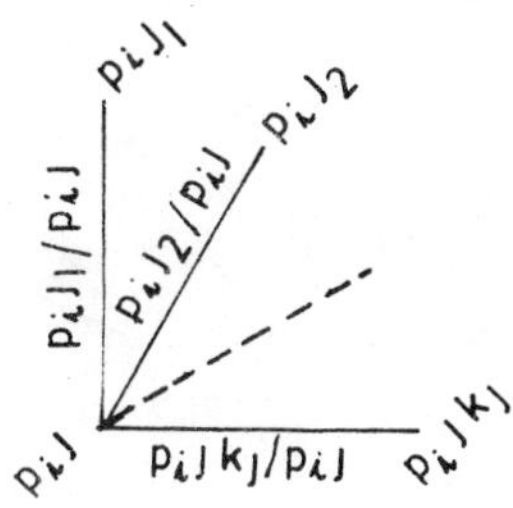

Fig. 6.2

and we shall get the result

$$\text{Final entropy} = \text{entropy between classes} + \sum_{i=1}^{n} p_i\ (\text{entropy within } i^{\text{th}} \text{ class})$$

$$+ \sum_{i=1}^{n} \sum_{j=1}^{m_i} p_{ij}\ \ (\text{entropy within the } ij^{\text{th}} \text{ sub class}) \tag{9}$$

Thus decomposition can be carried to any number of stages so that

Entropy in the whole world = entropy between countries + entropy within countries in the world.

Entropy within a country = entropy between states in the country + entropy within states in the country.

Entropy within a state = entropy between districts in the state + entropy within districts in the state,

and so on. This decomposition principle has a large number of applications (Thiel [5]).

6.3 ADDITIVITY PROPERTY

In the above discussion, we apparently talk of only one probability distribution and its outcomes and the sub outcomes.

However, the outcomes may be the outcomes of one experiment and the sub outcome may be the results of combining the results of two experiments. Thus the first experiment may give us the ith outcome and corresponding to the ith outcome of the first, the second experiment may give us the i–j^{th} sub outcome.

Let p_i be the probability of the ith outcome of the first experiment and q_{ij} be the conditional probability of the jth outcome of the second experiment, when it is known that the first experiment has given the ith outcome

$$(j = 1, 2, \ldots, m_i\,; \quad i = 1, 2, \ldots, n)$$

Figure 6.1 may now be relabelled as follows

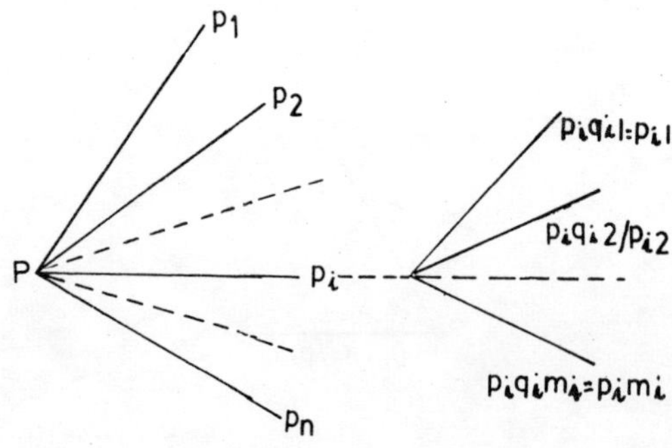

Fig. 6.3

and we rewrite the branching principle as

$$H_{m_1+m_2+\ldots+m_n}(p_1 q_{11}, p_1 q_{12}, \ldots, p_1 q_{1m_1}, \ldots, p_n q_{n_1}, p_n q_{n_2}, \ldots, p_n q_{nm_n})$$

$$= H_n(p_1, p_2, \ldots, p_n) + \sum_{i=1}^{n} p_i H_{m_i}(q_{i_1}, q_{i_2}, \ldots, q_{im_i}) \quad (10)$$

and we read it as follows

Entropy of the joint distribution = Entropy of the first distribution

$+ \sum_{i=1}^{n} p_i$ [conditional entropy of the second distribution when the first

distribution has the ith out-come] (11)

This is called the Additivity Property.

6.4 IDENTITY OF BRANCHING PRINCIPLE AND ADDITIVITY PROPERTY

It is obvious from the above discussion that there is no essential difference between these two properties. If there is any difference at all, it is of notation and terminology.

In the case of branching principle, we talk of one probability distribution of outcomes and sub outcomes and of entropy between and within outcomes.

In case of additivity, we break up outcomes into sub outcomes as the result of the second probability distribution and entropy within the classes becomes the expected value of conditional entropy of the second distribution.

In branching principle, we do not talk of why the branching is taking place.

The sub outcomes of the branching process become pairs of outcomes in the additivity case, i.e.

the $i-j^{th}$ sub outcome = $(i-j)^{th}$ outcome

and p_{ij} = probability of jth sub outcome of the ith outcome

= probability of $(i-j)^{th}$ outcome

= probability of ith outcome of the first distribution multiplied by the conditional probability of the jth outcome of the second distribution when it is known that the first outcome was the ith one

$$= p_i q_{ij} \tag{12}$$

6.5 RELATIONSHIP BETWEEN RECURSIVITY AND BRANCHING PRINCIPLES

The recursivity principle is stated as follows

$$H_n(p_1, p_2, \dots, p_n) = H_{n-1}(p_1 + p_2, p_3, \dots, p_n)$$

$$+ (p_1 + p_2) H_2\left(\frac{p_1}{p_1 + p_2}, \frac{p_2}{p_1 + p_2}\right), \quad n = 3, 4, 5, \dots \tag{13}$$

This is called recursivity property because if we know the expression for $H_2(p_1, p_2)$ and this recursivity formula, we can write in succession the formulae for H_3, H_4, $\dots, H_n, \dots$. Suppose we are given that

$$H_2(p_1, p_2) = -p_1 \ln p_1 - p_2 \ln p_2 \tag{14}$$

then (12) gives

$$H_3(p_1,p_2,p_3)=H_2(p_1+p_2,p_3)+(p_1+p_2)\,H_2\left(\frac{p_1}{p_1+p_2},\frac{p_2}{p_1+p_2}\right)$$

$$=-(p+p_2)\ln(p_1+p_2)-p_3\ln p_3$$

$$+(p_1+p_2)\left[-\frac{p_1}{p_1+p_2}\ln\frac{p_1}{p_1+p_2}-\frac{p_2}{p_1+p_2}\ln\frac{p_2}{p_1+p_2}\right]$$

$$=-p_1\ln p_1-p_2\ln p_2-p_3\ln p_3 \tag{15}$$

and if we assume that

$$H_n(p_1,p_2,\ldots,p_n)=-\sum_{i=1}^{n}p_i\ln p_i \tag{16}$$

then (13) will give us

$$H_{n+1}(p_1,p_2,\ldots,p_{n+1})=H_n(p_1+p_2,p_3,\ldots,p_{n+1})$$

$$+(p_1+p_2)\,H_2\left(\frac{p_1}{p_1+p_2},\frac{p_2}{p_1+p_2}\right)$$

$$=-(p_1+p_2)\ln(p_1+p_2)-\sum_{i=3}^{n+1}p_i\ln p_i$$

$$+(p_1+p_2)\left(-\frac{p_1}{p_1+p_2}\ln\frac{p_1}{p_1+p_1}-\frac{p_2}{p_1+p_2}\ln\frac{p_2}{p_1+p_2}\right)$$

$$=\sum_{i=1}^{n+1}p_i\ln p_i, \tag{17}$$

so that the general formula is proved by mathematical induction. Now we want to show that recursivity property implies branching property. From (13)

$$H_3(p_1,p_2,p_3)=H_2(p_1+p_2,p_3)+(p_1+p_2)\,H_2\left(\frac{p_1}{p_1+p_2},\frac{p_2}{p_1+p_2}\right) \tag{18}$$

which is the branching property when two sub outcomes are combined. Again from (13)

$$H_4(p_1,p_2,p_3,p_4)=H_3(p_1+p_2,p_3,p_4)$$

$$+(p_1+p_2)\,H_2\left(\frac{p_1}{p_1+p_2},\frac{p_2}{p_1+p_2}\right) \tag{19}$$

On using (13) again

$$H_4(p_1,p_2,p_3,p_4)=H_2(p_1+p_2+p_3,p_4)$$

$$+(p_1+p_2+p_3)\,H_2\left(\frac{p_1+p_2}{p_1+p_2+p_3},\frac{p_3}{p_1+p_2+p_3}\right)$$

$$+(p_1+p_2)\,H_2\left(\frac{p_1}{p_1+p_2},\frac{p_2}{p_1+p_2}\right)$$

$$=H_2(p_1+p_2+p_3,p_4)+(p_1+p_2+p_3)$$

$$\left[H_2\left(\frac{p_1+p_2}{p_1+p_2+p_3},\frac{p_3}{p_1+p_2+p_3}\right)\right.$$

$$\left.+\frac{p_1+p_2}{p_1+p_2+p_3}\,H_2\left(\frac{p_1}{p_1+p_2},\frac{p_2}{p_1+p_2}\right)\right] \tag{20}$$

Using (18) for probability distribution, $\frac{p_i}{p_1+p_2+p_3}, \frac{p_2}{p_1+p_2+p_3}, \frac{p_3}{p_1+p_2+p_3}$ the expression within square bracket comes out to be

$$H_3\left(\frac{p_1}{p_1+p_2+p_3},\frac{p_2}{p_1+p_2+p_3},\frac{p_3}{p_1+p_2+p_3}\right).$$

so that,

$$H_4(p_1,p_2,p_3,p_4)=H_2(p_1+p_2+p_3,p_4)$$

$$+(p_1+p_2+p_3)\,H_3\left(\frac{p_1}{p_1+p_2+p_3},\frac{p_2}{p_1+p_2+p_3},\frac{p_3}{p_1+p_2+p_3}\right) \tag{21}$$

so that the branching property holds when three outcomes are combined. Now we prove that it will hold for combining any number of sub outcomes by using mathematical induction. Let us assume that

$$H_{n+k}(p_1,p_2,\ldots,p_n,p_{n+1},p_{n+2},\ldots,p_{n+k})$$

$$=H_{k+1}(p_1+p_2+\ldots+p_n,p_{n+1},p_{n+2},\ldots,p_{n+k})$$

$$+(p_1+p_2+\ldots+p_n)\,H_n\left(\frac{p_1}{p_1+\ldots+p_n},\frac{p_2}{p_1+\ldots+p_n},\ldots,\frac{p_n}{p_1+\ldots+p_n}\right) \tag{22}$$

Since the principle holds for combining two sub outcomes, the first term on the R.H.S. is decomposed into two terms, viz.

$$H_k(p_1+p_2+\ldots+p_n+p_{n+1},p_{n+2},\ldots,p_{n+k})$$

$$+(p_1+p_2+\ldots+p_{n+1})\,H_2\left(\frac{p_1+p_2+\ldots+p_n}{p_1+\ldots+p_{n+1}},\frac{p_{n+1}}{p_1+\ldots+p_{n+1}}\right) \tag{23}$$

Combining the second terms on the R.H.S. of (22) and (23), we get

$$(p_1+p_2+\ldots+p_{n+1})\left[H_2\left(\frac{p_1+p_2+\ldots+p_n}{p_1+p_2+\ldots+p_{n+1}},\frac{p_{n+1}}{p_1+p_2+\ldots+p_{n+1}}\right)\right]$$

$$+H_n\left(\frac{p_1}{p_1+\ldots+p_n},\frac{p_2}{p_1+\ldots+p_n},\ldots,\frac{p_n}{p_1+\ldots+p_n}\right) \tag{24}$$

Using (19) for $k = 1$ for the distribution

$$\frac{p_1}{p_1+\ldots+p_{n+1}}, \frac{p_2}{p_1+\ldots+p_{n+1}}, \ldots, \frac{p_{n+1}}{p_1+\ldots+p_{n+1}},$$

we find that the expression within square branches of (24) is

$$H_{n+1}\left(\frac{p_1}{p_1+p_2+\ldots+p_{n+1}}, \frac{p_2}{p_1+p_2+\ldots+p_{n+1}}, \ldots, \frac{p_{n+1}}{p_1+p_2+\ldots+p_{n+1}}\right) \quad (25)$$

From (19), (20), (21) and (22), we find that if the branching principle holds for combining n outcomes, it also holds for combining $(n + 1)$ outcomes so that

$$H_{n+k}(p_1, p_2, \ldots, p_n, p_{n+1}, \ldots, p_{n+k}) = H_{k+1}(p_1+p_2+\ldots+p_n, p_{n+1}, p_{n+2}, \ldots, p_{n+k})$$

$$+(p_1+p_2+\ldots+p_n)H_n\left(\frac{p_1}{p_1+\ldots+p_n}, \frac{p_2}{p_1+\ldots+p_n}, \ldots, \frac{p_n}{p_1+\ldots+p_n}\right) \quad (26)$$

holds for all n, k

Just as we have combined the first set of n outcomes, we can combine other sets of outcomes, and so on, so that we get the branching property,

$$H_{m_1+m_2+\ldots+m_n}(p_{11}, \ldots, p_{1m_1}, p_{21}, \ldots, p_{2m_2}, \ldots, p_{n1}, p_{n2}, \ldots, p_{nm_n})$$

$$= H_n(p_1, p_2, \ldots, p_n) + \sum_{i=1}^{n} p_i H_{m_i}\left(\frac{p_{i1}}{p_i}, \frac{p_{i2}}{p_i}, \ldots, \frac{p_{im_i}}{p_i}\right) \quad (27)$$

Thus recursivity leads to branching property. Also recursivity is a special case of the branching property.

6.6 CONCLUDING REMARKS

We have established that,

RECURSIVITY ≡ BRANCHING ≡ ADDITIVITY ≡ DECOMPOSABILITY.

— Recursivity implies that entropy measure for n outcomes can be found when entropy measures for m outcomes are known where $m = 1, 2, \ldots, n-1$.

— Branching implies that entropy for $p_1, p_2; p_3, \ldots, p_n$ can be expressed in terms of entropy for $p_1 + p_2, p_3, \ldots, p_n$ and entropy of p_1, p_2.

— Additivity means that entropy of a joint distribution can be expressed as the sum of two entropies, one of which is the entropy of one marginal distribution and the other is the conditional distribution in the second distribution.

— Decomposability emphasises that the total entropy can be decomposed into entropy within classes and entropy between classes.

REFERENCES

1. J. Aczel and Z. Daroczy, (1975), On Measures of Information and their Characterisation, Academic Press, New York.
2. J.N. Kapur (1990), Maximum Entropy Models in Science and Engineering, John Wiley & Sons, New York.

3. A.M. Mathai & P.N. Rathie (1975), Basic Concepts of Information Theory and Statistics, John Wiley & Sons, New York.
4. C.E. Shannon (1948), ''A Mathematical Theory of Communication'', Bell Tech. J. Vol. 27, pp. 379–423, 623–659.
5. H. Theil (1974), Statistical Decomposition Analysis, North Holland, Amsterdam.

7

ON WEIGHTED ENTROPIES AND ENTROPIC MEANS

[A new justification is given for the use of weighted measures of entropy and directed divergence by showing that these are precisely the functions by maximizing or minimizing which we can get different weighted means of n positive numbers.]

7.1 INTRODUCTION

To measure the uncertainty of a probability distribution, $P = (p_1, p_2, \ldots, p_n)$, Shannon [10] gave his measure of entropy

$$S(P) = -\sum_{i=1}^{n} p_i \ln p_i \tag{1}$$

To take into account the possibility that different outcomes may have different weights $w_1, w_2, \ldots, w_n$ or different utilities $u_1, u_2, \ldots, u_n$ associated with them, Guiasu [4, 5] gave his measure of weighted entropy

$$H(P:W) = -\sum_{i=1}^{n} w_i p_i \ln p_i \tag{2}$$

and Belis and Guiasu [1] gave their measure of 'useful' information,

$$H(P:U) = -\sum_{i=1}^{n} u_i p_i \ln p_i \tag{3}$$

The measures (2) and (3) were obtained axiomatically on the basis of certain postulates, the plausibility of which was questioned by Longo [9] and Kapur [7].

In the present discussion, we give an alternative justification for the use of the measure (2) and other similar measures of Weighted Entropy and Weighted Directed Divergence. We obtain these as functions, by maximizing or minimizing which, we can obtain the various weighted means found useful in mathematics, science and engineering for the last three hundred years or more.

Ben Tal et al. [2] obtained these entropic means by minimizing entropy-like functions. We show that weighted entropy and weighted directed divergence measures are needed for getting weighted means and this can provide a strong motivation for introducing these weighted measures of information.

7.2 USE OF WEIGHTED ENTROPIES TO GET MEAN VALUES

Suppose positive numbers $a_1, a_2, \ldots, a_n$ are given and we want to find a number x which is as close to these as possible. For this we have to maximize a measure of equality of $a_1, a_2, \ldots, a_n, x$. Such a measure is given by the entropy of probability distribution,

$$\frac{a_1}{T+x}, \frac{a_2}{T+x}, \ldots, \frac{a_n}{T+x}, \frac{x}{T+x}; \quad T = \sum_{i=1}^{n} a_i \tag{4}$$

Shannon's entropy for this probability distribution is

$$-\sum_{i=1}^{n}\left(\frac{a_i}{T+x}\right)\ln\frac{a_i}{T+x} - \frac{x}{T+x}\ln\frac{x}{T+x} \tag{5}$$

To maximize it, we differentiate it with respect to x to get the estimate $\hat{x}$ of x as given by

$$\ln\hat{x} = \frac{\sum_{i=1}^{n} a_i \ln a_i}{\sum_{i=1}^{n} a_i} \tag{6}$$

This $\hat{x}$ is a mean of $a_1, a_2, \ldots, a_n$ and was called the 'entropic mean' by Kapur [6]. If instead of Shannon's measure, we use the more general measure,

$$S(P) = -\sum_{i=1}^{n} \phi(p_i) \tag{7}$$

where, $\phi(\cdot)$ is a convex function, we get

$$\phi'\left(\frac{\hat{x}}{T+\hat{x}}\right) = \frac{\sum_{i=1}^{n} a_i \phi'\left(\frac{a_i}{T+\hat{x}}\right)}{\sum_{i=1}^{n} a_i} \tag{8}$$

Thus, $\phi'\left(\frac{\hat{x}}{T+\hat{x}}\right)$ is a weighted mean of $\phi'\left(\frac{a_1}{T+\hat{x}}\right)$, $\phi'\left(\frac{a_2}{T+\hat{x}}\right)$, ..., $\phi'\left(\frac{a_n}{T+\hat{x}}\right)$. Since $\phi(x)$ is convex, $\phi'(x)$ is monotonic increasing and to every value of $\phi'(x)$, there is a unique value of x, so that

$\frac{\hat{x}}{T+\hat{x}}$ lies between the minimum and maximum values of

$\frac{a_1}{T+\hat{x}}, \frac{a_2}{T+\hat{x}}, \ldots, \frac{a_n}{T+\hat{x}}$, or $\hat{x}$ lies between minimum and maximum values of $a_1, a_2, \ldots, a_n$ so that $\hat{x}$ is a mean value for $a_1, a_2, \ldots, a_n$, since it also reduces to a when $a_1 = a_2 = \ldots = a_n = a$.

If instead of maximizing the measure (7), we maximize the weighted measure

$$H(P:W) = -\sum_{i=1}^{n+1} w_i \phi(p_i), \tag{9}$$

then instead of (8) we get

$$\phi'\left(\frac{\hat{x}}{T+\hat{x}}\right)=\frac{\sum_{i=1}^{n} a_i w_i \phi'\left(\frac{a_i}{T+\hat{x}}\right)}{\left(\sum_{i=1}^{n} a_i\right) w_{n+1}} \tag{10}$$

$\hat{x}$ will still be between min $(a_1, a_2, \ldots, a_n)$ and max $(a_1, a_2, \ldots, a_n)$ if we choose

$$w_{n+1}=\frac{\sum_{i=1}^{n} a_i w_i}{\sum_{i=1}^{n} a_i} \tag{11}$$

Thus, to get weighted mean, we can use weighted entropies, provided we associate with x the weight given by the weighted average of $w_1, w_2, \ldots, w_n$ as given by (11).

Thus we get the following result:

"If for obtaining a certain mean value of $a_1, a_2, \ldots, a_n$, we have to maximize the entropy measure (7) for the probability distribution (4), then to get the corresponding weighted mean, we have to maximize the corresponding weighted measure of entropy (9), where w_{n+1} is defined by (11)."

Thus weighted entropies are required as objective functions by maximizing which, we can obtain weighted means.

However for the weighted entropies, w_{n+1} has to be defined by (11).

7.3 RECURSIVE PROPERTY OF GUIASU'S MEASURE

If the measure (2) is to satisfy the recursive property,

$$-\sum_{i=1}^{n} w_i p_i \ln p_i = -w\,(p_1+p_2+\ldots+p_m)\ln(p_1+p_2+\ldots+p_m) - \sum_{i=m+1}^{n} w_i p_i \ln p_i$$
$$-(p_1+p_2+\ldots+p_m)\sum_{i=1}^{m} w_i \frac{p_i}{p_1+p_2+\ldots+p_m}\ln\frac{p_i}{p_1+p_2+\ldots+p_m}$$

or,

$$w\,(p_1+p_2+\ldots+p_m)\ln(p_1+p_2+\ldots+p_m) = -\left(\sum_{i=1}^{m} w_i p_i\right)\ln(p_1+p_2+\ldots+p_m)$$

or

$$w=\frac{w_1p_1+w_2p_2+\ldots+w_mp_m}{p_1+p_2+\ldots+p_m} \tag{12}$$

so that the weight to be associated with the incomplete probability distribution $(p_1, p_2, \ldots, p_m)$ is the weighted mean of the weights associated with each of the m probabilities.

It is interesting to compare (11) and (12) which look similar. Both (11) and (12) give mean weights; while (11) gives mean weight associated with the mean value of $a_1, a_2, \ldots, a_n$, (12) gives the mean weight to be associated with the incomplete probability distribution,

$$p_1, p_2, \ldots, p_m . \tag{13}$$

7.4 OBTAINING WEIGHTED DIRECTED DIVERGENCE

Ben Tal et al. minimized the entropy-like functions,

$$\sum_{i=1}^{n} a_i \phi\left(\frac{x}{a_i}\right) \text{ and } \sum_{i=1}^{n} w_i a_i \phi\left(\frac{x}{a_i}\right) \tag{14}$$

to get un-weighted and weighted means respectively. Kapur [8] minimized the entropy-like functions,

$$\sum_{i=1}^{n} x\phi\left(\frac{a_i}{x}\right) \text{ and } \sum_{i=1}^{n} w_i x\phi\left(\frac{a_i}{x}\right) \tag{15}$$

to obtain the same means. In any case, it is obvious that we require weighted measures to get weighted mean values.

7.5 SOME SPECIAL CASES

(*i*) To get the arithmetic mean $\sum_{i=1}^{n} w_i a_i \,/\, \sum_{i=1}^{n} w_i$, we have to maximize the weighted entropy measure,

$$\sum_{i=1}^{n+1} w_i \ln p_i \tag{16}$$

where,

$$p_i = \frac{a_i}{T+x}, i = 1, 2, \ldots, n; p_{n+1} = \frac{x}{T+x}, T = \sum_{i=1}^{n} a_i \tag{17}$$

and w_{n+1} is defined by (11).

(*ii*) To get the weighted entropic mean

$$\exp\left[\sum_{i=1}^{n} w_i a_i \ln a_i / \sum_{i=1}^{n} w_i a_i\right], \tag{18}$$

we require to maximize Guiasu's weighted entropy measure (2).

(*iii*) To get the weighted mean $\left[\sum_{i=1}^{n} w_i a_i^r / \sum_{i=1}^{n} w_i\right]^{1/r}$, we have to maximize

$$-\sum_{i=1}^{n+1} \frac{w_i}{a_i}\left(\frac{1-p_i^{r+1}}{r+1}\right), \; a_{n+1} = \frac{\sum_{i=1}^{n} w_i a_i}{\sum_{i=1}^{n} w_i} \tag{19}$$

As special cases of this, we find that

(*a*) To get the weighted harmonic mean, we put $r = -1$ and we have to maximize the measure

$$\sum_{i=1}^{n+1} \frac{w_i}{a_i} \ln p_i \tag{20}$$

(*b*) To get the weighted geometric mean, we put $r = 0$, and we have to maximize the measure

$$-\sum_{i=1}^{n+1} \frac{w_i}{a_i} (1 - p_i) \tag{21}$$

(*c*) To get the weighted arithmetic mean, we put $r = 1$, and have to maximize the measure

$$-\sum_{i=1}^{n+1} \frac{w_i}{a_i} \left(\frac{1 - p_i^2}{2} \right) = \sum_{i=1}^{n+1} \frac{w_i}{2a_i} (p_i^2 - 1) \tag{22}$$

We now make the following remarks:

(*i*) We get the very natural weighted arithmetic mean by maximizing (16) which is the weighted measure of entropy corresponding to Burg's [3] measure of entropy. If on the other hand, we maximize Guiasu's measure of entropy corresponding to Shannon's measure of entropy, we get the rather less natural weighted means given by (18). Thus while in practice Shannon's measure is supposed to be more natural than Burg's, from the point of view of weighted mean, weighted Burg's entropy gives a more natural weighted mean than the weighted Shannon entropy.

(*ii*) The weighted arithmetic mean can be obtained by either maximizing (16) or (19), for the special case $r = 1$.

7.6 CONCLUDING REMARKS

Over the last twenty years, many weighted measures of entropy and directed divergence have been given by either first defining them and then characterising them or by modifying the axioms for the classical measures to incorporate weights or utilities or by generalising the functional equations for classical measures and then solving them.

The present discussion provides a more natural way of getting measures of weighted entropy or directed divergence. To get ordinary means, we maximize or minimize entropy-like functions. We ask then as to what are the corresponding entropy like functions by maximizing or minimizing which we can get weighted means and these will give our weighted measures of entropy and directed divergence.

Using this approach, we have been able to get new measures of weighted entropy, not obtained by earlier methods.

REFERENCES

1. M. Belis and S. Guiasu. (1968). ''A Quantitative-qualitative Measures of Information in Cybernetic Systems. '' IEEE Trans. Inf. Th.. IT-4. pp. 593–594.
2. A. Ben Tal. A. Charnes and M. Taboulle (1989). ''Entropic Means'', Joun. Math Analysis and Appls. 139. 537–551.
3. J.P. Burg. (1972). ''The Relationship between Maximum Entropy Spectra and Maximum Likelihood spectra''. in Modern Spectral Analysis. ed. D.G. Childrers. pp. 130–131.
4. S. Guiasu. (1971). ''Weighted Entropy''. Reports on Math Physics. vol. 2. pp. 165-179.
5. S. Guiasu. (1977). ''Information Theory with Applications''. McGraw Hill. New York.
6. J.N. Kapur (1984). ''Maximum Entropy Principle and Search Theory''. Joun. Nat. Aca. of Mathematics 2(2) 99–104.
7. J.N. Kapur (1985). ''On the concept of useful information'' Joun. Org Behav Stat. 2 (3. 4). 147–162.
8. J.N. Kapur (1990). ''On Entropic Means'' MSTS Research report 555.
9. G. Longo (1972). ''Quantitative-Qualitative Measures of Information''. Springer-Verlag. New-York.
10. C.E. Shannon. (1948). ''A Mathematical Theory of Communication''. Bell-system Tech. J.. vol. 27. pp. 379–423. 623–659.

8

MEASURES OF ENTROPY FOR CONTINUOUS-VARIATE PROBABILITY DISTRIBUTIONS – I

[It is shown that defining an entropy measure of a probability distribution of a continuous random variate as a monotonic decreasing function of its directed divergence from the uniform distribution gives a better approach than defining it as the limit of the entropy measure of a discrete-variate distribution.]

8.1 INTRODUCTION

Literally scores of papers have been written on axiomatic derivation of Shannon's [12] measure of entropy

$$S(P) = -\sum_{i=1}^{n} p_i \ln p_i \tag{1}$$

of a discrete-variate probability distribution $P = (p_1, p_2, \ldots, p_n)$ from different sets of plausible axioms (Shannon [12], Khinchin [9], Mathai and Rathie [10], Kapur [8]). A great deal of mathematical sophistication and rigour is exercised in the process. However, when it comes to the derivation of the corresponding measure

$$S[f(x)] = -\int_a^b f(x) \ln f(x) dx \tag{2}$$

for the entropy of continuous-variate probability distribution with density function $f(x)$, all pretensions of rigour are given up and heuristic arguments are freely employed. The major argument in favour of (2) is not its rigorous derivation from clearly-stated axioms, but the fact that it gives useful results.

Similarly, rigorous arguments are used to deduce measures of entropy due to Havrda-Charvat [3], Kapur [7], Behara and Chawla [1] and others for the discrete case; but when measures for the continuous case are required, the sums are simply replaced by integrals, as if this operation is always valid. The weakness of this argument is demonstrated here. It is also shown that the weakness can be overcome by defining entropy as a monotonic decreasing function of the directed divergence of the given distribution from the uniform distribution.

8.2 DERIVATION OF SHANNON-WEINER MEASURE

Replacing p_i in (1) by $f(x_i)\,\Delta x_i$, we get

$$S[f(x)] = -\sum_{i=1}^{n} f(x_i)\Delta x_i \ln f(x_i)\Delta x_i$$

$$= -\sum_{i=1}^{n} f(x_i)\ln f(x_i)\Delta x_i - \sum_{i=1}^{n} f(x_i)\Delta x_i \ln \Delta x_i$$

If we take all intervals of equal length Δx, we get

$$S[f(x)] = -\sum_{i=1}^{n} f(x_i)\ln f(x_i)\,\Delta x - \ln \Delta x \sum_{i-i}^{n} f(x_i)\Delta x \tag{3}$$

If we put $\Delta x = \dfrac{b-a}{n}$ and let $n \to \infty$, we get

$$S[f(x)] \to -\int_a^b f(x)\ln f(x)dx - \underset{n\to\infty}{Lt}\ \ln\frac{b-a}{n} \tag{4}$$

Thus (1) does not approach (2): there appears to be a difference of an infinite constant. However, if $Q = (q_1, q_2, \ldots, q_n)$ is any other probability distribution and we replace q_i by $g(x_i)\,\Delta x_i$, we find on proceeding as above that

$$\left[\left(-\sum_{i=1}^{n} p_i \ln p_i\right) - \left(-\sum_{i=1}^{n} q_i \ln q_i\right)\right] \to \left[-\int_a^b f(x)\ln f(x)dx + \int_a^b g(x)\ln g(x)dx\right] \tag{5}$$

Thus while (2) is not a really a satisfactory measure of uncertainty (it can even be negative), it can still be used to find that density function $f(x)$ for which the entropy is greater than for any other density function. This gives us some justification for the use of (2) in implementing Jaynes [4] principle of maximum entropy for continuous-variate probability distributions.

8.3 DERIVATIONS OF OTHER MEASURES OF ENTROPY FOR A CONTINUOUS-VARIATE PROBABILITY DISTRIBUTION

(*a*) Renyi's [11] measure for a discrete probability distribution is given by

$$R_\alpha(P) = \frac{1}{1-\alpha} 1n \sum_{i=1}^{n} p_i^\alpha, \alpha \neq 1, \alpha > 0 \tag{6}$$

Replacing p_i by $f(x_i)\,\Delta x$, we get

$$R_\alpha[f(x)] = \frac{1}{1-\alpha}\ln \sum_{i=1}^{n} f^\alpha(x_i)(\Delta x)^\alpha = \frac{1}{1-\alpha}\left[\ln \sum_{i=1}^{n} f^\alpha(x_i) + \alpha \ln \Delta x\right] \tag{7}$$

$$\Delta x R_\alpha[f(x)] = \frac{1}{1-\alpha}\left[\ln \sum_{i=1}^{n} f^\alpha(x_i)\Delta x + \alpha \Delta x \ln \Delta x\right]. \tag{8}$$

Thus $R_\alpha f(x)$ does not approach a limit as $\Delta x \to 0$, but $\Delta x\, R_\alpha\,(f(x))$

approaches the limit $\frac{1}{1-\alpha}\ln\int_a^b f^{\alpha}(x)dx$ (9)

Similarly if $g(x)$ is any other density function, then $R_{\alpha}(g(x))$ does not approach a limit, but $\Delta x\, R_{\alpha}(g(x))$ approaches a limit

$$\frac{1}{1-\alpha}\ln\int_a^b g^{\alpha}(x)dx \tag{10}$$

Thus, we can say that Renyi's entropy for the density function $f(x) \gtrless$ Renyi's entropy for the density function $g(x)$ according as

$$\frac{1}{1-\alpha}\ln\int_a^b f^{\alpha}(x)dx \gtrless \frac{1}{1-\alpha}\ln\int_a^b g^{\alpha}(x)dx \tag{11}$$

Thus, we can use (9) in entropy maximization problems.

(*b*) Kapur's measure of entropy of order α and type β for a discrete probability distribution is given by

$$K_{\alpha,\beta}(P) = \frac{1}{\beta-\alpha}\ln\frac{\sum_{i=1}^{n} p_i^{\alpha}}{\sum_{i=1}^{n} p_i^{\beta}},\ \alpha, \beta > 0, \alpha \neq \beta \tag{12}$$

Proceeding as above, we find that $K_{\alpha,\beta}(P)$ does not approach a limit, but we can use

$$\frac{1}{\beta-\alpha}\ln\frac{\int_a^b f^{\alpha}(x)dx}{\int_a^b f^{\beta}(x)dx} \tag{13}$$

in entropy maximization processes

(*c*) Havrda-Charvat's [3] measure of entropy of order α for a given discrete-variate probability distribution is given by

$$H_{\alpha}(P) = \frac{1}{1-\alpha}\left(\sum_{i=1}^{n} p_i^{\alpha} - 1\right) \tag{14}$$

Replacing p_i by $f(x_i)\,\Delta x_i$ we get

$$H_{\alpha}[f(x)] = \frac{1}{1-\alpha}\left[\sum_{i=1}^{n} f^{\alpha}(x_1)(\Delta x)^{\alpha} - 1\right]$$

$$\frac{H_{\alpha}(f(x)) - H_{\alpha}(g(x))}{(\Delta x)^{1-\alpha}} \to \frac{1}{1-\alpha}\int [f^{\alpha}(x) - g^{\alpha}(x)]\,dx \tag{15}$$

so that
$$H_{\alpha}(f(x)) = \frac{1}{1-\alpha}\left[\int_a^b f^{\alpha}(x)dx - 1\right] \tag{16}$$

can be used in entropy maximization problems.

(*d*) We can similarly consider other measures of entropy. In every case, the limit of the entropy measure for a discrete variate distribution does not exist. However, in every case, by replacing the sum by an integral, we can find a measure whose maximisation will give us the maximum entropy probability distribution. This is not a satisfactory or rigorous approach. In the next section, we shall see that we have a much more satisfactory situation for the case of measures of directed divergence.

8.4 MEASURES OF DIRECTED DIVERGENCE FOR TWO CONTINUOUS-VARIATE PROBABILITY DISTRIBUTIONS

A general measure of directed divergence of a probability distribution $P = (p_1, p_2, \ldots, p_n)$ from a probability distribution $Q = (q_1, q_2, \ldots, q_n)$ is given by Csiszer's measure-[2]

$$D\,(P:Q) = \sum_{i=1}^{n} q_i \phi\,(p_i/q_i), \tag{17}$$

where ϕ (.) is a twice-differentiable convex function for which $\phi\,(1) = 0$. Replacing p_i by $f\,(x_i)\,\Delta x_i$, q_i by $g\,(x_i)\,\Delta x_i$ we get

$$D\,(f{:}g) = \sum_{i=1}^{n} g\,(x_i)\,\phi\left(\frac{f\,(x_i)}{g\,(x_i)}\right)\Delta x_i \tag{18}$$

which approaches

$$\int_a^b g\,(x)\,\phi\left[\frac{f\,(x)}{g\,(x)}\right]dx \tag{19}$$

Here the limit is approached even if we do not take the intervals of equal length. As special cases we have

(*i*) When $\phi\,(x) = x \ln x, D\,(f{:}g) = \int_a^b f\,(x) \ln \frac{f\,(x)}{g\,(x)}\,dx$ (20)

(*ii*) When $f\,(x) = \frac{x^\alpha - x}{\alpha - 1}, D\,(f:g) = \frac{1}{\alpha - 1}\left[\int_a^b f^a\,(x) g^{1-\alpha}(x) dx - 1\right], \alpha \neq 1$ (21)

the measure given by Havrda-Charvat [3]

(*iii*) When $f\,(x) = x \ln x - 1/c(1 + cx)\ln(1 + cx) + (x/c)(i + c)\ln(i + c)$

$$D\,(f{:}g) = \int_a^b f\,(x) \ln \frac{f\,(x)}{g\,(x)}\,dx - (1/c)\int_a^b [g\,(x) + cf\,(x)] \ln \frac{g\,(x) + cf\,(x)}{g\,(x)}\,dx$$

$$+ (1/c)(1 + c)\ln(1 + c), \tag{22}$$

the measure given by Kapur [6].

Thus we can derive measures of directed divergence axiomatically and rigorously for discrete variate probability distributions and then obtain the corresponding measures for continuous-variate probability distributions by considering a limiting process. In this case the replacement of sums by integrals can be justified in a rigorous manner.

This rigorous justification can be available to us for entropy measure if we define entropy measure as a monotonic decreasing function of a measure of directed divergence from the uniform distribution.

8.5 A NEW APPROACH TO DEFINING MEASURES OF ENTROPY

Since axiomatic derivation of measures of entropy for continuous-variate distributions creates difficulties, it is better to have axiomatic derivations of measures of directed divergence and then derive measures of entropy from these, since in this approach there is no distinction between discrete and continuous variate cases.

Putting
$$g(x) = \frac{1}{b-a}, \tag{23}$$

(*i*) in KL measure, we get

$$\int_a^b f(x) \ln \frac{f(x)}{\frac{i}{b-a}} dx = \ln(b-a) + \int_a^b f(x) \ln f(x) dx, \tag{24}$$

so that we take

$$-\int_a^b f(x) \ln f(x) dx \tag{25}$$

as a measure of entropy.

(*ii*) in HC measure, we get

$$\frac{1}{\alpha-1}\left[\int_a^b f^\alpha(x)\left(\frac{1}{b-a}\right)^{1-\alpha} dx - 1\right] = \left(\frac{1}{b-a}\right)^{1-\alpha}\left[\frac{1}{\alpha-1}\left\{\int_a^b f^\alpha(x)dx - 1\right\}\right.$$
$$\left.+\frac{1}{\alpha-1}\right] - \frac{1}{\alpha-1}, \tag{26}$$

so that we can take

$$\frac{1}{1-\alpha}\left[\int_a^b f^\alpha(x)dx - 1\right] \tag{27}$$

as a measure of entropy.

(*iii*) in Kapur's measure we get

$$\int_a^b f(x)\ln(f(x)(b-a))dx - 1/c\int_a^b\left[\left(\frac{1}{b-a}+cf(x)\right)\ln\{1+cf(x)(b-a)\}\right]dx$$

$$+1/c\,(1+c)\ln\,(1+c)\,, \qquad (28)$$

so that we can take

$$-\int_a^b f(x)\ln f(x)dx + 1/c\,\frac{1}{b-a}\int_a^b [\{1+c(b-a)f(x)\}$$

$$\ln\{1+c(b-a)f(x)\}]dx \qquad (29)$$

as a measure of entropy.

8.6 CONCLUDING REMARKS

(*i*) All the axioms of entropy measure for a discrete probability distribution cannot be taken over to the continuous case. In particular the limiting case of recursivity axiom does not make sense for continuous-variate probability distributions. Similarly all the logical arguments used for the discrete case cannot be taken over to the continuous case. As such, entropy measures have to be discussed separately for discrete and continuous variate cases.

(ii) However, we are more fortunate for the measure of directed divergence, since here the limit of measures for discrete probability distributions give measures for continuous variate distributions.

(iii) It is therefore desirable to consider the concept of directed divergence as more basic than that of entropy and to express the concept of entropy in terms of the concept of directed divergence.

REFERENCES

1. M. Behara, and J.S. Chawla, "Generalised Gamma Entropy, "In Selecta Statistica Canadiana, Vol. 2, pp. 15–38, 1974.
2. I. Csiszer, "A class of Measures of Informativity of Observation Channels, Periodica Math. Hungarica, Vol. 2, pp. 191–213, 1972.
3. Havrda, J.H. and F. Charvat, "Quantification Methods of Classification Processes: Concept of Structural Entropy," Kybernetica, Vol. 3, pp. 30–35, 1967.
4. Jaynes, E.T., "Information Theory and Statistical Mechanics," Physical Reviews, Vol. 106, pp. 620–630, 1957.
5. Kapur, J.N., "Twenty-five Years of Maximum Entropy," Jour. Math. Phy. Sci., Vol. 17, no. 2, pp. 103–106, 1983.
6. Kapur, J.N., "A Comparative Assessment of Various Measures of Directed Divergence," Advances in Management Studies, Vol. 3, pp. 1–16, 1984.
7. Kapur, J.N., "Four Families of Measures of Entropy," Ind. Jour. Pure and App. Maths., Vol. 17, no. 4, pp. 429–449, 1986.
8. Kapur, J.N., "Maximum-Entropy Models in Science and Engineering," Wiley Eastern, New Delhi, 1989.

9. Khinchin, A.I., Mathematical Foundations of Information Theory, Dover Publications, 1957.
10. Mathai, A.M. and P.N. Rathie, "Basic Concepts of Information Theory and Statistics, " John Wiley, New York, 1975.
11. Renyi, A., "On Measures of Entropy and Information," Proc. 4th Berkeley Symp. Maths. Stat. Prob., Vol. 1, pp. 547–561, 1961.
12. Shannon, C.E., "A Mathematical Theory of Communication," Bell System Tech. J., Vol. 27, pp. 379–423, 623–659, 1948.
13. Sharma, B.D. and I.J. Taneja, "Entropy of Type (α, β) and other Generalised Measures of Information," Metrika, Vol. 22, pp. 205–214, 1974.
14. Sharma, B.D. and D.P. Mittal, "New Non-additive Measures of Entropy for Discrete Probability Distributions," J. Math. Sci., Vol. 10, pp. 28–40, 1975.

9

MEASURES OF ENTROPY FOR CONTINUOUS-VARIATE PROBABILITY DISTRIBUTIONS-II

[The discussion given in Part I is farther elaborated and the two methods of obtaining measures of entropy for continuous-variate distributions are described and compared.]

9.1 THE FIRST METHOD

In this method, we first obtain a measure of entropy for a discrete-variate probability distribution $P = (p_1, p_2, \ldots, p_n)$ and then obtain the measure of entropy for a continuous variate probability distribution by using some limitating process as $n \to \infty$ [4].

To illustrate this process, we consider the measure of entropy,

$$K_c(P) = -\sum_{i=1}^{n} p_i \ln p_i + \frac{1}{C}\sum_{i=1}^{n}(1+cp_i)\ln(1+cp_i) - \frac{1}{c}(1+c)\ln(1+c) \tag{1}$$

where c is a real number ≥ -1. Two special cases of this measure corresponding to $c = -1$ and $c = 1$ were obtained by Kapur [3]. His object was to obtain those measures by maximizing which, subject to the constraints,

$$\sum_{i=1}^{n} p_i = 1, \quad \sum_{i=1}^{n} p_i \in_i = \hat{\in} \tag{2}$$

where $\varepsilon_1, \varepsilon_2, \ldots, \varepsilon_n$ are the n energy levels and $\hat{\varepsilon}$ is the expected energy of the system, we can get the distributions

$$p_i = \frac{1}{e^{\lambda+\mu\varepsilon_i}+1} \quad \text{and} \quad p_i = \frac{1}{e^{\lambda+\mu\varepsilon_i}-1} \tag{3}$$

where, λ and μ are obtained by using (2). These distributions correspond to Fermi-Dirac and Bose-Einstein distributions of statistical mechanics. The general measure (1) was given by Kapur [4] and by maximizing it subject to (2) one can also obtain the intermediate statistics distributions,

$$p_i = \frac{1}{e^{\lambda+\mu\varepsilon_i}+c} \tag{4}$$

To obtain the version of (1) applicable to continuous-variate distributions, we replace p_i by $f(x_i)\,\Delta x_1$, where we interpret $f(x)$ as the probability density function, to get

$$K = -\sum_{i=1}^{n} f(x_i)\,\Delta x_i \ln\left(f(x_i)\,\Delta x_i\right) + \frac{1}{c}\sum_{i=1}^{n} ((1 + c\,f(x_i)\,\Delta x_i)$$

$$\ln(1 + c\ f(x_i)\,\Delta x_i)) - \frac{1}{c}(1+c)\ln(1+c) \tag{5}$$

Let the variate x vary from a to b and let $x_1, x_2, \ldots, x_n$ divide (a, b) into n parts as shown in figure 9.1.

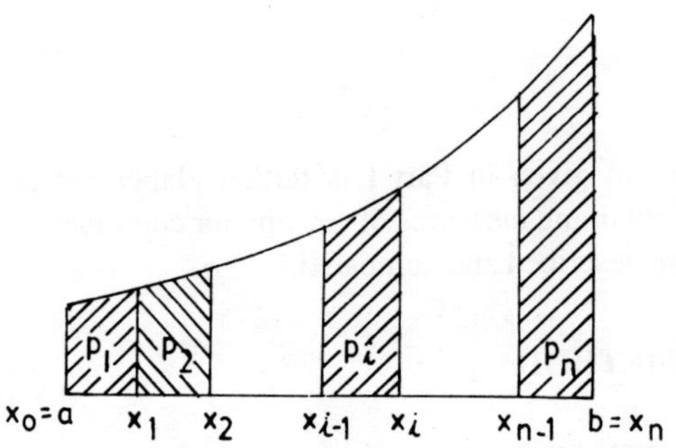

Fig. 9.1

Simplifying we get

$$K = -\sum_{i=1}^{n} f(x_i)\ln f(x_i)\,\Delta x_i - \sum_{i=1}^{n} f(x_i)\,\Delta x_i \ln \Delta x_i$$

$$+\frac{1}{c}\sum_{i=1}^{n}(1 + cf(x_i)\,\Delta x_i)(cf(x_i)\,\Delta x_i + \ldots) - \frac{1}{c}(1+c)\ln(1+c) \tag{6}$$

Let us take all sub-intervals of equal length so that $\Delta x_i = \dfrac{b-a}{n}$, so that

$$K = -\sum_{i=1}^{n} f(x_i)\ln f(x_i)\,\Delta x_i - \ln\frac{b-a}{n}\left(\sum_{i=1}^{n} f(x_i)\,\Delta x_i\right)$$

$$+\sum_{i=1}^{n} f(x_i)\,\Delta x_i - \frac{1}{c}(1+c)\ln(1+c) + 0\left(\frac{1}{n^2}\right) \tag{7}$$

Proceeding to the limit as $\Delta x_i \to 0, n \to \infty$, we get

$$K \cong -\int_a^b f(x)\ln f(x)dx - \ln\frac{b-a}{n} + 1 - \frac{1}{c}(1+c)\ln(1+c) \tag{8}$$

Now as n approaches ∞, the second term on the R.H.S. approaches ∞ and as such K does not approach a limit.

However, if $g(x)$ is any other density function over the same interval, then proceeding in the same way, we get

$$K' \equiv -\int_a^b g(x)\ln g(x)dx - \ln\frac{b-a}{n} + 1 - \frac{1}{c}(1+c)\ln(1+c) \tag{9}$$

so that

$$K - K' \to \left(-\int_a^b f(x)\ln f(x)dx\right) - \left(-\int_a^b g(x)\ln g(x)dx\right) \tag{10}$$

Thus while $K(P)$ does not approach a limit, $K(P) - K(Q)$, where Q is any other probability distribution approaches a limit.

Thus, while there maybe some difficulty in talking about

$$-\int_a^b f(x)\ln f(x)dx \tag{11}$$

as a measure of entropy for the continuous-variate distribution with density function $f(x)$, there should be no difficulty in speaking about

$$-\left(\int_a^b f(x)\ln f(x)dx\right) - \left(-\int_a^b g(x)\ln g(x)dx\right) \tag{12}$$

as the difference of entropies of two distributions with density functions $f(x)$ and $g(x)$.

A measure of entropy is used for two purposes:

(i) To measure uncertainty of a distribution, and

(ii) To maximize uncertainty, i.e. to get a distribution with maximum uncertainty out of all distributions satisfying some constraints.

There may be some doubt about the use of (11), i.e. we may or may not be able to use (11) as a measure of uncertainty, but we can use (11) for the second purpose, because in maximization we are comparing entropies of all distributions and finding the distribution whose entropy is greater than that of any other distribution satisfying the same constraints.

9.2 SOME REMARKS ON THIS DERIVATION

(*i*) We assume that $f(x)$ and $f(x)\ln f(x)$ are continuous functions so that definite integrals exist and we may take sub-intervals of equal length and prove that

$$\underset{\substack{n\to\infty \\ \Delta x_i \to 0 \\ \sum_{i=1}^{n}\Delta x_i = b-a}}{Lt}\left[-\sum_{i=1}^{n} f(x_i)\Delta x_i \ln f(x_i)\Delta x_i + \frac{1}{c}\sum_{i=1}^{n}(1+c\,f(x_i))\ln(1+c\,f(x_i)) - \frac{1}{c}(1+c)\ln(1+c)\right]$$

$$-\left[-\sum_{i=1}^{n} g(x_i)\,\Delta x_i \ln g(x_i)\,\Delta x_i + \frac{1}{c}\sum_{i=1}^{n}(1+c\,g(x_i))\ln(1+c\,g(x_i)) - \frac{1}{c}(1+c)\ln(1+c)\right]$$

$$=\left(-\int_a^b f(x)\ln f(x)dx\right)-\left(-\int_a^b g(x)\ln g(x)dx\right) \tag{13}$$

(*ii*) We find that while the measure of entropy for a discrete variate distribution can be interpreted as a measure of uncertainty, the corresponding measure of entropy for a continuous variate distribution cannot be so easily interpreted. It can even be negative. However, the differences in uncertainties can be interpreted.

(*iii*) In the present case, while the discrete-variate measures depends on c, the continuous-variate measure is independent of c. In fact as $c \to 0$, (1) approaches Shannon's [10] measure

$$S(P) = -\sum_{i=1}^{n} p_i \ln p_i \tag{14}$$

and (11) also corresponds to this measure.

(*iv*) This should not create the impression that all discrete-variate measures will approach Shannon's measure in the continuous-variate case.
Thus consider Renyi's [9] measure

$$R(P) = \frac{1}{\alpha(1-\alpha)} \ln \sum_{i=1}^{n} p_i^{\alpha}, \alpha \neq 0, 1 \tag{15}$$

Replacing p_i by $\ln f(x_i)\,\Delta x_i$, we get

$$R = \frac{1}{\alpha(1-\alpha)} \ln \sum_{i=1}^{n} f^{\alpha}(x_i)(\Delta x_i)^{\alpha} \tag{16}$$

$$= \frac{1}{\alpha(1-\alpha)} \ln \sum_{i=1}^{n} (\Delta x_i)^{\alpha-1} (f^{\alpha}(x_i)\,\Delta x_i \tag{17}$$

Replacing Δx_i by $(b-a)/n$, we get

$$R = \frac{1}{\alpha(1-\alpha)}\left[\ln\frac{(b-a)^{\alpha-1}}{n^{\alpha-1}} + \ln\left[\int_a^b f^{\alpha}(x)\,dx\right]\right] \tag{18}$$

Similarly, for any another density function $g(x)$ we shall get

$$R' = \frac{1}{\alpha(1-\alpha)}\left[\ln\frac{(b-a)^{\alpha-1}}{n^{\alpha-1}} + \ln\left[\int_a^b g^{\alpha}(x)\,dx\right]\right] \tag{19}$$

so that

$$R > R', \text{ if } \frac{1}{\alpha(1-\alpha)} \ln \int_a^b f^{\alpha}(x)\,dx > \frac{1}{\alpha(1-\alpha)} \ln \int_a^b g^{\alpha}(x)dx, \tag{20}$$

so that in entropy maximization problems, we can use

$$\frac{1}{\alpha(1-\alpha)}\ln\int_a^b f^{\alpha}(x)\,dx \tag{21}$$

as a measure of entropy. This may be regarded as the continuous variate version of Renyi [9] measure of entropy.

It appears from the above that we cannot get the measure

$$-\int_a^b f(x)\ln f(x)\,dx+\frac{1}{c}\int_a^b (c+df(x))\ln(c+df(x))dx \tag{22}$$

by the first method of transition from discrete probability distribution. However, this is a valid measure of entropy and we discuss in the next section another approach to get this measure.

9.3 THE SECOND METHOD

In this approach, we regard entropy as a measure of information of a distribution, i.e. we would like this measure to be maximum when the distribution is uniform. As such we want to choose our function $\phi(\cdot)$ in such a way that

$$\int_a^b \phi(f(x))dx \quad \text{is maximum subject to} \quad \int_a^b f(x)\,dx=1 \tag{23}$$

when $f(x)=\dfrac{1}{(b-a)}$

As such we want

$$\phi'(f(x))=\text{constant} \tag{24}$$

when $f(x)=\frac{1}{(b-a)}$, which is satisfied, we would also want this to give the global maximum value. For this reason, we would like to have $\phi(x)$ as a concave function of x. The function

$$\phi(x)=-x\ln x+\frac{1}{c}(1+cx)\ln(1+cx) \tag{25}$$

for which $$\phi'(x)=\ln\frac{1+cx}{x}$$

$$\phi''(x)=\frac{c}{1+cx}-\frac{1}{x}=-\frac{1}{x(1+cx)}, \tag{26}$$

is concave if $x>0$, $1+cx>0$

As such, we can use

$$-\int_a^b f(x)\ln f(x)\,dx+\frac{1}{c}\int_a^b (1+cf(x))\ln(1+cf(x))\,dx \tag{27}$$

as a measure of entropy provided

$$f(x) > 0, \quad 1 + cf(x) > 0 \tag{28}$$

The first condition is satisfied since $f(x)$ is to be a density function. The second condition is satisfied if $c > 0$. It will also be satisfied by negative value of c if we have some additional knowledge about $f(x)$. Thus if we know that $f(x)$ is always less than K, then we can choose $c = -1/K$.

Thus, in innovation diffusion models, we know that if $f(t)$, is the proportion of all possible adopters which has adopted a new innovation till time t, then $f(t) \leq 1$ and we can use the measure

$$-\int_a^b f(t) \ln f(t)\, dt - \int_a^b (1 - f(t)) \ln (1 - f(t))\, dt \tag{29}$$

In logistic models of population dynamics, we know that the population $N(t)$ at time t is less than or equal to the carrying capacity k of the environment and we can use the measure

$$-\int_a^b N(t) \ln N(t)\, dt - \int_a^b (k - N(t)) \ln (k - N(t))\, dt \tag{30}$$

(29) and (30) are essentially measures of 'equality' or 'uniformity' of $f(t)$ and $N(t)$, rather than of uncertainty.

Similarly, if the number of particles in n boxes are $x_1, x_2, \ldots, x_n$ and the maximum numbers of particles in these boxes are $m_1, m_2, \ldots, m_n$, then we can use

$$-\sum_{i=1}^{n} x_i \ln x_i - \sum_{i=1}^{n} (m_i - x_i) \ln (m_i - x_i) \tag{31}$$

as a measure of equality of the number of particles.
Again if we take,

$$\phi(x) = \frac{x^{\alpha} - x}{\alpha(1-\alpha)}, \quad \alpha \neq 0,\ \alpha \neq 1, \tag{32}$$

we find $f(x)$ is a concave function of x in [0, 1] and we can use,

$$\int_a^b \frac{f^{\alpha}(x) - f(x)}{\alpha(1-\alpha)}\, dx \quad \text{or} \quad \frac{1}{\alpha(1-\alpha)}\left(\int_a^b f^{\alpha}(x)\, dx - 1\right) \tag{33}$$

as a measure of entropy. This corresponds to Havrda-Charvat [2] measure of entropy.

9.4 A COMPARISON OF THE TWO METHODS

The basic philosophies of the two methods are different. In the first method, we first use the axioms of continuity, symmetry, expansionability, monotonocity, additivity, recursivity etc. to get a measure of entropy [1, 8] to represent 'uncertainty' or 'equality', and then take its limiting form. This limit involves an infinite constant which however cancels out if we take the limit of difference of measures of entropy for two probability distributions. This limiting form can represent difference in uncertainties rather than uncertainty itself.

In the second method, we do not use any axioms, we use mainly the fact that our measure should be maximum when the distribution is uniform.

Alternatively, the second method implies that we can deduce the measure of entropy from a measure of directed divergence since a measure of entropy has to be a monotonic decreasing function of the directed divergence of the distribution for the uniform distribution (Kapur [5, 7]).

REFERENCES

1. J. Aczel and Z. Daroczy (1975), On Measures of Information and their Characterization'', Academic Press, New York.
2. J.H. Havrda and F. Charvat (1967), ''Quantification Methods of Classification Processes. Concepts of Structural α Entropy'', Kybernetica 3, pp. 30–35.
3. J.N. Kapur (1972) 'Measures of Uncertainty, Mathematical Programming and Physics'', Journ. Ind. Soc. Agri. Stat, Vol. 24, pp. 47–66.
4. J.N. Kapur (1986), ''Four Families of Measures of Entropy'', Ind-Journ. Pure and App. Maths, Vol. 17, no. 4, pp. 429–449.
5. J.N. Kapur (1987), ''On the Relationship between Measures of Entropy and Directed Divergence,'' Proceedings Nat. Acad. Sci. **59 A** (4) 676–684.
6. J.N. Kapur (1988) ''On Generation of Measures of Directed Divergence from Measure of Entropy. Proc Nat. Acad. Sci. **58 A** (1) 113–121.
7. J.N. Kapur (1988), ''On the Basis for Relationship between Measures of Entropy and Directed Divergence'' Proc. Nat. Acad. Sci. **58A** (3), 375–387.
8. A.M. Mathai and P.N. Rathie, (1975), ''Basic Concepts of Information Theory and Statistics'', John Wiley, New York.
9. A. Renyi (1961), ''On Measures of Entropy and Information'', Proc. 4th Berkeley Symp. Maths Stat. Prob, Vol. 1, 547–561.
10. C.E. Shannon, (1948) ''A Mathematical Theory of Communication'', Bell system Tech, J., Vol. 27, pp. 379–423, 623–659.

10

DISCRETE ANALOGUES OF MEASURES OF ENTROPY FOR CONTINUOUS VARIATE DISTRIBUTIONS AND THEIR APPLICATIONS TO PARAMETER ESTIMATION

[Discrete analogues of measures of entropy for continuous variate distributions have been found and these have been applied to problems of parameter estimation. Alternative derivation of expressions for measures of entropy and directed divergence for continuous variate distributions are also given.]

10.1 DISCRETE ANALOGUE OF BOLTZMANN-SHANNON'S MEASURE OF ENTROPY

Shannon [14] gave his well-known measure of entropy for the discrete-variate probability distribution $P = (p_1, p_2, \ldots, p_n)$ as

$$S(P) = -\sum_{i=1}^{n} p_i \ln p_i \tag{1}$$

We have also the well-known Boltzmann-Shannon measure of entropy for a continuous-variate probability distribution with density function $f(x)$ as

$$S(F) = -\int_a^b f(x) \ln f(x) dx \tag{2}$$

Usually (1) is derived rigorously from axiomatic considerations (Aczel and Daroczy [1], Mathai and Rathie [13]) and then an effort is made to derive (4) by a limiting process (Kapur [8]). The derivation is however neither satisfactory nor rigorous. Its major merit to that it works. For the present discussion, we do not assume (2) and start from first principles to derive another measure for the entropy of a continuous-variate distribution.

Consider a partition of the interval $[a, b]$ into n subintervals by the set of points,

$$a = x_0 < x_1 < x_2 < \ldots < x_{n-1} < x_n = b \tag{3}$$

$$\text{and let } \int_{x_{i-1}}^{x_i} f(x)\, dx = p_i, \quad i = 1, 2, \ldots, n. \tag{4}$$

so that $p_i \geq 0$ for each i and $\sum_{i=1}^{n} p_i = 1$ and $P = (p_1, p_2, \ldots, p_n)$ is a probability distribution.

If instead of $f(x)$, we had the uniform distribution with density function,

$$g(x) = \frac{1}{b-a}, \tag{5}$$

we get the probability distribution $Q = (q_1, q_2, \ldots, q_n)$, where

$$\int_{x_{i-1}}^{x_i} g(x)dx = \int_{x_{i-1}}^{x_i} \frac{1}{b-a} dx = \frac{x_i - x_{i-1}}{b-a} = q_i, \quad i = 1, 2, \ldots, n \tag{6}$$

If $f(x)$ coincides with $g(x)$, then $P = Q$. The greater the discrepancy of $f(x)$ from the uniform distribution the greater will be the discrepancy or directed divergence of P from Q. Now the directed divergence of P from Q can be measured by the Kullback-Leibler [11] measure

$$D(P:Q) = \sum_{i=1}^{n} p_i \ln \frac{p_i}{q_i} \tag{7}$$

$$= \sum_{i=1}^{n} p_i \ln \frac{p_i}{\frac{x_i - x_{i-1}}{b-a}}$$

$$= \ln(b-a) - \left[-\sum_{i=1}^{n} p_i \ln \frac{p_i}{x_i - x_{i-1}} \right] \tag{8}$$

$$= \ln(b-a) - S_d(f), \tag{9}$$

where

$$S_d(f) = -\sum_{i=1}^{n} p_i \ln \frac{p_i}{x_i - x_{i-1}}; \quad p_i = \int_{x_{i-1}}^{x_i} f(x)\, dx \tag{10}$$

has been called by Forte and Hughes [4] as the discrete analogue of entropy measure (2) for the continuous-variate distribution with density function $f(x)$. This measure has been successfully applied by Forte and Sahoo [5] to solve a problem in image thresholding.

These authors of course did not proceed as above, but they used Boltzmann-Shannon entropy (2) and the principle of maximum entropy due to Jaynes [7]. We have not used either of these.

10.2 DISCUSSION OF THE DISCRETE ANALOGUE

We can have an infinity of partitions of the interval $[a, b]$ into n subintervals and we may be given some information to enable us to choose the partition. In the absence of any such information, we have no reason to choose any partition other than the uniform partition in which all the sub-intervals are of equal length (cf. Laplace's principle of insufficient reason), so that

$$x_i - x_{i-1} = \frac{b-a}{n}, q_i = \frac{1}{n} \tag{11}$$

and
$$D(P:Q) = \ln n - \left[-\sum_{i=1}^{n} p_i \ln p_i \right] \tag{12}$$

Now we define the entropy of any distribution, continuous variate or discrete variate, as any monotonic decreasing function of its directed divergence from the uniform distribution, so that we can consider

$$S(f) = -\sum_{i=1}^{n} p_i \ln p_i ; \quad p_i = \int_{x_{i-1}}^{x_i} f(x)\, dx \tag{13}$$

as a measure of entropy for the continuous-variate distribution with density function $f(x)$.

It may be noted that the expression (13) is the same as the expression (1), but it gives an entirely different measure. While (1) is a measure for a discrete-variate distribution, where the random variate takes a discrete set of n values with probabilities $p_1, p_2, \ldots, p_n$, (13) is a measure for a continuous-variate distribution where $p_1, p_2, \ldots, p_n$ are the probabilities of the random variate lying in n equal sub-intervals. Geometrically the two cases are represented in Figures 10.1 and 10.2.

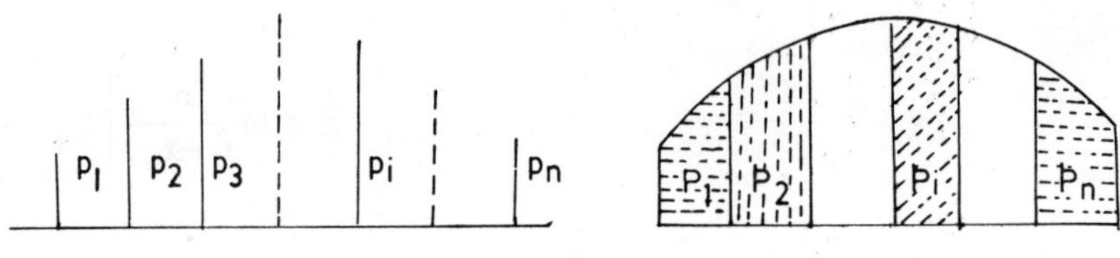

Fig. 10.1 **Fig. 10.2**

However, if we know $x_1, x_2, \ldots, x_{n-1}$, then we have to use the measure $S_d(f)$ given by (10) rather then $S(f)$ given by (13) and in that case the measure for the continuous variate case will not even look similar to the discontinuous variate case.

10.3 OTHER DISCRETE ANALOGUES

(*i*) The discrepancy between $f(x)$ and $g(x)$ can also be measured by $D(Q:P)$,

where $D(Q:P) = \sum_{i=1}^{n} q_i \ln \frac{q_i}{p_i} = \sum_{i=1}^{n} q_i \ln q_i - \sum_{i=1}^{n} q_i \ln p_i$

$$= \sum_{i=1}^{n} \frac{x_i - x_{i-1}}{b-a} \ln \frac{x_i - x_{i-1}}{b-a} - \sum_{i=1}^{n} \frac{x_i - x_{i-1}}{b-a} \ln \int_{x_{i-1}}^{x_i} f(x)\, dx , \tag{14}$$

so that for the given partition the discrete analogue is,

$$S_{d_1}(f) = \sum_{i=1}^{n} \frac{x_i - x_{i-1}}{b-a} \ln p_i \tag{15}$$

and if the sub-intervals are of equal length, we get the measure

$$S_1(f) = \frac{1}{n} \sum_{i=1}^{n} \ln p_i \tag{16}$$

Now (16) looks similar to Burg's [3] measure of entropy, but again p_i is not a probability mass concentrated at a point, but is a probability mass spread over the interval (x_{i-1}, x_i)

When the sub intervals are of unequal length and x_i's are known, we get (8) which differs from Burg's measure in two respects,

(*a*) it is a weighted measure, the weights being provided by the known lengths of the sub-intervals.

(*b*) p_i's are probability masses spread over intervals.

(*ii*) If we use Havrda-Charvat [6] measure of directed divergence,

$$D_\alpha (P:Q) = \frac{1}{\alpha(\alpha - 1)} \left(\sum_{i=1}^{n} p_i^\alpha q_i^{1-\alpha} - 1 \right), \tag{17}$$

the discrepancy of $f(x)$ from the uniform distribution is measured by

$$\frac{1}{\alpha(\alpha-1)} \left[\sum_{i=1}^{n} p_i^\alpha \left(\frac{x_i - x_{i-1}}{b-a} \right)^{1-\alpha} - 1 \right]; \quad p_i = \int_{x_{i-1}}^{x_i} f(x)\, dx \tag{18}$$

of which a monotonic decreasing function is

$$\frac{1}{\alpha(1-\alpha)} \sum_{i=1}^{n} p_i^\alpha (x_i - x_{i-1})^{1-\alpha} \tag{19}$$

When the intervals are of equal length this looks similar to (except for positive constants) to Havrda-Charvat [6] measure of entropy.

(*iii*) If appears that unlike the entropy measure for discrete variate distributions, the discrete analogue of a measure of entropy for a continuous-variate distribution need not be ≥ 0.

Thus $-\sum_{i=1}^{n} p_i \ln \frac{p_i}{x_i - x_{i-1}}$ need not be ≥ 0. What we can say is that it will be less than $\ln (b - a)$, but $\ln (b - a)$ can be negative. Similarly, the expression (19) > 0, if $\alpha < 1$ and is < 0 if $\alpha > 1$. Of course, Burg's entropy is negative even for a discrete variate distribution.

Thus, these discrete analogues can be used only if these are not required to be positive. In particular these can be used for entropy maximization problems.

(*iv*) It is obvious that to every measure of directed divergence for discrete probability distributions, there is a discrete analogue of a measure of entropy for a continuous variate distribution.

(*v*) Corresponding to Csiszer's family of measures

$$\sum_{i=1}^{n} q_i \phi \left(\frac{p_i}{q_i} \right) \tag{20}$$

where $\phi(\cdot)$ is a twice-differentiable convex function with $\phi(1) = \phi \;\; (0) = 0$, this gives the discrete analogue,

$$\sum_{i=1}^{n} \frac{x_i - x_{i-1}}{b-a} \phi\left(\frac{p_i(b-a)}{x_i - x_{i-1}}\right) \tag{21}$$

In general, we can get a discrete analogue from the discrete variate entropy measure by replacing p_i by $\frac{p_i}{(x_i - x_{i-1})}$ and then multiplying by $(x_i - x_{i-1})$ before summing from $i = 1$ to n. Thus,

$$-\sum_{i=1}^{n} p_i \ln p_i \quad \text{gives} \quad -\sum_{i=1}^{n} p_i \ln \frac{p_i}{x_i - x_{i-1}} \tag{22}$$

$$\sum_{i=1}^{n} \ln p_i \quad \text{gives} \quad \sum_{i=1}^{n} (x_i - x_{i-1}) \ln \frac{p_i}{(x_i - x_{i-1})} \tag{23}$$

$$\frac{1}{\alpha(1-\alpha)} \sum_{i=1}^{n} (p_i^{\alpha} - p_i) \quad \text{gives} \quad \frac{1}{\alpha(1-\alpha)} \sum_{i=1}^{n} (x_i - x_{i-1}) \left(\left(\frac{p_i}{x_i - x_{i-1}} \right)^{\alpha} - \frac{p_i}{x_i - x_{i-1}} \right)$$

$$= \frac{1}{\alpha(1-\alpha)} \left[\sum_{i=1}^{n} p_i^{\alpha} (x_i - x_{i-1})^{\alpha - 1} - 1 \right] \tag{24}$$

$$\frac{1}{\alpha(1-\alpha)} \ln \sum_{i=1}^{n} p_i^{\alpha} \quad \text{gives} \quad \frac{1}{\alpha(1-\alpha)} \ln \sum_{i=1}^{n} p_i^{\alpha} (x_i - x_{i-1})^{1-\alpha} \tag{25}$$

$$\frac{1}{\beta - \alpha} \ln \frac{\sum_{i=1}^{n} p_i^{\alpha}}{\sum_{i=1}^{n} p_i^{\beta}} \quad \text{gives} \quad \frac{1}{\beta - \alpha} \ln \frac{\sum_{i=1}^{n} p_i^{\alpha} (x_i - x_{i-1})^{1-\alpha}}{\sum_{i=1}^{n} p_i^{\beta} (x_i - x_{i-1})^{1-\beta}} \tag{26}$$

(*vii*) Two differences between entropy measure for a discrete variate probability distribution and the discrete analogue of the entropy for a continuous-variate distribution may be noted.

(*a*) In discrete variate case p_i's are probability masses concentrated at points, while in continuous-variate case, these are probability masses spread over intervals, and

(*b*) In the discrete case, entropy measure depends on p_i's only while in the continuous case these also depend on x_i's. This is the reason why Aczel [2] got an entropy measure for the continuous case by using the concept of an inset entropy, i.e. entropy measure depending on both p_i's and x_i's.

10.4 THE MAXIMUM ENTROPY PRINCIPLE

According to Jaynes [7] maximum entropy principle, of all probability distributions satisfying given constraints, we should choose that distribution which maximizes (1) in the discrete case, and (2) in the continuous-variate case subject to the given conditions. According to our discussion, the measure of entropy for the continuous-variate case is

$$-\sum_{i=1}^{n} p_i \ln \frac{p_i}{x_i - x_{i-1}}; \quad p_i = \int_{x_{i-1}}^{x_i} f(x)dx \tag{27}$$

We now appeal to Kapur and Kesavan's [10] generalized maximum entropy principle according to which we should maximize an appropriate measure of entropy subject to given conditions. Thus, our mathematical problem is:

Maximize,
$$-\sum_{i=1}^{n} \left(\int_{x_{i-1}}^{x_i} f(x)dx \right) \ln \frac{\int_{x_{i-1}}^{x_i} f(x)dx}{b-a} \tag{28}$$

subject to,
$$\int_a^b f(x)\,dx = \sum_{i=1}^{n} \int_{x_{i-1}}^{x_i} f(x)\,dx = 1 \tag{29}$$

$$\int_a^b f(x)\,\hat{g}_r(x)\,dx = \sum_{i=1}^{n} \int_{x_{i-1}}^{x_i} f(x)\,\hat{g}_r(x)\,dx = a_r, \quad r = 1, 2, \ldots, m \tag{30}$$

The major difference is that here instead of maximizing (2), we maximize (28). We would therefore like to examine the relationship between (2) and (28) more closely.

10.5 DERIVATION OF BOLTZMANN-SHANNON ENTROPY MEASURE

From (10) using mean-value theorem of integral calculus, we get

$$S_d(f) = -\sum_{i=1}^{n} \left(\int_{x_{i-1}}^{x_i} f(x)\,dx \right) \ln \frac{\int_{x_{i-1}}^{x_i} f(x)dx}{x_i - x_{i-1}}$$

$$= -\sum_{i=1}^{n} [x_i - x_{i-1}]\, f(\xi_i) \quad \ln \frac{(x_i - x_{i-1}) f(\xi_i)}{x_i - x_{i-1}}$$

$$= -\sum_{i=1}^{n} (x_i - x_{i-1}) f(\xi_i) \quad \ln (f(\xi_i)), \tag{31}$$

where
$$x_{i-1} < \xi_i < x_i \tag{32}$$

We now let $n \to \infty$, each $x - x_{i-1}$ tend to zero in such a way that the total length of the intervals remains $b - a$, then by the definition of Riemann integral and assuming $f(x)$ and $f(x) \ln f(x)$ to be continuous functions, we get when $n \to \infty$, $x_i - x_{i-1} \to 0$ for each i and $\sum_{i=1}^{n} (x_i - x_{i-1}) = b - a$

$$\underset{n \to \infty}{Lt} \quad S_d(f) = -\int_a^b f(x) \ln f(x)\, dx \tag{33}$$

which is Boltzmann-Shannon measure of entropy. In this derivation, what we have done is the following:

(*i*) Taken a partition of the interval $[a, b]$ into n subintervals.

(*ii*) Used the density function to get a probability distribution $P = (p_1, p_2, \ldots, p_n)$.

(*iii*) Used the density function $\frac{1}{b-a}$ to get another probability distribution

$$Q = (q_1, q_2, \ldots, q_n)$$

(*iv*) Found the measure of directed divergence $\sum_{i=1}^{n} p_i \ln \frac{p_i}{q_i}$ to get a measure of the discrepancy of $f(x)$ from the uniform distribution.

(*v*) Took the simplest monotonic decreasing function of this directed divergence as a measure of entropy.

(*vi*) Took the limit as the number of points of subdivision approaches infinity and all sub-interval tends to zero.

Thus we have here three distinct cases:

(*a*) $x_1, x_2, \ldots, x_{n-1}$ are specified. In this case we get the discrete analogue of the entropy measure.

(*b*) The interval $[a, b]$ is divided into n equal parts and we get a discrete analogue which appears similar to the entropy measure for the discrete case.

(*c*) When $n \to \infty$, $x_i - x_{i-1} \to 0$, we get the Boltzmann-Shannon entropy.

We may even say that if n is not specified, the objective way is to let it go to infinity and get the measure (2).

10.6 AN ALTERNATIVE DERIVATION OF BOLTZMANN-SHANNON ENTROPY MEASURE

We can use a similar procedure to find the discrete analogue for measure of directed divergence of the continuous-variate probability distributions. Let $f(x)$ and $g(x)$ be the density functions, and let

$$p_i = \int_{x_{i-1}}^{x_i} f(x)dx \quad q_i = \int_{x_{i-1}}^{x_i} g(x)\, dx \tag{34}$$

then $(p_1, p_2, \ldots, p_n)$ and $(q_1, q_2, \ldots, q_n)$ are probability distributions, so that,

$$D(P:Q) = \sum_{i=1}^{n} p_i \ln \frac{p_i}{q_i} \tag{35}$$

$$= \sum_{i=1}^{n} \left(\int_{x_{i-1}}^{x_i} f(x)\,dx \right) \ln \frac{\int_{x_{i-2}}^{x_i} f(x)dx}{\int_{x_{i-2}}^{x_i} g(x)dx} \tag{36}$$

Using the mean value theorem of integral calculus, we get

$$D(P:Q) = \sum_{i-1}^{n} (x_i - x_{i-1}) f(\xi_i) \ln \frac{(x_i - x_{i-1}) f(\xi_i)}{(x_i - x_{i-1}) g(\xi'_i)} \tag{37}$$

where,

$$x_{i-1} < \xi_i, \xi'_i < x_i \tag{38}$$

Now let $n \to \infty$, $x_i - x_{i-1} \to 0$, then taking the limit as $n \to \infty$, we get

$$D\,(f{:}g) = \int_a^b f(x) \ln \frac{f(x)}{g(x)}\,dx \tag{39}$$

If the second distribution is the uniform distribution, we get

$$D(f{:}U) = \int_a^b f(x) \ln \frac{f(x)}{\frac{i}{b-a}}\,dx = \ln (b-a) - \int_a^b f(x) \ln f(x)\,dx \tag{40}$$

and Boltzmann-Shannon measure (2) is obtained as a monotonic decreasing function of $D\,(f : U)$.

10.7 AN APPLICATION TO PARAMETER ESTIMATION

Here we are given the form of the density function $f(x, \theta)$ but we are not given the value of the parameter θ, and we have to estimate it in terms of a random sample x_1, $x_2, \ldots, x_n$ obtained from the population.

We choose θ so that the distribution is as uniform as possible subject to the knowledge of $x_1, x_2, \ldots, x_n$ and of the functional form f.

In this case p_i will be a function of θ, so that we have

$$p_i(\theta) = \int_{x_{i-2}}^{x_i} f(x, \theta)dx \tag{41}$$

and we have to choose θ to maximize

$$-\sum_{i=1}^{n} p_i(\theta) \ln \frac{p_i(\theta)}{x_i - x_{i-1}} \tag{42}$$

This can be compared with the method proposed by Lind and Solana [12] and various methods proposed by Kapur and Kesavan [10]. Lind and Solana assumed was that the random sample gives intervals in which the probabilities should be as equal as possible so that we choose θ to make $p_1(\theta), p_2(\theta), \ldots, p_n(\theta)$ as equal as possible and for this purpose they maximized

$$\sum_{i=1}^{n} \ln p_i(\theta), \quad \text{or} \quad -\sum_{i=1}^{n} p_i(\theta) \ln p_i(\theta) \tag{43}$$

We are assuming is that θ has such a value that $f(x, \theta)$ is as nearly constant as possible subject to the information given by the random sample so that in our case $p_i(\theta)$ is as nearly as possible proportional to $(x_i - x_{i-1})$, so that

$$\frac{p_i(\theta)}{x_1 - x_0}, \frac{p_2(\theta)}{x_2 - x_1}, \ldots, \frac{p_n(\theta)}{x_n - x_{n-1}} \tag{44}$$

are as nearly as equal as possible. This requires us to maximize either the entropy, $-\sum_{i=1}^{n} \frac{p_i(\theta)}{x_i - x_{i-1}} \ln \frac{p_i(\theta)}{x_i - x_{i-1}}$ or we have to minimize the directed divergence of $(p_1, \ldots, p_n)$ from

$$\left(\frac{x_1 - x_0}{b-a}, \frac{x_2 - x_1}{b-a}, \ldots, \frac{x_n - x_{n-1}}{b-a}\right)$$

i.e., we have to choose θ to minimize

$$\sum_{i=1}^{n} p_i(\theta) \ln \frac{p_i(\theta)}{\frac{x_i - x_{i-1}}{b-a}} \tag{45}$$

i.e., we have to choose θ to maximize

$$-\sum_{i=1}^{n} p_i(\theta) \ln \frac{p_i(\theta)}{x_i - x_{i-1}} \tag{46}$$

(43) and (46) are different and as such the estimates $\hat{\theta}_1$ and $\hat{\theta}_2$ obtained by these approaches will be different.

10.8 THE GENERAL PROBLEM OF STATISTICAL ESTIMATION

We have a number of approaches possible, depending on the information available.

(*i*) In the classical approach, we assume that we know the functional form $f(x, \theta)$ and we are given a random sample $x_1, x_2, \ldots, x_{n-1}, x_n$. In this case a number of methods have been proposed to estimate θ. These include Fisher's method of maximum likelihood, Pearson's method of moments, method of minimum chi-square, method of least squares, method of least information and a number of methods proposed in Kapur and Kesavan [10] and also the method proposed in section 7.

(*ii*) If we are not given $f(x, \theta)$ but we are given some characterizing moments as well as a random sample, we can first use the principle of maximum entropy, with Botlzmann-Shannon measure or any other classical measure of entropy and estimate $f(x, \theta)$ and then use the knowledge of the random sample to estimate the parameter θ (which may be a scalar or a vector, in terms of the sample values as in (i). In this approach we first use the moments and then use the random sample.

(*iii*) In this approach we first use the random sample to get the discrete analogue of entropy of the still unknown density function as

$$-\sum_{i-1}^{n} p_i \ln \frac{p_i}{x_i - x_{i-1}}, \quad p_i = \int_{x_{i-2}}^{x_i} f(x,\theta)dx \tag{47}$$

and then determine the unknown density function and the parameter by using the generalised principle of maximum entropy as suggested in section 4.

10.9 SUMMARY AND REMARKS

In the present discussion we have,

(*i*) given an alternative approach for finding discrete analogue of measure of entropy for a continuous random variate which is different from that proposed by Forte and Hughes [4]. In this derivation we have neither used the principle of maximum entropy nor the Boltzmann-Shannon measure of entropy.

(*ii*) discussed the interpretation of this discrete analogue is the light of this new derivation.

(*iii*) given some other discrete analogues corresponding to measures of directed divergence other than Kullback-Leibler's [11].

(*iv*) given an alternative derivation of Boltzmann-Shannon entropy measure for a continuous random variate without making explicit use of Shannon's measures of entropy.

(*v*) given discrete analogue of measure of directed divergence for probability distributions of a continuous random variate.

(*vi*) given another alternative derivation of Boltzmann-Shannon entropy measures, based on discrete analogue, of measure of directed divergence for a continuous random variate.

(*vii*) used the discrete analogue to give an alternative principle of parameter estimation.

(*viii*) discussed how the principle of maximum entropy can be used in conjunction with the new measures we have derived.

(*ix*) shown that if we are given informations in the form of both random sample and characterizing moments, different results will follow according to which information we use first.

REFERENCES

1. J. Aczel and Z. Daroczy (1975). On Measures of Information and their Characterization, Academic Press, New York.
2. J. Aczel (1984), "Measuring Information beyond Communication Theory". Inf. Proc. and Management, Vol. 20 no. 3, pp. 383–393.
3. J.P. Burg (1972), "The Relationship between Maximum Entropy Spectra and Maximum likelihood Spectra". In D.G. Childers, editor, Modern Spectral Analysis, pp. 130–131.

4. B. Forte and W. Hughes (1988), "The Maximum Entropy principle; a tool to define new entropies". Reports of Mathematical Physics, 26 (2), pp. 227–238.
5. B. Forte and P.K. Sahoo (1986), "Minimal loss of information and optimal threshold for digital image" . . . Manuscript.
6. J.H. Havrda and F. Charvat (1967), "Quantification Methods of Classification Processes: Concept of Structural α Entropy". Kybernatika, Vol. 3, pp. 30–35.
7. E. T. Jaynes (1957), "Information Theory and Statistical Mechanics". Physical Reviews, Vol. 106, pp. 620–630.
8. J.N. Kapur (1990), Maximum Entropy Models in Science and Engineering, Wiley Eastern, New Delhi and John Wiley, New York.
9. J.N. Kapur and H.K. Kesavan (1987), Generalized Maximum Entropy Principle (with Applications) Sandford Educational Press, University of Waterloo, Canada.
10. J.N. Kapur and H.K. Kesavan (1972), Entropy Optimization Principles and their Application. Academic Press, New York.
11. S. Kullback and R.A. Leibler (1951), "On Information and Sufficiency" Ann. Math. Stat. Vol 22, pp. 79–86.
12. N.C. Lind and V. Solana (1988), "Cross-entropy Estimation of Random Variables with Fractile Constraints." Institute of Risk Research, paper no. 11. University of Waterloo. Canada.
13. A. M. Mathai and P.N. Rathie (1975), Basic Concepts of Information Theory and Statistics, John Wiley, New York.
14. C.E. Shannon (1948), "Mathematical Theory of Communication". Bell System Tech. Journal, Vol. 27, No. 1–4, pp. 379–423, 623–656.

11

CORRECT MEASURES OF WEIGHTED DIRECTED DIVERGENCE

[We propose new measures of weighted directed divergence of a probability distribution $P = (p_1, p_2, \ldots, p_n)$ from another probability distribution $Q = (q_1, q_2, \ldots, q_n)$. Unlike some earlier measures, these are always ≥ 0 and vanish if and only if $P = Q$. These are also convex functions of $p_1, p_2, \ldots, p_n$ and can be used to determine that probability distribution, out of all those satisfying given constraints, for which the weighted directed divergence from a given distribution Q is least. It is also shown that these weighted measures arise naturally when we find measures of discrepancy between *a priori* and *a posteriori* distributions of numbers of balls in n boxes when balls are thrown at random in those boxes according to specified probability distributions].

11.1 INTRODUCTION

Let $P = (p_1, p_2, \ldots, p_n)$ be a probability distribution and let $W = (w_1, w_2, \ldots, w_n)$ be a set of weights associated with the n outcomes, then corresponding to Shannon's [12] measure of entropy,

$$S(P) = -\sum_{i=1}^{n} p_i \ln p_i \tag{1}$$

Guiasu [2] defind a measure of weighted entropy

$$S(P, W) = -\sum_{i=1}^{n} w_i p_i \ln p_i \tag{2}$$

and characterised it axiomatically.

Now if $Q = (q_1, q_2, \ldots, q_n)$ is another probability distribution, then the well-known Kullback-Leibler [10] number $D_1(P : Q)$ gives a measure of directed divergence of P from Q, where

$$D_1(P:Q) = \sum_{i=1}^{n} p_i \ln \frac{p_i}{q_i} \tag{3}$$

and Taneja and Tuteja [14] gave the corresponding measure of weighted directed divergence as

$$D_1(P:Q, W) = \sum_{i=1}^{n} w_i\, p_i \ln \frac{p_i}{q_i} \tag{4}$$

Measure (3) is a correct measure of directed divergence since it has the properties

(*i*) $D_1(P:Q) \geq 0$

(*ii*) $D_1(P:Q) = 0$ iff $P = Q$

(*iii*) $D_1(P:Q)$ is a convex function of P and Q.

However $D_1(P:Q, W)$ is not a correct measure of weighted directed divergence since it does not satisfy (*i*) and (*ii*). To see this,consider the simple case when

$$P = (1/3, 2/3), Q = (2/3, 1/3)$$

$$D_1(P:Q, W) = w_1 \frac{1}{3} \ln \frac{1}{2} + w_2 \frac{2}{3} \ln 2$$

$$= \frac{1}{3}(2w_2 - w_1) \ln 2 \tag{5}$$

This can be both greater than and less than zero. It can also vanish even when $P \neq Q$.

Thus, it is difficult to accept $D_1(P:Q, W)$ as a correct measure of weighted directed divergence.

One object of the paper is to find the correct measure of weighted directed divergence corresponding to measure (3). In fact we shall find a more general correct measure of weighted directed divergence corresponding to Csiszer's [1] measure of directed divergence,

$$D_2(P:Q) = \sum_{i=1}^{n} q_1 \phi\left(\frac{p_i}{q_i}\right), \tag{6}$$

where $\phi(\cdot)$ is a twice-differentiable convex function with $\phi(1) = 0$.

11.2 CORRECT MEASURE OF WEIGHTED DIRECTED DIVERGENCE CORRESPONDING TO CSISZER'S MEASURE

Since $\phi(x)$ is convex, $\phi'(x)$ is always increasing. Also $\phi(1) = 0$ At $x = 1$, $\phi'(x)$ may be positive, negative or zero (Figures 11.1 (*a*), (*b*), (*c*)).

In cases (*a*) and (*b*), $\phi(p_i/q_i)$ can be negative or positive. However $D_1(P:Q)$ is always ≥ 0, but $D_1(P:Q, W)$ can be negative or positive. In case (*c*), $\phi(p_i/q_i)$ is always ≥ 0 and vanishes only when $p_i = q_i$, so that in this case

$$D_2(P:Q, W) = \sum_{i=1}^{n} w_i q_i \phi\left(\frac{p_i}{q_i}\right) \tag{7}$$

would be ≥ 0 and vanishes if $p_i = q_i$ for each i. Thus the correct measure of weighted directed divergence corresponding to Csiszer's measure of directed divergence is given by (7), provided (*i*) $\phi(x)$ is convex and twice-differentiable

(*ii*) $\phi(1) = 0$

(*iii*) $\phi'(1) = 0$

The condition (*iii*) is the additional condition we are imposing on the function $\phi(x)$.

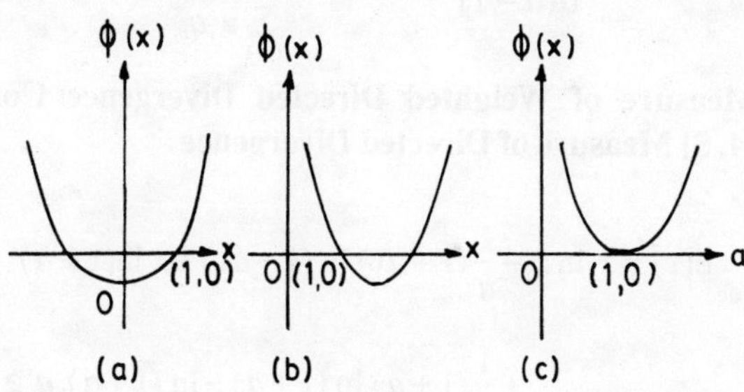

Fig. 11.1

11.3 SPECIAL CASES

(a) Correct Measure Corresponding to Kullback-Leibler Measure

Let $\phi(x) = x \ln x - x + 1$,
so that $\phi(x)$ is convex, $\phi(1) = 0$, $\phi'(1) = 0$, $\phi(x) \geq 0$ and,

$$D_3(P:Q,W) = \sum_{i=1}^{n} w_i q_i \left[\frac{p_i}{q_i} \ln \frac{p_i}{q_i} - \frac{p_i}{q_i} + 1 \right]$$

$$= \sum_{i=1}^{n} w_i \left[p_i \ln \frac{p_i}{q_i} - p_i + q_i \right] \tag{8}$$

(b) Correct Measure of Weighted Directed Divergence Corresponding to Havrda and Charvat's [3] Measure

Let
$$\phi(x) = \frac{x^{\alpha} - \alpha x + \alpha - 1}{\alpha(\alpha - 1)}, \alpha \neq 0, 1 \tag{9}$$

so that $\phi(x)$ is convex, $\phi(1) = 0, \phi'(1) = 0, \phi(x) \geq 0$

and
$$D_4(P:Q,W) = \frac{1}{\alpha(\alpha-1)} \sum_{i=1}^{n} w_i q_i \left[\left(\frac{p_i}{q_i} \right)^{\alpha} - \alpha \frac{p_i}{q_i} + \alpha - 1 \right]$$

$$= \frac{1}{\alpha(\alpha-1)} \sum_{i=1}^{n} w_i [p_i^{\alpha} q_i^{1-\alpha} - \alpha p_i + \alpha q_i - q_i] \tag{10}$$

(c) Correct Measure of Weighted Directed Divergence Corresponding to Renyi's [11] Measure of Directed Divergence

This measure is given by

$$D_5(P:Q,W)=\frac{\sum_{i=1}^{n} w_i q_i}{\alpha(\alpha-1)}\ln\frac{\sum_{i=1}^{n} w_i(p_i^{\alpha}q_i^{1-\alpha}-\alpha p_i+\alpha q_i)}{\sum_{i=1}^{n} w_i q_i},\alpha>1 \tag{11}$$

(d) Correct Measure of Weighted Directed Divergence Corresponding to Kapur's [4, 5] Measure of Directed Divergence

Let
$$\phi(x)=x\ln x-\frac{1}{a}(1+ax)\ln(1+ax)+x\ln(1+a)$$

$$+\frac{1}{a}(1+a)\ln(1+a)-\ln(1+a),a\geq 0 \tag{12}$$

so that $\phi(1)=0,\phi'(1)=0,\phi(x)\geq 0$

and

$$D_6(P:Q,W)=\sum_{i=1}^{n} w_i p_i \ln\frac{p_i}{q_i}+\sum_{i=1}^{n} w_i(p_i-q_i)\ln(1+a)$$

$$+\frac{1}{a}(1+a)\ln(1+a)\sum_{i=1}^{n} w_i q_i-\frac{1}{a}\sum_{i=1}^{n} w_i(q_i+ap_i)\ln\left(1+\frac{ap_i}{q_i}\right) \tag{13}$$

It may be noted that

(*i*) As $\alpha\to 1$ measures (10) and (11) approach measure (8),

(*ii*) As $\alpha\to 0$ measures (10) and (11) approach $D_3(Q:P,W)$,

(*iii*) As $a\to 0$ measure (13) approaches measure (8).

(*iv*) From (10)

$$D_{4,1-\alpha}(P:Q,W)=\frac{1}{(1-\alpha)(-\alpha)}\sum_{i=1}^{n} w_i(p_i^{1-\alpha}q_i^{\alpha}-(1-\alpha)(p_i-q_i)-q_i)$$

$$=\frac{1}{\alpha(\alpha-1)}\sum_{i=1}^{n} w_i(q_i^{\alpha}p_i^{1-\alpha}-\alpha q_i+\alpha p_i-p_i)$$

$$=D_{4,\alpha}(Q:P,W) \tag{14}$$

so that replacement of α by $1-\alpha$ in $D_4(P:Q:,W)$ is equivalent to interchanging Q and P.Similarly replacing α by $1-\alpha$ in $D_5(Q:P,W)$ is equivalent to interchanging P and Q.

(*v*) Putting $\alpha=0$ in (14), we get

$$D_{4,0}(Q:P,W)=D_{4,1}(P:Q,W)=D_3(P:Q,W) \tag{15}$$

Similarly $\qquad D_{5,0}(P:Q,W) = D_{5,1}(P:Q,W) = D_3(P:Q,W)$ (16)

(*vi*) In place of (7), we can use the measure

$$D_7(P:Q,W) = \frac{\sum_{i=1}^{n} w_i q_i \phi(p_i/q_i)}{\sum_{i=1}^{n} w_i q_i} \tag{17}$$

This is also ≥ 0, vanishes iff $p_i = q_i$ for each i and is a convex function of $p_1, p_2, \ldots, p_n$. This is also independent of the units in which weights are measured.

(*vii*) If weights w_i's are interpreted as 'utilities', then the above measures are called 'useful' measures of directed divergence. There is extensive literature on these measures.

11.4 ANOTHER CLASS OF MEASURES OF WEIGHTED DIRECTIVE DIVERGENCE

As a generalisation of (7), consider the measure

$$D_8(P:Q,W) = \sum_{i=1}^{n} w_i q_i \phi\left(\frac{p_i}{q_i}\right) + \sum_{i=1}^{n} w_i (1 + aq_i)\, \phi \frac{(1 + ap_i)}{(1 + aq_i)} \tag{18}$$

It is easily seen that if $\phi(\cdot)$ is a twice-differentiable convex function and $\phi(1) = 0$, $\phi'(1) = 0$ then $D_8(P:Q,W)$ is a convex function of $p_1, p_2, \ldots, p_n$ which attains its minimum value zero when each $p_i = q_i$ so that

$$D_8(P:Q,W) \geq 0 \quad \text{and vanishes iff} \quad P = Q \tag{19}$$

This result will be true whether $a > 0$ or $a < 0$. However, if $a < 0$, since we want $1 + ap_i > 0$, we would take $a \geq -1$ and if $a = -1$, we shall not consider degenerate distributions.

If $\phi(x) = \ln x - x + 1$, we get the measure

$$D_9(P:Q,W) = \sum_{i=1}^{n} w_i \left(p_i \ln \frac{p_i}{q_i} - p_i + q_i\right) + \sum_{i=1}^{n} w_i \left((1 + ap_i) \ln \frac{1 + ap_i}{1 + aqi} - ap_i + aq_i\right) \tag{20}$$

When $a = -1$, it gives

$$D_{10}(P:Q,W) = \sum_{i=1}^{n} w_i \left(p_i \ln \frac{p_i}{q_i} - p_i + q_i\right) + \sum_{i=1}^{n} w_i \left((1 - p_i) \ln \frac{1 - p_i}{1 - q_i} + p_i - q_i\right)$$

$$= \sum_{i=1}^{n} w_i \left(p_i \ln \frac{p_i}{q_i} + (1 - p_i) \ln \frac{1 - p_i}{1 - q_i}\right) \tag{21}$$

and if $a = 1$, this gives

$$D_{11}(P:Q,W)=\sum_{i=1}^{n} w_i\left(p_i \ln\frac{p_i}{q_i}-p_i+q_i\right)+\sum_{i=1}^{n} w_i\left((1+p_i)\ln\frac{1+p_i}{1+q_i}-p_i+q_i\right)$$

$$=\sum_{i=1}^{n} w_i\left(p_i \ln\frac{p_i}{q_i}+(1+p_i)\ln\frac{1+p_i}{1+q_i}-2p_i+2q_i\right) \tag{22}$$

If we take,

$$\phi(x)=\frac{x^{\alpha}-\alpha x+\alpha-1}{\alpha(\alpha-1)},$$

we get,

$$D_{12}(P:Q,W)=\frac{1}{\alpha(\alpha-1)}\sum_{i=1}^{n} w_i[p_i^{\alpha}q_i^{1-\alpha}-\alpha p_i+\alpha q_i-q_i$$

$$+(1+ap_i)^{\alpha}(1+aq_i)^{1-\alpha}-\alpha a(p_i-q_i)-q_i] \tag{23}$$

and when $a=-1, 1$, we get

$$D_{13}(P:Q,W)=\frac{1}{\alpha(\alpha-1)}\sum_{i=1}^{n} w_i[p_i^{\alpha}q_i^{1-\alpha}+(1-p_i)^{\alpha}(1-q_i)^{1-\alpha}-2q_i] \tag{24}$$

$$D_{14}(P:Q,W)=\frac{1}{\alpha(\alpha-1)}\sum_{i=1}^{n} w_i[p_i^{\alpha}q_i^{1-\alpha}+(1+p_i)^{\alpha}(1+q_i)^{1-\alpha}$$

$$-2\alpha p_i+2\alpha q_i-2q_i] \tag{25}$$

11.5 WEIGHTED MEASURES OF DIRECTED DIVERGENCE ARISING FROM CONSIDERATION OF 'MONKEYS THROWING BALLS' EXPERIMENTS

In a recent paper, Kapur [7] proved the following results:

(*i*) Let monkeys throw N balls at random in n boxes and let the *a priori* probability of a ball falling in the ith box be q_i, then the probability Π that the boxes receive proportions $p_1, p_2, \ldots, p_n$ of the balls, when N is large, is given by

$$\ln \Pi=-N\sum_{i=1}^{n} p_i \ln\frac{p_i}{q_i} \tag{26}$$

so that the proportions $p_1, p_2, \ldots, p_n$ which maximize Π is obtained by minimizing Kullback-Leibler measure

$$\sum_{i=1}^{n} p_i \ln\frac{p_i}{q_i} \tag{27}$$

subject to whatever information we have on $p_1, p_2, \ldots, p_n$. This is Kullback's [9] principle of minimum discrimination information.

(*ii*) Let n teams of monkeys throw balls in n boxes so that the probability distribution for the ith box is Poisson distribution with mean Nq_i, then the probability Π that $Np_1, Np_2, \ldots, Np_n$ balls are received in the n boxes, when N is large is given by

$$\ln \Pi = -N \sum_{i=1}^{n} \left(p_i \ln \frac{p_i}{q_i} - p_i + q_i \right) \tag{28}$$

This result was proved earlier by Skilling [13].

(*iii*) If the monkeys throw N_i balls in the ith box and the probability distribution for the number of balls in the ith box is binomial with probability of success q_i, then the probability that number of balls received in the ith box is $N_i\, p_i$ ($i = 1, 2, \ldots n$), when N_i's are large is given by

$$\ln \Pi = -\sum_{i=1}^{n} N_i \left[p_i \ln \frac{p_i}{q_i} + (1 - p_i) \ln \frac{1 - p_i}{1 - q_i} \right] \tag{29}$$

This corresponds to our weighted measure of directed divergence (21).

(*iv*) Let the probability distribution of the ith box be negative binomial with mean $N_i\, q_i$, then the probability Π that the number of balls in it is $N_i\, p_i$ is given by

$$\ln \Pi = -\sum_{i=1}^{n} N_i \left[p_i \ln \frac{p_i}{q_i} - (1 + p_i) \ln \frac{1 + p_i}{1 + q_i} \right] \tag{30}$$

This is slightly different from our weighted measure (22), but it is also seen that (30) is also a correct weighted measure of directed divergence.

(*v*) If in (11) we take the mean of the ith Poisson distribution to be $N_i\, q_i$ and we want the probability of getting $N_i\, p_i$ balls in the ith box, we get

$$\ln \Pi = -\sum_{i=1}^{n} N_i \left(p_i \ln \frac{p_i}{q_i} - p_i + q_i \right) \tag{31}$$

Thus in (29), (30), (31), we get weighted directed measures where the weights are given by the different numbers of balls available for different boxes.

11.6 SOME OTHER MEASURES OF WEIGHTED DIRECTED DIVERGENCE

Consider the measure,

$$D_{15}(P{:}Q, W) = \sum_{i=1}^{n} w_i \left[(p_i - a_i) \ln \frac{p_i - a_i}{q_i - a_i} + (b_i - p_i) \ln \frac{b_i - p_i}{b_i - q_i} \right] \tag{32}$$

This is easily seen to be a convex function of $p_1, p_2, \ldots, p_n$ whose minimum value zero arises when each $p_i = q_i$ so that (32) can be used as a valid measure of weighted directed divergence. This will be specially useful when we have to minimize $D_{15}(P : Q, W)$ subject to [8],

$$a_i \le p_i \le b_i, \quad i = 1, 2, \ldots n \tag{33}$$

Again while $\sum_{i=1}^{n} q_i \,\phi\left(\dfrac{p_i}{q_i}\right)$ a measure of directed divergence,

$\sum_{i=1}^{n} w_i q_i \phi\left(\dfrac{p_i}{q_i}\right)$ need not be a measure of directed divergence. However, we have got the following theorems:

Theorem 1. If $\phi(\cdot)$ is a convex twice-differentiable function such that $\phi(1) = 0$, then

$$\sum_{i=1}^{n} w_i \left[q_i \phi\left(\frac{p_i}{q_i}\right) + (1 - q_i)\,\phi\left(\frac{1 - p_i}{1 - q_i}\right)\right] \tag{34}$$

is a valid measure of directed divergence

Theorem 2. If $\phi(\cdot)$ is a convex twice-differentiable function such that $\phi(1) = 0$, $\phi'(1) = 0$, then

$$\sum_{i=1}^{n} w_i \left[q_i \,\phi\left(\frac{p_i}{q_i}\right) + (1 + a q_i)\,\phi\left(\frac{1 + a p_i}{1 + a q_i}\right)\right] \tag{35}$$

is a valid measure of weighted directed divergence.

By using (9), its limiting case as $\alpha \to 1$ or other suitable functions, we can get a large number of correct measures of weighted directed divergence.

11.7 CONCLUDING REMARKS

(*i*) If all the outcomes are equally important, we use Kullback's [9] principle of minimum directed divergence (or of minimum cross-entropy) according to which, out of all probability distributions satisfying given constraints, we choose that distribution P for which $D_1(P : Q)$ is minimum. If all the outcomes are not equally important, we use the modified principle of minimum weighted cross entropy, according to which, out of all probability distributions satisfying given constraints, we choose that distribution P which minimizes $D_1(P : Q, W)$.

(*ii*) We have given above correct measures of directed divergence for discrete random variate probability distribution. The corresponding measures for continuous-variate probability distributions can easily be obtained. Thus if $f(x)$ and $g(x)$ are the density functions for the probability distributions, then

$$D_3(f{:}g, w) = \int_a^b w(x)\left[f(x) \ln \frac{f(x)}{g(x)} - f(x) + g(x)\right] dx \tag{36}$$

$$D_4(f{:}g, w) = \int_a^b w(x)\,[f^{\alpha}(x) g^{1-\alpha}(x) - \alpha f(x) + \alpha\, g(x) - g(x)]\; dx \tag{37}$$

(*iii*) Instead of using the term weighted directed divergence, we may call it directed divergence with respect to weight function $w(x)$ or 'useful' directed divergence or quantitative-qualitative measure of directed divergence.

(*iv*) The measure $J_k(P : Q, W)$ of symmetric divergence of P from Q can be defined by

$$J_k(P{:}Q,W) = D_k(Q{:}P,W) + D_k(Q{:}P,W) \quad k = 1,2,3... \tag{38}$$

(*v*) The measure $I_k(P : Q, W)$ of weighted information improvement in going from Q to R when the true distribution is P and the weight function is $W(x)$ is defined by

$$I_k(P{:}Q{:}R,W) = D_k(P{:}Q,W) - D_k(P{:}R,W) \tag{39}$$

In particular we have

$$I_1(P{:}Q{:}R,W) = \sum_{i=1}^{n} w_i\left[p_i \ln\frac{p_i}{q_i} - p_i + q_i - p_i \ln\frac{p_i}{r_i} + p_i - r_i\right]$$

$$= \sum_{i=1}^{n} w_i\left[p_i \ln\frac{r_i}{q_i} + q_i - r_i\right] \tag{40}$$

It may be noted that this measure depends on four distributions viz., P, Q, R and W.

REFERENCES

1. I. Csiszer, (1972), "A class of Measures of Informativity of Observation Channels", Periodic Math. Hungarica, vol. 2 pp. 191–213.
2. S. Guiasu, (1971), "Weighted Entropy", Reports on Math. Physics, Vol. 2, pp. 165–179.
3. J.H. Havrda and F. Charvat, (1967), "Quantification Methods of Classification Processes : Concepts of Structural α Entropy, Kybernatica vol. 3, pp. 30–35.
4. J.N. Kapur, (1984), "A comparative assessment of various measures of directed divergence", Advances in Management Studies, vol. 3, pp. 1–16.
5. J.N. Kapur, (1986), "Four families of measures of entropy," Ind. Jour. Pure and App. Maths., vol 17, No. 4, pp. 429–449.
6. J.N. Kapur, (1990), *Maximum-Entropy Models in Science and Engineering*, Wiley Eastern, New Delhi.
7. J.N. Kapur, (1989), "Monkeys and Entropies", Bull. Math. Ass. India.
8. J.N. Kapur, (1989), "Maximum Entropy Probability Distributions when there are inequality constraints on probabilities and equality constraints on moments", Aligarh Jounral Stat 2, pp. 28–38.
9. S. Kullback, (1959), *Information Theory and Statistics*, John Wiley, New York.
10. S. Kullback and R.A. Leibler, (1951), "On Information and Sufficiency," Ann. Math. Stat., vol. 22, pp. 79–86.
11. A. Renyi, (1961), "On Measures of Entropy and Information." Proc. 4th Berkeley Symp. Maths. Stat. Prob., vol 1, pp. 547–561.
12. C.E. Shannon, (1948), "A Mathematical Theory of Communication", Bell System, Tech. J. vol 27, pp. 379–423, 623–659.
13. J.S. Skilling, (1990), "Quantified Maximum Entropy". In *Maximum Entropy and Bayesian Methods*, edited by P. Fourgere, pages 341–350, Kluwer Academic Press.
14. H.C. Taneja and R.K. Tuteja, (1984), "Characterisation of a qualitative-quantitative measures of relative information", Information Sciences, vol. 33, pp. 1–6.

12

SOME NEW MEASURES OF DIRECTED DIVERGENCE

[A number of new measures of directed divergence have been proposed and their properties have been discussed. These include Jensen's measures corresponding to various well-known entropies.]

12.1 INTRODUCTION

In a recent paper Lin [7] has proposed the following measures of directed divergence of a probability distribution $P = (p_1, p_2, \ldots, p_n)$ from another probability distribution $Q = (q_1, q_2, \ldots, q_n)$.

$$K(P:Q) = \sum_{i=1}^{n} p_i \, ln \frac{p_i}{(p_i + q_i)/2} \tag{1}$$

$$JS_\pi(P:Q) = H(\pi_1 P + \pi_2 Q) - \pi_1 H(P) - \pi_2 H(Q) \tag{2}$$

where, $$\pi = (\pi_1, \pi_2), \pi_1 \geq 0, \pi_2 \geq 0, \pi_1 + \pi_2 = 1 \tag{3}$$

and, $$H(P) = -\sum_{i=1}^{n} p_i \ln p_i \tag{4}$$

is Shannon's [9] measure of entropy.
He has also defined a symmetric measure corresponding to (1) by,

$$L(P:Q) = K(P:Q) + K(Q:P) \tag{5}$$

and generalised Jensen-Shannon measure corresponding to (2) by

$$JS_\pi(P_1, P_2, \ldots, P_n) = H(\pi_1 P_1 + \pi_2 P_2 + \ldots + \pi_n P_n) - \sum_{i=1}^{n} \pi_i H(P_i) \tag{6}$$

In the present discussion, we generalise Lin's measures in two directions:

(*i*) Instead of using $(P + Q)/2$, we use $\lambda P + (1 - \lambda)Q, 0 \leq \lambda \leq 1$

(*ii*) Instead of using only Shannon's [9] measure of entropy or Kullback Leibler's [6] measure of directed divergence, we use other generalised measures of entropy and directed divergence [5].

12.2 FIRST GENERALISATION

Consider the measure,

$$D\,(P{:}Q) = \sum_{i=1}^{n} (\lambda p_i + (1-\lambda) q_i)\, \phi\left(\frac{p_i}{\lambda p_i + (1-\lambda)\, q_i}\right); \quad 0 \le \lambda \le 1\,, \tag{7}$$

where, $\phi\,(\cdot)$ is a twice - differentiable convex function for which $\phi\,(1) = 0$. This is a generalisation of Csiszer's [2] measure of directed divergence which comes out as special case if we put $\lambda = 0$, i.e. it is a generalisation of

$$D_1\,(P{:}Q) = \sum_{i=1}^{n} q_i\, \phi\left(\frac{p_i}{q_i}\right) \tag{8}$$

We note the following properties of $D\,(P:Q)$

(*i*) $D\,(P:Q)$ is a continuous function of $p_1, p_2, \ldots, p_n$. Except when $\lambda = 0$, this is defined even if a $q_i = 0$ and the corresponding $p_i \neq 0$ or if $p_i = 0$ and corresponding $q_i \neq 0$. It is not defined only when both p_i and q_i are zero, but then the ith outcome can be easily ignored.

(*ii*) $D\,(P:Q)$ is permutationally symmetric, i.e it does not change if the pairs $(p_1, q_1), (p_2, q_2), \ldots, (p_n, q_n)$ are permuted among themselves.

(*iii*) Let $D\,(P:Q) = f\,(p_1, p_2, \ldots, p_n\,;\, q_1, q_2, \ldots, q_n)$

then,

$$\frac{\partial f}{\partial p_i} = \lambda\, \phi\left(\frac{p_i}{\lambda p_i + (1-\lambda)\, q_i}\right) + \frac{(1-\lambda)\, q_i}{(\lambda p_i + (1-\lambda)\, q_i)}\, \phi'\left(\frac{p_i}{\lambda p_i + (1-\lambda)\, q_i}\right) \tag{9}$$

$$\frac{\partial^2 f}{\partial p_i^2} = \frac{(1-\lambda)^2 q_i^2}{(\lambda p_i + (1-\lambda) q_i)^3}\, \phi''\left(\frac{p_i}{\lambda p_i + (1-\lambda)\, q_i}\right) \tag{10}$$

$$\frac{\partial^2 f}{\partial p_i\, \partial p_j} = 0 \tag{11}$$

Since $\phi\,(\cdot)$ is a convex function , $\partial^2 f/\partial p_i^2 > 0$ and so $D\,(P:Q)$ is a convex function of $p_1, p_2, \ldots, p_n$.

The global minimum of $D\,(P:Q)$ subject to $\sum_{i=1}^{n} p_i = 1$ occurs when

$$\frac{\partial f / \partial p_1}{1} = \frac{\partial f / \partial p_2}{1} = \cdots = \frac{\partial f / \partial p_n}{1} \tag{12}$$

i.e., when $p_i = q_i\ \forall_i$ and from (7), its minimum value is

$$\sum_{i=1}^{n} p_i\, \phi\,(1) = 0, \tag{13}$$

so that,

$$D\,(P:Q) \ge 0 \tag{14}$$

and vanishes iff $P = Q$, so that it can be used as a measure of directed divergence.

12.3 SOME SPECIAL CASES

(*i*) Let $\phi(x) = x \ln x$, (15)

then (7) gives,

$$D_{2,\lambda}(P:Q) = \sum_{i=1}^{n} p_i \ln \frac{p_i}{\lambda p_i + (1-\lambda) q_i}, \qquad 0 \le \lambda \le 1 \tag{16}$$

We may call it generalised Lin's measure

If $\lambda = 0$, it gives Kullback - Leibler [6] measure

$$D_{2,0}(P:Q) = \sum_{i=1}^{n} p_i \ln p_i / q_i \tag{16a}$$

If $\lambda = 1/2$ it gives Lin's measure

$$D_{2,1/2}(P:Q) = \sum_{i=1}^{n} p_i \ln \frac{p_i}{(p_i + q_i)/2} \tag{16b}$$

(*ii*) Let $\qquad \phi(x) = \dfrac{x^{\alpha} - x}{\alpha(\alpha - 1)}, \qquad \alpha \ne 0, 1$ (17)

then (7) gives

$$D_{3,\lambda}(P:Q) = \frac{1}{\alpha(\alpha-1)} \sum_{i=1}^{n} (\lambda p_i + (1-\lambda) q_i) \left[\left(\frac{p_i}{\lambda p_i + (1-\lambda) q_i} \right)^{\alpha} - \frac{p_i}{\lambda p_i + (1-\lambda) q_i} \right]$$

$$= \frac{1}{\alpha(\alpha-1)} \left[\sum_{i=1}^{n} p_i^{\alpha} (\lambda p_i + (1-\lambda) q_i)^{1-\alpha} - 1 \right] \tag{18}$$

If we put $\lambda = 0$, it gives Havrda- Charvat [3] measure of directed divergence. As such (18) gives a generalisation of this measure.

If $\alpha \to 1$, it gives $D_{2\lambda}(P:Q)$. If $\lambda = 1/2$ and $\alpha \to 1$, it gives Lin's measure, and if $\lambda = 0$, $\alpha \to 1$, it gives Kullback-Leibler [6] measure. If $\alpha \to 0$, it gives

$$D_{4,\lambda}(P:Q) = -\sum_{i=1}^{n} (\lambda p_i + (1-\lambda) q_i) \ln \frac{p_i}{\lambda p_i + (1-\lambda) q_i} \tag{19}$$

$$= D_{2,\lambda}((\lambda P + (1-\lambda) Q) : P) \tag{20}$$

so that this is another version of generalised Lin's measure.

(*iii*) Expression (18) gives a generalisation of Renyi's [8] measure viz

$$D_{5,\lambda}(P:Q) = \frac{1}{\alpha(\alpha-1)} \ln \sum_{i=1}^{n} p_i^{\alpha} (\lambda p_i + (1-\lambda) q_i)^{1-\alpha}, \quad \alpha \ne 0, \quad \alpha \ne 1, \tag{21}$$

(*iv*) Let $\qquad \phi(x) = x \ln x - \dfrac{1}{a}(1 + ax) \ln(1 + ax) + \dfrac{x}{a}(1 + a) \ln(1 + a)$ (22)

then (7) gives,

$$D_{6,\lambda}(P:Q)=\sum_{i=1}^{n}[\lambda p_i+1-\lambda)q_i]\{\frac{p_i}{\lambda p_i+(1-\lambda)q_i}\ln\frac{p_i}{\lambda p_i+(1-\lambda)q_i}$$

$$-\frac{1}{a}\left(1+\frac{ap_i}{\lambda p_i+(1-\lambda)q_i}\right)\ln\left(1+\frac{ap_i}{\lambda p_i+(1-\lambda)q_i}\right)+\frac{p_i}{a(\lambda p_i+(1-\lambda)q_i)}\quad(1+a)\ln(1+a)\}$$

$$=\sum_{i=1}^{n}p_i\ln\frac{p_i}{\lambda p_i+(1-\lambda)q_i}-\frac{1}{a}\sum_{i=1}^{n}(\lambda p_i+ap_i+(1-\lambda)q_i)\ln\left(1+\frac{ap_i}{\lambda p_i+(1-\lambda)q_i}\right)$$

$$+\frac{1}{a}(1+a)\ln(1+a) \tag{23}$$

If we put $\lambda=0$, we get Kapur's [4] measure of directed divergence.

12.4 SOME OTHER GENERALISED MEASURES

(*i*) Corresponding to Burg's [1] measure of directed divergence,

$$\sum_{i=1}^{n}\left(\frac{p_i}{q_i}-\ln\frac{p_i}{q_i}-1\right), \tag{24}$$

we get the generalised measure,

$$D_{7,\lambda}(P:Q)=\sum_{i-1}^{n}\left(\frac{p_i}{\lambda p_i+(1-\lambda)q_i}-\ln\frac{p_i}{\lambda p_i+(1-\lambda)q_i}-1\right) \tag{25}$$

(*ii*) Corresponding to generalised Burg's measure of directed divergence,

$$\sum_{i=1}^{n}\left(\frac{(1+ap_i)}{1+aq_i}-\ln\frac{1+ap_i}{1+aq_i}-1\right), \tag{26}$$

we get the generalised measure,

$$D_{8,\lambda}(P:Q)=\sum_{i=1}^{n}\left(\frac{1+ap_i}{1+a(\lambda p_i+(1-\lambda)q_i)}-\ln\frac{1+ap_i}{1+a(\lambda p_i+(1-\lambda)q_i)}-1\right) \tag{27}$$

It can easily be verified that both $D_7(P:Q)$ and $D_8(P:Q)$ are convex functions of $p_1, p_2, \ldots, p_n$ and have thus minimum value zero when $p_i=q_i\ \forall_i$

12.5 JENSEN-SHANNON MEASURE AND ITS GENERALISATION

We have already shown that the generalised measure $D_2(P:Q)$ is a convex function of $p_1, p_2, \ldots, p_n$. It is also a convex function of $q_1, q_2, \ldots, q_n$ since

$$f(p_1,p_2,\ldots,p_n;q_1,q_2,\ldots,q_n)=\sum_{i=1}^{n}p_i\ln\frac{p_i}{\lambda p_i+(1-\lambda)q_i}$$

$$=-\sum_{i=1}^{n}p_i\ln(\lambda p_i+(1-\lambda)q_i)+\sum_{i=1}^{n}p_i\ln p_i \tag{28}$$

$$\Rightarrow \frac{\partial f}{\partial q_i} = -\frac{p_i(1-\lambda)}{\lambda p_i + (1-\lambda)\, q_i} \Rightarrow \frac{\partial^2 f}{\partial q_i^2} = \frac{(1-\lambda)^2 p_i^2}{(\lambda p_i + (1-\lambda)\, q_i)^2} > 0\,, \quad \frac{\partial^2 f}{\partial q_i \partial q_j} = 0 \qquad (29)$$

Now consider the measure,

$$D_{9,\lambda}(P{:}Q,\lambda) = \lambda \sum_{i=1}^{n} p_i \ln \frac{p_i}{\lambda p_i + (1-\lambda)\, q_i} + (1-\lambda) \sum_{i=1}^{n} q_i \ln \frac{q_i}{\lambda p_i + (1-\lambda)\, q_i} \qquad (30)$$

$$= \lambda \sum_{i=1}^{n} p_i \ln p_i + (1-\lambda) \sum_{i=1}^{n} q_i \ln q_i$$

$$- \sum_{i=1}^{n} (\lambda p_i + (1-\lambda)\, q_i) \ln (\lambda p_i + (1-\lambda)\, q_i) \qquad (31)$$

$$= H\,(\lambda P + (1-\lambda)\,(Q) - \lambda\, H(P) - (1-\lambda)\, H\,(Q)), \qquad (32)$$

where,

$$H\,(P) = -\sum_{i=1}^{n} p_i \ln p_i \qquad (33)$$

is Shannon's measure of entropy. The measure (32) is a generalisation of Jensen-Shannon measure defined by Lin [7] whose measure arises when $\lambda = 1/2$. It is obvious that

$$D_{9,\lambda}\,(P{:}Q;\lambda) = D_{9,\lambda}\,(Q{:}P, 1-\lambda) \qquad (34)$$

Thus generalised Jensen-Shannon measure need not be defined separately. It arises from the generalisation of Kullback-Leibler measure in a natural way.

Let us now consider the variation of this measure with respect to λ.

Let

$$g\,(\lambda) = \lambda \sum_{i=1}^{n} p_i \ln p_i + (1-\lambda) \sum_{i=1}^{n} q_i \ln q_i$$

$$- \sum_{i=1}^{n} (\lambda p_i + (1-\lambda)\, q_i) \ln (\lambda p_i + (1-\lambda)\, q_i) \qquad (35)$$

Then,

$$g'(\lambda) = \sum_{i=1}^{n} p_i \ln p_i - \sum_{i=1}^{n} q_i \ln q_i$$

$$- \sum_{i=1}^{n} (1 + \ln (\lambda p_i + (1-\lambda) q_i))\,(p_i - q_i)$$

$$= \sum_{i=1}^{n} p_i \ln p_i - \sum_{i=1}^{n} q_i \ln q_i - \sum_{i=1}^{n} (p_i - q_i) \ln (\lambda p_i + (1-\lambda) q_i) \qquad (36)$$

$$g''(\lambda) = -\sum_{i=1}^{n} \frac{(p_i - q_i)^2}{\lambda p_i + (1-\lambda) q_i} < 0, \qquad (37)$$

so that $g\,(\lambda)$ is a concave function of λ and $g'\,(\lambda)$ is always decreasing. Now,

$$g\,(0) = \sum_{i=1}^{n} q_i \ln q_i - \sum_{i=1}^{n} q_i \ln q_i = 0 \qquad (38)$$

$$g(1) = \sum_{i=1}^{n} p_i \ln p_i - \sum_{i=1}^{n} p_i \ln p_i = 0 \tag{39}$$

$$g'(0) = \sum_{i=1}^{n} p_i \ln p_i - \sum_{i=1}^{n} q_i \ln q_i - \sum_{i=1}^{n} (p_i - q_i) \ln q_i = \sum_{i=1}^{n} p_i \frac{p_i}{q_i} \geq 0 \tag{40}$$

$$g'(1) = \sum_{i=1}^{n} p_i \ln p_i - \sum_{i=1}^{n} q_i \ln q_i - \sum_{i=1}^{n} (p_i - q_i) \ln p_i = \sum_{i=1}^{n} q_i \ln \frac{q_i}{p_i} \leq 0 \tag{41}$$

so that the graph of g(λ) is as in Figure 12.1.

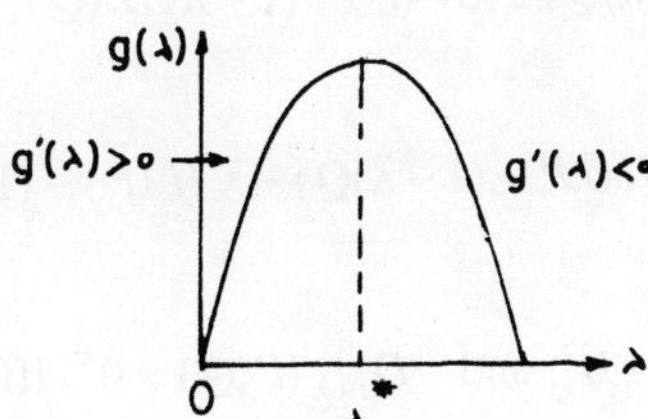

Fig. 12.1

Now $g(\lambda)$ is maximum at λ^* where

$$\sum_{i=1}^{n} p_i \ln p_i - \sum_{i=1}^{n} q_i \ln q_i = \sum_{i=1}^{n} (p_i - q_i) \ln (\lambda^* p_i + (1-\lambda^*) q_i) \tag{42}$$

Let $\lambda^* P + (1 - \lambda^*) Q$ be denoted by R,

then $$D_{2,\lambda}(P:R) = \sum_{i=1}^{n} p_i \ln \frac{p_i}{\lambda^* p_i + (1-\lambda^*) q_i} \tag{43}$$

and $$D_{2,\lambda}(Q:R) \equiv \sum_{i=1}^{n} q_i \frac{\ln q_i}{\lambda^* p_i + (1-\lambda^*) q_i} \tag{44}$$

Equation (42) shows that

$$D_{2,\lambda}(P:R) = D_{2,\lambda}(Q:R) \tag{45}$$

so that for given P and Q, Jensen Shannon generalised measure is maximum for that value of λ which makes $D_{2,\lambda}(P:R) = D_{2,\lambda}(Q:R)$ i.e. for a distribution which is in some sense, midway between P and Q.

By choosing $\lambda = 1/2$, Lin made his measure symmetrical, but he did not give it the maximum value. In fact from (36)

$$g'\left(\frac{1}{2}\right) = \sum_{i=1}^{n} p_i \ln \frac{p_i}{\frac{p_i+q_i}{2}} - \sum_{i=1}^{n} q_i \ln \frac{q_i}{\frac{p_i+q_i}{2}} \tag{46}$$

$$= D_{2,1/2}\left(P:\frac{P+Q}{2}\right) - D_{2,1/2}\left(Q:\frac{P+Q}{2}\right) \tag{47}$$

so that $D_{2,1/2}(P:(P+Q)/2) \gtrless D_{2,1/2}(Q:(P+Q)/2) \rightarrow g'(1/2) \gtrless 0$

Whether $1/2 \lessgtr \lambda^*$, g (1/2) < g (λ^*) and Lin's measure < maximum value of generalised Lin's measure.

12.6 OTHER JENSEN'S MEASURES OF DIRECTED DIVERGENCE

Let $H(P)$ be any concave measure of entropy, then by the definition of a concave function

$$H(\lambda P + (1-\lambda)Q) \geq \lambda H(P) + (1-\lambda) H(Q), \quad 0 \leq \lambda \leq 1, \tag{48}$$

so that if we define

$$D_{10,\lambda}(P:Q) = H(\lambda P + (1-\lambda)Q) - \lambda H(P) - (1-\lambda) H(Q) \tag{49}$$

then,

$$D_{10,\lambda}(P:Q) \geq 0 \quad \text{and} \quad D_{10,\lambda}(P:Q) = 0 \quad \text{iff} \quad P = Q \tag{50}$$

Special Cases:

(i) $H(P) = -\sum_{i=1}^{n} p_i \ln p_i$ gives Jensen-Shannon measure (31) of directed divergence.

(ii) $H(P) = (1-\alpha)^{-1}\left(\sum_{i=1}^{n} p_i^{\alpha} - 1\right)$ gives Jensen-Havrda-Charvat measure of directed divergence,

$$D_{11,\lambda}(P:Q) = (1-\alpha)^{-1} \sum_{i=1}^{n} [(\lambda p_i + (1-\lambda) q_i)^{\alpha} - \lambda p_i^{\alpha} - (1-\lambda) q_i^{\alpha}] \tag{51}$$

(iii) $H(P) = \sum_{i=1}^{n} \ln p_i$ gives Jensen-Burg measure of directed divergence,

$$D_{12,\lambda}(P:Q) = \sum_{i=1}^{n} \ln(\lambda p_i + (1-\lambda) q_i) - \lambda \sum_{i=1}^{n} \ln p_i - (1-\lambda) \sum_{i=1}^{n} \ln q_i$$

$$= \ln \prod_{i=1}^{n} \frac{\lambda p_i + (1-\lambda) q_i}{p_i^{\lambda} q_i^{1-\lambda}} \tag{52}$$

(iv) $H(P) = \sum_{i=1}^{n} \ln(1 + a p_i) - \ln(1+a)$ gives Jensen - generalised Burg measure of directed divergence,

$$D_{13,\lambda}(P:Q) = \ln \prod_{i=1}^{n} \frac{1+a\,(\lambda p_i + (1-\lambda)q_i)}{(1+a\,p_i)^{\lambda}\,(1+a\,q_i)^{1-\lambda}} \tag{53}$$

(*v*) $$H(P) = -\sum_{i=1}^{n} p_i \ln p_i + \frac{1}{a}\sum_{i=1}^{n}(1+ap_i)\ln(1+ap_i) - \frac{1}{a}(1+a)\ln(1+a) \tag{54}$$

gives Jensen-Kapur measure of directed divergence,

$$D_{14,\lambda}(P:Q) = -\sum_{i=1}^{n}(\lambda p_i + (1-\lambda)q_i)\ln(\lambda p_i + (1-\lambda)q_i) + \frac{1}{a}\sum_{i=1}^{n}(1+a\lambda p_i + a(1-\lambda)q_i)$$

$$\ln(1+a\lambda p_i + a(1-\lambda)q_i) + \lambda\sum_{i=1}^{n} p_i \ln p_i - \frac{\lambda}{a}\sum_{i=1}^{n}(1+ap_i)\ln(1+ap_i)$$

$$+ (1-\lambda)\sum_{i=1}^{n} q_i \ln qi - \frac{(1-\lambda)}{a}\sum_{i=1}^{n}(1+aq_i)\ln(1+aq_i) \tag{55}$$

(*vi*) Renyi's measure $(1-\alpha)^{-1}\ln\sum_{i=1}^{n} p_i^{\alpha}$ is concave if $\alpha < 1$ and this gives Jensen-Renyi's measure of directed divergence,

$$D_{15,\lambda}(P:Q) = \frac{1}{1-\alpha}\left[\ln\sum_{i=1}^{n}(\lambda p_i + (1-\lambda)\,q_i)^{\alpha} - \lambda\ln\sum_{i=1}^{n} p_i^{\alpha} - (1-\lambda)\ln\sum_{i=1}^{n} q_i^{\alpha}\right]$$

$$= \frac{1}{1-\alpha}\ln\frac{\left(\sum_{i=1}^{n}\lambda p_i + (1-\lambda)\,q_i\right)^{\alpha}}{\left(\sum_{i=1}^{n} p_i^{\alpha}\right)^{\lambda}\left(\sum_{i=1}^{n} q_i^{\alpha}\right)^{1-\lambda}}, \quad 0<\alpha<1 \tag{56}$$

12.7 CONCAVITY OF JENSEN'S MEASURES WITH RESPECT TO λ

Let $H(P) = \sum_{i=1}^{n} h(p_i)$

where $h(\bullet)$ is a concave function, and let

$$f(\lambda) = H(\lambda P + (1-\lambda)\,Q) - \lambda H(P) - (1-\lambda)\,H(Q)$$

$$= \sum_{i=1}^{n} h(\lambda p_i + (1-\lambda)q_i) - \lambda\sum_{i=1}^{n} h(p_i) - (1-\lambda)\sum_{i=1}^{n} h(q_i) \tag{58}$$

so that,

$$f'(\lambda) = \sum_{i=1}^{n}(p_i - q_i)\,h'(\lambda p_i + (1-\lambda)\,q_i) - \sum_{i=1}^{n} h(p_i) + \sum_{i=1}^{n} h(q_i) \tag{59}$$

$$f''(\lambda) = \sum_{i=1}^{n}(p_i - q_i)^2\,h''(\lambda h_i + (1-\lambda)\,q_i) < 0 \tag{60}$$

so that $f(\lambda)$ is a concave function of λ. Also,

$$f(0) = 0, \qquad f(1) = 0 \tag{61}$$

so that $f(\lambda)$ increases from 0 to some maximum value as λ goes to λ^* and decreases from maximum value to 0 as λ goes from λ^* to 0, where λ^* is given by

$$\sum_{i=1}^{n} h'(\lambda^* p_i + (1-\lambda^*) q_i) = \lambda^* \sum_{i=1}^{n} h(p_i) + (1-\lambda^*) \sum_{i=1}^{n} h(q_i) \tag{62}$$

Thus the graph of (58) has the same shape as in Figure 12.1.

The generalised Jensen measure is

$$f(\lambda_1, \lambda_2, \ldots, \lambda_m) = \sum_{i=1}^{n} h\left(\sum_{j=1}^{m} \lambda_j p_{ij}\right) - \sum_{j=1}^{m} \lambda_j \sum_{i=1}^{n} h(p_{ij}) \tag{63}$$

$$\frac{\partial f}{\partial \lambda_j} = \sum_{i=1}^{n} h'\left(\sum_{j=1}^{m} \lambda_j p_{ij}\right) p_{ij} - \sum_{i=1}^{n} h(p_{ij}) \tag{64}$$

$$\frac{\partial^2 f}{\partial \lambda_j^2} = \sum_{i=1}^{n} h''\left(\sum_{j=1}^{m} \lambda_j p_{ij}\right) p_{ij}^2 \tag{65}$$

$$\frac{\partial^2 f}{\partial \lambda_j \partial \lambda_k} = \sum_{i=1}^{n} h''\left(\sum_{j=1}^{m} \lambda_j p_{ij}\right) p_{ij} p_{ik} \tag{66}$$

so that $f(\lambda_1, \lambda_2, \ldots, \lambda_m)$ is a concave function of $\lambda_1, \lambda_2, \ldots, \lambda_m$ which attains its maximum value when $\lambda = \lambda^*$, where $\lambda^* = (\lambda_1^*, \lambda_2^*, \ldots, \lambda_m^*)$ is determined from the m equations

$$\sum_{i=1}^{n} h'\left(\sum_{j=1}^{m} \lambda_j^* p_{ij}\right) p_{ij} = \sum_{i=1}^{n} h(p_{ij}), \quad j = 1, 2, \ldots, m \tag{67}$$

REFERENCES

1. J.P Burg (1972), "The Relationship between Maximum Entropy Spectra and Maximum likelihood Spectra." In Modern Spectral Analysis, ed D.G. childers, pp. 130–131, M.S.A.
2. I. Csiszer (1972), "A class of Measures of Informativity of Observation Channels," Periodica Math. Hungarica, Vol. 2 pp. 191–213.
3. J.H. Havrda and F. Charvat, (1967), " Quantification Methods of Classification Processes: Concept of Structural α Entropy", Kybernetica, Vol. 3, pp. 30–35.
4. J.N. Kapur (1986), "Four Families of Measures of Entropy", Ind. Journ Pure and Applied Maths., Vol. 17, No.4, pp. 429–449.
5. J.N. Kapur (1990), Maximum Entropy Models in Science and Engineering, Wiley Eastern, New Delhi and John Wiley, New York.
6. S. Kullback and R.A. Leibler (1951), "On Information and Sufficiency", Ann. Math. Stat., Vol. 22, pp. 79–86.
7. J. Lin (1991), Divergence Measures based on the Shannon Entropy, IEEE, Trans, INF, Th. Vol. 17, No. 1. pp. 145–151.
8. A. Renyi (1961), "On Measures of Entropy and Information", Proc. 4th Berkeley Symp. Math. Stat. Prob. Vol. 1. pp. 547–561.
9. C.E. Shannon (1948), "A Mathematical Theory of Communication", Bell System Tech. J. Vol. 27, pp. 379–423, 623–659.

13

TWO NEW MEASURES OF ENTROPY AND DIRECTED DIVERGENCE

[Two new two-parameter measures of entropy have been obtained by simultaneous use of the generalisations of Shannon's measure of entropy due to Kapur and Havrda and Charvat.]

13.1 INTRODUCTION

Let $P = (p_1, p_2, \ldots, p_n)$; $\sum_{i=1}^{n} p_i = 1, \quad p_i \geq 0, \quad i = 1, 2, \ldots, n$ (1)

be a probability distribution. Then Shannon [6] obtained the function

$$H_0(P) = -\sum_{i=1}^{n} p_i \ln p_i \tag{2}$$

for measuring the uncertainty or entropy of the probability distribution P. Later, Kapur [4] generalised this measure to get the parametric measure of entropy;

$$H_a(P) = -\sum_{i=1}^{n} p_i \ln p_i + \frac{1}{a} \sum_{i=1}^{n} (1 + ap_i) \ln(1 + ap_i) - \frac{1}{a}(1 + a) \ln(1 + a);\ a \geq -1, \tag{3}$$

It is easily verified that $H_a(P) \to H_0(P)$ as $a \to 0$. $H_a(P)$ also includes the following two special cases [3].

(i) Bose-Einstein entropy,

$$H_1(P) = -\sum_{i=1}^{n} p_i \ln p_i + \sum_{i=1}^{n} (1 + p_i) \ln(1 + p_i) - 2 \ln 2 \tag{4}$$

(ii) Fermi-Dirac entropy,

$$H_{-1}(P) = -\sum_{i=1}^{n} p_i \ln p_i - \sum_{i=1}^{n} (1 - p_i) \ln(1 - p_i) \tag{5}$$

Now Havrda-Charvat [2] generalised (1) to another parametric measure of entropy,

$$H^{\alpha}(P) = \frac{1}{1 - \alpha}\left(\sum_{i=1}^{n} p_i^{\alpha} - 1\right), \quad \alpha \neq 1, \quad \alpha > 0 \tag{6}$$

It is easily verified that $H^{\alpha}(P) \to H_0(P)$ as $\alpha \to 1$ so that $H_0(P)$ may also be written as $H'(P)$.

We want to combine the two generalisations and get two two-parameter measures of entropy.

13.2 FIRST NEW MEASURE OF ENTROPY

The forms of (2) and (3) suggest that we consider,

$$H_a^\alpha(P) = \frac{1}{1-\alpha}\left(\sum_{i=1}^n p_i^\alpha - \sum_{i=1}^n p_i\right) - \frac{1}{a}\frac{1}{1-\alpha}\left(\sum_{i=1}^n (1+ap_i)^\alpha - \sum_{i=1}^n (1+ap_i)\right)$$

$$+\frac{1}{a}\frac{1}{1-\alpha}((1+a)^\alpha - (1+a)) \tag{7}$$

This approaches Kapur's measure (3) as $\alpha \to 1$.
It approaches Havrda-Charvat measure (6) as $a \to 0$.
It approaches Bose-Einstein entropy (4) as $\alpha \to 1, a \to 1$
It approaches Fermi-Dirac entropy (5) as $\alpha \to 1, a \to -1$

Thus (7) includes all the five measures of entropy as special or limiting cases, but that does not ensure that (7) is itself a measure of entropy, unless it itself satisfies the conditions expected from a measure of entropy.

It is easily seen that (7) is continuous, permutationally symmetric and reduces to zero for the degenerate distributions,

$$\Delta_1 = (1,0,...,0),\quad \Delta_2 = (0,1,0,\ldots,0),\ldots,\Delta_n = (0,0,\ldots 0,1) \tag{8}$$

Since (2) is neither additive nor recursive, we do not expect (7) to be additive or recursive. However, all of (2) , (3), (4), (5) and (6) are concave and therefore we examine whether (7) is concave. Its concavity will ensure that it will be maximum for the uniform distribution.

$$U = \left(\frac{1}{n},\frac{1}{n},\ldots,\frac{1}{n}\right) \tag{9}$$

and it will have its minimum value at the degenerate distributions and it will always be ≥ 0. Now, since both

$$\frac{1}{1-\alpha}\left(\sum_{i=1}^n p_i^\alpha - \sum_{i=1}^n p_i\right) \quad\text{and}\quad \frac{1}{1-\alpha}\left(\sum_{i=1}^n (1+ap_i)^\alpha - \sum_{i=1}^n (1+ap_i)\right) \tag{10}$$

are concave functions, (7) will be a concave function if $a < 0$, since the sum of two concave functions is concave. However, if $a > 0$, (7) represents the difference of two concave functions and this difference may or may not be concave. To examine its concavity, we consider the function,

$$\phi(x) = \frac{x^\alpha - x}{1-\alpha} - \frac{1}{a}\frac{(1+ax)^\alpha - (1+ax)}{1-\alpha} \tag{11}$$

This gives

$$\phi''(x) = -\alpha x^{\alpha-2} + a\alpha(1+ax)^{\alpha-2} = -\alpha(x^{\alpha-2} - a(1+ax)^{\alpha-2}) \tag{12}$$

From (7) and (11)

$$H_a^\alpha(P) = \sum_{i=1}^n \phi(p_i) - \phi(1) \tag{13}$$

so that, $H_a^\alpha(P)$ will be concave if

$$x^{\alpha-2} \geq a\,(1+ax)^{\alpha-2} \qquad \text{when} \quad 0 \leq x \leq 1. \tag{14}$$

We consider now three cases :-

Case (i) If $\alpha > 2$, this requires

$$x \geq a^{\frac{1}{\alpha-2}}\,(1+ax) \quad \text{or} \quad x\left(1-a^{\frac{a-1}{\alpha-2}}\right) > a^{\frac{1}{\alpha-2}} \tag{15}$$

This condition is not satisfied if $a \geq 1$. For $a < 1$, it requires

$$x \geq \frac{a^{\frac{1}{\alpha-2}}}{1-a^{\frac{\alpha-1}{\alpha-2}}} \tag{16}$$

The R.H.S. of (16) has a definte positive value and (16) will not be satisfied for those values of probability which are less than this value.
Thus when $\alpha > 2$, (7) is not a concave function except when $a = 0$.

Case (ii)

If $\alpha = 2$, (14) becomes

$$1 \geq a \tag{17}$$

so that when $\alpha > 2$, $H_a^{\alpha}(P)$ is concave whenever $a \leq 1$, so that

$$H_a^2(P) = \left(1 - \sum_{i=1}^{n} p_i^2\right) + \sum_{i=1}^{n} p_i\,(1+ap_i) - (1+a)$$

$$= (1-a)\left(1 - \sum_{i=1}^{n} p_i^2\right), \qquad a \leq 1 \tag{18}$$

is a measure of entropy. However, it is just a positive multiple of Havrda-Charvat [2] entropy of order 2.

Case (iii)

$\alpha < 2$. In this case (14) gives,

$$\frac{1}{x^{2-\alpha}} \geq a\frac{1}{(1+ax)^{2-\alpha}} \quad \text{or} \quad \frac{1+ax}{x} \geq a^{\frac{1}{2-\alpha}} \quad \text{or} \quad \frac{1}{x} \geq a^{\frac{1}{2-\alpha}} - a$$

$$\text{or} \quad a^{\frac{1}{2-\alpha}} - a \leq \frac{1}{x} \tag{19}$$

Since $1/x$ lies between 1 and ∞, condition (19) will be satisfied if,

$$a^{\frac{1}{2-\alpha}} - \; a \leq 1 \quad \text{or} \quad a\left(a^{\frac{\alpha-1}{2-\alpha}} - 1\right) \leq 1 \tag{20}$$

We now consider two sub-cases

Sub-case (a) : $0 < \alpha < 1$

If $a > 1$, $a^{\frac{\alpha-1}{2-\alpha}} < 1$ and L.H.S. of (20) is negative and (20) is satisfied.

If $a < 1$, $a^{\frac{1}{2-\alpha}} < 1$ and $a^{\frac{1}{2-\alpha}} - a < 1$ so that (20) is again satisfied. Thus when $0 < \alpha < 1$, $H_a^\alpha(P)$ is a concave function for all $a \geq -1$. When $\alpha = 1$, it reduces to Kapur's measure (2) which is known to be concave.

Sub-case (b): $1 < \alpha < 2$. In this case, let

$$f(a) = a^{\frac{1}{2-\alpha}} - a - 1 = a\left(a^{\frac{\alpha-1}{2-\alpha}} - 1\right) - 1 \tag{21}$$

so that,

$$f(0) = -1, f(1) = -1, \quad f(\infty) = \infty \tag{22}$$

and,

$$f'(a) = a^{\frac{\alpha-1}{2-\alpha}} - 1 \tag{23}$$

$$f''(a) = \frac{\alpha - 1}{2-\alpha} a^{\frac{2\alpha-3}{2-\alpha}} > 0 \tag{24}$$

so that $f(a)$ is convex function and has the shape given in Figure 13.1.

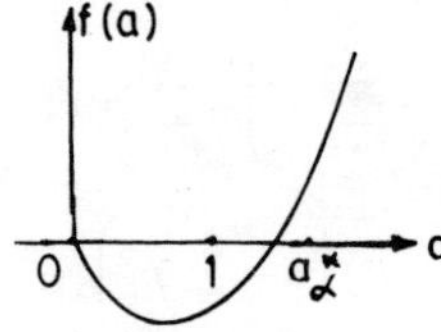

Fig. 13.1

Thus for every value of α between 1 and 2,

$$f(a) < 0 \quad \text{when} \quad a < a_\alpha^* \quad \text{where} \quad a_\alpha^* > 1 \tag{25}$$

Thus the condition of concavity is not satisfied for all values of a, but for all values of $a <$ a critical value a_α^* which is greater than unity and depends on α. It is easily seen that

$$\begin{aligned} a_\alpha^* &= \infty \quad \text{when} \quad \alpha = 1 \\ &= \frac{1+\sqrt{5}}{2} \quad \text{when} \quad \alpha = 3/2 \\ &= 1 \quad \text{when} \quad \alpha = 2 \end{aligned} \tag{26}$$

$$a_\alpha^{*\frac{1}{2-\alpha}} - a_\alpha^* - 1 = 0 \Rightarrow \left(\frac{1}{2-\alpha} a_\alpha^{*\frac{\alpha-1}{2-\alpha}} - 1\right)\frac{da_\alpha^*}{d\alpha} + a_\alpha^{*\frac{1}{2-\alpha}}\left(\frac{1}{2-\alpha}\right)^2 \ln a_\alpha^* = 0 \tag{27}$$

Since, $a_\alpha^* > 1, \quad 1 < \alpha < 2$, (27) gives $\dfrac{da_\alpha^*}{d\alpha} < 0$ (28)

so that a_α^* decreases with α and as α goes for 1 to 2, a_α^* decreases from α to 1. Its graph is shown in Figure 13.2.

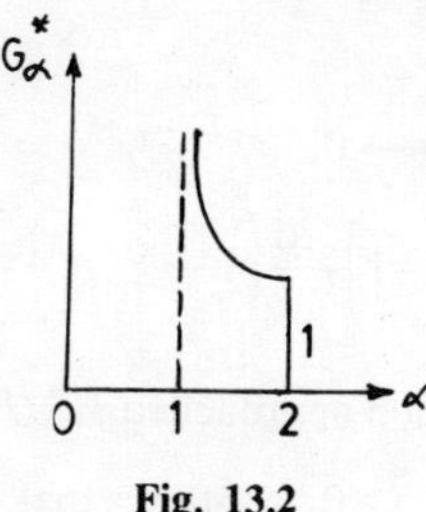

Fig. 13.2

The above discussion leads to the theorem:

Theorem: $H_a^\alpha(P)$ is concave function of P for all values of $a \geq -1$ if $\alpha \leq 1$. If α lies between 1 and 2, $H_a^\alpha(P)$ in a concave function of P for all values of a lying between -1 and a_α^*, where a_α^* is always ≥ 1 and decreases from ∞ to unity as α goes from 1 to 2. In any case $H_a^\alpha(P)$ is always concave for all values of α between 1 and 2 when a lies between -1 and 1. When $\alpha > 2$, $H_a^\alpha(P)$ is not a concave function of P for any value of a other than zero.

13.3 A SECOND MEASURE OF ENTROPY

Kapur [4] also gave a second generalisation of Shannon's entropy, viz.

$$H_b(P) = -\sum_{i=1}^n p_i \ln p_i + \frac{1}{b^2}\sum_{i=1}^n (1+bp_i)\ln(1+bp_i) - \frac{1}{b^2}(1+b)\ln(1+b), \; b \geq -1 \quad (29)$$

This suggests that we consider the corresponding generalisation of Havrda-Charvat measure,

$$H_b^\alpha(P) = \frac{1}{1-\alpha}\left(\sum_{i=1}^n p_i^\alpha - \sum_{i=1}^n p_i\right) - \frac{1}{b^2}\frac{\sum_{i=1}^n (1+bp_i)^\alpha - \left(\sum_{i=1}^n (1+bp_i)\right)}{1-\alpha}$$
$$+\frac{1}{b^2}\frac{(1+b)^\alpha - (1+b)}{1-\alpha} \quad (30)$$

This is always the difference between two concave functions.

Proceeding as before, the condition for its concavity corresponding to (14) is

$$x^{\alpha-2} \geq (1+bx)^{\alpha-2} \quad \text{for } 0 \leq x \leq 1 \quad (31)$$

If $\alpha > 2$, it becomes $x \geq 1 + bx$ which cannot be satisfied. If $\alpha = 2$ it becomes $1 \geq 1$, it is satisfied, but it gives a multiple of Havrda-Charvat measure of 2nd order,

If $$\alpha < 2, \quad \frac{1}{x} \geq \frac{1}{1+bx}, x \leq 1+bx, x\,(1-b) \leq 1 \tag{32}$$

This is satisfied when $b \geq 0$
We thus get theorem 2.

Theorem 2: $H_b^\alpha(P)$ is a concave function of P for all values of $b \geq -1$ if $\alpha \leq 2$, and is not a concave function of P if $\alpha > 2$.

Remark: It may be noted that as $b \to 0$,

$$H_b^\alpha(P) \to \frac{1}{1-\alpha}\left(\sum_{i=1}^{n} p_i^\alpha - 1\right) - \frac{\alpha}{2}\left(1 - \sum_{i=1}^{n} p_i^2\right) = H^\alpha(P) - \frac{\alpha}{2} H^{(2)}(P) \tag{33}$$

This is unlike the case $H_a^\alpha(P)$ which approached $H^\alpha(P)$ as $a \to 0$. We also find that $H_b^2(P) \to 0$ as $b \to 0$. Since $H_b^\alpha(P) \geq 0$, it follows that

$$H^\alpha(P) \geq \frac{1}{2}\alpha H^{(2)}(P) \quad \text{for} \quad \alpha \leq 2 \tag{34}$$

This is obvious otherwise since $H^\alpha(P)$ is a monotonic decreasing function of α (Kapur [5]) and as such in the range $0 < \alpha \leq 2$, its minimum value is $H^2(P)$ so that

$$H^\alpha(P) \geq H^{(2)}(P) \geq \frac{1}{2}\alpha H^{(2)}(P) \quad \text{since} \quad \alpha \leq 2. \tag{35}$$

13.4 PAIRED ENTROPIES

Let $\varphi(\alpha)$ be a convex twice-differentiable function, then

$$S_1(P) = -\sum_{i=1}^{n} \phi(p_i) + \phi(1) + (n-1)\,\phi(0) \tag{36}$$

can be used as a measure of entropy, since it is permutationally symmetric, continuous, concave, has its maximum value when the distribution is uniform and minimum value zero at degenerate distributiors. It will also be expansible if $\phi(0) = 0$. Its maximum value will also increase with n if

$$\phi'\left(\frac{1}{n}\right) - n\,\phi\left(\frac{1}{n}\right) + n\phi(0) > 0 \tag{37}$$

It is also easily verified that

$$S_2(P) = -\sum_{i=1}^{n} \phi(1 + a p_i) + \phi(1+a) + (n-1)\,\phi(1) \tag{38}$$

is also a measure of entropy whether $a > 0$ or $a < 0$. It will be expansible if $\phi(1) = 0$ and its maximum value will increase with n if

$$a\phi'\left(1 + \frac{a}{n}\right) - n\,\phi\left(1 + \frac{a}{n}\right) + n\,\phi(1) > 0 \tag{39}$$

If b is any positive number, then

$$S_3(P) = S_1(P) + bS_2(P) = -\sum_{i=1}^{n} \phi(p_i) - b\sum_{i=1}^{n} \phi(1 + ap_i) + b\,\phi(1 + a)$$
$$+ ((n-1)b + 1)\,\phi(1) + (n-1)\,\phi(0) \qquad (40)$$

can also be used as a measure of entropy. In particular if $b = 1$, $a = -1$, this gives

$$-\sum_{i=1}^{n} \phi(p_i) - \sum_{i=1}^{n} \phi(1 - p_i) + n\,(\phi(0) + \phi(1)) \qquad (41)$$

This has sometimes been called a paired measure of entropy.

The interesting case arises when

$$S_4(P) = S_1(P) - bS_2(P) = -\sum_{i=1}^{n} \phi(p_i) + b\sum_{i=1}^{n} \phi(1 + ap_i) - b\,\phi(1 + a)$$
$$+ (1 - (n-1)b)\,\phi(1) + (n-1)\,\phi(0) \qquad (42)$$

Kapur had investigated $S_4(P)$ when,

$$\phi(x) = x\ln x,\; b = \frac{1}{a}, \frac{1}{a^2} \qquad (43)$$

We have investigated above $S_4(P)$ when,

$$\phi(x) = \frac{x^{\alpha} - x}{\alpha - 1}, \quad b = \frac{1}{a}, \frac{1}{a^2} \qquad (44)$$

Our present discussion includes the earlier discussion when $\alpha \to 1$.

In the same way, we can discuss the other measures arising from other convex functions and other values of b.

13.5 MESURES OF DIRECTED DIVERGENCE

(i) Corresponding to the measure of entropy (3), Kapur [4] proposed measure of directed divergence,

$$D(P:Q) = \sum_{i=1}^{n} p_i \ln\frac{p_i}{q_i} - \frac{1}{a}\sum_{i=1}^{n} (1 + ap_i)\ln\frac{(1 + ap_i)}{(1 + aq_i)}, \quad a \geq -1 \qquad (45)$$

This is permutationally symmetric, continuous convex function of $p_1, p_2, \ldots, p_n$ and vanishes iff $p_i = q_i\ \forall_i$. However, it is not in general a convex function of $q_1, q_2, \ldots, q_n$.

We now generalise (45) to consider the measure,

$$D(P:Q) = \sum_{i=1}^{n} p_i \ln\frac{p_i}{q_i} + A\sum_{i=1}^{n} (1 + ap_i)\ln\frac{1 + ap_i}{1 + q_i} \qquad (46)$$

This will be a convex function of $p_1, p_2, \ldots, p_n$ if

$$\frac{1}{p_i}+\frac{Aa^2}{1+ap_i}>0 \tag{47}$$

This will always be satisfied if $A>0$. If A is negative, it will still be satisfied if

$$Aa^2>-\left(\frac{1}{p_i}+a\right) \quad \text{or} \quad Aa^2+a>-\frac{1}{p_i} \tag{48}$$

Now, $1/p_i$ varies from 1 to ∞, so that (48) will be satisfied if

$$Aa^2+a>-1 \quad \text{or} \quad A>-\frac{1}{a}-\frac{1}{a^2} \tag{49}$$

The graph of $A=-1/a-1/a^2$ when $a\geq -1$ is given in Figure 13.3,

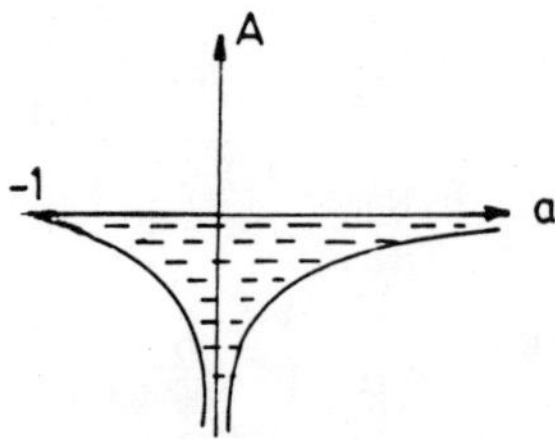

Fig. 13.3

where all points inside the shaded region give permissible values of A, a. Now,

$$\underset{a\to 0}{Lt} \sum_{i=1}^{n}(1+ap_i)\ \ln\frac{(1+ap_i)}{1+aq_i}=0 \tag{50}$$

$$\underset{a\to 0}{Lt}\ \frac{1}{a}\sum_{i=1}^{n}(1+ap_i)\ \ln\frac{1+ap_i}{1+aq_i}=\sum_{i=1}^{n}(p_i-q_i)=0 \tag{51}$$

so that for all finite values of A, positive or negative which are independent of a, (46) approaches K.L. measure [5] as $a\to 0$. Also, we can use

$$D(P:Q)=\sum_{i=1}^{n}p_i\ln\frac{p_i}{q_i}-\left(\frac{c}{a}+\frac{d}{a^2}\right)\sum_{i=1}^{n}(1+ap_i)\ln\frac{1+ap_i}{1+aq_i} \tag{52}$$

where, c and d are any positive numbers less than unity. We get interesting particular cases when $c=1, d=0$ or $c=0, d=1$ or $c=1, d=1$.

The measure (52) is again in general not a convex function of $q_1, q_2, \ldots, q_n$.

(*ii*) A measure which is a convex function of both P and Q is obtained from Csiszer's [1] measure

$$\sum_{i=1}^{n} q_i \, \phi\left(\frac{p_i}{q_i}\right) \tag{53}$$

where $\phi\,(\bullet)$ is a twice-differentiable convex function with $\phi\,(1) = 0$ by taking,

$$\phi(x) = x \ln x - \frac{1}{a}(1+ax)\ln\frac{1+ax}{1+a}, \quad a > 0, \tag{54}$$

which gives,

$$D(P:Q) = \sum_{i=1}^{n} p_i \ln\frac{p_i}{q_i} - \frac{1}{a}\sum_{i=1}^{n}(q_i + a p_i)\ln\frac{q_i + a p_i}{q_i\,(1+a)} \tag{55}$$

We can generalise it by considering

$$\phi\,(x) = x \ln x + A\,(1+ax)\ln\frac{1+ax}{1+a} \tag{56}$$

It will be convex if

$$\frac{1}{x} + \frac{Aa^2}{1+ax} > 0, \quad Aa^2 + a > -\frac{1}{x} \tag{57}$$

Now x can vary from 0 to ∞, so that $-1/x$ can vary from $-\infty$ to 0, so that our condition is

$$A\,a^2 + a > 0 \quad \text{or} \quad A > -\frac{1}{a} \tag{58}$$

Thus our generalised measure of directed divergence which is a convex function of both P and Q is

$$D(P:Q) = \sum_{i=1}^{n} p_i \ln\frac{p_i}{q_i} - A\sum_{i=1}^{n}(q_i + ap_i)\ln\frac{q_i + ap_i}{q_i\,(1+a)} \tag{59}$$

where A is any positive number or a negative number $\geq -1/a$.

(*iii*) Now consider,

$$\phi\,(x) = \frac{x^{\alpha} - x}{\alpha - 1} + A\,\frac{(1+ax)^{\alpha} - (1+ax)}{\alpha - 1} - A\,\frac{(1+a)^{\alpha} - (1+a)}{\alpha - 1} \tag{60}$$

This will be covex if

$$\alpha x^{\alpha-2} + A\,\alpha\,(1+ax)^{\alpha-2} \geq 1 \tag{61}$$

If A is positive, this is always satisfied. If $A = -B$, this gives

$$B \leq \left(\frac{x}{1+ax}\right)^{\alpha-2} = \left(\frac{1+ax}{x}\right)^{2-\alpha} \tag{62}$$

As x goes from 0 to ∞, $x\,/\,(1 + ax)$ goes from 0 to $1/a$.

If $\alpha > 2$, this requires $B \leq 0$ or $B = 0$

If $\alpha = 2$, $B \leq 1$ (63)

If $\alpha < 2$, $(1 + \text{ax}) / x$ can vary from a to ∞,
Expression (62) gives,

Also,

$$B \leq a^{2-\alpha} \tag{64}$$

$$D(P:Q) = \frac{1}{\alpha - 1} [(\sum_{i=1}^{n} p_i^{\alpha} q_i^{1-\alpha} - 1 + A\ [(q_i + ap_i)^{\alpha} q_i^{1-\alpha} - (q_i + a p_i)]$$

$$- A\ (1 + a)^{\alpha} + A\ (1 + a)] \tag{65}$$

gives a valid measure of directed divergernce for all non-negative values of A. Also,

$$D(P:Q) = \frac{1}{\alpha - 1}\left[\sum_{i=1}^{n} p_i^{\alpha} q_i^{1-\alpha} - 1 - B\ (q_i + a p_i)^{\alpha} q_i^{1-\alpha} - (q_i + a p_i) + B\ (1 + a)^{\alpha} - B\ (1 + a)\right]$$

gives a valid measure of directed divergence if $B \leq a^{2-\alpha}$ when $0 < a < 2$, and $B = 0$ when $\alpha > 2$.

REFERENCES

1. I. Csiszer (1972). "A class of measures of informativity of observation channels", Periodica Math. Hangarica. Vol. 2. pp. 191–213.
2. J.H. Havrda and F. Charvat (1967). "Quantification methods of classification processes: Concept of Structural α Entropy", Kybernatica. Vol. 3. pp. 30–35.
3. J.N. Kapur (1972). "Measures of uncertainty. mathematical programming and physics", Journ. Ind. Soc. Agri. Stat. Vol. 24, pp. 47–66.
4. J.N. Kapur (1986). "Four families of measures of entropy", Ind. Jour. Pure and App. Maths; Vol. 17. No. 4. pp. 429–449.
5. J.N. Kapur (1987). "Monotonocity and concavity of some parametric measures of entropy. Tamkang. Journal of Mathematics 18 (3). 25–40.
6. S. Kullback and R.A. Leibler (1951). "On information and sufficiency", Ann. Math. Stat; Vol. 22. pp. 79–86.
7. C.E. Shannon (1948). "A mathematical theory of communication", Bell System Tech. Journ. Vol. 27. pp. 379–423. 623–659.

14

SOME NEW ADDITIVE MEASURES OF ENTROPY AND DIRECTED DIVERGENCE

[Some new additive measures of entropy and directed divergence are obtained. It is explained why the entropy and directed divergences of type β which are obtained as limiting cases of author's entropy and directed divergence of order α and type β when $\alpha \to \beta$, are not valid measures, inspite of their being additive.]

14.1 INTRODUCTION

Additivity is considered an important property for a measure of entropy. Inspite of the importance of this property, only three measures due to Shannon [9], Renyi [8] and Aczel-Daroczy [1]/Kapur [3] were obtained till 1967 and no new additive measures have been given during the last twenty-five years.

Of course, some authors have modified the definition of additivity itself and obtained measures satisfying these modified definitions. The measures they thus obtain are not additive in the original sense.

The known additive measures of entropy for a probability distribution $P = (p_1, p_2, \ldots, p_n)$ are given by

$$H_1(P) = -\sum_{i=1}^{n} p_i \ln p_i \qquad \text{(Shannon [9])} \qquad (1)$$

$$H_2(P) = \frac{1}{1-\alpha} \ln \sum_{i=1}^{n} p_i^{\alpha}, \quad \alpha > 0,\ \alpha \neq 1 \qquad \text{(Renyi [8])} \qquad (2)$$

$$H_3(P) = \frac{1}{\beta - \alpha} \ln \frac{\sum_{i=1}^{n} p_i^{\alpha}}{\sum_{i=1}^{n} p_i^{\beta}}, \alpha > 1,\ \beta < 1 \text{ or } \alpha < 1, \beta > 1 \qquad \text{(Kapur [3])} \qquad (3)$$

$$\underset{\alpha \to 1}{Lt} \quad H_2(P) = H_1(P) \qquad (4)$$

$$(H_3(P))_{\beta=1} = H_2(P) \qquad (5)$$

$$\underset{\substack{\alpha \to 1 \\ \beta \to 1}}{Lt} \quad H_3(P) = H_1(P) \qquad (6)$$

By taking the limit of $H_3(P)$ as $\alpha \to \beta$, we get the 'entropy' of type β, viz.

$$H_4(P) = -\frac{\sum_{i=1}^{n} p_i^\beta \ln p_i}{\sum_{i=1}^{n} p_i^\beta} \tag{7}$$

It was soon found that this measure was not maximum

when, $$p_1 = p_2 = \dots = p_n = \frac{1}{n} \tag{8}$$

for all values of β. Since, entropy measures uncertainty, a measure which does not give maximum uncertainty when all the outcomes are equally likely, cannot be considered a valid measure of uncertainty or entropy. Stolarsky [10] and Clausing [2] therefore spent considerable effort in finding those values of β for which H_β (*P*) will have its maximum value for the uniform distribution.

It is easily verified that H_4 (P) is additive, since if $P = (p_1, p_2, \dots, p_n)$ and $Q = (q_1, q_2, \dots, q_n)$ are two independent probability distributions and $P * Q$ is their joint distribution,

$$\begin{aligned}
H_4(P^*Q) &= -\frac{\sum_{j=1}^{m}\sum_{i=1}^{n} p_i^\beta q_j^\beta \ln p_i q_j}{\sum_{j=1}^{m}\sum_{i=1}^{n} p_i^\beta q_j^\beta} \\
&= -\frac{\sum_{j=1}^{m} q_j^\beta \sum_{i=1}^{n} p_i^\beta \ln p_i - \sum_{i=1}^{n} p_i^\beta \sum_{j=1}^{m} q_j^\beta \ln q_j^\beta}{\sum_{i=1}^{n} p_i^\beta \sum_{j=1}^{m} q_i^\beta} \\
&= -\frac{\sum_{i=1}^{n} p_i^\beta \ln p_i}{\sum_{i=1}^{n} p_i^\beta} - \frac{\sum_{j=1}^{m} q_i^\beta \ln q_j^\beta}{\sum_{i=1}^{n} q_j^\beta} \\
&= H_4(P) + H_4(Q)
\end{aligned} \tag{9}$$

It is also verified that H_3 (*P*) is maximum when $P = U$, where *U* is the uniform distribution given by (8).

One object of the present discussion is to explain the reason why H_4 (*P*) is not maximum when $P = U$, although H_3 (*P*) is, though H_4 (*P*) is only a limiting form of H_3 (*P*).

A second object is to find measures of directed divergence D_3 (*P* : *Q*) and D_4 (*P* : *Q*) corresponding to H_3 (*P*) and H_4 (*P*) and explain why D_4 (*P* : *Q*) may not be a valid measure inspite of its being a limiting case of the valid measure D_3 (*P* : *Q*).

A third object is to find new additive measures of entropy and directed divergence in addition to the three measures given above.

14.2 SOME NEW ADDITIVE MEASURES OF ENTROPY

Consider the measure,

$$H_5(P) = \frac{\lambda}{1-\alpha} \ln \sum_{i=1}^{n} p_i^{\alpha} + \frac{\mu}{1-\beta} \ln \sum_{i=1}^{n} p_i^{\beta}, \quad \alpha \neq 1,\ \beta \neq 1 \tag{10}$$

where $\lambda, \mu, \alpha, \beta$ are > 0 but α and β may be less than or greater than unity. Since both

$$\frac{1}{1-\alpha} \ln \sum_{i=1}^{n} p_i^{\alpha} \quad \text{and} \quad \frac{1}{1-\beta} \ln \sum_{i=1}^{n} p_i^{\beta} \tag{11}$$

are maximum when $P = U$ and λ, μ are > 0, $H_5(P)$ is also maximum when $P = U$, whenever $\alpha > 0,\ \beta > 0$

Now let,

$$\frac{\lambda}{1-\alpha} = -\frac{k\mu}{1-\beta}, \tag{12}$$

where $k > 0$. Since λ, μ, k are > 0, $1 - \alpha$ and $1 - \beta$ have to have opposite signs. If $\alpha < 1,\ \beta > 1$ or if $\alpha > 1,\ \beta < 1$, using (10) and (12), we get

$$H_5(P) = \frac{\lambda}{1-\alpha}\left[\ln \sum_{i=1}^{n} p_i^{\alpha} - k \ln \sum_{i=1}^{n} p_i^{\beta}\right]$$

$$= \frac{\lambda}{1-\alpha} \ln \frac{\sum_{i=1}^{n} p_i^{\alpha}}{\left(\sum_{i=1}^{n} p_i^{\beta}\right)^k}, \qquad \begin{matrix} \alpha < 1, & \beta > 1 \\ \text{or} \quad \alpha > 1, & \beta < 1 \end{matrix} \tag{13}$$

Whether $\alpha < 1,\ \beta > 1$ or $\alpha > 1,\ \beta < 1,\ 1 - \alpha$ has the same sign as $\beta - \alpha$, so we consider the measure,

$$H_6(P) = \frac{1}{\beta - \alpha} \ln \frac{\sum_{i=1}^{n} p_i^{\alpha}}{\left(\sum_{i=1}^{n} p_i^{\beta}\right)^k} \tag{14}$$

This is maximum when $P = U$ and so is its special case $H_3(P)$ when $k = 1$. It is easily seen that if P and Q are two independent distributions,

$$H_6(P^*Q) = H_6(P) + H_6(Q) \tag{15}$$

so that the new measure $H_6(P)$ is additive and it is a valid measure since it is maximum when $P = U$.

This is a three-parameter measure of entropy which includes the earlier three measures as special or limiting cases, since

$$(H_6(P))_{k=1} = H_3(P), \quad (H_6(P))_{\substack{k=1\\ \beta=1}} = H_2(P), \quad \underset{\substack{\beta\to 1\\ k\to 1\\ \alpha\to 1}}{Lt} H_6(P) = H_1(P) \tag{16}$$

By giving specific values like $k = 1/4\ ,\ 1/2\ ,\ 2,\ 3,\ \ldots$, we can get two-parameter measures, different from $H_3\ (P)$.

By giving specific values to k different from unity and by giving specified values to β (> 1 if $\alpha < 1$ and < 1 if $\beta > 1$), we can get one-parametric measure, different from $H_2\ (P)$.

By giving k, α, β specified values ($k \neq 1, \beta > 1, \alpha < 1$ or $\alpha < 1, \beta > 1$) we get a non-parametric measure different from $H_1\ (P)$.

All these measures of entropy are additive and are valid measures of entropy. Thus we have got an infinity or new additive measures of entropy.

14.3 WHY H_4 (P) IS NOT A VALID MEASURE OF ENTROPY?

$H_3\ (P)$ is defined when $\alpha < 1, \beta > 1$ or when $\alpha > 1, \beta < 1$. Of course α and β may be as near to unity as we like, provided they satisfy, these conditions but α can approach β only in case $\beta = 1$ and in this case (7) becomes,

$$-\sum_{i=1}^{n} p_i \ln p_i \,/\, \sum_{i=1}^{n} p_i = -\sum_{i=1}^{n} p_i \ln p_i\,, \tag{17}$$

which is a perfectly valid measure of entropy.

Now suppose $\beta < 1$, then as $\alpha \to \beta, \alpha < 1$ and in this case (10) gives difference of two entropies of type H_3 (P) and while each is maximum when $P = U$, their difference may not be maximum at $P = U$.

However, it is also possible that in some cases, the difference may also be maximum when $P = U$, but these cases have to be specifically determined and this is what Stolarsky [10] and Clausing [3] did.

14.4 SOME NEW MEASURES OF DIRECTED DIVERGENCE

Since $\dfrac{1}{\alpha-1} \ln \sum_{i=1}^{n} p_i^{\alpha} q_i^{1-\alpha}$ and $\dfrac{1}{\beta-1} \ln \sum_{i=1}^{n} p_i^{\beta} q_i^{1-\beta}$ $\quad \alpha \neq 1, \beta \neq 1; \alpha; \beta > 0$ (18)

are valid measures of directed divergence, we consider the measure,

$$\frac{\lambda}{\alpha-1} \ln \sum_{i=1}^{n} p_i^{\alpha} q_i^{1-\alpha} + \frac{\mu}{(\beta-1)} \ln \sum_{i=1}^{n} p_i^{\beta} q_i^{1-\beta} \tag{19}$$

and, let

$$\frac{\mu}{\beta-1} = -\frac{k\lambda}{\alpha-1}, \quad \mu > 0\,,\ \lambda > 0,\ k > 0 \tag{20}$$

so that $\alpha - 1$ and $\beta - 1$ have opposite signs. We get by using arguments similar to those used above, the measure,

$$D_6\,(P{:}Q) = \frac{1}{\alpha-\beta} \ln \frac{\sum_{i=1}^{n} p_i^{\alpha} q_i^{1-\alpha}}{\left(\sum_{i=1}^{n} p_i^{\beta} q_i^{1-\beta}\right)^{k}}, \tag{21}$$

which is a new measure of directed divergence. If P_1, P_2 and Q_1, Q_2 are two pairs of independent distributions, we get

$$D_6(P_1^* P_2 : Q_1^* Q_2) = \frac{1}{\alpha-\beta} \ln \frac{\sum_{j=1}^{m}\sum_{i=1}^{n} (p_{i1} p_{i2})^{\alpha} (q_{j1} q_{j2})^{1-\alpha}}{\left[\sum_{j=1}^{m}\sum_{i=1}^{n} (p_{i1} p_{i2})^{\beta} (q_{j1} q_{j2})^{1-\beta}\right]^k}$$

$$= \frac{1}{\alpha-\beta} \ln \frac{\sum_{j=1}^{m}\sum_{i=1}^{n} p_{i1}^{\alpha} q_{j1}^{1-\alpha} \sum_{j=1}^{m}\sum_{i=1}^{n} p_{i2}^{\alpha} q_{j2}^{1-\alpha}}{\left[\sum_{j=1}^{m}\sum_{i=1}^{n} p_{i1}^{\beta} q_{j1}^{1-\beta} \sum_{j=1}^{m}\sum_{i=1}^{n} p_{i2}^{\beta} q_{j2}^{1-\beta}\right]^k}$$

$$= \frac{1}{\alpha-\beta} \ln \frac{\sum_{j=1}^{m}\sum_{i=1}^{n} p_{i1}^{\alpha} q_{j1}^{1-\alpha}}{\left[\sum_{j=1}^{m}\sum_{i=1}^{n} p_{i1}^{\beta} q_{j1}^{1-\beta}\right]^k} + \frac{1}{\alpha-\beta} \ln \frac{\sum_{j=1}^{m}\sum_{i=1}^{n} p_{i2}^{\alpha} q_{j2}^{1-\alpha}}{\left[\sum_{j=1}^{m}\sum_{i=1}^{n} p_{i2}^{\beta} q_{j2}^{1-\beta}\right]^k}$$

$$= D_6(P_1 : Q_1) + D_6(P_2 : Q_2) \tag{22}$$

so that the new measure is additive.

When $k = 1$, it gives the measure,

$$D_3(P:Q) = \frac{1}{\alpha-\beta} \ln \frac{\sum_{i=1}^{n} p_i^{\alpha} q_i^{1-\alpha}}{\sum_{i=1}^{n} p_i^{\beta} q_i^{1-\beta}} \tag{23}$$

Letting $\alpha \to \beta$, we get the measure,

$$D_4(P:Q) = \frac{\sum_{i=1}^{n} p_i^{\beta} q_i^{1-\beta} \ln \frac{p_i}{q_i}}{\sum_{i=1}^{n} p_i^{\beta} q_i^{1-\beta}} \tag{24}$$

When $\beta = 1$, this gives Kullback-Leibler [7] measure of directed divergence,

$$D(P:Q) = \sum_{i=1}^{n} p_i \ln \frac{p_i}{q_i} \tag{25}$$

and $D(P : Q) \geq 0$ for all probability distributions.

Also,

$$\frac{d}{d\beta} D_4(P:Q) = \frac{d}{d\beta}\left[\frac{\sum_{i=1}^{n} q_i \left(\frac{p_i}{q_i}\right)^{\beta} \ln \frac{p_i}{q_i}}{\sum_{i=1}^{n} q_i \left(\frac{p_i}{q_i}\right)^{\beta}}\right]$$

$$= \frac{\left(\sum_{i=1}^{n} q_i \left(\frac{p_i}{q_i}\right)^{\beta}\right)\left(\sum_{i=1}^{n} q_i \left(\frac{p_i}{q_i}\right)^{\beta} \left(\ln \frac{p_i}{q_i}\right)^2\right) - \left[\sum_{i=1}^{n} q_i \left(\frac{p_i}{q_i}\right)^{\beta} \ln \frac{p_i}{q_i}\right]^2}{\left[\sum_{i=1}^{n} q_i \left(\frac{p_i}{q_i}\right)^{\beta}\right]^2}, \tag{26}$$

so that by using Cauchy-Schwarz inequality, we find that

$$\frac{d}{d\beta} D_4 (P:Q) \geq 0 \tag{27}$$

so that for given (P, Q), $D_4 (P : Q)$ is a monotonic increasing function of β,

$$(D_4 (P:Q))_{\beta > 1} > (D_4(P:Q))_{\beta = 1} = D\ (P,Q) \geq 0 \tag{28}$$

For all values of $\beta \geq 1$, D_4 (P : Q) ≥ 0 and it also vanishes when $P = Q$. When $\beta = 1$, it vanishes only when $P = Q$. Also,

$$(D_4 (P:Q))_{\beta < 1} < (D_4 (P:Q))_{\beta = 1} = D\ (P:Q) \geq 0 \tag{29}$$

so that for $\beta < 1$, $D_4 (P : Q)$ may or may not always be ≥ 0, though it will still vanish when $P = Q$. It will be worth while to investigate whether there a β_0 such that for $\beta_0 < \beta < 1$; also $D_4 (P : Q) \geq 0$ for all P, Q.

14.5 CONCLUDING REMARKS

We have given an infinity of non-parametric, one parametric, two-parametric and three parametric measures of entropy and directed divergence which are additive and which are different from the known measures and which satisfy the true additivity property. We can use these if we are specially fascinated by the additivity property.

However, being maximum at U is an even important property and should not be sacrificed for the sake of additivity. In fact, in entropy optimization problems, we would prefer concavity to additivity because concave measures give global maximum values, while non-concave functions may not do so.

In fact, if we consider both sum function property or logarithm of sum function property as well as additivity, we will find that we have exhausted all possibilities and the only other measures can be positive linear combinations of the measures we have given.

Another important point is that generalisation is always achieved at some cost. A special measure has some properties and the generalised measure will not have all these properties and whenever we generalise, we should investigate what property or properties we have lost and how valuable these are.

Thus, in going from $H_1 (P)$ to $H_2 (P)$ we lost recursivity and sub-additivity. We also lost concavity when $\alpha > 1$ and replaced it by pseudo-concavity. In going from $H_2 (P)$ to H_3 (P) we lost even pseudo-concavity and we retained the property that the measure was maximum at $P = U$. In going to $H_4 (P)$ we lost even this property. In going to $H_6 (P)$ we did not lose the property of being maximum at $P = U$ and at same time we were able to generalise.

Moreover, we would like to have the minimum number of parameters and at the same time we should like our measures to be flexible for better fit to data and for taking into account intangible factors which cannot be taken into account in the absence of parameters except by complicating the constraints significantly.

There is always a trade-off between gains and losses in generalisation and one should use one's judgement and knowledge of the situation to draw the line at the right point.

The relative merits of some of these properties from the point of view of optimization principles are discussed in Kapur are Kesavan [6].

REFERENCES

1. J. Aczel and Z. Daroczy (1963), "*Uber ver all gemeinerte quásilinare Mittel woret, die mit Genichts funktunon gebildett sind*", Publicatunes Mall **10**, pp. 171-190.
2. A Clausing (1983), "Type t entropy and majorisation", Siam J. Math. Analysis, Vol. 14, No. 1, pp. 203-208.
3. J.N. Kapur (1967), "On some properties of generalised entropies", Ind. J. Math. Vol. 9, pp. 427-442.
4. J.N. Kapur (1987), "On the range of validity of certain measures of inaccuracy", Mathematics Today **5**, pp. 54-62
5. J.N. Kapur (1986), "On the range of validity of Sharma and Taneja's measure of entropy", Gujarat Statistical Reviews, **17**, pp. 33-40.
6. J.N. Kapur and H.K. Kesavan (1992), "Entropy Optimization Principles and their Application", Academic Press, New York.
7. S. Kullback & R.A. Leibler (1951), "On information and sufficiency", Ann. Math. Stat. Vol. 22, pp. 79-86.
8. A. Renyi (1961), "On Measures of Entropy and Information", Proc. 4th Berkeley Symp. Math. Stat. Prob. Vol. 1. pp. 547-561.
9. C.E. Shannon (1948), "A Mathematical Theory of Communication", Bell System, Tech J., Vol. 27, pp. 373-423, 623-659.
10. K.B. Stolarsky (1980), "A stronger logarithm inequality suggested by the entropy inequality", SIAM, J. Math Analysis, Vol. 11, pp. 242-247.

15

ON EQUIVALENT SETS OF MEASURES OF ENTROPY AND CROSS-ENTROPY

[Measures of entropy (cross-entropy) which give same maximising (minimising) probability distributions when these are maximized (minimized) subject to a given set of constraints are said to be equivalent to one another. It is shown here that the four measures of entropy (and the corresponding measures of cross-entropy) which were independently proposed by Renyi [15] Harvada and Charvat [2], Sharma and Mittal [16] and Behara and Chawla [1] from different considerations, are in fact equivalent. Some more measures which are equivalent to these are also introduced. While some measures of an equivalance class may be additive, recursive, concave / convex, others may not have some of these properties.]

15.1 INTRODUCTION

Let $P = (p_1, p_2, \ldots, p_n)$ be a probability distribution then Shannon [17] gave the measure of entropy

$$S(P) = -\sum_{i=1}^{n} p_i \ln p_i \tag{1}$$

to measure the uncertainty, diversity or equality represented by P. Later Havrda-Charvat [2], Renyi [5], Behara-Chawla [1] and Sharma-Mittal [16] introduced the measures

$$H(P) = \frac{1}{1-\alpha}\left(\sum_{i=1}^{n} p_i^{\alpha} - 1\right), \qquad \alpha \neq 1 \qquad \alpha > 0 \tag{2}$$

$$R(P) = \frac{1}{1-\alpha} \ln \sum_{i=1}^{n} p_i^{\alpha}, \qquad \alpha \neq 0 \qquad \alpha > 0 \tag{3}$$

$$B(P) = \frac{\left(\sum_{i=1}^{n} p_i^{\alpha}\right)^{1/\alpha - 1}}{1-\alpha}, \qquad \alpha \neq 1 \qquad \alpha > 0 \tag{4}$$

$$M(P) = \frac{\left(\sum_{i=1}^{n} p_i^{\alpha}\right)^{\beta} - 1}{(1-\alpha)\beta}, \qquad \alpha \neq 1, \quad \alpha > 0, \quad \beta \neq 0 \tag{5}$$

These measures were obtained independently and from quite different considerations.

Now suppose only some partial information about $p_1, p_2, \ldots, p_n$ is available in the form,

$$\sum_{i=1}^{n} p_i = 1\,,\ \sum_{i=1}^{n} p_i\, g_{ri} = a_r\,,\ r = 1,2,\ldots,m,\ m+1<n;\quad p_1 \geq 0,\ldots,\ p_n \geq 0 \tag{6}$$

These are in general not sufficient to determine $p_1, p_2, \ldots, p_n$ uniquely. Accordingly Jaynes [3] proposed his maximum entropy principle according to which we should choose $p_1, p_2, \ldots, p_n$ to maximize $S(P)$ subject to constraints (6). Later Kapur and Kesavan [9, 10] proposed that we can as well choose $p_1, \ldots, p_n$ to maximize any other measure of entropy subject to (6). They have argued the case for the generalised principle of maximum entropy in [10, 11]. Of course, the generalised measures of entropy and the generalied principle of maximum entropy had been earlier used implicitly by many workers.

The measure $H(P)$, $R(P)$, $B(P)$, $M(P)$ are generalised measures because these involve parameters α, β, for special or limiting values of which the measures reduce to Shannon's measure $S(P)$. Thus it is easily shown that

$$\underset{\alpha\to 1}{Lt}\ H(P) = S(P)\quad \underset{\alpha\to 1}{Lt}\ R(P) = S(P),\quad \underset{\alpha\to 1}{Lt}\ B(P) = S(P)$$

$$\underset{\beta\to 0}{Lt}\ M(P) = R(P),\ \underset{\alpha\to 1}{Lt}\ M(P) = S(P),\quad (M(P))_{\beta=1} = H(P) \tag{7}$$

In general, different measures of entropy arrange a given set of probability distribution in different order for their uncertainty or diversity [14]. Thus one measure may give the result that P is more uncertain than Q, while another measure may give the result that P is less uncertain than Q. In spite of this, different measures of entropy have been used in economics, genetics, sociology, ecology, linguistics and other fields [11], because of the appropriateness of different measures for different situations.

However, some measures, inspite of being different, may lead to same arrangements and in particular they may lead to the same probability distribution as the maximum entropy probability distribution. We call these measures as equivalent from the generalised maximum entropy point of view or from the point of view of arranging probability distributions according to their entropies or equalities or diversities.

In sections 2, 3 and 4 we show that the measures $H(P)$, $R(P)$, $B(P)$ and $M(P)$ are equivalent in this sense. In section 5, we give an alternative proof of this important result. In section 6 we discuss the concavity of these measures and show that while these are equivalent, these need not all be concave and these need not even share the properties of additivity and recursivity. In section 7, we give some relations between these measures and in section 8 we give some new measures belonging to this equivalence class. In section 9, we give some other equivalence classes of measures, disjoint from the earlier set and in section 10, we give equivalence classes of measures of cross-entropy. In section 11, we give a case where equivalent measures can give different results, and finally, in section 12, we give a summary of the results obtained in the present discussion.

15.2 EQUIVALENCE OF HAVRDA-CHARVAT AND RENYI'S MEASURES OF ENTROPY

It is easily seen that

$$R(P)=\frac{1}{1-\alpha}\ln\sum_{i=1}^{n}p_i^{\alpha}=\frac{1}{1-\alpha}\ln[(1-\alpha)H(P)+1] \tag{8}$$

When $\alpha > 1$, then

$$H(P)\underset{<}{>}H(Q) \Rightarrow (1-\alpha)H(P)+1 \underset{>}{<} (1-\alpha)H(Q)+1$$

$$\Rightarrow \ln[(1-\alpha)H(P)+1] \underset{>}{<} \ln[(1-\alpha)H(Q)+1]$$

$$\Rightarrow \frac{1}{1-\alpha}\ln[(1-\alpha)H(P)+1] \underset{<}{>} \frac{1}{1-\alpha}\ln[(1-\alpha)H(Q)+1]$$

$$\Rightarrow R(P) \underset{<}{>} R(Q) \tag{9}$$

Similarly, when $\alpha < 1$

$$H(P) \underset{<}{>} H(Q) \Rightarrow R(P) \underset{<}{>} R(Q) \tag{10}$$

and whether $\alpha > 1$ or < 1

$$R(P) \underset{<}{>} R(Q) \Leftrightarrow H(P) \underset{<}{>} H(Q) \tag{11}$$

Thus Havrda-Charvat and Renyi's entropy measures arrange probabilty distributions in the same order of magnitude of the entropies. Alternatively from (8)

$$\frac{d(R(P))}{d(H(P))}=\frac{1}{1-\alpha}\,\frac{1-\alpha}{(1-\alpha)H(P)+1}=\frac{1}{(1-\alpha)H(P)+1}=\frac{1}{\sum_{i=1}^{n}p_i^{\alpha}}>0, \tag{12}$$

so that $R(P)$ and $H(P)$ increase or decrease together.

It also follows that the maximum entropy probability distributions obtained by using $R(P)$ or $H(P)$, will be the same.

15.3 EQUIVALENCE OF HAVRDA-CHARVAT AND BEHARA-CHAWLA MEASURES OF ENTROPY

It is easily shown from (2) and (4) that

$$B(P)=\frac{1}{1-\alpha}[(1-\alpha)H(P)+1)^{\frac{1}{\alpha}}] \tag{13}$$

so that, $$\frac{dB(P)}{dH(P)} = \frac{1}{\alpha}[(1-\alpha)H(P)+1]^{\frac{1}{\alpha}-1} = \frac{1}{\alpha}\left(\sum_{i=1}^{n} p_i^{\alpha}\right)^{\frac{1}{\alpha}-1} > 0, \tag{14}$$

so that B (P) and H (P) increase and decrease together and

$$B(P) \underset{<}{\overset{>}{}} B(Q) \Leftrightarrow H(P) \underset{<}{\overset{>}{}} H(Q) \tag{15}$$

Again, $$H(P) = H(Q) \Leftrightarrow \sum_{i=1}^{n} p_i^{\alpha} = \sum_{i=1}^{n} q_i^{\alpha}$$

$$\Leftrightarrow B(P) = B(Q) \tag{16}$$

Thus Havrda-Charvat and Behara-Chawla measures are equivalent and lead to the same MEPD.

15.4 EQUIVALENCE OF HAVRDA CHARVAT AND SHARMA-MITTAL MEASURE'S OF ENTROPY

From (2) and (15), we find that

$$M(P) = \frac{1}{(1-\alpha)\beta}[((1-\alpha)H(P)+1)^{B} - 1] \tag{17}$$

so that, $$\frac{d(M(P))}{d(H(P))} = [(1-\alpha)H(P)+1]^{\beta-1} = \left(\sum_{i=1}^{n} p_i^{\alpha}\right)^{\beta-1} > 0, \tag{18}$$

so that M (P) and H (P) again increase and decrease together and lead to the same maximum entropy probability distribution.

15.5 ALTERNATIVE PROOF OF EQUIVALENCE OF THE FOUR MEASURES

Maximizing (2), (3), (4), (5) subject to (6), we get

$$\frac{\alpha}{1-\alpha} p_i^{\alpha-1} = \lambda_0 + \lambda_1 g_{1i} + \lambda_2 g_{2i} + \dots + \lambda_m g_{mi} \tag{19}$$

$$\frac{\alpha}{1-\alpha} \frac{1}{\sum_{i=1}^{n} p_i^{\alpha}} p_i^{\alpha-1} = \lambda'_0 + \lambda'_1 g_{1i} + \lambda'_2 g_{2i} + \dots + \lambda'_m g_{mi} \tag{20}$$

$$\frac{1}{1-\alpha}\left(\sum_{i=1}^{n} p_i^{\alpha}\right)^{\frac{1}{\alpha}-1} p_i^{\alpha-1} = \lambda''_0 + \lambda''_1 g_{1i} + \lambda''_2 g_{2i} + \dots + \lambda''_m g_{mi} \tag{21}$$

$$\frac{\alpha}{1-\alpha}\left(\sum_{i=1}^{n} p_i^{\alpha}\right)^{\beta-1} p_i^{\alpha-1} = \lambda'''_0 + \lambda'''_1 g_{1i} + \dots + \lambda m''' g_{mi} \tag{22}$$

where in each case λ's are obtained by using the constraints (6).

Each of the four equations can be written as

$$p_i = (\mu_0 + \mu_1 g_{1i} + \ldots + \mu_m g_{mi})^{\frac{1}{\alpha}-1} \tag{23}$$

where, μ's are determined in each case by using the same constraints (6) and as such have the same values in all four cases.

Thus, we find that the maximum entropy probability distribution is the same and is independent of which of the four measures we use. In particular, it is also independent of the value of β we use.

15.6 CONCAVITY OF MEASURES

In entropy maximization problems, it is desirable that entropy function should be a concave function of $p_1, p_2, \ldots, p_n$.

Though all these four measures increase or decrease together, it does not follow that if one measure is concave, others will also be concave functions.

Since $-x \ln x$ is a concave function and x^α is convex if $\alpha > 1$ and concave function if $\alpha < 1$, it easily follows that $S(P)$ and $H(P)$ are concave functions of P.

Let us now examine the nature of $B(P)$ and $H(P)$. Now

$$\frac{\partial}{\partial p_i} B(P) = \frac{1}{1-\alpha}\left[\left(\sum_{i=1}^n p_i^\alpha\right)^{\frac{1}{\alpha}-1}\right]$$

$$\frac{\partial^2}{\partial p_i^2} B(P) = -p_i^{\alpha-1}\left(\sum_{i=1}^n p_i^\alpha\right)^{\frac{1}{\alpha}-2}\left(\sum_{i=1}^n p_i^\alpha - p_i^\alpha\right) \tag{24}$$

$$\frac{\partial^2}{\partial p_i \partial p_j} B(p) = p_i^{\alpha-1} p_j^{\alpha-1}\left(\sum_{i=1}^n p_i^\alpha\right)^{\frac{1}{\alpha}-2} \tag{25}$$

so that,

$$\sum_{j=1}^n \sum_{i=1}^n \left[\frac{\partial^2}{\partial p_i \partial p_j} B(p)\right] x_i x_j = -\left(\sum_{i=1}^n p_i^\alpha\right)^{(1/\alpha)-2}\left[\sum_{i=1}^n p_i^\alpha \sum_{i=1}^n p_i^{\alpha-2} x_i^2 - \left(\sum_{i=1}^n p_i^{\alpha-1} x_i\right)^2\right]$$

$$= -\left(\sum_{i=1}^n p_i^\alpha\right)^{\frac{1}{\alpha}-2}\left[\sum_{i=1}^n u_i^2 \sum_{i=1}^n v_i^2 - \left(\sum_{i=1}^n u_i v_i\right)^2\right] \tag{26}$$

where $$u_i = p_i^{\alpha/2} \quad , \quad v_i = p_i^{\frac{\alpha-2}{2}} x_i \quad , \quad i = 1, 2, \ldots, n \tag{27}$$

By using Cauchy-Schwartz inequality, we find that the Hessian matrix of the second order derivatives of $B(P)$ is negative definite, so that $B(P)$ is a concave function of $p_1, p_2, \ldots, p_n$.

Similarly,

$$\frac{\partial}{\partial p_i}(M(P)) = \frac{1}{(1-\alpha)\beta}\left[\alpha\beta\left(\sum_{i=1}^n p_i^\alpha\right)^{\beta-1} p_i^{\alpha-1}\right]$$

Similarly,

$$\frac{\partial^2}{\partial p_i^2} M(P) = \frac{\alpha}{1-\alpha}\left(\sum_{i=1}^{n} p_i^{\alpha}\right)^{\beta-2} p_i^{\alpha-2}\left\{\left(\sum_{i=1}^{n} p_i^{\alpha}\right)(\alpha-1)+\alpha p_i^{\alpha}(\beta-1)\right] \tag{28}$$

and,

$$\frac{\partial^2}{\partial p_i \partial p_j} M(P) = \frac{\alpha}{1-\alpha}(\beta-1)\left(\sum_{i=1}^{n} p_i^{\alpha}\right)^{\beta-2} \alpha p_i^{\alpha-1} p_j^{\alpha-1} \tag{29}$$

so that,

$$\sum_{j=1}^{n}\sum_{i=1}^{n}\frac{\partial^2}{\partial_i \partial p_j} M(P)\, x_i x_j = \left(\sum_{i=1}^{n} p_i^{\alpha}\right)^{\beta-\alpha}\frac{\alpha}{1-\alpha}\left[(\alpha-1)\sum_{i=1}^{n} p_i^{\alpha}\sum_{i=1}^{n} p_i^{\alpha-2}x_i^2+\alpha(\beta-1)\left(\sum_{i=1}^{n} p_i^{\alpha-1}x_i\right)^2\right]$$

$$= \left(\sum_{i=1}^{n} x_i^2\right)^{p-2}\alpha\left[-\sum_{i=1}^{n} u^2\sum_{i=1}^{n} v_i^2+\frac{\alpha(\beta-1)}{1-\alpha}\left(\sum_{i=1}^{n} u_i v_i\right)^2\right] \tag{30}$$

so that $M(P)$ will be a concave function of P if

$$\frac{\alpha(\beta-1)}{1-\alpha} \le 1$$

i.e. if $\beta \le \frac{1}{\alpha}$ when $\alpha < 1$ or $\beta \ge \frac{1}{\alpha}$ when $\alpha > 1$ (31)

Behara and Chawla chose $\beta = 1/\alpha$ but we can choose other pairs of values of α, β satisfying the conditions (31).

If the equation (31) are violated, i.e. if $\alpha(\beta-1)/(1-\alpha) > 1$, then $M(P)$ is not necessarily concave function of $p_1, p_2, \ldots, p_n$. However, we can still use it as a measure of entropy for the simple reason that its maximization gives the same distribution as Havrda-Charvat measure which is a concave function.

In Renyi's measure, we note that if $\alpha < 1$, $\sum_{i=1}^{n} p_i^{\alpha}$ is a concave function, its logarithm is also a concave function and as such $(1-\alpha)^{-1}\ln\sum_{i=1}^{n} p_i^{\alpha}$ is also a concave function. However, when $\alpha > 1$, $\sum_{i=1}^{n} p_i^{\alpha}$ is a convex function, its logarithm is not necessarily a convex function and Renyi's measure is not a concave function. In fact, it is pseudo-concave function and it can still be used in entropy maximization problems.

In the measure $M(P)$, β is allowed to be negative. If $\beta = -\beta'$ the condition (31) is satisfied if $\alpha < 1$. If $\beta = -\beta'$, $\alpha > 1$ (31) becomes

$$\frac{\alpha(-\beta'-1)}{1-\alpha} < 1 \quad \text{or} \quad \frac{\alpha(1+\beta')}{\alpha-1} < 1 \quad \text{or} \quad \beta' < \frac{1}{\alpha} \quad \text{or} \quad -\beta < \frac{1}{\alpha} \tag{32}$$

Thus the measure $M(P)$ is concave if

(i) $\beta > 0 \quad \beta < \frac{1}{\alpha}, \quad \alpha < 1$ (33)

(ii) $\beta > 0 \quad \beta > \frac{1}{\alpha}, \quad \alpha > 1$ (34)

(iii) $\beta > 0 \quad \beta < \frac{1}{\alpha}, \quad \alpha < 1$ (35)

(iv) $\beta < 0 \quad \alpha < 1$ (36)

(v) $\beta < 0 \quad \beta < \frac{1}{\alpha}, \quad \alpha > 1$ (37)

15.7 RELATIONS AMONG THESE ENTROPY MEASURES

(a) It has been proved earlier [4, 5] that for a given probability distribution P,

(i) $H(P)$ is a monotonic decreasing function of α and $H(P) \to 0$ as $\alpha \to \infty$.

(ii) $R(P)$ is a monotonic decreasing function of α and $R(P) \to -\ln p_{\max}$ as $\alpha \to \infty$, where

$$p_{\max} = \max(p_1, p_2, \ldots, p_n) \tag{38}$$

(iii) For a given β, $M(P)$ is a monotonic decreasing function of α.

(b) It has also been conjectured that $B(P)$ is also monotonic decreasing function of α.

(c) $$R(P) - H(P) = \frac{1}{1-\alpha}\left[\ln \sum_{i=1}^{n} p_i^\alpha - \sum_{i=1}^{n} p_i^\alpha + 1\right] = \frac{1}{1-\alpha}[\ln x - x + 1]\ ; \quad x = \sum_{i=1}^{n} p_i^\alpha \tag{39}$$

However, $\ln x - x + 1 \leq 0$ when $x > 0$ and the equality sign holds only when $x = 1$, so that

$$R(P) \leq H(P) \quad \text{when} \quad \alpha < 1$$
$$R(P) \geq H(P) \quad \text{when} \quad \alpha > 1 \tag{40}$$

In fact, $R(P) = H(P)$ when $\alpha = 1$ or when P is a degenerate distribution.

(*d*) Now $$\alpha > 1 \Rightarrow \sum_{i=1}^{n} p_i^\alpha < 1 \Rightarrow \left(\sum_{i=1}^{n} p_i^\alpha\right)^{\beta-1} \begin{array}{l} < 1 \text{ if } \beta > 1 \\ > 1 \text{ if } \beta < \end{array}$$

$$\Rightarrow \left(\sum_{i=1}^{n} p_i^\alpha\right)^{\beta} \begin{array}{c} < \\ > \end{array} \left(\sum_{i=1}^{n} p_i^\alpha\right) \text{ according as } \beta \begin{array}{c} > \\ < \end{array} 1$$

$$\Rightarrow \frac{\left(\sum_{i=1}^{n} p_i^\alpha\right)^{\beta} - 1}{1-\alpha} \begin{array}{c} > \\ < \end{array} \frac{\sum_{i=1}^{n} p_i^\alpha - 1}{1-\alpha} \text{ according as } \beta \begin{array}{c} > \\ < \end{array} 1$$

$$\Rightarrow \beta M(P) \begin{array}{c} > \\ < \end{array} H(P) \text{ according as } \beta \begin{array}{c} > \\ < \end{array} 1 \tag{41}$$

Similarly, $\alpha < 1 \quad \Rightarrow \sum_{i=1}^{n} p_i^{\alpha} > 1 \Rightarrow \left(\sum_{i=1}^{n} p_i^{\alpha}\right)^{\beta-1} \overset{>}{<} 1$ according as $\beta \overset{>}{<} 1$

$$\Rightarrow \left(\sum_{i=1}^{n} p_i^{\alpha}\right)^{\beta} \overset{>}{<} \sum_{i=1}^{n} p_i^{\alpha} \quad \text{according as } \beta \overset{>}{<} 1$$

$$\Rightarrow \frac{\left(\sum_{i=1}^{n} p_i^{\alpha}\right)^{\beta} - 1}{1-\alpha} \overset{>}{<} \frac{\sum_{i=1}^{n} p_i^{\alpha} - 1}{1-\alpha} \quad \text{according as } \beta \overset{>}{<} 1$$

$$\Rightarrow \beta M(P) \overset{>}{<} H(P) \quad \text{according as } \beta \overset{>}{<} 1 \tag{42}$$

Thus whether $\alpha \gtreqless 1$, $\beta M(P) \gtreqless H(P)$ according as $\beta \gtreqless 1$ (43)

(e) $\alpha > 1 \Rightarrow \sum_{i=1}^{n} p_i^{\alpha} < 1 \Rightarrow \left(\sum_{i=1}^{n} p_i^{\alpha}\right)^{\frac{1}{\alpha}-1} > 1$

$$\Rightarrow \frac{\left(\sum_{i=1}^{n} p_i\right)^{\frac{1}{\alpha}-1}}{1-\alpha} < \frac{\sum_{i=1}^{n} p_i^{\alpha} - 1}{1-\alpha} \Rightarrow B(P) < H(P) \tag{44}$$

and, $\alpha < 1 \Rightarrow \sum_{i=1}^{n} p_i^{\alpha} > 1 \Rightarrow \left(\sum_{i=1}^{n} p_i^{\alpha}\right)^{\frac{1}{\alpha}-1} > 1$

$$\Rightarrow \frac{\sum_{i=1}^{n} p_i^{\alpha} - 1}{1-\alpha} > \frac{\sum_{i=1}^{n} p_i^{\alpha} - 1}{1-\alpha} \Rightarrow B(P) > H(P) \tag{45}$$

(f) Let $\beta > \frac{1}{\alpha}$, then

$$\alpha > 1 \Rightarrow \sum_{i=1}^{n} p_i^{\alpha} = 1 \Rightarrow \left(\sum_{i=1}^{n} p_i^{\alpha}\right)^{\beta-\frac{1}{\alpha}} < 1 \Rightarrow \left(\sum_{i=1}^{n} p_i^{\alpha}\right)^{\beta} < \left(\sum_{i=1}^{n} p_i^{\alpha}\right)^{\frac{1}{\alpha}}$$

$$\Rightarrow \frac{\left(\sum_{i=1}^{n} p_i^{\alpha}\right)^{\beta} - 1}{1-\alpha} > \frac{\left(\sum_{i=1}^{n} p_i^{\alpha}\right)^{\frac{1}{\alpha}} - 1}{1-\alpha} \Rightarrow \beta M(P) > B(P) \tag{46}$$

$$\alpha < 1 \Rightarrow \sum_{i=1}^{n} p_i^{\alpha} > 1 \Rightarrow \left(\sum_{i=1}^{n} p_i^{\alpha}\right)^{\beta-\frac{1}{\alpha}} > 1 \Rightarrow \left(\sum_{i=1}^{n} p_i^{\alpha}\right)^{\beta} > \left(\sum_{i=1}^{n} p_i^{\alpha}\right)^{\frac{1}{\alpha}}$$

$$\Rightarrow \frac{\left(\sum_{i=1}^{n} p_i^{\alpha}\right)^{\beta} - 1}{1-\alpha} > \frac{\left(\sum_{i=1}^{n} p_i^{\alpha}\right)^{\frac{1}{\alpha}} - 1}{1-\alpha} \Rightarrow \beta M(P) > B(P) \tag{47}$$

Thus when $\beta > \frac{1}{\alpha}$ then whether $\alpha >$ or < 1 $\beta M(P) > B(P)$ (48)

Similarly, when $\beta < \frac{1}{\alpha}$, then whether $\alpha > 1$ or < 1, $\beta M(P) < B(P)$ (49)

(*g*) combining the results of (c) and (e)

$$\alpha < 1 \quad \Rightarrow \quad R(P) \le H(P) \le B(P) \tag{50}$$

$$\alpha > 1 \quad \Rightarrow \quad R(P) \ge H(P) \ge B(P) \tag{51}$$

$$\alpha = 1 \quad \Rightarrow \quad R(P) = H(P) \ge B(P) \tag{52}$$

Also when $\alpha = 0, \quad R(P) = \ln n, \quad H(P) = (n-1), \quad R(P) = \infty$ (53)

when $\alpha = \infty, \quad R(P) = -\ln p_{\max}, \quad H(P) = 0, \quad B(P) = 0$ (54)

Using the results, we get the graphs of the type given in Figure 15.1

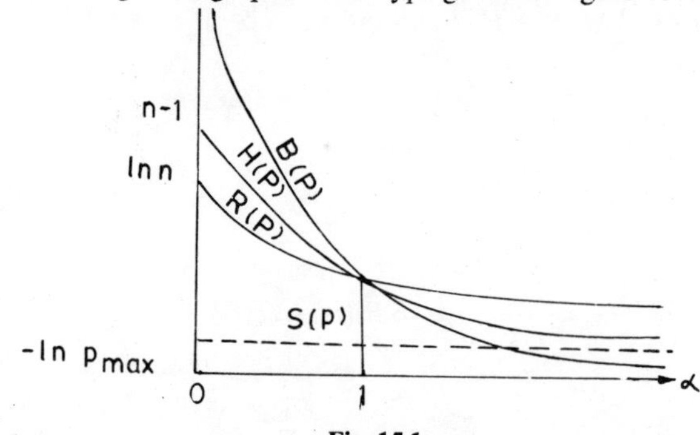

Fig. 15.1

The graph of *B* (*P*) differs from the graph of *H* α (*P*) given by Nayak [15], since

$$B(P) = \frac{\left(\sum_{i=1}^{n} p_i^{\alpha}\right)^{\frac{1}{\alpha}} - 1}{1-\alpha}, \quad H_\alpha(P) = \frac{\left(\sum_{i=1}^{n} p_i^{\alpha}\right)^{\frac{1}{\alpha}} - 1}{2^{\frac{1}{\alpha}-1} - 1} \tag{55}$$

so that,

$$\underset{\alpha \to 1}{Lt} \quad B(P) = -\sum_{i=1}^{n} p_i \ln p_i = S(P)$$

$$\underset{\alpha \to 1}{Lt} \quad H_\alpha(P) = \left(-\sum_{i=1}^{n} p_i \ln p_i\right) \underset{\alpha \to 1}{Lt} \frac{1-\alpha}{2^{\frac{1}{\alpha}-1} - 1} = -\frac{\sum_{i=1}^{n} p_i \ln p_i}{\ln 2} = -\sum_{i=1}^{n} p_i \log_2 p_i$$

$$= \frac{S(P)}{\ln 2} \tag{56}$$

(h) Monotonicity of *M* (*P*) with respect to β

Let $\sum_{i=1}^{n} p_i^{\alpha} = a$ so that $a \lessgtr 1$ when $\alpha \gtrless 1$,

$$\text{then} \quad M(P) = \frac{1-a^{\beta}}{(\alpha-1)\beta} \tag{57}$$

so that,

$$\beta(\alpha-1)\frac{d}{d\beta}M(P) = -\beta a^{\beta}\ln a + a^{\beta} - 1 = g(B) \quad \text{(say)} \tag{58}$$

then,

$$g'(\beta) = -\beta a^{\beta}(\ln a)^2, \quad g(0) = 0, \quad g'(0) = 0 \tag{59}$$

so that the graph of g (β) is as given in Figure 15.2,

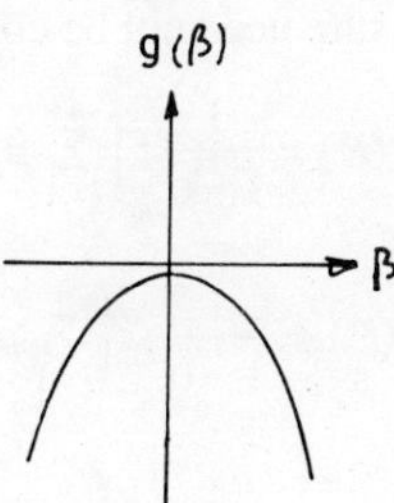

Fig. 15.2

so that $$(\alpha-1)\frac{d}{d\beta}M(P) < 0 \tag{60}$$

Thus if $\alpha > 1$ then for fixed α, P M (P) decreases with β
and if $\alpha < 1$ then for fixed α, P M (P) increases with β

Again,

$$\frac{d}{d\beta}(\beta M(P)) = \frac{d}{d\beta}\left(\frac{\left(\sum_{i=1}^{n} p_i^{\alpha}\right)^{\beta} - 1}{1-\alpha}\right) = \frac{1}{1-\alpha}\left(\sum_{i=1}^{n} p_i^{\alpha}\right)^{\beta}\ln\left(\sum_{i=1}^{n} p_i^{\alpha}\right) = \frac{1}{1-\alpha}a^{\beta}\ln a \tag{61}$$

If $\alpha > 1,\ a < 1$ and if $\alpha < 1\ \ a > 1$ so that from (61),

$$\frac{d}{d\beta}\beta(M(P)) > 0 \tag{62}$$

so that for fixed α and β, $\beta\,M$ (P) always increases with β.

15.8 A GENERAL MEASURE EQUIVALENT TO THE GIVEN CLASS OF MEASURES

Consider the measure,

$$F(P) = \frac{1}{1-\alpha}\left[f\left(\sum_{i=1}^{n} p_i^{\alpha}\right) - f(1)\right] \tag{63}$$

where $f(x)$ is a positive monotonic increasing function of x, so that

$$F(P) = \frac{1}{1-\alpha}[f((1-\alpha)H(P)+1) - f(1)] \tag{64}$$

and,

$$\frac{d(F(P))}{d(H(P))} = f'((1-\alpha)\,H(P)+1) = f'\left(\sum_{i=1}^{n} p_i^{\alpha}\right) > 0 \tag{65}$$

so that $F(P)$ and $H(P)$ increase or decrease together and attain the maximum value for the same probability distribution.

When $f(x) = x$, we get $H(P)$, when $f(x) = \ln x$, we get $R(P)$ when $f(x) = x^{1/\alpha}$, we get $B(P)$ and when $f(x) = x^{\beta}/\beta$, we get $M(P)$. We can however now take any monotonic increasing function of x which is positive when x is positive and get a measure equivalent to the four given measures.

However the general measure need not be concave. In fact,

$$\frac{\partial}{\partial p_i} F(P) = \frac{1}{1-\alpha} f'\left(\sum_{i=1}^{n} p_i^{\alpha}\right) \alpha p_i^{\alpha-1}$$

$$\frac{\partial^2}{\partial p_i^2} F(P) = \frac{1}{1-\alpha} f''\left(\sum_{i=1}^{n} p_i^{\alpha}\right) \alpha^2 p_i^{2\alpha-1} - f'\left(\sum_{i=1}^{n} p_i^{\alpha}\right) \alpha\, p_i^{\alpha-2}$$

$$\frac{\partial^2}{\partial p_i \partial p_j} F(P) = \frac{1}{1-\alpha} f''\left(\sum_{i=1}^{n} p_i^{\alpha}\right) \alpha^2 p_i^{\alpha-1} p_j^{\alpha-1}$$

so that

$$\sum_{j=1}^{n}\sum_{i=1}^{n} \frac{\partial^2}{\partial p_i \partial p_j} F(P) x_i x_j = \frac{x^2}{1-\alpha} f''\left(\sum_{i=1}^{n} p_i^{\alpha}\right)\left(\sum_{i=1}^{n} p_i^{\alpha-1} x_i\right)^2$$

$$-\alpha f'\left(\sum_{i=1}^{n} p_i^{\alpha}\right) \sum_{i=1}^{n} p_i^{\alpha-2} x_i^2 \tag{66}$$

The general measure would be concave if either,

(i) $\alpha < 1$ $f(x)$ is a concave monotonic increasing function.
(ii) $\alpha > 1$ $f(x)$ is a convex monotonic increasing function.

These conditions are sufficient to ensure concavity, these are not necessary

When $f(x) = x$, these show that Harvda-Charvat measure is concave whether $\alpha \gtrless 1$.

When $f(x) = \ln x$ these show that Renyi's measure is concave when $\alpha < 1$
When $f(x) = x^{1/\alpha}$, these conditions are not satisfied, yet $B(P)$ is concave.
When $f(x) = x^{\beta}/\beta$ these conditions are satisfied if $\beta < 1, \alpha < 1$ or $\beta > 1, \alpha > 1$

and in both these cases, $\alpha(\beta - 1)/(1 - \alpha)$ is negative and $\alpha < 1$ and thus $M(P)$ is concave.

These conditions show that we can get an infinite number of concave measures equivalent to these four given measures. Thus we can consider the trignometric measure

$$\frac{1}{1-\alpha}\left[\tan^{-1} \sum_{i=1}^{n} p_i^{\alpha} - \pi/4\right], \quad \alpha < 1 \tag{67}$$

This is concave for all probability distributions and is equivalent to the four given measure and approaches half the value of Shannon entropy as $\alpha \to 1-0$.

Since all the measures equivalent to these four measures lead to the same *MEP*, we can use any one of these. In practice, we should use the simplest which appears to be the Havrda-Charvat measure.

15.9 OTHER EQUIVALENT CLASSES OF MEASURES OF ENTROPY

From our definition of equivalent measures, it follows that equivalence has the following properties.

(i) it is reflexive, i.e. every measure is equivalent to itself.

(ii) it is symmetric, i.e. if $X(P)$ is equivalent to $Y(P)$, then $Y(P)$ is equivalent to $X(P)$.

(iii) It is transitive, i.e. if $X(P)$ is equivalent to $Y(P)$ and $Y(P)$ is equivalent to $Z(P)$ than $X(P)$ is equivalent to $Z(P)$.

Thus our definition gives an equivalence relation and this result gives a set of equivalence classes of measures of entropy since all measures belonging to a class are equivalent to one another and no member of one class can be equivalent to a member of another class.

We have discussed above the equivalence class of measures of entropy which are equivalent to Havrada-Charvat measure of entropy.

Similarly, we can have an equivalent class of measures of entropy which are equivalent to Shannon entropy. These can include

$$-\prod p_i \ln p_i, \quad \prod_{i=1}^{n} p_i^{-p_i}, \quad \left(-\sum_{i=1}^{n} p_i \ln p_i\right)^2, \quad \ln\left(-\sum_{i=1}^{n} p_i \ln p_i\right) \tag{68}$$

and in general any measure,

$$f\left(-\sum_{i=1}^{n} p_i \ln p_i\right) \tag{69}$$

where $f(x)$ is positive and monotonic increasing when $x \geq 0$. Of course, in practice, we use the simplest of these, viz. Shannon measure. Similarly,

$$-\sum_{i=1}^{n} p_i \ln p_i - \sum_{i=1}^{n} (1-p_i) \ln (1-p_i) \quad \text{and} \quad \prod_{i=1}^{n} p_i^{-p_i} (1-p_i)^{(1-p_i)} \tag{70}$$

are equivalent, but we use the former in preference to the latter.

In our first class of equivalent measures, Havrda-Charvat measure is most convenient, but it is non-additive, while Renyi' measure is less convenient, but it is additive. Some persons who consider additivity as desirable may like to use Renyi entropy, but for entropy maximization problems, additivity property makes no difference.

15.10 EQUIVALENT MEASURES OF DIRECTED DIVERGENCE

According to Kullback's [12] principle of minimum cross-entropy, given constraints (6) and a priori probability distribution Q, we shall choose that probability distribution P which minimizes Kullback-Leibler [13] measure of directed divergence,

$$D(P,Q)=\sum_{i=1}^{n} p_i \ln\frac{p_i}{q_i} \tag{71}$$

However according to the generalised principle of minimum cross-entropy [11]' we can choose P to minimize any appropriate measures of cross-entropy subject to (6). The measures of cross-entropy corresponding to the five measures of entropy given by (2), (3), (4), (5) and (6) are:

$$\frac{1}{\alpha-1}\left(\sum_{i=1}^{n} p_i^{\alpha} q^{1-\alpha}-1\right), \quad \alpha\neq 1 \quad \alpha>0 \tag{72}$$

$$\frac{1}{\alpha-1}\left(\ln\sum_{i=1}^{n} p_i^{\alpha} q^{1-\alpha}\right) \quad \alpha\neq 1, \quad \alpha>0 \tag{73}$$

$$\frac{1}{\alpha-1}\left(\left(\sum_{i=1}^{n} p_i^{\alpha} q^{1-\alpha}\right)^{\frac{1}{\alpha}}-1\right) \quad \alpha\neq 1 \quad \alpha>0 \tag{74}$$

$$\frac{1}{(\alpha-1)\beta}\left[\left(\sum_{i=1}^{n} p_i^{\alpha} q^{1-\alpha}\right)^{\beta}-1\right) \quad \alpha\neq 1 \quad \alpha>0 \quad \beta\neq 0 \tag{75}$$

$$\frac{1}{\alpha-1}\left[f\left(\sum_{i=1}^{n} p_i^{\alpha} q^{1-\alpha}\right)-f(1)\right], \quad f(x) \text{ is monotonic} \tag{76}$$

All these are equivalent measures since these increase or decrease together and lead to the same probability distribution when minimized subject to (6). These five given equivalence class of measures of directed divergence. Similarly,

$$\sum_{i=1}^{n} p_i \ln\frac{p_i}{q_i}, \quad \prod_{i=1}^{n}\left(\frac{p_i}{q_i}\right)^{p_i}, \quad \left(\sum_{i=1}^{n} p_i \ln\frac{p_i}{q_i}\right)^2, \quad f\left(\sum_{i=1}^{n} p_i \ln\frac{p_i}{q_i}\right) \tag{77}$$

give another equivalence class of measures of cross-entropy.

Since, entropy measure of dependence [6] are also measures of directed divergence of cross-entropy of the joint probability distribution from the product probability distribution of marginal distributions, equivalent measures of cross-entropy lead to equivalent measures of dependence.

15.11 A CASE WHERE USE OF EQUIVALENT MEASURES MAY GIVE DIFFERENT RESULTS

Consider the problem of thresholding in image reconstruction. Kapur et. al [8] proposed that we should choose that threshold value which maximizes sum of the entropies due to (*i*), pixels with gray levels $\leq s$ and (*ii*) pixels with gray levels $> s$. If $g_1(s)$, $g_2(s)$ denote these entropies, we should choose s to maximize $g_1(s)+g_2(s)$. If we use another equivalent measures of entropy, we would have to maximize $f(g_1(s)) + (f(g_2(s))$. The first will be maximum when $g'_1(s)+g'_2(s)=0$ and the second will be maximum when $f'(g_1(s)\,g'_1(s)+f'_1(g_2(s)\,g'_2(s)=0$ and in general these two will give different threshold values. It may be noted that here we are not using the principle or maximum entropy.

15.12 SUMMARY OF RESULTS

(i) The four measures of entropy proposed independently by Havrda-Charvat, Renyi, Behara-Chawla, Sharma-Mittal are equivalent, i.e. when these are maximized subject to some given constraints, these lead to the same maximum entropy probability distribution.

(ii) The measure of entropy $(1-\alpha)^{-1}\left[f\left(\sum_{i=1}^{n} p_i^{\alpha}\right)-f(1)\right]$ where $f(x)$ is a monotonic increasing function of x is a general measure which includes these four measures as special cases and includes an infinity of other measures equivalent to these, an infinity of which can be concave. In fact the measure will be' concave if $\alpha < 1$ and $f(x)$ is concave or $\alpha > 1$ and $f(x)$ is convex.

(iii) $H(P)$ and $B(P)$ are always concave functions, $R(P)$ is concave if $\alpha < 1$ and $M(P)$ is concave if $\alpha(\beta-1)/(1-\alpha) < 1$

(iv) $\alpha < 1 \Rightarrow R(P) \le H(P) \le B(P)$

$\alpha > 1 \Rightarrow R(P) \ge H(P) \ge B(P)$

$\beta > \frac{1}{\alpha} \Rightarrow \beta M(P) > H(P)$, whether $\alpha > 1$ or < 1

$\beta < \frac{1}{\alpha} \Rightarrow \beta M(P) < H(P)$, whether $\alpha > 1$ or < 1

(v) For fixed α, P, $\alpha > 1 \Rightarrow M(P)$ decreases with β, $\alpha < 1 \Rightarrow M(P)$ increases.

(vi) All measure of entropy can be divided into disjoint equivalence class such that all members of a class are equivalent to every member of the same class and no memebr of a class is equivalent to a member of another class.

(vii) The measures of cross-entropy due to Havrda-Charvat, Renyi and those corresponding to the measure of entropy due to Behara and Charvat and Sharma-Mittal are equivalent to the measure,

$$\frac{1}{\alpha-1}\left[f\left(\sum_{i=1}^{n} p_i^{\alpha} q^{1-\alpha}\right)-f(1)\right]$$

where $f(x)$ is a monotonic increasing function including all these as special cases and an infinity of others, an inifinity of which can be convex functions of P.

(viii) In particular,

$$\frac{1}{1-\alpha}\left[\tan^{-1}\left(\sum_{i=1}^{n} p_i^{\alpha}\right)-\frac{\pi}{4}\right],\quad \alpha < 1 \qquad \frac{1}{\alpha-1}\left[\tan^{-1}\left(\sum_{i=1}^{n} p_i^{\alpha} q_i^{1-\alpha}\right)-\frac{\pi}{4}\right],\quad \alpha > 1 \tag{78}$$

are trignometric measures of entropy and directed divergence.

(ix) When the object is not to find maximizing or minimizing probability distribution but it is to find maximizing or minimizing parameter values, equivalent measures may not give the same results.

REFERENCES

1. M. Behara and J.S Chawla (1974), ''Generalised Gamma Entropy'' Selecta Statistica Canadiana, **2**, 15-38.
2. J.H. Havrda and F.C. Charvat (1967), ''Quantification Methods of Classification Processes , concept of structural α entropy'' Kybernetika, vol. 3, pp. 30-35.
3. E.T. Jaynes (1957), ''Information Theory and Statistical Mechanics, Physical Reviews, vol. 106, pp. 620-630.
4. J.N. Kapur (1986), ''On monotonocity property of various measures of entropy and directed divergence'' Jonrn. Math. Phy. Sci. 20(3), 235-253.
5. J.N. Kapur (1987), ''Monotonocity and Concavity of some parametric measures of entropy, directed divergence and releated functions.'' Tamkang Journal of Mathematics, 18 (3), 21-40.
6. J.N. Kapur (1990), Information Theoretic-measures of stochastic dependence'' Bull Math. Ind 22, 43-58.
7. J.N. Kapur (1990), ''Maximum Entropy Models in Science and Engineering'', Wiley Eastern, New Delhi and John Wiley, New York.
8. J.N. Kapur, P.K. Sahaio and K.C. Wong (1985)'', A new method of gray level picture thresholding using entropy of the histogram'' Computer Vision, Graphics and Image Processing, vol. 29, pp. 273-288.
9. J.N. Kapur and H.K. Kesavan (1987), ''Generalised Maximum Entropy Principle (with Applications)'', Sand Ford Educational Press, University of Waterloo, Canada.
10. J.N. Kapur and H.K. Kesavan (1992), '' Entropy Optimization Principles with Applications, Academic Press, New York.
11. H.K. Kesavan and J.N. Kapur (1989), The generalised maximum entropy principle, IEEE Sys Man Cyb. 19(9) 1042-1052.
12. S. Kullback (1959), ''Information Theory and Statistics'', John Wiley, New york.
13. S. Kullback and R.A. Leibler (1951), ''On Information and Sufficienty'' Ann Math. Stat. vol. 22, pp. 79-86.
14. T.K. Nayak (1985), ''On Diversity Measures based on Entropy Function'', Communications in Statistics 14(11) 223.
15. A. Renyi (1961), on Measures of Entropy and Information'' Proc. First. Berkeley Symp. Stat. 1, 547-561
16. B.D. Sharma and D. P. Mittal (1975), New Non additive Measures of Entropy for Discreate Probability Distributions,'' J. Math Sci. vol 10, pp 28-40.
17. C.E. Shannon (1968), ''A Mathematical Theory of Communications'' Bell System Tech Journ. 27, pp. 379-423, 623-659.

16

ENTROPY AND DIRECTED DIVERGENCE MEASURES WHICH LEAD TO POSITIVE PROBABILITIES ON OPTIMISATION SUBJECT TO LINEAR CONSTRAINTS

[More than half a dozen measures of entropy are proposed which when maximized subject to linear moment constraints always lead to positive maximizing probabilities. Corresponding measures of directed divergence are also given.]

16.1 INTRODUCTION

Let $P = (p_1, p_2, \ldots, p_n)$ be a probability distribution about which only partial information is available in the form of linear moment constraints.

$$\sum_{i=1}^{n} p_i = 1, \sum_{i=1}^{n} p_i\, g_{ri} = a_r, r = 1, 2, ..., m, \tag{1}$$

where $m + 1 \ll n$ so that (1) cannot, in general determine $(p_1, p_2, \ldots, p_n)$ uniquely. In other words, there may, in general, be an infinity of probability distributions satisfying (1). According to the principle of generalised maximum entropy [5, 10, 11], we should choose that one of these which has maximum value for a measure of uncertainty or entropy of the form $H(p_1, p_2, \ldots, p_n)$. We shall like $H((p_1, p_2, \ldots, p_n)$ to satisfy the following conditions:

(*i*) It should be continuous and permutationally symmetric, i.e. it should not change when $p_1, p_2, \ldots, p_n$ are interchanged among themselves.

(*ii*) It should be a concave or pseudo-concave function of $(p_1, p_2, \ldots, p_n)$, so that when it is maximized subject to (1), its local maximum is a global maximum.

(*iii*) when it is maximized subject to linear constraints, the maximising probabilities should be positive (at least non-negative) since negative probabilites have no physical meaning.

(*iv*) It should be a monotonic decreasing function of some 'distance' from the uniform distribution, since the uniform distribution represents the distribution of maximum uncertainty.

We are not interested in the measure being positive, since in maximization, we are interested in relative values of entropy rather than in the absolute values. We are not interested in additivity, recursivity, sub-additivity etc. because these properties are not of much relevance to the maximization problem.

We have not listed its being maximum for the uniform distribution separately since it is ensured by (ii) and (iv). However, by (iii) we will like each maximizing probability to be > 0, because if it is zero, we can as well leave the corresponding outcome out of our consideration.

So far the measure which has been used in such an overwhelming number of cases that it is considered by many as the only measure of entropy is Shannon's [15] measure,

$$S(P) = -\sum_{i=1}^{n} p_i \ln p_i \tag{2}$$

It satisfies all the four properties given above. Many other proposed measures of entropy like those of Renyi's measure $R_\alpha(P)$ [14], Havrada and Charvat's measure $H_\alpha(P)$ [4] Vajda's entropy [16] given respectively by

$$R_\alpha(P) = \frac{1}{1-\alpha} \ln \sum_{i=1}^{n} p_i^\alpha, \quad H_\alpha(P) = \frac{1}{1-\alpha}\left(\sum_{i=1}^{n} p_i^\alpha - 1\right), \quad \alpha > 0 \quad \alpha \neq 1$$

$$V(P) = \left(1 - \sum_{i=1}^{n} p_i^2\right) \tag{3}$$

do not satisfy all these conditions. In particular, these do not satisfy condition (iii). This was used as a strong argument against these, since if we use these, we have to explicitly use the non-negatively constraints,

$$p_1 \geq 0, p_2 \geq 0, \ldots, p_n \geq 0 \tag{4}$$

and this will make the mathematical problem rather complicated. As such, we have to search for measures which satisfy all the four conditions and in particular those which satisfy (iii), i.e. which are such that when these are maximized subject to linear constraints, the maximizing probabilities are all > 0.

16.2 DERIVATION OF SHANNON'S MEASURE SATISFYING THE ABOVE CONDITIONS

Let $\phi(x)$ be twice-differentiable convex function with $\phi(1) = 0$ then Csiszer's [3] measure of directed divergence of the probability distribution P from another distribution $Q = (q_1, q_2, \ldots, q_n)$ is given by

$$D(P:Q) = \sum_{i=1}^{n} q_i \, \phi(p_i/q_i) \tag{5}$$

If U is the uniform distribution, we get

$$D(P:U) = \sum_{i=1}^{n} \frac{1}{n} \phi(np_i) \tag{6}$$

A monotonic decreasing function of this is $-\sum_{i=1}^{n} \phi(np_i)$. Thus, the function,

$$H(P) = -\sum_{i=1}^{n} \phi(p_i) \tag{7}$$

satisfies the condition (i), (ii) and (iv). We have now to consider how we can choose the convex function $\phi(p_i)$ so that the condition (iii) is also satisfied.

Maximizing (7) subject to (1), we get

$$\phi'(p_i) = \lambda_0 + \lambda_1 g_{1i} + \lambda_2 g_{2i} + \dots + \lambda_m g_{mi} \tag{8}$$

We have to solve (8) for p_i and we would like the value of p_i to be positive,

We get
$$p_i = \phi'^{-1}(\lambda_0 + \lambda_1 g_{1i} + \dots + \lambda_m g_{mi}) \tag{9}$$

We would also like $\phi'(x)$ to be one-one.

We therefore consider such functions which are defined for only positive values of the arguments.

One function which satisfies these conditions is

$$\phi'(p_i) = \ln p_i \tag{10}$$

This is defined for only positive values of p_i. For every positive value of p_i we get only one value of $\ln p_i$ and for every value $\ln p_i$, we get only one value p_i. This gives

$$\phi(p_i) = p_i \ln p_i - p_i \tag{11}$$

and this gives us Shannon's measure of entropy [15].

16.3 SOME OTHER MEASURES SATISFYING THE ABOVE CONDITIONS

(i) We consider the function,

$$\phi'(p_i) = \frac{(\ln p_i)^m}{p_i} \tag{12}$$

This is defined only for positive value of p_i. For every positive value of p_i ($0 \le p_i \le 1$) there is a unique value of $\phi'(p_i)$ and for every value of $\phi'(p_i)$ (positive if m is even and negative if m is odd) there is a unique value of p_i. This is easily seen to be true since the graphs of

$$y = p_i \quad \text{and} \quad y = (\ln p_i)^m \tag{13}$$

intersect in one point only whether m is odd or even (Figures 16.1 and 16.2).

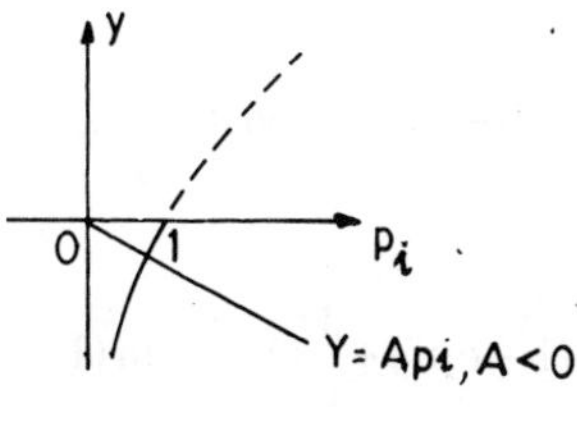

Fig. 16.1

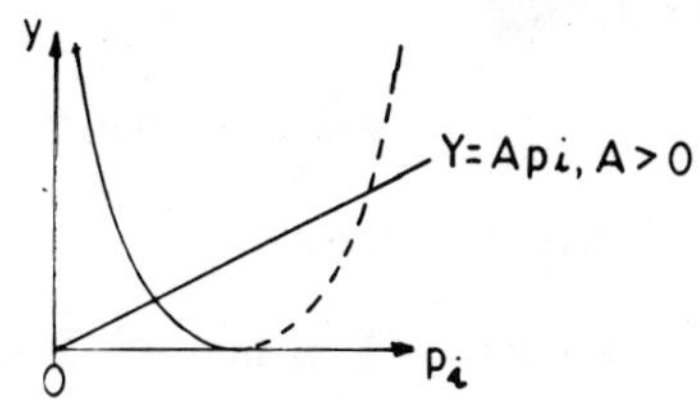

Fig. 16.2

Expression (12) gives us the measure,

$$B \sum_{i=1}^{n} (\ln p_i)^{m+1} \tag{14}$$

Let
$$\psi(p_i) = B\,(\ln p_i)^{m+1}, \tag{15}$$

so that
$$\psi'(p_i) = B\,(m+1)\,(\ln p_i)^m \frac{1}{p_i}$$

$$\psi''(p_i) = B(m+1)\left[m\,(\ln p_i)^{m-1}\frac{1}{p_i^2} - (\ln p_i)^m \frac{1}{p_i^2}\right]$$

$$= B\frac{m+1}{p_i^2}(\ln p_i)^{m-1}\,[m - \ln p_i] \tag{16}$$

We choose B to be < 0 when m is odd and to be > 0 when m is even, so that our measure of entropy is concave. This gives,

$$H_2(P) = -\sum_{i=1}^{n} (\ln p_i)^{m+1}, \quad m \text{ odd}$$

or
$$H_3(P) = \sum_{i=1}^{n} (\ln p_i)^{m+1}, \quad m \text{ even} \tag{17}$$

In either case the measure is negative but this negativity does not matter, since we are interested in positivity of resulting probabilities, rather than in the positivity of measures of entropy.

(ii) When $m = 0$, it gives Burg's measure of entropy, which we shall discuss further below. For other vlaues of m, it gives the measures

$$-\sum_{i=1}^{n} (\ln p_i)^2, \quad \sum_{i=1}^{n} (\ln p_i)^3, \quad -\sum_{i=1}^{n} (\ln p_i)^4, \ldots \tag{18}$$

The maximum value of our measure is
$$n\,(\ln n)^{m+1} \tag{19}$$
which always increases with n. This is a desirable property of these measures.

(ii) Another function which satisfies the two conditions of being defined only for positive values and being one-one is

$$\phi'(p_i) = \ln\frac{p_i}{1 - c\,p_i}, \quad c \le 1 \tag{20}$$

This gives,

$$\phi(p_i) = p_i \ln p_i + \frac{1}{c}(1 - c\,p_i)\ln(1 - c\,p_i) + \text{const} \tag{21}$$

which gives the measure,

$$H_4(P) = -\sum_{i=1}^{n} p_i \ln p_i - \frac{1}{C}\sum_{i=1}^{n}(1 - c\,p_i)\ln(1 - cpi) + \frac{1}{c}(1-c)\ln(1-c) \tag{22}$$

This measure was originally proposed by Kapur [8]. As a special case of it when $c = 1$, if

$$H_5(P) = -\sum_{i=1}^{n} p_i \ln p_i - \sum_{i=1}^{n}(1 - p_i)\ln(1 - p_i) \tag{23}$$

has been found specially useful [7].

(iii) $m = 0$ in (14) leads to Burg's entropy

$$H_0(P) = \sum_{i=1}^{n} \ln p_i \tag{24}$$

In this case $\phi'(p_i) = 1/_{p_i}$ does not imply that p_i is always positive. However, if we maximize (24) subject to (1), we get

$$\frac{1}{p_i} = \lambda_0 + \lambda_1 g_{1i} + \ldots + \lambda_m g_{mi}, \tag{25}$$

where $\lambda_0, \lambda_1, \ldots, \lambda_m$ depend on the values of the parameter $a_1, a_2, \ldots, a_m$. If we take,

$$a_r = \sum_{i=1}^{n} \frac{1}{n} g_{ri}, \quad r = 1, 2, \ldots m. \tag{26}$$

these will give the uniform distribution where p_i's are positive. As we change a_r's, p_i's will change continuously, but if these have to become negative, they have to go through zero values. If p_i has to pass through zero value, some of the λ's will have to change from $+\infty$ to $-\infty$ and this cannot happen for continuously changing a_r's. As such with (25), p_i's will always be positive and there will be one-one correspondence.

(iv) The same argument applies if we take

$$H_7(P) = -\sum_{i=1}^{n} \frac{1}{p_i^m} \tag{27}$$

or even

$$H_8(P) = \sum_{i=1}^{n} p_i{}^{1}/_{2} \tag{28}$$

which give positive probabilities on maximization subject to known constraints. However, (28) gives

$$p_i = \frac{1}{(\lambda_0 + \lambda_1 g_{1i} + \ldots + \lambda_m g_{mi})^2} \tag{29}$$

This will make p_i's positive but can give more than one solution and care will have to be used in this case.

16.4 SOME DERIVED MEASURES

(i) Consider the measure,

$$H_9(P) = -\sum_{i=1}^{n} p_i \ln p_i - \sum_{j=1}^{k} \frac{\mu_j}{\varepsilon_j} \sum_{i=1}^{n} (1 - \varepsilon_j p_i) \ln (1 - \varepsilon_j p_i) \tag{30}$$

where

$$\mu_1 \geq 0, \mu_2 \geq 0, \ldots, \mu_k \geq 0, \quad \sum_{j=1}^{k} \mu_j = 1, \qquad 0 \leq \varepsilon_j \leq 1 \quad \forall j \tag{31}$$

Now, $$\frac{\partial}{\partial p_i} H_9(P) = -(1+\ln p_i) + \sum_{j=1}^{k} \mu_j \,(1+\ln(1-\varepsilon_j p_j))$$

$$= \sum_{j=1}^{k} \mu_j \ln\left(\frac{1-\varepsilon_j p_i}{p_i}\right) \tag{32}$$

$$\frac{\partial^2 H_9(P)}{\partial p_i^2} = -\sum_{j=1}^{k} \frac{\mu_j \varepsilon_j}{1-\varepsilon_j p_i} - \frac{1}{p_i} \tag{33}$$

$$\frac{\partial^2 H_9(P)}{\partial p_i \,\partial p_i} = 0 \tag{34}$$

Thus $H_9(P)$ is a concave function of $(p_1, p_2, \ldots, p_n)$ and when it is maximized subject to linear constraints (1), we get

$$\prod_{j=1}^{k} \frac{(1-\varepsilon_j p_i)^{\mu_j}}{p_i^{\mu_j}} = \exp(\lambda_1 + \lambda_1 g_{1i} + \ldots + \lambda_m g_{mi}) > 0 \tag{35}$$

Negative p_i's can not satisfy (35). As such all maximizing probabilities have to be positive.

(ii) Similarly the measure,

$$H_{10}(P) = -\sum_{i=1}^{n} p_i \ln p_i + a \sum_{i=1}^{n} \ln p_i\,, \quad a > 0 \tag{36}$$

is a concave function which when maximized subject to (1) gives

$$\ln p_i + \frac{a}{p_i} = \lambda_0 + \lambda_1 g_{1i} + \ldots + \lambda_m g_{mi} \tag{37}$$

which can be satisfied by only positive values of p_i's.

16.5 CORRESPONDING MEASURES OF DIRECTED DIVERGENCE

Let $\phi(x)$ be a convex twice differentiable function of x for which $\phi(1) = 0$, $\phi(0) = 0$ and let $-\sum_{i=1}^{n} \phi(p_i)$ be a measure of entropy, then the corresponding measure of directed divergence is $\sum_{i=1}^{n} q_i \,\phi(p_i/q_i)$, then from (II),

$$D_1(P:Q) = \sum_{i=1}^{n} q_i \,p_i/q_i \ln p_i/q_i = \sum_{i=1}^{n} p_i \ln p_i/q_i \tag{38}$$

which is the measure due to Kullback and Leibler [12]. Using (21) we get the measure,

$$D_4(P:Q) = \sum_{i=1}^{n} p_i \ln p_i/q_i + \frac{1}{c} \sum_{i=1}^{n} (q_i - c p_i) \ln \frac{q_i - c p_i}{q_i}$$

$$-\frac{1}{c}(1-c)\ln(1-c), \tag{39}$$

which is a measure due to Kapur [6, 9]. Corresponding to (23) we get,

$$D_5(P:Q) = \sum_{i=1}^{n} p_i \ln p_i/q_i + \sum_{i=1}^{n} (q_i - p_i) \ln \frac{q_i - p_i}{q_i} \tag{40}$$

Corresponding to (24), we have the measure,

$$D_6(P:Q) = \sum_{i=1}^{n} \left(\frac{p_i}{q_i} - \ln \frac{p_i}{q_i} + 1 \right) \quad (40)$$

since, $$D_6(P:U) = \sum_{i=1}^{n} \left(\frac{p_i}{1/n} - \ln \frac{p_i}{1/n} + 1 \right) = -n \ln n + 2n - \sum_{i=1}^{n} \ln p_i \tag{42}$$

and $H_6(P)$ is monotonic decreasing function of $D_6(P:U)$.
Corresponding to (27), we get

$$D_7(P:Q) = \sum_{i=1}^{n} q_i \left(\frac{p_i}{q_i} \right)^m - 1 = \sum_{i=1}^{n} p_i^{-m} q_i^{-m+1} - 1 \tag{43}$$

This correspondes to Havrda-Charvat [4] measure of directed divergence and can be used where none of the q_i's vanish. Corresponding to (20), we get

$$D_8(P:Q) = \sum_{i=1}^{n} (\sqrt{p_i} - \sqrt{q_i})^2 \tag{44}$$

which correspond to Bhattacharya's [1] measure .
Corresponding to (30), we get

$$D_9(P:Q) = \sum_{i=1}^{n} p_i \ln p_i/q_i + \sum_{j=1}^{k} \frac{\mu_i}{a_j} \sum_{i=1}^{n} (q_i - a_j p_i) \ln \frac{(q_i - a_j p_i)}{q_i}$$

$$- \sum_{j=1}^{k} \frac{\mu_j}{a_j} (1 - a_j) \ln (1 - a_j) \tag{45}$$

Corresponding to (36) we get

$$D_{10}(P:Q) = \sum_{i=1}^{n} p_i \ln \frac{p_i}{q_i} + a \sum_{i=1}^{n} q_i \ln \frac{q_i}{p_i} \tag{46}$$

which for $a = 1$, reduces to Kullback's [13] symmatric divergence measure,

$$J(P:Q) = D(P:Q) + D(Q:P) \tag{47}$$

All these measures of directed divergence, when minimized subject to linear constraints will give rise to positive probabilities.

REFERENCES

1. A. Bhattacharya (1943), "On a measure of divergence between two statistical populations defined by their probability distributions" Bull Cal. Math Soc. Vol 35, pp. 99–109.
2. J.P. Burg (1972), "The Relationship between Maximum Entropy Spectra and Maximum Likelihood Spectra". In Modern Spectral Analysis ed. D.G. Childers, pp. 130–131 MSA.
3. I. Csiszer (1972), "A class of measures of Information of observation channels" Persodica Mathematica Hungarica vol 2, 191–202.

4. J.H. Havrda and F. Charvat (1967), "Quantification Methods of Classification Processes" concept of structural α entropy" Kybernatica vol 3 pp. 30–35.
5. E.T. Jaynes (1957), "Information Theory and Statistical Mechanics" Physical Reviews. vol. 106, pp. 620–630.
6. J.N. Kapur (1967), "On Information of order α and type β" Proc. Ind. Accd. Sci. 48A, 65–73.
7. J.N. Kapur (1972), "Measures of uncertainty mathematical programming and physics" J. Ind. Soc. Ag. Stat. vol 24, 47–66.
8. J.N. Kapur (1986), "Four families of measures of entropy" Ind. Jour. Pure. App. Maths, vol 17, No. 4 pp. 429–449.
9. J.N. Kapur (1990), "Maximum Entropy Models in Science and Engineering," Wiley Eastern New Delhi and John Wiley, New York.
10. J.N. Kapur & H.K. Kesavan (1987), "Generalised Maximum Entropy Principle (with Applications)," Sand ford Education Press University of Waterloo.
11. J.N. Kapur and H.K. Kesavan (1992), "Entropy Optimization Principles with Applications," Academic Press, New York.
12. S. Kullback and R.A. Lubler (1951), "On Information and Sufficiency" Ann. Math Stat. vol 22 pp. 79–86.
13. S. Kullback (1959), "Information Theory ant Statistics", John Wiley New York.
14. A. Renyi (1961), "On Measures of Entropy and Information" Pre 1st Berkeley Synp. Prob. Stat. 1, 547–561.
15. C.E. Shannon (1948), "A Mathematical Theory; of Communication" Bell System Tech J. Vol 27, 379–423.
16. S. Vajda (1969), "A contribution to information Analysis of Patterns", In S. Watanbe (ed) Methodologies of Pattern Recognition.

17

GENERATING MEASURES OF CROSS-ENTROPY BY USING MEASURES OF WEIGHTED ENTROPY

[A number of new measures of cross-entropy have been generated by using the concepts of weighted entropy and weighted directed divergence and choosing the weights appropriately. Conversely some new measures of weighted entropy have been suggested on the basis of the new measures of cross-entropy generated.]

17.1 INTRODUCTION

Consider the measure of weighted entropy,

$$H_1(P:W) = -\sum_{i=1}^{n} w_i\,\phi\,(p_i), \tag{1}$$

where, $\phi\,(\bullet)$ is a continuous convex twice-differentiable function. Its maximum subject to the natural constraint,

$$\sum_{i=1}^{n} p_i = 1 \tag{2}$$

occurs when,

$$w_1\,\phi'\,(p_1) = w_2\phi'(p_2) = \ldots\ldots\ldots = w_n\phi'(p_n) \tag{3}$$

If the maximum occurs when $p_i = q_i$, $i = 1, 2, \ldots, n$, then we get

$$\frac{w_1}{[\phi'(q_1)]^{-1}} = \frac{w_2}{[\phi'(q_2)]^{-1}} = \ldots\ldots\ldots = \frac{w_n}{[\phi'(q_n)]^{-1}}, \tag{4}$$

so that (1) becomes,

$$-\sum_{i=1}^{n} \frac{\phi(p_i)}{\phi'(q_i)} \quad \text{or} \quad \sum_{i=1}^{n} \frac{\phi(p_i)}{\phi'(q_i)} \tag{5}$$

according as $\phi'(q_i)$ is positive or negative for each $i = 1, 2, 3, \ldots, n$

Now consider the measure,

$$\sum_{i=1}^{n} \frac{\phi(p_i) - \phi(q_i)}{\phi'(q_i)} \quad \text{or} \quad \Sigma \frac{\phi(q_i) - \phi(p_i)}{\phi'(q_i)} \tag{6}$$

according as $\phi'(q_i)$ is positive or negative.

In either case, the measure is a continuous twice-differentiable convex function of $p_1, p_2, \ldots, p_n$ which has its minimum value at $(q_1, q_2, \ldots, q_n)$. As such, this can be used as a measure of directed divergence or cross-entropy of P from Q, where

$$P = (p_1, p_2, ..., p_n)\,; \quad Q = (q_1, q_2, ..., q_n) \tag{7}$$

provided the condition on $\phi'(q_i)$ is satisfied for all i,i.e. provided $(q_1, q_2, \ldots, q_n)$ are such that either $\phi'(q_i) > 0$ or < 0 for all i's. $\phi(x)$ being monotonic increasing or monotonic decreasing is sufficient, but not necessary for this purpose.

Thus corresponding to every measure of weighted entropy of type (1), we can get a measure of cross-entropy. However, this measure is not applicable for all probability distributions Q, but only for a certain class of apriori distributions. We examine three special cases in the next three-sections.

17.2 MEASURE OF CROSS-ENTROPY CORRESPONDING TO GUIASU'S MEASURE OF WEIGHTED ENTROPY

Guiasu's [1] gave the measure,

$$H_2(P:W) = -\sum_{i=1}^{n} w_i\, p_i \ln p_i, \tag{8}$$

corresponding to Shannon's [4] measure of entropy,

$$-\sum_{i=1}^{n} p_i \ln p_i \tag{9}$$

This gives us the measure of cross-entropy,

$$\sum_{i=1}^{n} \frac{q_i \ln q_i - p_i \ln p_i}{1 + \ln q_i}, \tag{10}$$

provided $\qquad 1 + \ln\, q_i < 0 \;$ or $\; \ln\, q_i < -1 \;$ or $\; q_i < e^{-1} \qquad (11)$

This is not a serious restriction if n is large. Thus our first new measure is

$$D_1(P:Q) = \sum_{i=1}^{n} \frac{q_i \ln q_i - p_i \ln p_i}{1 + \ln q_i}, \quad q_i < e^{-1}, i = 1, 2, ..., n \tag{12}$$

17.3 MEASURE OF CROSS-ENTROPY CORRESPONDING TO HAVRDA-CHARVAT MEASURE OF ENTROPY

Havrda-Charvat [2] measure of entropy is given by

$$\sum_{i=1}^{n} \frac{(p_i^{\alpha} - p_i)}{\alpha(1-\alpha)}, \quad \alpha \neq 1, \quad \alpha > 0 \tag{13}$$

and the corresponding measure of weighted entropy is

$$H_3\,(P:W) = \sum_{i=1}^{n} w_i \frac{p_i^{\alpha} - p_i}{\alpha\,(1-\alpha)}, \qquad \alpha \neq 1 \tag{14}$$

The corresponding measure of cross-entropy is

$$D_2(P:Q) = \sum_{i=1}^{n} \frac{p_i^{\alpha} - p_i - q_i^{\alpha} + q_i}{\alpha(1-\alpha)(\alpha q_i^{\alpha-1} - 1)} \tag{15}$$

provided,

$$\alpha q_i^{\alpha-1} < 1 \quad \text{or} \quad q_i < \left(\frac{1}{\alpha}\right)^{\frac{1}{\alpha-1}} \tag{16}$$

Now we have the Table:

α:	1/10	1/9	1/4	1/3	1/2	2/3	3/4	.9	.99	1	2	3	5
$\alpha^{\frac{1}{1-\alpha}}$:	0.077	.134	.137	.192	.25	.296	.316	.348	.3667	.368	.5	.577	.669
α	10	100	α										
$\alpha^{\frac{1}{1-\alpha}}$:	.774	.926	1										

Thus we can use the measure for a suitable value of α unless one of the components of Q is large. In this case we can find $D\ (Q : P)$ unless one of the components of P is also very large.

Even if one of the component of Q or P is large, we can find a suitable value of α unless the component happens to be unity, in which we cannot use even Kullback-Leibler measure.

17.4 MEASURE OF CROSS-ENTROPY CORRESPONDING TO KAPUR'S MEASURE OF ENTROPY

This measure of entropy is given by [3]

$$-\sum_{i=1}^{n} p_i \ln p_i + \frac{1}{a} \sum_{i=1}^{n} (1 + a\, p_i) \ln (1 + a\, p_i) - \sum_{i=1}^{n} \frac{1}{a} (1+a) \ln (1+a)\, p_i \tag{17}$$

and the corresponding measure of weighted entropy is

$$H_u(P:W) = -\sum_{i=1}^{n} w_i \left[p_i \ln p_i - \frac{1}{a}(1 + a\, p_i) \ln (1 + a\, p_i) + \frac{1}{a}(1+a) \ln (1+a)\, p_i \right] \tag{18}$$

The corresponding measure of cross-entropy is

$$D_3(P:W) = \sum_{i=1}^{n} \frac{(q_i \log q_i - p_i \ln p_i) - \frac{1}{a}(1 + a\, q_i) \ln (1 + a q_i) + \frac{1}{a}(1 + a p_i) \ln (1 + a\, p_i)}{\ln q_i - \ln (1 + a\, q_i)}, \tag{19}$$

provided $a > 0$

If $a < 0$, the above formulas will be valid if

$$q_i < 1 + a\, q_i\,, \quad \text{i.e. if} \quad q_i(1-a) < 1 \quad \text{or} \quad q_i < \frac{1}{1-a} \tag{20}$$

For the above formulae to be valid, we also require that

$$1 + a\, q_i > 0 \quad \text{or} \quad q_i < -\frac{1}{a}, \quad p_i < -\frac{1}{a} \tag{21}$$

It will be better to let $a > -1 \text{and} q_i < \dfrac{1}{1-\alpha}$

17.5 SOME OTHER MEASURES OF CROSS-ENTROPY

The measures (12), (15) and (19) suffer from the weakness that these are not applicable for all probability distribution Q. To overcome this weakness, we choose measures,

$$\sum_{i=1}^{n} \frac{\phi(q_i) - \phi(p_i)}{\phi'(q_i)} \quad \text{or} \quad \sum_{i=1}^{n} \frac{\phi(q_i) - \phi(p_i)}{-\phi'(q_i)} \tag{23}$$

with only such convex functions $\phi(x)$ such that $\phi'(q_i)$ has always one sign. We now consider some special cases,

(i) $\phi(x) = x \ln x - x\,, \quad \phi'(q_i) = \ln q_i < 0$ (24)

This gives,
$$D_4(P:Q) = \sum_{i=1}^{n} \frac{q_i \ln q_i - q_i - p_i \ln p_i + p_i}{\ln q_i} \tag{25}$$

We note that $D_4(P:Q)$ is a convex function of P which has its minimum value zero when $p_i = q_i$ for each i.

(ii) $\phi(x) = \dfrac{1}{\alpha(\alpha-1)} x^{\alpha}, \quad \alpha \neq 0, \quad \alpha \neq 1, \quad \phi'(q_i) = \dfrac{q_i^{\alpha-1}}{\alpha - 1}$ (26)

This gives,

$$D_5(P:Q) = \frac{1}{\alpha}(1-\alpha) \sum_{i=1}^{n} \frac{q_i^{\alpha} - p_i^{\alpha}}{q_i^{\alpha-1}}, \quad \alpha \neq 0, \quad \alpha \neq 1 \tag{27}$$

This is again a convex function of $p_1, p_2, \ldots, p_n$ which attains minimum value zero only when $p_i = q_i$ for each i.

(iii) $\phi(x) = -\ln(1 + ax)\,, \quad a > -1, \quad \phi'(q_i) = \dfrac{-a}{(1 + a\, q_i)}$ (28)

This gives,

$$D_6(P:Q) = \sum_{i=1}^{n} \frac{\ln(1 + a\, q_i) - \ln(1 + a\, p_i)}{\frac{a}{1 + a\, q_i}}, \quad a > 0 \tag{29}$$

$$D_7(P:Q) = \sum_{i=1}^{n} \frac{\ln(1 + a\, q_i) - \ln(1 + a\, p_i)}{\frac{-a}{1 + a\, q_i}}, \quad -1 < a < 0 \tag{30}$$

(iv) $$\phi(x) = x \ln x - (x+a)\ln(x+a), \quad a > 0 \tag{31}$$

$$\phi'(q_i) = \ln \frac{q_i}{q_i + a} < 0 \tag{32}$$

This gives,

$$D_8(P:Q) = \sum_{i=1}^{n} \frac{q_i \ln q_i - (q_i + a)\ln(q_i + a) - p_i \ln p_i + (p_i + a)\ln(p_i + a)}{\ln \frac{q_i}{q_i + a}}, \quad a > 0 \tag{33}$$

17.6 SOME POSSIBLE MEASURES OF WEIGHTED ENTROPY

The above discussion suggests that the following measures are possible candidates for measures of weighted entropy.

(i) $$H_5(P:W) = \sum_{i=1}^{n} w_i (p_i - p_i \ln p_i) \tag{34}$$

(ii) $$H_6(P:W) = \frac{1}{\alpha(1-\alpha)} \sum_{i=1}^{n} w_i p_i^{\alpha} \quad \alpha \neq 0, 1 \tag{35}$$

(iii) $$H_7(P:W) = \sum_{i=1}^{n} w_i \ln(1 + a p_i), \quad a > 0 \tag{36}$$

(iv) $$H_8(P:W) = \sum_{i=1}^{n} w_i (-p_i \ln p_i - (1 - p_i)\ln(1 - p_i)) \tag{37}$$

Each of the above is a concave function of $p_1, p_2, p_3, \ldots, p_n$ and is permutationally symmetric, i.e. it does not change when the pairs $(p_1, w_1), (p_2, w_2), \ldots, (p_n, w_n)$ are permuted among themselves. Each is maximum when p_i is a function of w_i only. Each has a local minimum at a degenerate distribution and the global minimum is also at a degenerate distribution. The globally minimum value need not be zero, but can be made zero by adding a suitable constant.

Let,
$$w_{\min} = \min(w_1, w_2, \ldots, w_n) \tag{38}$$

then we get,

(i) $$H'_1(P:W) = \sum_{i=1}^{n} w_i (p_i - p_i \ln p_i) - w_{\min} \tag{39}$$

(ii) $$H'_2(P:W) = \frac{1}{\alpha(1-\alpha)} \left(\sum_{i=1}^{n} w_i p_i^{\alpha} - w_{\min} \right) \tag{40}$$

(iii) $$H'_3(P:W) = \sum_{i=1}^{n} w_i \ln(1 + a p_i) - w_{\min} \ln(1 + a) \tag{41}$$

(iv) $$H'_4(P:W) = H_4(P:W) \tag{42}$$

The first three of these measures violate one of the conditions considered desirable for an appropriate measure of a weighted entropy namely that it should be ≥ 0 and should vanish for all degenerate distributions. However, the lack of symmetry with respect to different degenerate distributions is understandable because the weights are different.

REFERENCES

1. S. Guiasu (1971), "Weighted Entropy", Reports on Math. Physics, vol. 2. pp. 165–179.
2. J.H. Havrda, F. Charvat (1967), "Quantification Methods of Classification Processes: Concepts of Structural α Entropy", Kybernetica, vol. 3, pp. 30–35.
3. J.N. Kapur (1986), "Four Families of Measures of Entropy, " Ind. Jour. Pure and App. Maths, vol. 17, No. 4, pp. 429–449.
4. C.E. Shannon (1948), "A Mathematical Theory of Communication", Bell System Tech. J., vol. 27, pp. 379–423, 623–659.

18

NORMALISED MEASURES OF INFORMATION

[Different motivations for normalising measures of entropy, directed divergence, and inaccuracy are discussed and about half a dozen normalised measures of each type are obtained.]

18.1 NEED FOR NORMALISING MEASURES OF ENTROPY

$$\text{Let } P = (p_1, p_2, \ldots, p_n)\ ;\quad p_i \geq 0\ \ (i = 1, 2, \ldots, n),\ \sum_{i=1}^{n} p_i = 1 \tag{1}$$

be a probability distribution, then Shannon's [16] measure of entropy is given by

$$H(P) = -\sum_{i=1}^{n} p_i \ln p_i \tag{2}$$

where $0 \times \ln 0$ is defined as zero. This measures the degree of equality among the probabilities, i.e. the greater the equality among, $p_1, p_2, \ldots, p_n$, the greater is the value of $H(P)$ and this measure has its maximum value $\ln n$ when all the probabilities are equal, i.e. when

$$p_1 = p_2 = \ldots\ldots = p_n = 1/n\ ;\quad P = U\ ,\quad H(U) = \ln n, \tag{3}$$

where U is the uniform distribution.

The entropy (2) also measures the uniformity of P or the 'closeness' of P to the uniform distribution U, since according to Kullback-Leibler [14] measure, the directed divergence of P from U is given by

$$D(P:U) = \sum_{i=1}^{n} p_i \ln \frac{p_i}{1/n} = \ln n + \sum_{i=1}^{n} p_i \ln p_i$$

$$= \ln n - H(P) \tag{4}$$

Thus greater the value of entropy $H(P)$, the nearer is P to U.

This entropy provides a measure of equality or uniformity of the probabilities $p_1, p_2, \ldots, p_n$ among themselves

Now consider the probability distributions,

$$P = (1/3,\ 1/3,\ 1/3)\ ,\quad Q = (1/8,\ 1/8,\ 1/4,\ 1/4,\ 1/4) \tag{5}$$

Which is more uniform or in which the probabilities are more equal? The obvious answer is that probabilities of P are more equal than the probabilities of Q. However

$$H(P) = \ln 3, \quad H(Q) = \ln 4 + 1/4 \ln 2 > \ln 3 \tag{6}$$

P is perfectly uniform yet $H(P) \leq H(Q)$. From the value of the two entropies, it appears that Q is more uniform than P. This is obviously wrong.

The fallacy arises due to the fact that the entropy depends not only on the degree of equality among the probabilities; it also depends on the value of n.

So long as n is the same, entropy can be used to compare the uniformity (or lack of it) of the distributions, but if the numbers of outcomes are different, then entropy is not a satisfactory measure of uniformity.

In this case, we try to eliminate the effect of n by normalising the entropy, i.e. by defining a normalised measure of entropy by

$$\overline{H}(P) = \frac{H(P)}{\max\limits_{P} H(P)} \tag{7}$$

It is obvious that $\quad 0 \leq \overline{H}(P) \leq 1$ (8)

For Shannon's measure (2),

$$\overline{H}(P) = \frac{-\sum_{i=1}^{n} p_i \ln p_i}{\ln n} \tag{9}$$

Thus for the distribution given in (5), we have

$$\overline{H}(P) = \frac{\ln 3}{\ln 3} = 1, \quad \overline{H}(Q) = \frac{\ln 4 + 1/4 \ln 2}{\ln 5} = .9566 \tag{10}$$

This gives the correct result that Q is less uniform than P.

Thus to compare the uniformity, or equality or uncertainty of two distributions, we should compare their normalised measures of entropy.

18.2 SOME NORMALISED MEASURES OF ENTROPY

Every measure of entropy is a concave function which has its maximum value when P is the uniform distribution, so that

$$\overline{H}(P) = \frac{H(P)}{H(U)} \tag{11}$$

As special cases, we get the following normalised measures of entropy

(*i*) Shannon's [16] normalised measure of entropy,

$$\frac{-\sum_{i=1}^{n} p_i \ln p_i}{\ln n} \tag{12}$$

(*ii*) Renyi's [15] normalised measure of entropy,

$$\frac{\frac{1}{1-\alpha} \ln \sum_{i=1}^{n} p_i^{\alpha}}{\frac{1}{1-\alpha} \ln (n\, 1/n^{\alpha})} = \frac{\ln \sum_{i=1}^{n} p_i^{\alpha}}{(1-\alpha) \ln n}; \quad \alpha \neq 1 \tag{13}$$

(*iii*) Havrda-Charvat's [3] normalised measure of entropy,

$$\frac{\frac{1}{1-\alpha}\left(\sum_{i=1}^{n} p_i^{\alpha}-1\right)}{\frac{1}{1-\alpha}(n\ 1/n^{\alpha}-1)}=\frac{\left(\sum_{i=1}^{n} p_i^{\alpha}-1\right)}{n^{1-\alpha}-1} \tag{14}$$

In fact Havrda-Charvat [3] themselves gave the measure

$$\frac{\sum_{i=1}^{n} p_i^{\alpha}-1}{2^{1-\alpha}-1}, \tag{15}$$

which is a normalised measure when $n = 2$. Again

(*iv*) Sharma-Taneja [17] normalised measure of entropy

$$=\frac{\frac{1}{\beta-\alpha}\left(\sum_{i=1}^{n} p_i^{\alpha}-\sum_{i=1}^{n} p_i^{\beta}\right)}{\frac{1}{\beta-\alpha}(n(1/n^{\alpha})-n(1/n^{\beta}))}$$

$$=\frac{\sum_{i=1}^{n} p_i^{\alpha}-\sum_{i=1}^{n} p_i^{\beta}}{n^{1-\alpha}-n^{1-\beta}},\quad \begin{array}{ll} 0<\alpha<1, & \beta>1 \\ \text{or}\quad \alpha>1, & 0<\beta<1 \end{array} \tag{16}$$

(*v*) Kapur's [4] first normalised measure of entropy is

$$\frac{\frac{1}{\beta-\alpha}\ln\frac{\sum_{i=1}^{n} p_i^{\alpha}}{\sum_{i=1}^{n} p_i^{\beta}}}{\frac{1}{\beta-\alpha}\ln\frac{n^{1-\alpha}}{n^{1-\beta}}}=\frac{\ln\frac{\sum_{i=1}^{n} p_i^{\alpha}}{\sum_{i=1}^{n} p_i^{\beta}}}{(\beta-\alpha)\ln n},\quad \begin{array}{ll} 0<\alpha<1, & \beta>1 \\ \text{or} & \\ \alpha>1, & 0<\beta<1 \end{array} \tag{17}$$

(*vi*) Kapur's [6] second normalised measure of entropy is

$$\frac{-\sum_{i=1}^{n} p_i \ln p_i + 1/a \sum_{i=1}^{n} (1+ap_i)\ln(1+ap_i)_{-1/a}(1+a)\ln(1+a)}{\ln n+\frac{(n+a)}{n}\ln\frac{(n+a)}{n}-\frac{1}{a}(1+a)\ln(1+a)},\quad a\geq -1 \tag{18}$$

(*vii*) Sharma-Mittal's [18] normalised measure of entropy is

$$=\frac{\frac{1}{(\alpha-1)\beta}\left[\left(\sum_{i=1}^{n} p_i^{\alpha}\right)^{\beta}-1\right]}{\frac{1}{(\alpha-1)\beta}[(n^{1-\alpha})^{\beta}-1]}=\frac{\left(\sum_{i=1}^{n} p_i^{\alpha}\right)^{\beta}-1}{n^{(1-\alpha)\beta}-1},\quad \alpha\neq 1,\quad \beta\neq 0 \tag{19}$$

Kapur, Kumar and Kumar [7] drew the graphs of the normalised measures (12), (13), (14), and (17) for the distribution $(p, 1-p)$ and showed that the normalised measures are much closer to one another than the non-normalised measures. All these represent concave functions with minimum value zero at $p = 0$ and $p = 1$, and maximum value unity at $p = 1/2$.

18.3 NEED FOR NORMALISING MEASURES OF DIRECTED DIVERGENCE

Let $P = (p_1, p_2, ..., p_n)$, $Q = (q_1, q_2, ..., q_n)$ (20)

be two probability distributions, Then a measure of directed divergence, $D(P:Q)$ has the following properties:

(*i*) $D(P:Q) \geq 0$ (*ii*) $D(P:Q) = 0$ *iff* $P = Q$ (*iii*) $D(P:Q)$ is a convex function of both $p_1, p_2, \ldots, p_n$; $q_1, q_2, \ldots, q_n$

We can consider three possible normalisations, viz.

$$\overline{D}_1(P:Q) = \frac{D(P:Q)}{\max\limits_{P,Q} D(P:Q)}, \overline{D}_2(P:Q) = \frac{D(P:Q)}{\max\limits_{P} D(P:Q)}, \overline{D}_3(P:Q) = \frac{D(P:Q)}{\max\limits_{Q} D(P:Q)} \tag{21}$$

In general, the first and third denominators are infinite and as such the first and third normalisations are meaningless. We therefore consider the second normalisation.

For this, we should also assume that none of the components of Q is zero. Keeping Q fixed, we find that distribution P for which $D(P:Q)$ is maximum. Since $D(P:Q)$ is a convex function of P and $p_i \geq 0$, the maximum value of $D(P:Q)$ will arise when P is one of the degenerate distributions.

$$\Delta_1 = (1, 0, 0..., 0), \Delta_2 = (0, 1, 0, 0...., 0), ..., \Delta_n = (0, 0, 0, ..., 1), \tag{22}$$

so that $\max\limits_{P} D(P:Q) = \max\limits_{r} D(\Delta_r:Q)$, (23)

so that
$$\overline{D}_2(P:Q) = \frac{D(P:Q)}{\max\limits_{r} D(\Delta_r:Q)} \tag{24}$$

The normalised measure lies between 0 and 1. It is zero when $P = Q$ and it is unity when P is that degenerate distribution for which $D(\Delta_r:Q)$ is maximum. Let $r = k$ in this case.

Thus given any two distributions P and Q, to find $\overline{D}_2$, we keep Q fixed and

(*i*) find directed divergence of P from Q,

(*ii*) find maximum possible directed divergence of any distribution from Q,

(*iii*) find the ratio of (*i*) to (*ii*).

Thus the purpose of the normalisation of a measure of directed divergence is quite distinct from the purpose of normalisation of a measure of entropy.

In the case of measures of entropy, the object was to eliminate the effect of having different numbers of outcomes for different distributions. In the case of directed divergence, the object is to get an idea of the ratio of directed divergence of P from Q to the maximum possible directed divergence of any distribution from Q.

18.4 NORMALISED MEASURES OF DIRECTED DIVERGENCE

The most important measure of directed divergence is due to Csiszer [2], viz.

$$D\,(P:Q)=\sum_{i=1}^{n} q_i\, f\,(p_i/q_i) \tag{25}$$

where $f\,(\bullet)$ is any convex twice-differentiable function which is such that $f\,(1)=0$, so that

$$\begin{aligned} D(\Delta_r:Q) &= q_1 f(0)+q_2 f(0)+\ldots..+q_{r-1} f(0) \\ &\quad + q_r f(1/q_r)+q_{r+1} f(0)+\ldots..+q_n f(0) \\ &= q_r f\,(1/q_r)+(1-q_r)\,f\,(0) \end{aligned} \tag{26}$$

Let $$\phi\,(q)=q\,f(1/q)+(1-q)\,f\,(0), \tag{27}$$

so that, $$\phi'(q)=f(1/q)-1/q\,f'(1/q)-f(0) \tag{28}$$

and, $$\phi''(q)=1/q^3\,f''(1/q) \tag{29}$$

Since $f\,(\bullet)$ is a convex function, $f''\,(1/q)>0$, so that from (29), $\phi''\,(q)>0$ and $\phi\,(q)$ is a convex function of q, so that $\phi'\,(q)$ is a monotonic increasing function of q. Now from (28)

$$\begin{aligned} \phi'(1) &= f(1)-f'(1)-f(0) \\ &= -f'(1)-f(0), \end{aligned} \tag{30}$$

since $f\,(1)=0$. We assume that $f\,(\bullet)$ is such that

$$f'(1)+f\,(0)>0 \tag{31}$$

so that $\phi'(1)<0$. In this case $\phi'(q)$ is a monotonic increasing function of q which increases as q increases from 0 to 1, and its final value $\phi'(1)$ is negative. Therefore, $\phi'(q)$ is always negative in the interval [0, 1] and $\phi(q)$ is a monotonic decreasing function of q.

Now from (26) and (27), $$D(\Delta_r:Q)=\phi\,(q_r) \tag{32}$$

Since $\phi\,(q)$ is a monotonic decreasing function of q,

$$\max\,[\phi(q_1),\quad \phi\,(q_2),\ldots,\phi\,(q_n)]=\phi\,(q_k), \tag{33}$$

where, $$q_k=\min\,(q_1,q_2,\ldots,q_n) \tag{34}$$

From (24), (26), (27), (33), (34)

$$\overline{D}_2\,(P:Q)=\frac{\sum_{i=1}^{n} q_i\, f\,(p_i/q_i)}{q_k\, f(1/q_k)+(1-q_k)\,f(o)} \tag{35}$$

Now we discuss some special cases:

(*i*) Kullback-Leibler [14] normalised measure of directed divergence is obtained by putting $f\,(x)=x\ln x$. In this case

$$f'(1)+f(0)=1+\ln 1+0=1>0, \tag{35a}$$

$$\text{and} \quad \overline{D}_2(P:q) = \frac{\sum_{i=1}^{n} p_i \ln (p_i/q_i)}{-\ln q_k} \tag{36}$$

(*ii*) Havrda:Charvat [3] normalised measure of directed divergence is obtained by putting $f(x) = \frac{x^\alpha - x}{\alpha - 1}$, $\alpha \neq 1$. In this case

$$f'(1) + f(0) = 1 > 0 \tag{37}$$

and,
$$\overline{D}_2(P:Q) = \frac{\sum_{i=1}^{n} p_i^\alpha q_i^{1-\alpha} - 1}{q_k^{1-\alpha} - 1}, \quad \alpha \neq 1 \tag{38}$$

(*iii*) Sharma-Taneja [17] normalised measure of directed divergence is obtained by putting $f(x) = \frac{x^\alpha - x^\beta}{\alpha - \beta}$, $0 < \alpha < 1,\ \beta > 1$ or $\alpha > 1,\ 0 < \beta < 1$

In this case $f'(1) + f(0) = 1 > 0,$ (39)

$$\overline{D}_2(P:Q) = \frac{\sum_{i=1}^{n} p_i^\alpha q_i^{1-\alpha} - \sum_{i=1}^{n} p_i^\beta q_i^{1-\beta}}{q_k^{1-\alpha} - q_k^{1-\beta}} \tag{40}$$

(*iv*) Kapur's [6] second normalised measure of directed divergence is obtained by putting $f(x) = x \ln x - 1/a\,(1 + ax) \ln \frac{1+ax}{1+a}$, $a > -1$ In this case,

$$f'(1) + f(0) = 1/a \ln (1 + a) > 0 \tag{41}$$

$$D_2(P:Q) = \frac{\sum_{i=1}^{n} p_i \ln (p_i/q_i) - 1/a \sum_{i=1}^{n} (q_i + ap_i) \ln \frac{1 + ap_i/q_i}{1+a}}{1/a\,[q_k \ln q_k - (q_k + a) \ln (q_k + a) + (1 + a) \ln (1 + a)]} \tag{42}$$

(*v*) Renyi's [15] measure of directed divergence is defined by

$$R\,(P:Q) = \frac{1}{(\alpha - 1)} \ln \sum_{i=1}^{n} p_i^\alpha q_i^{1-\alpha} = \frac{1}{\alpha - 1} \ln [(\alpha - 1) H\,(P:Q) + 1], \quad \alpha \neq 1, \tag{43}$$

where $H\,(P:Q)$ is Havrda-Charvat [3] measure of directed divergence. This shows that whether $\alpha < 1$ or $\alpha > 1$, for a given Q, $R\,(P:Q)$ is maximum for the same P for which $H\,(P:Q)$ is maximum so that

$$\max_P R(P:Q) = \frac{1}{\alpha - 1} \ln\left[(\alpha - 1) \max_P H\,(P:Q) + 1\right], \tag{44}$$

$$\max_P R\,(P:Q) = \frac{1}{\alpha - 1} \ln\left[(\alpha - 1)\left\{\frac{1}{\alpha - 1}(q_k^{1-\alpha} - 1)\right\} + 1\right]$$

$$= \frac{1}{\alpha - 1} \ln q_k^{1-\alpha} = -\ln q_k,$$

so that,

$$\overline{D}_2(P:Q)=\frac{\frac{1}{\alpha-1}\ln\sum_{i=1}^{n}p_i^{\alpha}q_i^{1-\alpha}}{-\ln q_k} \tag{45}$$

Alternatively, if $\alpha > 1$, $\sum_{i=1}^{n} p_i^{\alpha} q_i^{1-\alpha}$ is maximum when $P = \Delta_k$ and if $\alpha < 1$, $\sum_{i=1}^{n} p_i^{\alpha} q_i^{1-\alpha}$ is maximum when $P = \Delta_k$. In either case

$$\max_p\left[\frac{1}{1-\alpha}\ln\sum_{i=1}^{n}p_i^{\alpha}q_i^{1-\alpha}\right]=\frac{1}{\alpha-1}\ln q_k^{1-\alpha}=-\ln q_k \tag{46}$$

(*v*) For Kapur's [4] first measure of directed divergence,

$$K(P:Q)=\frac{1}{\alpha-\beta}\ln\frac{\sum_{i=1}^{n}p_i^{\alpha}q_i^{1-\alpha}}{\sum_{i=1}^{n}p_i^{\beta}q_i^{1-\beta}},\quad \begin{matrix}0<\alpha<1, & \beta>1\\ 0<\beta<1, & \alpha>1\end{matrix} \tag{47}$$

In both cases, the minimum value occurs when $P = \Delta_k$ so that

$$\overline{D}_2(P:Q)=\frac{\ln\left(\sum_{i=1}^{n}p_i^{\alpha}q_i^{1-\alpha}\right)-\ln\left(\sum_{i=1}^{n}p_i^{\beta}q_i^{1-\beta}\right)}{\ln q_k^{1-\alpha}-\ln q_k^{1-\beta}}$$

$$=\frac{\ln\left(\sum_{i=1}^{n}p_i^{\alpha}q_i^{1-\alpha}\right)-\ln\left(\sum_{i=1}^{n}p_i^{\beta}q_i^{1-\beta}\right)}{(\beta-\alpha)\ln q_k} \tag{48}$$

(*vi*) For Sharma-Mittal [18] measure of directed divergence,

$$S(P:Q)=\frac{1}{(\alpha-1)\beta}\left[\left(\sum_{i=1}^{n}p_i^{\alpha}q_i^{1-\alpha}\right)^{\beta}-1\right]\quad \begin{matrix}\alpha\neq 1\\ \beta\neq 1\end{matrix} \tag{49}$$

$$\max_p S(P:Q)=\frac{1}{(\alpha-1)\beta}\left(q_k^{(1-\alpha)\beta}-1\right) \tag{50}$$

so that,

$$\overline{D}_2(P:Q)=\frac{\left(\sum_{i=1}^{n}p_i^{\alpha}q_i^{1-\alpha}\right)^{\beta}-1}{q_k^{(1-\alpha)\beta}-1}. \tag{51}$$

In the case $n = 2$, Kapur, Kumar and Kumar [8] have drawn the graphs in three dimensional space of $\overline{D}_2(p, 1-p; q, 1-q)$ against p and q for the measures (36), (38) and (42).

18.5 NEED FOR NORMALISING MEASURES OF INACCURACY

Kerridge [12] defined a measure of inaccuracy of the probability distribution P relative to Q as

$$I(P:Q) = -\sum_{i=1}^{n} p_i \ln q_i \tag{52}$$

$$= \sum_{i=1}^{n} p_i \ln p_i/q_i - \sum_{i=1}^{n} p_i \ln p_i$$

$$\text{or} \quad I(P:Q) = D(P:Q) + H(P) \tag{53}$$

where, $D(P:Q)$ is Kullback-Leibler [15] measure of directed divergence and $H(P)$ is the entropy of the distribution P.

Even if P is known, there is an uncertainty about the outcome of P and this uncertainty is measured by $H(P)$. If instead of P, we know Q, then uncertainty is increased by $D(P:Q)$. The uncertainty of P relative to Q

= uncertainty of P + uncertainty due to P being different from Q. (54)

Kapur defined [17] a general measure of inaccuracy by using (53) as the sum of two corresponding measures of entropy and directed divergence. Thus Havrda and Charvat's measure of inaccuracy

$$= \frac{1}{\alpha-1}\left(\sum_{i=1}^{n} p_i^{\alpha} q_i^{1-\alpha} - 1\right) + \frac{1}{1-\alpha}\left(\sum_{i=1}^{n} p_i^{\alpha} - 1\right)$$

$$= \frac{1}{\alpha-1}\left[\sum_{i=1}^{n} p_i^{\alpha}(q_1^{1-\alpha} - 1)\right] \tag{55}$$

Similarly, some other measures of inaccuracy are Renyi's [15]:

$$\frac{1}{\alpha-1} \ln \frac{\sum_{i=1}^{n} p_i^{\alpha} q_i^{1-\alpha}}{\sum_{i=1}^{n} p_i^{\alpha}}, \qquad \alpha \neq 1 \tag{56}$$

Kapur's first [4]:

$$\frac{1}{\alpha-\beta} \ln \left[\frac{\sum_{i=1}^{n} p_i^{\alpha} q_i^{1-\alpha}}{\sum_{i=1}^{n} p_i^{\alpha} q_i^{1-\beta}} \, \frac{\sum_{i=1}^{n} p_i^{\beta}}{\sum_{i=1}^{n} p_i^{\alpha}} \right]$$

$$= \frac{1}{\alpha-\beta} \ln \left[\frac{\sum_{i=1}^{n} p_i^{\alpha} q_i^{1-\alpha}}{\sum_{i=1}^{n} p_i^{\alpha}} \, \frac{\sum_{i=1}^{n} p_i^{\beta}}{\sum_{i=1}^{n} p_i^{\beta} q_i^{1-\beta}} \right] \tag{57}$$

Kapur's second [6]:

$$\sum_{i=1}^{n} p_i \ln p_i/q_i - 1/a \sum_{i=1}^{n} (q_i + a p_i) \ln \frac{1 + a p_i/q_i}{1+a}$$

$$-\sum_{i=1}^{n} p_i \ln p_i + 1/a \quad \sum_{i=1}^{n} (1+ap_i)\ln(1+ap_i) - 1/a\,(1+a)\ln(1+a) \tag{58}$$

Sharma and Taneja's [17]:

$$\frac{1}{\alpha-\beta}\left[\sum_{i=1}^{n} p_i^{\alpha} q_i^{1-\alpha} - \sum_{i=1}^{n} p_i^{\beta} q_i^{1-\beta}\right] + \frac{1}{\beta-\alpha}\left[\sum_{i=1}^{n} P_i^{\alpha} - \sum_{i=1}^{n} P_i^{\beta}\right]$$

$$= \frac{1}{\alpha-\beta}\left[\sum_{i=1}^{n} p_i^{\alpha}(q_i^{1-\alpha}-1) - \sum_{i=1}^{n} p_i^{\beta}(q_i^{1-\beta}-1)\right] \tag{59}$$

Sharma-Mittal's [18]:

$$\frac{1}{(\alpha-1)\beta}\left[\left(\sum_{i=1}^{n} p_i^{\alpha} q_i^{1-\alpha}\right)^{\beta} - 1\right] + \frac{1}{(1-\alpha)\beta}\left[\left(\sum_{i=1}^{n} p_i^{\alpha}\right)^{\beta} - 1\right] \tag{60}$$

$$= \frac{1}{(\alpha-1)\,\beta}\left[\left(\sum_{i=1}^{n} p_i^{\alpha} q_i^{1-\alpha}\right)^{\beta} - \left(\sum_{i=1}^{n} p_i^{\alpha}\right)^{\beta}\right] \tag{61}$$

We normalise these measures in the same spirit in which we normalise measures of directed divergence. We define.

$$I(P:Q) = \frac{I(P:Q)}{\max\limits_{P} I(P:Q)} = \frac{D(P:Q)+H(P)}{\max\limits_{p}(D(P:Q)+H(P))} \tag{62}$$

This normalised measure represents the ratio of Inaccuracy of P relative to Q to maximum possible inaccuracy of any distribution relative to Q.

(63)

Now, $D(P:Q)$ is a convex function of P and $H(P)$ is a concave function of p and their sum may be a convex function or a concave function or it may be neither convex nor concave. As such maximization of inaccuracy for variation of P present different problems from those encountered in maximization of $D(P:Q)$ alone.

18.6 NORMALISED MEASURES OF INACCURACY

(*i*) Kerridge's [13] normalised measure of inaccuracy,

$$= \frac{-\sum_{i=1}^{n} p_i \ln q_i}{\max\limits_{P}\left(-\sum_{i=1}^{n} p_i \ln q_i\right)} = \frac{\sum_{i=1}^{n} p_i \ln q_i}{\ln q_k} \tag{64}$$

(*ii*) Havrda-Charvat [3] normalised measure of inaccuracy.
The measure (55) is a concave function of P if $\alpha > 1$ and is convex function if $\alpha < 1$, so that the measure is maximum when $P = U$ when $\alpha > 1$ and it is maximum when $P = \Delta_k$ if $\alpha < 1$, so that the normalised measure,

$$= \frac{\sum_{i=1}^{n} p_i^{\alpha}(q_i^{1-\alpha}-1)}{n^{-\alpha}\sum_{i=1}^{n}(q_i^{1-\alpha}-1)} \quad \text{if} \quad \alpha > 1 \tag{65}$$

$$= \frac{\sum_{i=1}^{n} p_i^{\alpha} (q_i^{1-\alpha} - 1)}{q_k^{1-\alpha} - 1} \quad \text{if} \quad \alpha < 1 \tag{66}$$

(*iii*) Sharma-Taneja's normalised measure of inaccuracy:
From (59)

$$\frac{\partial^2 S}{\partial p_i^2} = \frac{1}{\alpha - \beta} [\alpha(1-\alpha) p_i^{\alpha-2} (q_i^{1-\alpha} - 1) - \beta(\beta - 1) p_i^{\beta-2} (q_i^{1-\beta} - 1)] \tag{67}$$

If $\alpha > 1, 0 < \beta < 1$ or $0 < \alpha < 1$, $\beta > 1$, the second order derivative is throughout negative so that S is concave function of P and we can easily normalise S.

(*iv*) Kapur's second normalised measure of inaccuracy:
From (58),

$$\frac{\partial^2 k}{\partial p_i^2} = -\frac{a}{q_i + ap_i} + \frac{a}{1 + ap_i} = \frac{a}{(1 + ap_i)(q + ap_i)} (q_i - 1) \tag{68}$$

This measure is concave if $a > 0$ and is convex if $a < 0$. Therefore normalised measure of inaccuracy,

$$= \frac{-\sum_{i=1}^{n} p_i \ln q_i + \frac{1}{a} \sum_{i=1}^{n} (1 + ap_1) \ln (1 + ap_i) - \frac{1}{a} \sum_{i=1}^{n} (q_i + ap_i) \ln \frac{q_i + ap_i}{q_i(1+a)} - \frac{1}{a}(1+a) \ln (1+a)}{-\frac{1}{n} \sum \ln q + \frac{1}{a} \sum \left(1 + \frac{a}{n}\right) \ln \left(1 + \frac{a}{n}\right) - \frac{1}{a} \sum \left(q_i + \frac{a}{n}\right) \ln \frac{q_i + \frac{a}{n}}{q_i(1+a)} - \frac{1}{a}(a+1) \ln (a+1)}$$

$$\text{if } a > 0 \tag{69}$$

$$= \frac{-\sum_{i=1}^{n} p_i \ln q_i + 1/a \sum_{i=1}^{n} (1 + ap_i) \ln (1 + ap_i) - 1/a \sum_{i=1}^{n} (q_i + ap_i) \ln \frac{q_i + ap_i}{q_i (1+a)} - 1/a\,(1+a) \ln (1+a)}{-\ln q_k - 1/a\,(q_k + a) \ln \frac{q_k + a}{q_k(1+a)}}$$

$$\text{if } a < 0 \tag{70}$$

18.7 NORMALISED MEASURES OF 'USEFUL' INFORMATION

(*a*) Belis and Guiasu [1] gave the measure of 'useful' information as

$$-\sum_{i=1}^{n} u_i p_i \ln p_i, \tag{71}$$

where $u_1, u_2, \ldots, u_n$ are the utilities of n outcomes and these are assumed to be positive. In this case the entropy is a concave function whose minimum value is zero and whose maximum value is obtained when

$$\frac{1 + \ln p_1}{1/u_1} = \frac{1 + \ln p_2}{1/u_2} = \ldots\ldots = \frac{1 + \ln p_n}{1/u_n} = A \text{ (say)}, \tag{72}$$

so that,

$$p_i = e^{A/u_i - 1}, \qquad i = 1.2 \ldots\ldots n \tag{73}$$

where A is determinèd from the equation,

$$\sum_{i=1}^{n} e^{A/u_i} = e, \tag{74}$$

so that the maximum value of this useful information,

$$= -\sum_{i=1}^{n} u_i e^{(A/u_i - 1)} (A/u_i - 1) = -\sum_{i=1}^{n} A e^{A/u_i - 1} + \sum_{i=1}^{n} u_i e^{A/u_i - 1}$$

$$= -A + \sum_{i=1}^{n} u_i e^{A/u_i - 1} \tag{75}$$

and the normalised measure of useful information is obtained by dividing (71) by (75).

(*b*) A measure proposed for measuring useful directed divergence is,

$$\sum_{i=1}^{n} u_i p_i \ln p_i/q_i \tag{76}$$

This does not strictly represent a directed divergence since it can be negative. It vanishes when $p_i = q_i$ for each i, but it can also vanish when $p_i \neq q_i$ for each i.

However, we can construct a normalised measure for this which is always ≥ 0 by subtracting its minimum value from it and then dividing by the maximum value of this difference, so that the normalised measure is,

$$\frac{\sum_{i=1}^{n} u_i p_i \ln p_i/q_i - \min_P \sum_{i=1}^{n} u_i p_i \ln p_i/q_i}{\max_P \sum_{i=1}^{n} u_i p_i \ln p_i/q_i - \min_P \sum_{i=1}^{n} u_i p_i \ln p_i/q_i} \tag{77}$$

Since, $\sum_{i=1}^{n} u_i p_i \ln p_i / q_i$ is a convex function of P, its global minimum value can be found by minimizing it subject to $\sum_{i=1}^{n} p_i = 1$ and its maximum value will occur at one of the degenerate distributions.

For finding minimum value we have,

$$\frac{1 + \ln p_i/q_i}{1/u_1} = \frac{1 + \ln p_i/q_i}{1/u_2} = \ldots\ldots = \frac{1 + \ln p_n/q_n}{1/u_n} = B \tag{78}$$

where, B is obtained by solving

$$\sum_{i=1}^{n} q_i e^{B/u_i - 1} = 1, \tag{79}$$

and then $p_1, p_2, \ldots, p_n$ can be found by using (78). The maximum value of

$$\sum^{n} u_i p_i \ln p_i/q_i = \max (u_1 \ln 1/q_1, u_2 \ln 1/q_2, \ldots, u_n \ln 1/q_n) \tag{80}$$

18.8 CONCLUDING REMARKS

Normalised measures of entropy, directed divergence and useful information have been discussed by Kapur, Kumar and Kumar [7, 8, 9], but they discussed only the special cases when there are two outcomes. Here, we have discussed normalised measures for the general value, of n. We have also discussed the normalised measures of inaccuracy [10]. Normalised measures of stochastic interdependence have been separately discussed by Kapur [11], and Kapur and Medha [12].

REFERENCES

1. Belis, M. and S. Guisau, "A Quantitative-Qualitative Measure of Information in Cybernatic Systems", IEEE, Trans. Inf. Th. vol. IT. 4, pp 593–594, 1968.
2. Csiszer, I., "A Class of Measures of Informativity of Observation Channels" Periodica Math. Hungarica, vol. 2. pp. 191–213, 1972.
3. Havrda, J.H. and F. Charvat, "Quantification Methods of Classification Processes: Concept of Structural α Entropy, "Kybernatica, vol 3, pp. 30–35, 1967.
4. Kapur, J.N., "On Information of Order α and type β, "Proc. Ind. Acad. Sci., vol 48A pp. 65–75, 1968.
5. Kapur, J.N., "Comparative Assesment of Various Measures of Entropy" Jour. Inf, and Opt. Sci vol 4, no 1, pp. 207–232, 1983.
6. Kapur, J.N., "Four families of measures of entropy," Ind. Jour. Pure and App. Maths., Vol 17, no 4. pp 429–449, 1986.
7. Kapur, J.N., V, Kumar and U, Kumar, "Normalised Measures of Entropy", Int. Jour Gen Systems 1, pp 55–59, 1986.
8. Kapur, J.N., V, Kumar and U, Kumar, "Normalised Measures of Directed Divergence" Int Jour Gen Systems. 13, pp 5–16, 1986.
9. Kapur, J.N., V, Kumar and U, Kumar, "Normalised Measures of Useful Information" Operations Research (Adm Science A Canada) 7 (12), 1986.
10. Kapur, J.N., "Inaccuracy, Entropy, and Coding Theory", Tamkang Jour Math 18 (1), 35–48, 1987.
11. Kapur, J.N., "On Normalised Measures of Statistical Dependence", Proc Nat. Acad Sciences (to appear).
12. Kapur, J.N. and M, Dhande, "Families of Normalised Measures of Dependence" Acta. Ciencia Indica 16 M (2), 193–198, 1990.
13. Kerridge, D.E.K., "Inaccuracy and Inference", J. Roy. Stat. Soc, vol. 23 pp 184–194, 1961.
14. Kullback, S. and R.A. Leibler, "On Information and Sufficiency", Ann. Math. Stat. Vol. 22, pp. 79–86, 1951.
15. Reyni, A., "On Measures of Entropy and Information", Proc. 4th Berkley Symp. Math. Stat. Prob., Vol. 1, pp. 547–561, 1961.
16. Shannon, C.E., "A Mathematical Theory of Communication", Bell System Tech. J., Vol. 27, pp 379–423, 623–659, 1948.
17. Sharma, B.D. and I.J. Taneja, "Entropy of type (α, β) and other generalised Measures of Information", Matrika vol 22, pp. 205–214, 1974.
18. Sharma, B.D. and D.P. Mittal, "New Non-additive Measures of Entropy for Discrete Probability Distributions", J. Math. Sci., vol 10, pp. 28–40, 1975.

19

GENERATING FUNCTIONS FOR INFORMATION MEASURES

[Golomb [2] and Guiasu and Reischer [3] have given generating functions for Shannon's [17] measure of entropy and Kullback-Leibler's [15] measure of relative information. Here we obtain generating functions for other measures of entropy and relative information as well as for other measures of information based on two, three, four or m probability distributions.]

19.1 INTRODUCTION

Let,

$$P = (p_1, p_2, ..., p_n), Q = (q_1, q_2, ..., q_n) \tag{1}$$

be two probability distributions, then Golomb [2] defined an information generating function,

$$f(t) = -\sum_{i=1}^{n} p_i^t \tag{2}$$

with the property that,

$$f'(1) = -\sum_{i=1}^{n} p_i \ln p_i = S(P), \tag{3}$$

where $S(P)$ is Shannon's [17] measure of entropy for the probability distribution P. Also from (2), we get

$$f^{(r)}(1) = (-1)^{r-1} \sum_{i=1}^{n} p_i(-\ln p_i)^r, r = 1, 2, ... \tag{4}$$

Now $-\ln p_i$ is the self-information of the ith outcome, so that from (4), $(-1)^{r-1} f^r(1)$ gives the expected value of the rth power of the self-information or the average value of the rth powers of the self-informations of the n outcomes: Also, $-f^{(2)}(1) - (f'(1))^2$ gives the variance of the self-information.

Golomb [2] obtained simple expressions for the information generating function $f(t)$ for the uniform (discrete and continuous) distributions, geometric distribution, Zeta distribution, exponential distribution, Pareto distribution and normal distribution.

Later Guiasu and Reisher [3] defined the generating function $g(t)$ for relative information or cross-entropy or directed divergence of P from Q by

$$g(t) = \sum_{i=1}^{n} q_i (p_1/q_i)^t \tag{5}$$

with the property,

$$g'(1) = \sum_{i=1}^{n} p_i \ln p_i/q_i \tag{6}$$

$$g^{(r)}(1) = \sum_{i=1}^{n} p_i (\ln p_i/q_i)^r, \ r = 1,2,3,... \tag{7}$$

Since $(-\ln q_i) - (-\ln p_i) = \ln (p_i/q_i)$ is the change in self-information of the ith outcome when probability distribution P changes to Q, we deduce from (6) and (7) that

(*a*) average value of the change in self-information is given by $g'(1)$

(*b*) average value of the rth power of change in self-information is given by $g^{(r)}(1)$.

Using (6) and (7), we can find not only the variance of the change of self-information, as suggested by Guiasu and Reisher [3], but we can find all the moments about the mean of the distribution of self-information.

Now, it is well-known that,

$$D(P:Q) = \sum_{i=1}^{n} p_i \ln p_i/q_i \tag{8}$$

is always ≥ 0 and vanishes iff $P = Q$. It is known as Kullback-Leibler [15] measure of directed divergence or of relative information or of cross-entropy.

Thus, $f(t)$ and $g(t)$ give generating functions for two important measures of information, viz. Shannon's measure of entropy and Kullback-Leibler's measure of directed divergence.

However, during the last forty years, after these two measures were discovered, other useful measures of information have been given and it will be useful to have generating functions for all the measures.

We propose to give such generating functions in the present discussion for most measures of entropy, directed divergence and information improvement, generalised information improvement, measures of dependence and other measures of information based on m probability distribution. We would also prove some general theorems about generating functions.

19.2 GENERATING FUNCTIONS FOR MEASURES OF INFORMATION BASED ON ONE PROBABILITY DISTRIBUTION

Let,

$$f_\alpha(t) = \frac{1}{1-\alpha}\left(\sum_{i=1}^{n} (p_i^\alpha)^t - 1\right), \alpha \neq 1 \tag{9}$$

then,

$$f_\alpha(1) = \frac{1}{1-\alpha}\left(\sum_{i=1}^{n} p_i^\alpha - 1\right), \alpha \neq 1 \tag{10}$$

so that $f_\alpha(1)$ gives Harvda-Charvat's [4] measure of entropy. Also

$$f'_\alpha(0) = \frac{1}{1-\alpha} \ln \sum_{i=1}^{n} p_i^\alpha, \quad \alpha \neq 1 \tag{11}$$

which is Renyi's [16] measure of entropy so that $f_\alpha(t)$ is the generating function for Renyi's measure of entropy. Now

$$f_1(t) = \underset{\alpha \to 1}{Lt}\, f_\alpha(t) = -t \sum_{i=1}^{n} p_i \ln p_i \tag{12}$$

which is a trivial generating function for Shannon's measure of entropy. Now, as

$$f_1(1) = -\sum_{i=1}^{n} p_i \ln p_i, f'_1(1) = -\sum_{i=1}^{n} p_i \ln p_i, \tag{13}$$

so that both $f_1(1)$ and $f'_1(1)$ give Shannon's measure of entropy. Now define,

$$f_{\alpha,\beta}(t) = \frac{1}{\beta-\alpha}\left[\left(\frac{\sum_{i=1}^{n} p_i^\alpha}{\sum_{i=1}^{n} p_i^\beta}\right)^t - 1\right] \tag{14}$$

so that,

$$f_{\alpha,\beta}(1) = \frac{1}{\beta-\alpha}\frac{\sum_{i=1}^{n} p_i^\alpha - \sum_{i=1}^{n} p_i^\beta}{\sum_{i=1}^{n} p_i^\beta}, \alpha \neq \beta \tag{15}$$

which is $\left(\sum_{i=1}^{n} p_i^\beta\right)^{-1}$ times Sharma's and Taneja's [18] measure of entropy.

Now,

$$f_{\alpha,\beta}'(0) = \frac{1}{\beta-\alpha}\ln\frac{\sum_{i=1}^{n} p_i^\alpha}{\sum_{i=1}^{n} p_i^\beta}, \quad \beta \neq \alpha \tag{16}$$

$$\alpha > 1, \beta < 1$$

$$\text{or } \alpha < 1, \beta > 1$$

which is the measure of entropy due to Aczel and Daroczy [1] and Kapur [5] so that $f_{\alpha,\beta}(t)$ is the generating function for Aczel-Darozy-Kapur's measure of entropy.

Again we define

$$\bar{f}_{\alpha,\beta}(t) = \frac{1}{\beta-\alpha}\left[\left(\sum_{i=1}^{n} p_i^\alpha\right)^t - \left(\sum_{i=1}^{n} p_i^\beta\right)^t\right], \quad \alpha \neq \beta \tag{17}$$

so that,

$$\bar{f}_{\alpha,\beta}(1) = \frac{1}{\beta-\alpha}\left(\sum_{i=1}^{n} p_i^\alpha - \sum_{i=1}^{n} p_i^\beta\right), \quad \alpha \neq \beta \tag{18}$$

$$\alpha > 1, \beta < 1$$

$$\alpha < 1, \beta > 1$$

so that $\bar{f}_{\alpha,\beta}(1)$ is Sharma and Taneja [18] measure of entropy, and

$$\bar{f}'_{\alpha,\beta}(0) = \frac{1}{\beta-\alpha}\left(\ln\frac{\sum_{i=1}^{n} p_i^{\alpha}}{\sum_{i=1}^{n} p_i^{\beta}}\right), \quad \beta \neq \alpha \tag{19}$$

$$\beta > 1, \alpha < 1$$

$$\beta < 1, \alpha > 1$$

which is the same as (16), so that $\bar{f}_{\alpha,\beta}(t)$ can also be regarded as generating function for the measure given by (16).
Again we define

$$\bar{f}_a(t) = -\sum_{i=1}^{n} p_i^t + 1/a \sum_{i=1}^{n} (1+ap_i)^t - 1/a \sum_{i=1}^{n} (1+a)^t p_i, \tag{20}$$

so that,

$$\bar{f}_a'(1) = -\sum_{i=1}^{n} p_i \ln p_i + 1/a \sum_{i=1}^{n} (1+ap_i)\ln(1+ap_i) - 1/a(1+a)\ln(1+a). \tag{21}$$

which is Kapur's [8] measure of entropy so that $\bar{f}_a(t)$ is the generating function for Kapur's measure of entropy.

19.3 GENERATING FUNCTIONS FOR MEASURES OF INFORMATION BASED ON TWO PROBABILITY DISTRIBUTIONS

(*a*) *Measure of Directed Divergence*

(*i*) Let,

$$g_\alpha(t) = \frac{1}{\alpha-1}\left(\left(\sum_{i=1}^{n} p_i^{\alpha} q_i^{1-\alpha}\right)^t - 1\right), \alpha \neq 1 \tag{22}$$

so that,

$$g_\alpha(1) = \frac{1}{\alpha-1}\left(\sum_{i=1}^{n} p_i^{\alpha} q_i^{1-\alpha} - 1\right), \alpha \neq 1 \tag{23}$$

which is Havrda-Charvat [4] measure of directed divergence. Also,

$$g_\alpha'(1) = \frac{1}{\alpha-1}\ln\sum_{i=1}^{n} p_i^{\alpha} q_i^{1-\alpha}, \quad \alpha \neq 1 \tag{24}$$

so that $g_\alpha(t)$ is the generating function for Renyi's [16] measure of directed divergence.
Also,

$$g_1(1) = \underset{\alpha\to 1}{Lt}\, g_\alpha(1) = \sum_{i=1}^{n} p_1 \ln p_1/q_i; g_1'(1) = \sum_{i=1}^{n} p_i \ln p_i/q_i \tag{25}$$

give Kullback-Leibler's [15] measure of divergence.

(*ii*) Let

$$g_{\alpha,\beta}(t) = \frac{1}{\alpha-\beta}\left(\left(\sum_{i=1}^{n} p_i^{\alpha} q_i^{1-\alpha}\right)^t - \left(\sum_{i=1}^{n} p_i^{\beta} q_i^{1-\beta}\right)^t\right) \tag{26}$$

so that,

$$g'_{\alpha,\beta}(0) = \frac{1}{\alpha-\beta} \ln \frac{\sum_{i=1}^{n} p_i^{\alpha} q_i^{1-\alpha}}{\sum_{i=1}^{n} p_i^{\beta} q_i^{1-\beta}}, \quad \alpha \neq \beta \tag{27}$$

so that $g_{\alpha,\beta}(t)$ can be regarded as a generating function for Kapur's [6] measure of directed divergence

(*iii*) Let,

$$g_a(t) = \sum_{i=1}^{n} [p_i (p_i/q_i)^t - 1/a (1 + a p_i/q_i)^{t+1}] \tag{28}$$

so that,

$$\overline{g}_a'(0) = \sum_{i=1}^{n} p_i \ln p_i/q_i = 1/a \sum_{i=1}^{n} (1 + a p_i/q_i) \ln (1 + a p_i/q_i) \tag{29}$$

which gives Kapur's [6] measure of directed divergence so that $\overline{g}_a(t)$ can be regarded as a generating function of Kapur's measure of directed divergence.

(*b*) *Measures of Inaccuracy*

Let,

$$h_{\alpha}(t) = \frac{1}{1-\alpha} \left[\left(\frac{\sum_{i=1}^{n} p_i^{\alpha}}{\sum_{i=1}^{n} p_i^{\alpha} q_i^{1-\alpha}} \right)^t - 1 \right], \quad \alpha \neq 1, \tag{30}$$

so that,

$$h_{\alpha}(1) = \frac{1}{1-\alpha} \left[\left(\frac{\sum_{i=1}^{n} p_i^{\alpha}}{\sum_{i=1}^{n} p_i^{\alpha} q_i^{1-\alpha}} \right) - 1 \right], \quad \alpha \neq 1, \tag{31}$$

and,

$$h_{\alpha}'(0) = \frac{1}{1-\alpha} \ln \frac{\sum_{i=1}^{n} p_i^{\alpha}}{\sum_{i=1}^{n} p_i^{\alpha} q_i^{1-\alpha}}, \quad \alpha \neq 1, \tag{31a}$$

which is Renyi's measure of inaccuracy so that $h_{\alpha}(t)$ can be regarded as a generating function for Renyi's measure of inaccuracy. Again let

$$h(t) = -\sum_{i=1}^{n} p_i q_i^t, \tag{32}$$

so that,

$$h'(0) = -\sum_{i=1}^{n} p_i \ln q_i \tag{33}$$

which is Kerridge's [14] measure of inaccuracy. Thus $h(t)$ can be regarded as a generating function for Kerridge's measure of inaccuracy.

19.4 GENERATING FUNCTIONS FOR MEASURES OF INFORMATION BASED ON THREE OR FOUR PROBABILITY DISTRIBUTIONS

(*a*) *Measure of Information Improvement*

$$K_\alpha(t) = \frac{1}{\alpha - 1}\left[\left(\sum_{i=1}^{n} p_i^\alpha q_i^{1-\alpha}\right)^t - \left(\sum_{i=1}^{n} p_i^\alpha r_i^{1-\alpha}\right)^t\right], \quad \alpha \neq 1 \tag{34}$$

so that,

$$K'_\alpha(0) = \frac{1}{\alpha - 1}\ln\frac{\sum_{i=1}^{n} p_i^\alpha q_i^{1-\alpha}}{\sum_{i=1}^{n} p_i^\alpha r_i^{1-\alpha}} \quad \alpha \neq 1 \tag{35}$$

and,

$$K_1'(0) = \sum_{i=1}^{n} p_i \ln\frac{r_i}{q_i} \tag{36}$$

(36) gives Theil's [19] measure of information improvement and (35) gives a generalization due to Kapur [9]. $K_\alpha(t)$ can be regarded as a generating function for both these measures of information improvement. Similarly,

$$K_{\alpha,\beta}(t) = \frac{1}{\alpha - 1}\left[\left(\sum_{i=1}^{n} p_i^{\alpha+\beta-1} q_i^{1-\alpha}\right)^t - \left(\sum_{i=1}^{n} p_i^{\alpha+\beta-1} r_i^{1-\alpha}\right)^t\right], \alpha \neq 1 \tag{37}$$

is a generating function for the measure of information improvement given by,

$$\frac{1}{\alpha - 1}\ln\frac{\sum_{i=1}^{n} p_i^{\alpha+\beta-1} q_i^{1-\alpha}}{\sum_{i=1}^{n} p_i^{\alpha+\beta-1} r_i^{1-\alpha}}, \alpha \neq 1 \tag{38}$$

(*b*) *Measure of Generalised Information Improvement*

Let the true distribution be P and let the probability distribution Q be revised to R in course of time. In the same time, let the target distribution be revised to S, then the measure of generalised information improvement is,

$$1/2\left(\sum_{i=1}^{n} p_i \ln r_i/q_i + \sum_{i=1}^{n} s_i \ln r_i/q_i\right) \tag{39}$$

Now consider the generating function,

$$u_\alpha(t) = \frac{1}{2(\alpha - 1)}\left(\sum_{i=1}^{n} p_i^\alpha q_i^{1-\alpha}\right)^t + \left(\sum_{i=1}^{n} p_i^\alpha r_i^{1-\alpha}\right)^t - \left(\sum_{i=1}^{n} s_i^\alpha q_i^{1-\alpha}\right)^t - \left(\sum_{i=1}^{n} s_i^\alpha r_i^{1-\alpha}\right)^t \tag{40}$$

so that,

$$u_\alpha'(0) = \frac{1}{\alpha - 1} \ln \frac{\sum_{i=1}^{n} p_i^\alpha q_i^{1-\alpha} \sum_{i=1}^{n} s_i^\alpha q_i^{1-\alpha}}{\sum_{i=1}^{n} p_i^\alpha r_i^{1-\alpha} \sum_{i=1}^{n} s_i^\alpha r_i^{1-\alpha}} \tag{41}$$

$$u'_1(0) = 1/2 \sum_{i=1}^{n} [p_i \ln r_i/q_i + s_i \ln r_i/q_i], \tag{42}$$

so that $u_\alpha(t)$ can be regarded as a generating function for measure of generalised information improvement given by (39) and (40).

19.5 GENERATING FUNCTION FOR MEASURES OF INFORMATION BASED ON *m* PROBABILITY DISTRIBUTIONS

Let $P_1, P_2, \ldots\ldots P_m$ be m probability distributions and let P be the distribution whose i^{th} component is proportional to the geometrical mean of the ith components of $P_1, P_2, \ldots, P_m$, so that let

$$p_i^* = (p_{1i} p_{2i} \ldots p_{mi})^{1/m} \tag{43}$$

and,

$$A = \sum_{i=1}^{n} (p_{1i} p_{2i} \ldots p_{mi})^{1/m}, \tag{44}$$

then a measure of divergence of the m probability distributions [12, 13] is given by

$$-\ln A = -\ln[p_1^* + p_2^* + \ldots + p_m^*] \tag{45}$$

and the generating function is

$$v(t) = -(p_1^* + p_2^* + \ldots p_m^*)^t, \tag{46}$$

since

$$v'(0) = -\ln(p_1^* + p_2^* + \ldots + p_m^*) \tag{47}$$

19.6 SOME GENERAL PROPERTIES OF GENERAL FUNCTIONS

Theorem 1. The Shannon entropy generating function for the joint probability distribution of m independent probability distributions is equal to the product of the generating functions of the m probability distribution of m variates.

Proof: Let,

$$P(x_1 = i_1) = p_{1i_1}, P(x_2 = i_2) = p_{2i_2}, \ldots P(x_m = i_m) = p_{mi_m}, \tag{48}$$

then,

$$P(x_1 = i_1, x_2 = i_2, \ldots, x_m = i_m)$$

$$= p_{1i_1} p_{2i_2} \ldots p_{mi_m} \tag{49}$$

and
$$f(t) = \sum_{i_m=1}^{n_m} \dots \sum_{i_1=1}^{n_1} (p_{1i_1} p_{2i_2} \dots p_{mi_m})^t$$

$$= \sum_{i_1=1}^{n_1} (p_{1i_1})^t \sum_{i_2=1}^{n_2} (p_{2i_2})^t \dots \sum_{i_m=1}^{n_m} (p_{mi_m})^t$$

$$= f_1(t) f_2(t) \dots f_m(t) \tag{50}$$

Theorem 2. The Shannon entropy for the joint probability of m independent variates is equal to the sum of the Shannon entropies of the individual variates.

Proof: Taking logarithmic differentiation of (50)

$$\frac{f'(t)}{f(t)} = \frac{f_1'(t)}{f_1(t)} + \dots + \frac{f_m'(t)}{f_m(t)} \tag{51}$$

Putting $t = 1$ and remembering that

$$f_1(1) = f_2(1) = \dots = f_m(1) = 1 \tag{52}$$

and,

$$f_1'(1), f_2'(1), \dots, f_m'(1), f'(1) \tag{53}$$

are the entropies of the m variates and of the joint distribution, we get the required result.

Theorem 3. For Renyi's entropy generating function $f_\alpha(t)$, we have for m independent variates

$$(1 + (1-\alpha) f_\alpha(t)) = (1 + (1-\alpha) f_{1\alpha}(t)) \cdot (1 + (1-\alpha) f_{2\alpha}(t)) \dots (1 + (1-\alpha) f_{m\alpha}(t)) \tag{54}$$

where, $f_{1\alpha}(t), f_{2\alpha}(t) \dots f_{m\alpha}(t)$ are the generating functions for probability distributions for the m individual independent variates and $f_\alpha(t)$ is the generating function for the joint probability distribution.

Proof: This easily follows from (9).

Theorem 4. A result similar to that the theorem 2 holds for Renyi's entropy measure

Proof: Taking the logarithmic differentiation of (54) and putting $t = 0$ we get

$$\frac{f_\alpha'(0)}{1 + (1-\alpha) f_\alpha(0)} = \frac{f_{1\alpha}'(0)}{1 + (1-\alpha) f_{1\alpha}(0)} + \dots + \frac{f_{m\alpha}'(0)}{1 + (1-\alpha) f_{m\alpha}(0)}$$

or $$f'_\alpha(0) = f'_{1\alpha}(0) + \dots + f'_{m\alpha}(0) \tag{55}$$

which proves the theorem.

Theorem 5. For Aczel-Daroczy-Kapur entropy generating function, we have

$$(1 + (\beta - \alpha) f_{\alpha,\beta}(t)) = (1 + (\beta - \alpha) f_{1\alpha,\beta}(t)) \dots (1 + (\beta - \alpha) f_{m\alpha,\beta}(t)) \tag{56}$$

Proof: This follows easily from (14).

Theorem 6. A result similar to those of theorems 2 and 4 holds for Aczel-Daroczy-Kapur measure of entropy.

Proof: This follows from logarithmic differentiation of (56) and putting $t = 0$.

19.7 CONCLUDING REMARKS

(*i*) We have discussed the results for a random variate which takes a finite set of discrete values. The discussion can be easily extended to the cases when the number of values taken is infinite or the variate is continuous, provided the series and integrals which arise are all convergent.

(*ii*) The discussion can also be easily extended to multi-variate distributions.

(*iii*) Guiasu and Reisher [3] showed that their generating function for relative information can be used when both the distributions are binomial (with same N) or both are Poisson. It can be easily shown that their measure applies equally well when both distributions are negative binomial or both are generalised negative binomial or both belong to the class of generalised Lagrangian distributions, uni variate or multivariate [7].

Guiasu and Reisher [3] also showed that their generating function can also be used to find Watanbe's [20] measure of interdependence. However, the generating function can also be used to obtain many more measures of interdependence [10, 11], since every measure of directed divergence of the joint probability distribution of a number of random variates $x_1, x_2, \ldots, x_m$ from the probability distribution obtained as the product of marginal distributions gives a measure of interdependence.

Theorems 2, 4 and 6 about additive property of Shannon, Renyi and Kapur's measure of entropy are well known, but their proofs given here make use of the concept of generating function and as such are different from the earlier proofs.

REFERENCES

1 J. Aczel & Z. Daroczy. On Measures of Information and their Characterization. Academic Press. New York. 1975.

2. S. Golomb, "The Information Generating Function of a Probability Distribution" IEEE Trans Inf. Theory I T–12, 75–79 (1966).

3. S. Guiasu and C. Reisher, "The Relative Information Generating Function" Information Sciences 35, 235–241 (1985).

4. J.H. Havrda and F. Charvat, "Quantification Methods of Classification Processes: Concept of Structural α Entropy," Kybernatica, Vol. 3, pp. 30–35 (1967).

5. J.N. Kapur, 'Generalised Entropy of order α and type β", Maths Seminar 4, 78–94 (1967).

6. J.N. Kapur, "On Information of order α and type β", Proc Ind Acd Sci 48A, 65–74 (1968).

7. J.N. Kapur, "Maximum Entropy Formalism for some univariate and multivariate Lagrangian Distributions, Aligarh Jour. Stat 2, 1–16 (1982).

8. J.N. Kapur, "Four families of measures of Entropy", Ind. Jour Pure. App Maths 17 (4), 429–446 (1986).

9. J.N. Kapur, "Some Measures of Information Improvement" J C I S S 11 (1), 39–41 (1986).

10. J.N. Kapur, "Information Theoretic Measures of Stochastic Interdependence", Bull Math Ass Ind.22,43–58,1990.
11. J.N. Kapur and M. Dhande, "On Entropic Measures of Stochastic Dependence" Ind Jour Pure App Maths 17 (5), 581–591 (1986).
12. J.N. Kapur and V. Kumar, "A measure of mutual divergence among a number of probability distributions", Int Jour Math and Math Sci 10 (3), 597-600 (1987).
13. J.N. Kapur, and G.P. Tripathi, "New measures of directed and symmetric divergence based on *m* probability distributions", Ind Jour Pure App Math 19 (7), 617–626 (1988).
14. D.E, F. Kerridge, "Inaccuracy and Inference "J. Roy. Stat. Soc, vol. 23B, pp. 1 184–194 (1961).
15. S. Kullback and R.A. Liebler, "On Information and Sufficiency" Ann Math, Stat, Vol. 22, pp 79–86, (1951).
16. A Renyi, "On Measures of Entropy and Information", Proc 4th Berkley Symp. Maths, Prob. Vol. 1, pp 547–561, (1961).
17. C.E. Shannon, "A Mathematical Theory of Communication" Bell System Tech J, Vol. 27, pp 379–423, 623–659 (1948).
18. B.D. Sharma and I.J. Taneja, "Entropy of Type (α, β) and other generalised measures of Information" Metrika vol. 27, pp. 205–214, (1974).
19. H. Theil, "Economics and Information Theory", North Holland, Amesterdom (1967).
20. S. Watanabe, "Knowing and Guessing," John Wiley, New York (1969).

20

ON GENERATING APPROPRIATE MEASURES OF WEIGHTED ENTROPY AND WEIGHTED DIRECTED DIVERGENCE

[Appropriate measures of weighted entropy and weighted directed divergence have been defined and a number of these have been generated. The appropriate unweighted measures have been obtained as special cases.]

20.1 INTRODUCTION

Let $q_1, q_2, \ldots, q_n$ be the apriori probabilities of n possible outcomes and let $p_1, p_2, \ldots, p_n$ be their aposteriori probabilities. Let $w_1, w_2, \ldots, w_n$ denote the weights assigned to these outcomes and let $u_1, u_2, \ldots, u_n$ be their utilities, then Guiasu [4] gave his measures of weighted entropy,

$$H(P:W) = -\sum_{i=1}^{n} w_i p_i \ln p_i \tag{1}$$

and characterised it axiomatically. Belis and Gruiasu [1] give their qualitative-quantitative measure of information,

$$H(P:U) = -\sum_{i=1}^{n} u_i p_i \ln p_i \tag{2}$$

Taneja and Tuteja [8] gave the measure of weighted directed divergence,

$$D(P,Q:W) = \sum_{i=1}^{n} w_i p_i \ln \frac{p_i}{q_i} \tag{3}$$

Since then a large number of measures of weighted entropy and weighted directed divergence have been obtained in the literature [5] and have been characterised axiomatically.

Some of these are always positive, but some of these can also be negative for special weight distributions.

None of these is additive in the general case, but some of these can be additive in the special case when the weights are equal.

None of them is strictly recursive, but some are recursive in a limited sense and some are recursive when weights are equal.

Some of these are concave/convex, but others can be neither of these.

One major implicit consideration appears to have been that these weighted measures should be 'generalisations' of the well-known unweighted measures in the sense that these should reduce to the corresponding unweighted measures when the weights are equal and these should satisfy some plausible looking axioms. Similarly, unweighted parametric measures themselves are obtained as generalisations of Shannon's [7] measure of entropy,

$$H(P) = -\sum_{i=1}^{n} p_i \ln p_i \tag{4}$$

or of Kullback–Leibler's [6] measure of directed divergence,

$$D(P:Q) = \sum_{i=1}^{n} p_i \ln \frac{p_i}{q_i} \tag{5}$$

in the sense that for some special or limiting values of the parameters involved, the parametric measure should reduce to Shannon's or Kullback–Leiblers measure; otherwise different authors have assumed different conditions for their measures.

The object of the present discussion is to define these measures precisely and then to proceed to obtain some measures satisfying these stated conditions.

20.2 DEFINITION OF APPROPRIATE MEASURES OF WEIGHTED ENTROPY

An appropriate measure of $H(P:W)$ of probability distribution $P = (p_1, p_2, \ldots, p_n)$ $(p_i \geq 0\ \forall_i,\ \sum_{i=1}^{n} p_i = 1)$ and weight distribution $(w_1, w_2, \ldots, w_n)$ $(w_i \geq 0\ \forall_i)$ must satisfy the following conditions:

(*i*) It must be a continuous, permutationally symmetric function of $p_1, p_2, \ldots, p_n$; $w_1, w_2, \ldots, w_n$ in the sense that it must not change when the pairs (p_1, w_1), $(p_2, w_2), \ldots, (p_n, w_n)$ are permuted among themselves.

(*ii*) It must be maximum subject to $\sum_{i=1}^{n} p_i = 1$ when each p_i is same function of w_i, i.e. when

$$p_1 = g(w_1), p_2 = g(w_2), \ldots, p_n = g(w_n) \tag{6}$$

In particular, when the weights are equal, it must be maximum when,

$$p_1 = p_2 = \ldots = p_n = 1/n \tag{7}$$

and in this case, the maximum value should be a monotonic increasing function of n.

(*iii*) Its minimum value should be zero and this should arise for every degenerate distribution,

$$\Delta_1 = (1, 0, \ldots, 0), \Delta_2 = (0, 1, 0, \ldots, 0), \ldots, \Delta_n = (0, 0, 0, \ldots, 0, 1) \tag{8}$$

In other words

$$H(P:W) \geq 0 \tag{9}$$

and the equality sign holds when $P = \Delta_i$ for some i.

(*iv*) $H(P:W)$ should be a concave function of $p_1, p_2, \ldots, p_n$ so that when it is maximized subject to $\sum_{i=1}^{n} p_i = 1$ and other linear constraints of the form,

$$\sum_{i=1}^{n} p_i g_r(x_i) = a_r, r = 1, 2, ..., m, m \ll n, \tag{10}$$

the local maximum is a global maximum.

(v) When $H(P : W)$ is maximized subject to linear constraints, the maximizing probabilities should be ≥ 0, so that non-negativity constraints have not to be imposed separately.

We shall call such a measure of weighted entropy as 'appropriate for entropy maximization purposes'. Moreover, appropriateness here includes both conceptual appropriateness as well as mathematical-convenience appropriateness.

20.3 SOME APPROPRIATE MEASURES OF WEIGHTED ENTROPY

(i) *First category of appropriate measures of weighted entropy*

Consider,

$$H_\alpha(P:W) = -\sum_{i=1}^{n} w_i p_i^\alpha \ln p_i, \alpha \geq 0 \tag{11}$$

This is obviously continuous, permutationally symmetric, maximum when p_i is a function of w_i only, same for each i, always ≥ 0 and vanishes only for degenerate distributions. Now, let

$$f(p) = -p^\alpha \ln p, f'(p) = -p^{\alpha-1}(1+\alpha \ln p), f''(p) = -p^{\alpha-2}(\alpha(\alpha-1)\ln p + (2\alpha-1)) \tag{12}$$

Since $\ln p < 0, f''(p) < 0$ if $1/2 \leq \alpha \leq 1$. Thus $f(p)$ is concave and therefore $H_\alpha(P : W)$ is concave for $1/2 \leq \alpha \leq 1$.

Again maximizing (11) subject to

$\sum_{i=1}^{n} p_i = 1$ and (10), we get

$$-w_i p_i^{\alpha-1}(1+\ln p_i) = \lambda_0 + \lambda_1 g_1(x_i) + ... + \lambda_m g_m(x_i) \tag{13}$$

Since, L.H.S. involves In $p_i$and In p_i is defined only when $p_i > 0$, the maximizing probabilities will all be positive.

Thus,

$$H_\alpha(P:W) = -\sum_{i=1}^{n} w_i p_i^\alpha \ln p_i, \frac{1}{2} \leq \alpha \leq 1 \tag{14}$$

gives a family of appropriate measures of weighted entropy. When $\alpha = 1$, it gives Guiasu's measure of weighted entropy.

If $\alpha = 0$, (11) gives $-\sum_{i=1}^{n} w_i \ln p_i$ which is a convex function of $p_1, p_2, \ldots, p_n$. As such

$$H_0(P:W) = \sum_{i=1}^{n} w_i \ln p_i \tag{15}$$

gives an appropriate measure of weighted entropy, since it is concave and since it is easily shown that it always gives positive maximizing probabilities. This may be called Burg's [2] measure of weighted entropy. Of course, this is not positive, but this does not prevent us from using it for entropy-maximization purposes.

(*ii*) *Second category of appropriate measures of weighted entropy*

Consider,

$$H_a(P:W) = -\sum_{i=1}^{n} w_i p_i \ln p_i + \frac{1}{a}\sum_{i=1}^{n} w_i(1+ap_i)\ln(1+ap_i)$$

$$-w_{\min}\sum_{i=1}^{n}(1+a)\ln(1+a)p_i \tag{16}$$

where,

$$w_{\min} = \min(w_1, w_2, \ldots, w_n), \tag{17}$$

This is also continuous, permutationally symmetric and is maximum when p_i is same function of w_i only. Now consider,

$$g(p) = -wp\ln p + \frac{1}{a}w(1+ap)\ln(1+ap) - \frac{w_{\min}}{a}(1+a)\ln(1+a)p \tag{18}$$

so that,

$$g'(p) = -w(\ln p - \ln(1+ap)) - \frac{w_{\min}}{a}(1+a)\ln(1+a) \tag{19}$$

$$g''(p) = -w\left(\frac{1}{p} - \frac{a}{1+ap}\right) = \frac{-w}{p(1+ap)} < 0, \tag{20}$$

so that, $g(p)$ is concave if $a \geq -1$

Its minimum value is zero at one of the degenerate distributions, though at all other degenerate distributions, its value must be > 0. Thus, $H_a(P: W) \geq 0$ and the equality sign holds at one of the degenerate distributions. If $a = -1$, we get,

$$H_{-1}(P:W) = -\sum_{i=1}^{n} w_i p_i \ln p_i - \sum_{i=1}^{n} w_i(1-p_i)\ln(1-p_i) \tag{21}$$

and this has the minimum value zero at every degenerate distribution. Also,

$$H_0(P:W) = -\sum_{i=1}^{n} w_i p_i \ln p_i + \underset{a\to 0}{Lt}\frac{\sum_{i=1}^{n} w_i(1+ap_i)\ln(1+ap_i)}{a}$$

$$-\underset{a\to 0}{Lt}\, w_{\min}\frac{(1+a)\ln(1+a)}{a}$$

$$= -\sum_{i=1}^{n} w_i p_i \ln p_i + \sum_{i=1}^{n} w_i p_i - w_{\min}, \tag{22}$$

so that the family (16) includes Guiasu's measure as a special case since in maximization problem, the last two terms make no difference.

Thus, when a ≥ -1, (16) gives a family of appropriate measure of weighted entropy.

(iii) Third Category of appropriate measure of weighted entropy.
consider the measure,

$$H_F(P:W) = \sum_{i=1}^{n} w_i F\,(p_i), \tag{23}$$

where,

$$F'(p) = G(\ln p), F''(p) = G'(\ln p)\frac{1}{p}, \tag{24}$$

so that $F(p)$ is a concave function of p if $G'(x) < 0$ when $x < 0$. Moreover, maximizing (23) subject to linear constraints, we get

$$w_i G\,(\ln p_i) = \lambda_0 + \lambda_1 g_1(x_i) + ... + \lambda_m g_m(x_i), \tag{25}$$

so that all maximizing probabilities are positive.

As a special case, let

$$F_m'(p) = (-\ln p)^m, F_m''(p) = m\,(-\ln p)^{m-1}\left(-\frac{1}{p}\right) < 0 \tag{26}$$

When $m = 1$,

$$F_1'(p) = -\ln p$$

$$F_1(p) = -p \ln p + p = p\,(1 - \ln p) > 0 \tag{27}$$

When $m = 2$

$$F_2'(p) = (-\ln p)^2 = (\ln p)^2,$$

so that,

$$F_2(p) = p\,(\ln p)^2 - 2\int \ln p\,dp = p\,(\ln p)^2 + 2F_1(p)$$

$$= (p \ln p)^2 + 2p\,(1 - \ln p) > 0 \tag{28}$$

When $m = 3$

$$F_3'(p) = -(\ln p)^3$$

$$F_3(p) = -p\,(\ln p)^3 + 3\int (\ln p)^2 dp = -p\,(\ln p)^3 + 3F_2(p) > 0 \tag{29}$$

In the same way,

$$F_4(p) = p\,(\ln p)^4 + 4F_3(p) > 0 \tag{30}$$

$$F_5(p) = -p\,(\ln p)^5 + 5F_4(p) > 0 \tag{31}$$

and in general,

$$F_m(p) = p(-\ln p)^m + mF_m(p)$$

$$= p\,(-\ln p)^m + m[p\,(-\ln p)^{m-1} + (m-1)F_{m-2}(p)]$$

$$= p\,[(-\ln p)^m + m(-\ln p)^{m-1} + m\,(m-1)(-\ln p)^{m-2}$$

$$+ m\,(m-1)(m-2)(-\ln p)^{m-3} + ... + m\,(m-1)...1 \tag{32}$$

Thus,

$$H_m(P:W) = \sum_{i=1}^{n} p_i w_i[(-\ln p_i)^m + m(-\ln p_i)^{m-1} + \ldots + m(m-1)\ldots 1] \quad (33)$$

gives an appropriate measure of weighted entropy for every positive value of m. For $m = 1$, it gives

$$H_1(P:W) = \sum_{i=1}^{n} w_i p_i - \sum_{i=1}^{n} w_i p_i \ln p_i \quad (34)$$

which is only slightly different from Gauisu's measure of weighted entropy, However, its maximum subject to $\sum_{i=1}^{n} p_i = 1$ occurs when $p_i = q_i$, where

$$\frac{\ln q_i}{\frac{1}{w_1}} = \frac{\ln q_2}{\frac{1}{w_2}} = \ldots = \frac{\ln q_n}{\frac{1}{w_n}} \text{ or } w_i = -\frac{k}{\ln q_i}, i = 1, 2, \ldots, n \quad (35)$$

so that our measure becomes,

$$\sum_{i=1}^{n} \frac{p_i \ln p_i - p_i}{\ln q_i} \quad (36)$$

(36) gives a measure of weighted entropy whose maximum subject to $\sum_{i=1}^{n} p_i = 1$ occurs when $p_i = q_i$ where $(q_1, q_2, \ldots, q_n)$ is a specified probability distribution, so that

$$\sum_{i=1}^{n} \frac{q_i \ln q_i - q_i - p_i \ln p_i + p_i}{\ln q_i} \quad (37)$$

is a convex function of $p_1, p_2, \ldots, p_n$ which is always ≥ 0 and attains its minimum value when $p_i = q_i$ for each i so that (37) can be used as a measure of directed divergence of $P = (p_1, p_2, \ldots, p_n)$ from $Q = (q_1, q_2, \ldots, q_n)$.

If we maximize (33) subject to linear constraints, we get

$$w_i(-\ln p_i)^m = \lambda_0 + \lambda_1 g_1(x_i) + \ldots + \lambda_m g_m(x_i) \quad (38)$$

or,

$$-\ln p_i = [((\lambda_0 + \lambda_1 g_1(x_i) + \ldots + \lambda_m g_m(x_i)) \mid w_i]^{1/m} \quad (39)$$

If m is odd, the R.H.S has only one real value which has to be positive and we equate it to $-\ln p_i$ to get p_i.

If m is even, the R.H.S. can have two real values, one of which has to be positive and we have to equate it to $-\ln p_i$

In either case, we get a unique value of p_i.

20.4 DEFINITION OF AN APPROPRIATE MEASURE OF WEIGHTED DIRECTED DIVERGENCE

A measure $D(P : Q, W)$ will be said to be an appropriate measure of directed divergence if

(*i*) It is a continuous functions of $p_1, p_2, \ldots, p_n; q_1, q_2, \ldots, q_n; w_1, w_2, \ldots, w_n$.

(*ii*) It is permututionally symmetric, i.e. it does not change when the triplets $(p_1, q_1, w_1), (p_2, q_2, w_2), \ldots, (p_n, q_n, w_n)$ are permuted among themselves.

(*iii*) It is always ≥ 0 and vanishes when $p_i = q_i$ for each i.

(*iv*) It is a convex function of $p_1, p_2, \ldots, p_n$ which has its minimum value zero when $p_i = q_i$ for each i.

(*v*) It reduces to an ordinary measure of directed divergence when all the weights are equal.

20.5 SOME APPROPRIATE MEASURES OF WEIGHTED DIRECTED DIVERGENCE

(*i*) *First category of appropriate measures of weighted directed divergence*

Csiszer's [3] general class of measure of directed divergence are given by

$$D(P:Q) = \sum_{i=1}^{n} q_i \phi\left(\frac{p_i}{q_i}\right) \tag{40}$$

where, $\phi(\cdot)$ is a twice-differentiable convex function with $\phi(1) = 0$. We are naturally interested in knowing whether

$$D(P:Q;W) = \sum_{i=1}^{n} w_i q_i \, \phi\left(\frac{p_i}{q_i}\right) \tag{41}$$

can be an appropriate measure of weighted directed divergence. It satisfies conditions (*i*) and (*ii*) above. It also vanishes when $p_i = q_i$ for each i. It is also a convex function of $p_1, p_2, \ldots, p_n$. Its minimum value subject to $\sum_{i=1}^{n} p_i = 1$ arises when

$$\frac{\phi'\left(\frac{p_i}{q_i}\right)}{\frac{1}{w_1}} = \frac{\phi'\left(\frac{p_2}{q_2}\right)}{\frac{1}{w_2}} = \ldots = \frac{\phi'\left(\frac{p_n}{q_n}\right)}{\frac{1}{w_n}} \tag{42}$$

and when $w_1, w_2, \ldots, w_n$ are equal, its minimum value arises when $p_i = q_i$ for each i. However, if $w_1, w_2, \ldots, w_n$ are not equal, then this can arise when $p_i = q_i$ for each i only when,

$$\phi'(1) = 0 \tag{43}$$

As such in this case of weighted measure $\phi(x)$ has to satisfy the condition (43) in addition to its begin convex, twice differentiable and vanishing at $x = 1$.

A function satisfying all these conditions is

$$\phi(x) = x \ln x - x + 1 \tag{44}$$

and gives appropriate measure,

$$D_1(P:Q;W) = \sum_{i=1}^{n} w_i q_i \left[\frac{p_i}{q_i} \ln \frac{p_i}{q_i} - \frac{p_i}{q_i} + 1\right] = \sum_{i=1}^{n} w_i \left[p_i \ln \frac{p_i}{q_i} - p_i + q_i\right] \tag{45}$$

Another function which satisfies all our conditions is

$$\phi(x) = \frac{x^{\alpha} - \alpha x + \alpha - 1}{\alpha(\alpha - 1)}, \alpha \neq 0, 1 \tag{46}$$

and this gives us the appropriate measure of weighted directed divergence,

$$D_2(P:Q;W) = \frac{1}{\alpha(\alpha-1)} \sum_{i=1}^{n} w_i [p_i^{\alpha} q_i^{1-\alpha} - \alpha p_i + (\alpha-1) q_i], \alpha \neq 0, \alpha \neq 1 \tag{47}$$

When, $\alpha \to 1$, this gives,

$$D_1(P:Q;W) = \sum_{i=1}^{n} w_i \left[p_i \ln \frac{p_i}{q_i} - p_i + q_i \right], \tag{48}$$

which is the same as (45) and when $\alpha \to 0$, it gives

$$D_1(P:Q;W) = -\sum_{i=1}^{n} w_i \left[q_i \ln \frac{p_i}{q_i} - p_i + q_i \right] = \sum_{i=1}^{n} w_i \left[q_i \ln \frac{q_i}{p_i} - q_i + p_i \right], \tag{49}$$

which is essentially the same as (45).

A third function which satisfies our conditions is

$$\phi(x) = x \ln x - \frac{1}{a}(1+ax) \ln(1+ax) + \frac{1}{a} \ln(1+a) + x \ln(1+a) \tag{50}$$

This gives the third appropriate measure of weighted directed divergence,

$$D_3(P:Q;W) = \sum_{i=1}^{n} w_i \left[p_i \ln \frac{p_i}{q_i} - \frac{1}{a}(q_i + a p_i) \ln\left(1 + a\frac{p_i}{q_i}\right) + \frac{1}{a} \ln(1+a) q_i + p_i \ln(1+a) \right] \tag{51}$$

(ii) A Second Category of Appropriate Measures of Weighted Directed Divergence

Consider the measure (17) of unweighted directed divergence and consider the corresponding measure,

$$\sum_{i=1}^{n} w_i \left[\frac{q_i \ln q_i - q_i - p_i \ln p_i + p_i}{\ln q_i} \right] \tag{52}$$

This is a convex function of $p_1, p_2, \ldots, p_n$ which vanishes when $p_i = q_i$ for each i. However its minimum subject to $\sum_{i=1}^{n} p_i = 1$ occurs when

$$w_1 \frac{\ln p_i}{\ln q_i} = w_2 \frac{\ln p_i}{\ln q_i} = - \ldots = w_n \frac{\ln p_n}{\ln q_n} \tag{53}$$

and this will not occur when $p_i = q_i$ for each i unless the weights are all equal. As such the measure (52) can be negative. We now modify it to

$$D_4(P:Q;W) = \sum_{i=1}^{n} w_i \frac{p_i \ln q_i - p_i \ln p_i - q_i + p_i}{\ln q_i} \tag{54}$$

This is a convex function of $p_1, p_2, \ldots, p_n$ which vanishes when $p_i = q_i$ for each i and its minimum occurs when

$$\frac{w_1(\ln q_1 - \ln p_1)}{\ln q_i} = \frac{w_2(\ln q_2 - \ln p_2)}{\ln q_2} = \ldots = \frac{w_n(\ln q_n - \ln p_n)}{\ln q_n} \tag{55}$$

i.e. when $q_i = p_i$ for each i, so that (54) gives an appropriate measure of weighted directed divergence.

Again corresponding to (23), we get the measure,

$$D_5(P:Q;W) = \sum_{i=1}^{n} w_i \left[\frac{F(q_i) - F(p_i)}{F'(q_i)} + p_i - q_i \right] \tag{56}$$

This is appropriate when $F'(q_i) > 0$ since it is convex ($\because F(p_i)$ is concave and $F'(q_i) > 0$), it vanishes when $p_i = q_i$ for each i and it is minimum when,

$$w_1\left[-\frac{F'(p_1)}{F'(q_1)} + 1 \right] = w_2\left[-\frac{F'(p_2)}{F'(q_2)} + 1 \right] = \ldots = w_n\left[-\frac{F'(p_n)}{F'(q_n)} - 1 \right] \tag{57}$$

i.e. when $p_i = q_i$ for each i.

In fact (56) gives a formula for generating appropriate measures of weighted directed divergence from any measure of entropy of the form,

$$\sum_{i=1}^{n} F(p_i) \tag{58}$$

where $F(\cdot)$ is a continuous twice-differentiable concave function and $F'(q) > 0$. Thus,

$F(x) = x - x \ln x$ gives

$$D_6(P:Q;W) = \sum_{i=1}^{n} w_i \left[\frac{p_i - p_i \ln p_i - q_i + q_i \ln q_i}{\ln q_i} + p_i - q_i \right] \tag{59}$$

$F(x) = \frac{x^\alpha}{\alpha(1-\alpha)}$ gives

$$D_7(P:Q,W) = \sum_{i=1}^{n} \frac{w_i}{1-\alpha} \left[\frac{q_i^\alpha - p_i^\alpha}{\alpha\, q_i^{\alpha-1}} + p_i - q_i \right], \alpha \neq 0, 1$$

$$= \sum_{i=1}^{n} w_i \left[\frac{q_i^\alpha - p_i^\alpha + \alpha p_i q_i^{\alpha-1} - \alpha q_i^\alpha}{\alpha(1-\alpha) q_i^{\alpha-1}} \right]$$

$$= \sum_{i=1}^{n} \frac{w_i}{\alpha(1-\alpha)} [q_i - p_i^\alpha q_i^{1-\alpha} + \alpha p_i - \alpha q_i] \tag{60}$$

20.6 SOME NEW MEASURES OF DIRECTED DIVERGENCE

In the new measures of weighted directed divergence, we have obtained in section 5, we can take w_i as any positive function of q_i since the measure will still be always ≥ 0, would vanish iff $p_i = q_i$ and would remain a convex function of $p_1, p_2, \ldots, p$. In particular we can always choose $w_i = q^\beta{}_i$, to get new measure of directed divergence where in general $\beta > 0$, but if none of q_i's is 0, β can be taken to be negative. Thus, from (48), we get the measure,

$$D_1(P:Q) = \sum_{i=1}^{n} \left[\frac{p_i}{q_i} \ln \frac{p_i}{q_i} - \frac{p_i}{q_i} + 1 \right] \tag{59}$$

From (49) we get

$$D_2(P:Q) = \sum_{i=1}^{n} \left[\ln \frac{q_i}{p_i} - 1 + \frac{p_i}{q_i} \right] = \sum_{i=1}^{n} \left(\frac{p_i}{q_i} - \ln \frac{p_i}{q_i} - 1 \right) \tag{60}$$

From (50) we get

$$D_3(P:Q) = \sum_{i=1}^{n} \left[\frac{p_i}{q_i} \ln \frac{p_i}{q_i} - \frac{1}{a}\left(1 + \frac{ap_i}{q_i}\right) \ln \left(1 + \frac{ap_i}{q_i}\right) + \frac{p_i}{q_i} \ln(1+a) + \frac{1}{a} \ln(1+a) \right] \tag{61}$$

From (54) we get

$$D_4(P:Q) = \sum_{i=1}^{n} [p_i \ln p_i - p_i - p_i \ln q_i + q_i] \tag{62}$$

which is essentially Kullback-Liebler measure.

From (48) we get

$$D_5(P:Q) = \frac{1}{\alpha(\alpha-1)} \sum_{i=1}^{n} [p_i^{\alpha} - \alpha p_i q_i^{\alpha-1} + (\alpha-1) q_i^{\alpha}], \alpha \neq 0, \alpha \neq 1 \tag{63}$$

In the general case if we put $w_i = q_i^{\beta}$, $\beta > 0$ in each of the appropriate measure of directed divergence, we shall get measures of directed divergence containing one more parameter than the original measure.

20.7 CONCLUDING REMARKS

(*i*) Earlier very often measures of entropy, weighted entropy, directed divergence, weighted directed divergence have been obtained without stating explicitly the conditions they are expected to satisfy. Here the conditions are explicitly stated and measures satisfying these conditions are obtained. Many of the earlier measure satisfy these conditions. The two conditions not considered here are those of additivity and recursivity. These do not play an important role in optimization problems. Thus when a measure is maximized, every monotonic increasing function of it is also maximized, but while some of these may be additive or recursive, others may not be, though all are maximized by the same probability distribution. It may be noted here that in all entropy maximization problems, we are not so much interested in the maximum value of the entropy, as in the maximizing probability distribution.

(*ii*) We have not discussed here how to assign the weights partly because such an assignment will depend on the specific applications we have in view. Thus the weights may depend on the penalty to be paid due to our not being able to meet the demands, these may be due to different returns from different outcomes in stock market or the different losses caused by earthquakes of different intensities and so on. Some of these matters may be discussed in another communications. There is no doubt however that deeper research is needed in this area and research will have to be oriented to specific applications.

REFERENCES

1. M. Belis and S. Guiasu (1968), "A Quantitative-Qualitative Measures of Information in Cybernatic Systems". IEEE. Trans-inf. Th., vol. IT–4, pp. 593–594.

2. J.P. Burg (1972), ''The relationship between Maximum Entropy Spectra and Maximum Likelihood Spectra'', in Modern Spectral Analysis, ed. D.G. Childers pp. 130–131.
3. I. Csiszer (1972), ''A class of Measures of Informativity of Observations Channels,'' Periodic Math. Hungarica, Vol. 2, pp. 191–213.
4. S. Guiasu (1971), ''Weighted Entropy, ''Reports on Math. Physics, Vol. 2. pp. 165–179.
5. J.N. Kapur (1986), ''New qualitative-quantitative measure of information'', Nat. Acad. Sci. Lelters , vol. 6 (2), 51–54.
6. S. Kullback and R.A. Leibler (1951), ''On Information and sufficiency,'' Ann. Math. Stat., vol. 22, pp. 79–86.
7. C.E. Shannon (1948), ''A Mathematical Theory of Communication, ''Bell System Tech. J., Vol. 27, pp. 379–423, 623–659.
8. H.C. Taneja and R.K. Tuteja (1984), ''Characterisation of a qualitative quantitative measure of relative information'' Information Science, vol. 33, pp. 106.

21

MEASURING THE UNCERTAINTY OF A SET OF PROBABILITY DISTRIBUTIONS

[Two alternative ways are given for measuring the uncertainty of a set of probability distributions, which satisfy a given set of constraints.]

21.1 INTRODUCTION

Uncertainty is a very important phenomenon in modern life and there have been many efforts to understand it, to describe it, to measure it, to regulate it and to maximize or minimize it.

There is a great struggle between probability theorists (Cheesman [1] Lindley [9]) and fuzzy-set theorists (Klir [7] Kosko [9]) for possessing the soul of uncertainty. While probability theorists claim that all aspects of uncertainty can be encompassed by probability theory, the fuzzy-set theorists feel that there are concepts like possibility, plausibility, feasibility, specificity etc. which require fuzzy-set theory for their description. There are still others who feel that there are many aspects of uncertainty which are encompassed neither by probability theory nor by fuzzy-set theory.

Shannon [11] gave an important measure for measuring the uncertainty of the probability distribution,

$$P = (p_1, p_2, \ldots, p_n)$$

viz.,

$$S(P) = -\sum_{i=1}^{n} p_i \ln p_i \tag{1}$$

This measure was used by Jaynes [3] in his principle of maximum entropy. This principle has had tremendous applications in a variety of fields like statistical mechanics, thermodynamics, regional and urban planning, spatial structures, business, economics statistics, non-linear spectral analysis, computerised tomography, pattern recognition, risk analysis, queueing theory, reliability theory, marketing, data structures (Kapur [4, 5], Kapur and Kesavan [6]).

There is another type of uncertainty corresponding to a set of probability distributions which satisfy the set of constraints,

$$\sum_{i=1}^{n} p_i = 1, \sum_{i=1}^{n} p_i g_{ri} = a_r, r = 1, 2, \ldots m; p_i \geq 0 \ldots p_n \geq 0 \tag{2}$$

where, $m << n$.

This uncertainty arises because in general, there is an infinity of probability distributions satisfying (2) and there is uncertainty as to which one of these we should choose.

This is conceptually different from the uncertainty of a single probability distribution.

Thus, there is an uncertainty as to which outcome will be realised. This can be removed only by performing an experiment and finding out the outcome. The uncertainty is about the outcome, i.e. whether it is outcome 1 or outcome 2 or . . . outcome n. We have to make a choice out of a finite number of alternatives.

In the present case, there is an infinity of probability distributions satisfying (2) and there is an uncertainty about which probability distribution is the 'real' one. This uncertainty can be removed by getting more constraints of the type (2). Every independent constraint consistent with the earlier constraints will give us information which will reduce this uncertainty and hopefully we can remove this uncertainty altogether by getting sufficient additional constraints.

Our problem is to find a measure for this uncertainty of the set and to find how much of this uncertainty can be removed by each additional constraint which represents an additional piece of information.

21.2 THE MEASUREMENT OF UNCERTAINTY OF A SET OF DISTRIBUTIONS

Since the entropy measure (1) is a concave function, it has a unique global maximum value when it is maximized subject to (2). Let this value be S_{max}. There will be a unique probability distribution corresponding to this maximum value and this is called the maximum entropy probability distribution.

Again there will be a globally minimum value S_{min} when it is minimized subject to the same constraint (2), but there can be many minimum-entropy probability distributions having this minimum value S_{min}.

Every other probability distribution satisfying (2) has its entropy S between S_{min} and S_{max}, i.e. it lies in the interval (S_{min}, S_{max}).

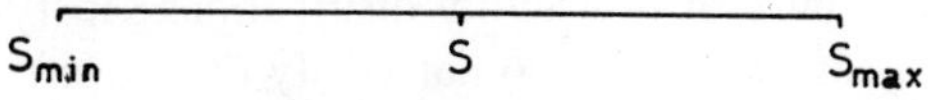

S_{min} S S_{max}

Fig. 21.1

When there is only one constraint, viz. the natural constraint

$$\sum_{i=1}^{n} p_i = 1. \tag{3}$$

then,
$$S_{max} = \ln n, \; S_{min} = 0, \; S_{max} - S_{min} = \ln n \tag{4}$$

As more and more constraints are imposed S_{max} will in general decrease and S_{min} will in general increase, so that with every additional constraint $S_{max} - S_{min}$ will in general decrease.

It will not decrease and retain its earlier value only if the additional constraint is not independent of the earlier ones and can be obtained as a linear combination of the earlier constraints.

We now define

$$U = \frac{S_{max} - S_{min}}{\ln n} \tag{5}$$

and call it the uncertainty index of the set of probability distributions defined by (2). It lies between 0 and 1 and goes on decreasing with every additional independent constraint. It becomes zero when sufficient constraints have been introduced to give a unique probability distribution since in that case $S_{max} = S_{min}$ and there will be no uncertainty about the probability distribution since it would be uniquely known. Thus the second type of uncertainty about which we are talking, will be removed, though the first type of uncertainty about which Shannon talked will remain, showing that the two types of uncertainty are distinct:

If our set consists of one probability distribution only, there is only the first type of uncertainty.

If our set consists of many, possibly infinity of probability distributions, we have both types of uncertainties. First we have to remove the second type by choosing one distribution and then we have to remove the first type by choosing one outcome.

21.3 TWO EXAMPLES TO ILLUSTRATE THE USE OF THIS MEASURE

Suppose we have a six-faced dice, then in the beginning, we have only one constraint, viz.

$$p_1 + p_2 + \ldots + p_6 = 1, \quad p_1 \geq 0, \ldots, p_6 \geq 0 \tag{6}$$

There is an infinity of probability distributions satisfying it. The maximum uncertainty is ln 6, the minimum uncertainty is 0 and the length of the interval of uncertainty is ln 6,.

Now suppose we are given the additional constraint

$$p_1 + 2p_2 + 3p_3 + 4p_4 + 5p_5 + 6p_6 = 3.5 \tag{7}$$

There will still be an infinity of probability distributions satisfying it, though an infinity of these which satisfy (6) and do not satisfy (7) are left out. The uniform distribution still satisfies it and so S_{max} is still ln 6, None of the degenerate distribution (i.e. those in which one component is unity) satisfies it and as such $S_{min} > 0$. The length of the interval of uncertainty < ln 6 and the uncertainty index is less than unity.

Similarly, if we impose is succession, the constraints,

$$p_1 + 2^2 p_2 + 3^2 p_3 + 4^2 p_4 + 5^2 p_5 + 6^2 p_6 = \frac{91}{6} \tag{8}$$

$$p_1 + 2^3 p_2 + 3^3 p_3 + 4^3 p_4 + 5^3 p_5 + 6^3 p_6 = \frac{441}{6} \tag{9}$$

$$p_1+2^4p_2+3^4p_3+4^4p_4+5^4p_5+6^4p_6=\frac{2275}{6} \tag{10}$$

$$p_1+2^5p_2+3^5p_3+4^5p_4+5^5p_5+6^5p_6=\frac{12201}{6} \tag{11}$$

the length of the interval of uncertainty goes on decreasing at each stage. Since the uniform distribution satisfies all these constraints, S_{max} is always ln 6, but S_{min} goes on increasing till at the last stage, the uniform distribution is the only distribution satisfying all the constraints and as such $S_{min} = S_{max} = \ln 6$, so that the length of the interval is continuously decreasing and ultimately has become zero.

The Figure 21.2 illustrates how S_{max} is always ln 6, S_{min} always increases, the length $S_{max}-S_{min}$ decreases, U decreases and finally it becomes zero and we get only one probability distribution, viz. the uniform distribution

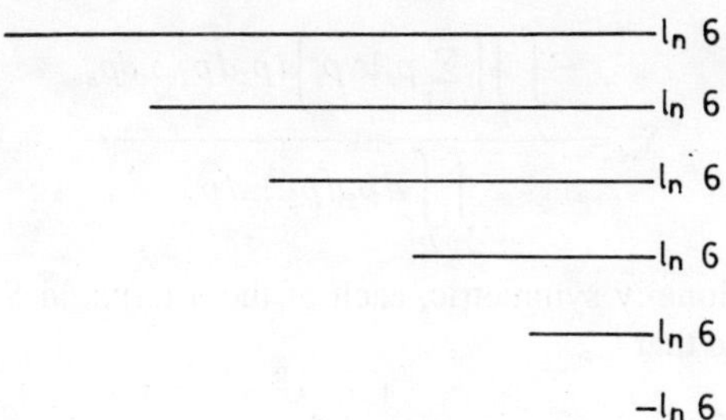

Fig. 21.2

As a second example, consider the game of finding a number thought of by a person by asking him a sequence of questions. Suppose we ask him to choose a number between 1 and 1024. Let p_i be the probability of his having thought of the number i, then we get the probability distribution $(p_1, p_2, \ldots, p_{1024})$ with $S_{max} = \ln 1024$, $S_{min} = 0$. If we use logarithms to the base 2, then $S_{max} = 10$, $S_{min} = 0$. Now we ask him to tell us whether his number lies between 1 and 512 or between 513 and 1024. Whatever be his answer, the new probability distribution will have $S_{max} = \ln_2 512 = 9$ but $S_{min} = 0$. Thus if we ask him 10 questions of the same type, S_{max} will reduce from 10 bits to 0 bits and S_{min} will always remain 0 so that U will change from 1 to 0 and every question reduces U by 1/10.

21.4 AN ALTERNATIVE WAY OF MEASURING THE UNCERTAINTY OF THE SET OF DISTRIBUTIONS

In this method we attempt to find the average of the Shannon entropies of all the feasible distributions.

The constraint set (2) defines a region in an n-dimensional probability space bounded by $n + m + 1$ hyper planes, viz.

$$p_1=0, p_2=0, \ldots, p_n=0, \sum_{i=1}^{n} p_i=1, \sum_{i=1}^{n} p_i g_{ri}=a_r, r=1,2\ldots m \tag{12}$$

Each feasible probability distribution is represented by a point in this space. Since we have no reason to believe otherwise on the basis of the given constraints, we assume that all points in this region are equally likely or of $(p_1, p_2, \ldots, p_n)$ falling in a sub-region of this space is proportional to the hyper volume of this space so that the average uncertainty is given by

$$S_{av} = \frac{\int\int_D S(P) dp_1 dp_2 \ldots dp_n}{\int\int_D dp_1 dp_2 \ldots dp_n} \tag{13}$$

where D is the region bounded by the $n + m + 1$ hyper planes,

We first consider the simple problem when there is only one constraint, the natural constraint so that D is the region bounded by the hyper planes.

$$p_1 = 0, p_2 = 0, \ldots, p_n = 0, \quad p_1 + p_2 + \ldots + p_n = 1 \tag{14}$$

so that

$$S_{av} = \frac{-\int\int_D \left(\sum_{i=1}^{n} p_i \ln p_i\right) dp_1 dp_2 \ldots dp_n}{\int\int_D dp_1 dp_2 \ldots dp_n} \tag{15}$$

Since S is permutationally symmetric, each of the n terms in S (P) will make the same contribution, so that

$$S_{av} = -n \frac{\int_0^1 p_1 \ln p_1 dp_1 \ldots \int_0^{1-p_i \cdots p_{n-2}} dp_{n-1} \int_0^{1-p_1 \cdots p_{n-1}} dp_n}{\int_0^1 dp_1 \int_0^{1-p_1 \cdots p_{n-2}} dp_{n-1} \int_0^{1-p_1 \cdots p_{n-1}} dp_n} \tag{16}$$

$$= -n \frac{\int_0^1 p_1 \ln p_1 \frac{(1-p_i)^{n-1}}{(n-1)!} dp_1}{\int_0^1 \frac{(1-p_1)^{n-1}}{(n-1)!} dp_1}$$

$$= -n^2 \int_0^1 p_1 \ln p_1 (1-p_1)^{n-1} dp_1 \tag{17}$$

Now using the definition of beta function

$$B(m,n) = \int_0^1 x^{m-1}(1-x)^{n-1} dx \tag{18}$$

Differentiating with respect to m

$$\frac{\partial}{\partial m} \ln B(m,n) = \frac{\int_0^1 x^{m-1} \ln x (1-x)^{n-1} dx}{\int_0^1 x^{m-1}(1-x)^{n-1} dx}$$

or,
$$\frac{\Gamma'(m)}{\Gamma(m)} - \frac{\Gamma'(m+n)}{\Gamma(m+n)} = \frac{\int_0^1 x^{m-1} \ln x (1-x)^{n-1} dx}{\Gamma(m)\Gamma(n)/\Gamma(m+n)} \tag{19}$$

Putting $m = 2$

$$\int_0^1 x \ln x (1-x)^{n-1} dx = \frac{1}{n(n+1)} \left[\frac{\Gamma'(2)}{\Gamma(2)} - \frac{\Gamma'(n+2)}{\Gamma(n+2)} \right] \tag{20}$$

Now,
$$\Gamma(n+2) = (n+1)\Gamma(n+1)$$

so that,
$$\frac{\Gamma'(n+2)}{\Gamma(n+2)} = \frac{1}{n+1} + \frac{\Gamma'(n+1)}{\Gamma(n+1)}$$
$$= \frac{1}{n+1} + \frac{1}{n} + \frac{\Gamma'(n)}{\Gamma(n)}$$
$$= \ldots$$
$$= \frac{1}{n+1} + \frac{1}{n} + \ldots + \frac{1}{2} + \frac{\Gamma'(2)}{\Gamma(2)} \tag{21}$$

From (20) and (21)

$$-\int_0^1 x \ln x \, (1-x)^{n-1} dx = \frac{1}{n(n+1)} \left[\frac{1}{2} + \frac{1}{3} + \ldots + \frac{1}{n+1} \right] \tag{22}$$

From (17) and (22)

$$S_{av} = \frac{n}{n+1} \left[\frac{1}{2} + \frac{1}{3} + \ldots + \frac{1}{n+1} \right] \tag{23}$$

Also,
$$S_{max} = \ln n \tag{24}$$

so that
$$\frac{S_{av}}{S_{max}} = \frac{n}{n+1} \frac{1}{\ln n} \left[\frac{1}{2} + \frac{1}{3} + \ldots + \frac{1}{n+1} \right] \tag{25}$$

From (23), (24) and (25) we get Table 1.

Table 1

n	2	3	4	5	6	7	8
S_{av}	0.55555	.8124999	1.0266666	1.208333	1.385306	1.503625	1.6257623
S_{max}	0.693	1.0986	1.386	1.609438	1.79176	1.94592	2.07944
$\frac{S_{av}}{S_{max}}$	.8086	.7398	.742	.7509	.73316	.77206	0.7818

from this we deduce that,

(*i*) S_{av} increases with n

(*ii*) S_{max} increases with n

(*iii*) $\frac{S_{av}}{S_{max}}$ also increases with n

Now for large n

$$1+\frac{1}{2}+\frac{1}{3}+\ldots+\frac{1}{n}\approx \ln n+\gamma_n \tag{26}$$

where, $\gamma_n \to$ Euler's constant < 1 as $n \to \infty$

$$\frac{S_{av}}{S_{max}} \approx \frac{n}{n+1}\frac{1}{\ln n}[\ln(n+1)-1+\gamma_{n+1}]$$

$$=\frac{n}{n+1}\frac{\ln(n+1)}{\ln n}-\frac{n}{n+1}\frac{1-\gamma_{n+1}}{\ln n} \tag{27}$$

so that,

$$\frac{S_{av}}{S_{max}} \to 1 \text{ as } n \to \infty. \tag{28}$$

This shows that even for moderate values of n, the average entropy may be about 80% of maximum entropy so that most of the feasible probability distributions have entropy values near the maximum value and as n becomes large, almost all the distributions are in a very narrow neighbourhood of the uniform distribution. This argument can be used as some justification for the maximum entropy principle.

21.5 USE OF ANOTHER MEASURE OF ENTROPY

This is sometimes also taken as a justification for the use of Shannon's measure as a measure of entropy. Let us see whether the above result will hold if we use some other measure as a measure of entropy. Let us consider the measure due to Havrda and Charvat [2],

$$H(P)=\frac{1}{1-\alpha}\sum_{i=1}^{n}(p_i^{\alpha}-p_i), \alpha>0, \alpha\neq 1 \tag{29}$$

then proceeding as before

$$H_{av}=\frac{n^2}{1-\alpha}\int_0^1 (p_1^{\alpha}-p_1)(1-p_1)^{n-1}dp_1$$

$$=\frac{n^2}{1-\alpha}\left[\int_0^1 p_1^{\alpha}(1-p_1)^{n-1}dp_1-\int_0^1 p_1(1-p_1)^{n-1}dp_1\right]$$

$$=\frac{n^2}{1-\alpha}\left[\frac{\Gamma(\alpha+1)\Gamma(n)}{\Gamma(\alpha+n+1)}-\frac{\Gamma(2)\Gamma(n)}{\Gamma(n+2)}\right] \tag{30}$$

$$H_{\max} = \frac{1}{1-\alpha}(n^{1-\alpha}-1) \tag{31}$$

$$\frac{H_{av}}{H_{\max}} = \frac{n^2(n-1)!}{n^{1-\alpha}-1}\left[\frac{\Gamma(\alpha+1)}{\Gamma(\alpha+n+1)} - \frac{\Gamma(2)}{\Gamma(n+1)}\right] \tag{32}$$

$$= \frac{n^2}{n^{1-\alpha}-1}(n-1)!\left[\frac{1}{(\alpha+n)(\alpha+n-1)\dots(\alpha+1)} - \frac{1}{(n+1)\dots 2}\right] \tag{33}$$

If α is an integer,

$$\frac{H_{av}}{H_{\max}} = \frac{n^{2+\alpha-1}}{1-n^{\alpha-1}}(n-1)!\,\frac{\alpha(\alpha-1)\dots 2-(\alpha+n)(\alpha+n-1)\dots(\alpha+1)}{(\alpha+n)(\alpha+n-1)\dots 2}$$

$$= \frac{n^{\alpha+1}}{n^{\alpha-1}-1}\,\frac{(n+2)(n+1)\dots(\alpha+n)-2.3\dots\alpha}{n(n+1)\dots(n+\alpha)}$$

$$= \frac{n^{\alpha+1}}{n^{\alpha-1}-1}\left[\frac{1}{n(n+1)} - \frac{2.3\dots\alpha}{n(n+1)\dots(n+\alpha)}\right]$$

$$= \frac{n^{\alpha}}{n^{\alpha}+n^{\alpha-1}-n-1} - \frac{2.3\dots\alpha}{n(n+1)-(n+\alpha)}\,\frac{n^{\alpha+1}}{n^{\alpha-1}-1} \tag{34}$$

which approaches unity as $n \to \infty$

This shows that there is nothing special about Shannon's entropy. Other measures of entropy have also the property that

$$\underset{n\to\infty}{Lt}\ \frac{H_{av}}{H_{\max}} = 1 \tag{35}$$

21.6 VARIANCE OF ENTROPIES

This is give by

$$E(S(P))^2 - (E(S(P)))^2 \tag{36}$$

or by

$$E(H(P))^2 - (E(H(P)))^2 \tag{37}$$

Let us calculate the second measure since it includes the first as a special case when $\alpha \to 1$.

$$(H(P))^2 = +\frac{1}{(1-\alpha)^2}\left[\sum_{i=1}^{n}(p_i^{\alpha}-p_i)^2 + 2\sum_{i=1}^{n}\ \sum_{j>i\ j=1}^{n}(p_i^{\alpha}-p_i)(p_j^{\alpha}-p_i)\right] \tag{38}$$

$$E[H(P)]^2 = \frac{1}{(1-\alpha)^2}\left[nE(p_1^{\alpha}-p_1)^2 + n(n-1)E(p_1^{\alpha}-p_1)(p_2^{\alpha}-p_2)\right] \tag{39}$$

$$E(p_1^{\alpha}-p_1)^2 = n\int_0^1 (p_1^{\alpha}-p_1)^2(1-p_1)^{n-1}dp_1 \tag{40}$$

$$E(p_1^\alpha - p_1)(p_2^\alpha - p_2) = n(n-1)\int_0^1\int_0^{1-p_1} (p_1^\alpha - p_1)(p_2^\alpha - p_2)(1-p_1-p_2)^{n-2} dp_1 dp_2 \quad (41)$$

so that,

$$E(p_1^\alpha - p_1)^2 = n\left[\frac{\Gamma(2\alpha+1)\Gamma(n)}{\Gamma(2\alpha+n+1)} - 2\frac{\Gamma(\alpha+2)\Gamma(n)}{\Gamma(\alpha+2+n)} + \frac{\Gamma(3)\Gamma(n)}{\Gamma(n+3)}\right] \quad (42)$$

$$E(p_1^\alpha - p_1)(p_2^\alpha - p_2) = n(n-1)\int_0^1 (p_1^\alpha - p_1)(1-p_1)^{\alpha+n-1}(dp_1)\int_0^1 x^\alpha(1-x)^{n-2}dx$$

$$-n(n-1)\int_0^1 (p_1^\alpha - p_1)(1-p_i)^n (dp_1)\int_0^1 x^1(1-x)^{n-2}dx$$

$$= n(n-1)\frac{\Gamma(\alpha+1)\Gamma(n-1)}{\Gamma(\alpha+n)}\int_0^1 (p_1^\alpha - p_1)(1-p_1)^{\alpha+n-1}dp_1$$

$$-n(n-1)\frac{\Gamma(2)\Gamma(n-1)}{\Gamma(n+1)}\int_0^1 (p_1^\alpha - p_1)(1-p_1)^n dp_1$$

$$= n(n-1)\frac{\Gamma(\alpha+1)\Gamma(n-1)}{\Gamma(\alpha+n)}\left[\frac{\Gamma(\alpha+1)\Gamma(\alpha+n)}{\Gamma(2\alpha+n+1)} - \frac{\Gamma(2)\Gamma(\alpha+n)}{\Gamma(\alpha+n+2)}\right]$$

$$-n(n-1)\frac{\Gamma(2)\Gamma(n-1)}{\Gamma(n+1)}\left[\frac{\Gamma(\alpha+1)\Gamma(n+1)}{\Gamma(\alpha+n+2)} - \frac{\Gamma(2)\Gamma(n+1)}{\Gamma(n+3)}\right] \quad (43)$$

Substituting from (42) and (43) in (39) we get $E(H(P))^2$, and then using (30) and (38), we get $Var(H(P))$. Making $\alpha \to 1$, we shall get $Var(S(P))$.

21.7 CONCLUDING REMARKS

In this paper, we have defined a measure for the uncertainty of the set of probability distributions, all of which satisfy a given set of constraints. In the first approach, we defined it as the ratio of the difference between maximum and minimum entropies of individual distributions of this set divided by ln n. As we get more and more information, this uncertainty becomes smaller and smaller. When it becomes zero, a unique probability distribution is identified and there is no uncertainty about which probability distribution of the set we are choosing. The final probability distribution we get will also have an uncertainty, about possible outcomes, but this is different from the uncertainty of the set of distributions.

In the second approach, we find the average of the uncertainties of the individual probability distributions of the set. Each probability distribution is represented as a point in n-dimensional space and different volume elements will correspond to different sets of probability distributions. We assume a uniform distribution of probability distributions in the space and with this assumption we are able to find the mean of the uncertainties when the only constraint is the natural constraint.

In the two approaches we have used two different types of probabilistic uncertainties.

REFERENCES

1. P. Cheeseman, (1985), "In defense of probability", Proc. 9th Intern. Joint Conf. on Artificial Intelligence, Los Angeles, pp. 1002–1009.
2. J.H. Havrda and F. Charvat (1967), "Quantification Methods of Classification Processes: Concept of Structural α Entropy", Kybernetica, vol. 3, pp. 30–35.
3. E.T. Jaynes., (1957), "Information Theory and Statistical Mechanics", Physical Reviews, vol. 106, pp. 620–30.
4. J.N. Kapur (1983), "Twenty-Five Years of Maximum Entropy", Jour. Math Phy. Sci. vol. 17, No. 2, pp. 103–156.
5. J.N. Kapur (1990), Maximum Entropy Models in Science and Engineering. John Wiley Eastern.
6. J.N. Kapur & H.K. Kesavan (1987), Generalised Maximum Entropy Principle with Applications, Sandford Educational Press, University of, Waterloo.
7. George. J. Klir (1989), Is there More to Uncertainty than some probability Theorists might have us believe ? Int J. General Systems vol. 15 pp. 347–78.
8. G.J. Klir and T.A. Folger, (1988), Fuzz-sets, Uncertainty and Informations, Prentice–Hall, Eaglewood Cliffs, NJ.
9. B. Kosko (1987), "Fuzziness *vs* Probability" unpublished Manuscript, E.E. Department University of Southern California.
10. D.V. Lindley (1982), "The probability approach to the treatment of uncertainty in artificial intelligence and-expert systems." Statistical Science 2 No. 1, pp. 17–24.
11. C.E. Shannon (1948), "A Mathematical Theory of Communication," Bell System Tech. J., vol. 27, pp. 379–423, 623–659.

22

MEASURING THE UNCERTAINTY OF A BIRTH DEATH-PROCESS

[Measures of uncertainty for the probability distribution of population size at time t in a birth-death process are obtained as functions of birth and death rates λ and μ and time t. It is shown that if birth rate = death rate = λ, then uncertainty is maximum when $t = 1/\lambda$ and if $\lambda < \mu$, then in the steady state, the uncertainty increases monotonically with $\rho = \lambda/\mu$. In the nonsteady case, a conjecture is made that the uncertainty is maximum when $t = (\lambda - \mu)^{-1} \ln (\lambda/\mu)$. The effect of a constant rate of immigration on the variation of uncertainty is also studied.]

22.1 INTRODUCTION

We consider a simple birth-death process under the following assumptions:

(*i*) The probability of a birth in a small time-interval Δt per unit individual of a population is $\lambda\Delta t + 0(\Delta t)$, where $0(\Delta t)$ is an infinitesimal of a higher order than Δt.

(*ii*) The probability of a death in the same time interval per unit individual of the population is $\mu\,\Delta t + 0\,(\Delta t)$.

(*iii*) The probability of more than one birth or death in the interval is $0\,(\Delta t)$. Let $p_n\,(t)$ denotes the probability of there being n persons in the population at time t and let n_0 denote the number of persons at time $t = 0$, then using the theorems of total and compound probabilities, an expression for $p_n\,(t)$ is obtained in most text-books on theory of probability and stochastic processes [1, 2, 7, 9].

In fact, if we define the probability generating function by

$$\phi(s,t) = \sum_{n=0}^{\infty} p_n(t)s^n, \tag{1}$$

we get the result,

$$\phi(s,t) = \left[\frac{(\lambda - \mu x)s + \mu(x-1)}{(\lambda - \lambda x)s + (\lambda x - \mu)}\right]^{n_0}, \lambda \neq \mu, \tag{2}$$

$$= \left[\frac{\lambda t - (\lambda t - 1)s}{1 + \lambda t - \lambda t s}\right]^{n_0}, \lambda = \mu, \tag{3}$$

$$\text{where} \quad x = \exp(\lambda - \mu)t \tag{4}$$

By expanding $\phi(s, t)$ in power series of s, we can find $p_n(t)$.

In a queuing system, let λ and μ denote arrival and service rates in the steady state, i.e., when $p_n(t)$ is independent of t, then we get

$$p_n = (1-\rho)\rho^n, \quad n = 0, 1, 2, \ldots\ldots; \rho = \lambda/\mu \tag{5}$$

At any time t, the number of persons in the system (whether birth-death or queuing) can be 0, 1, 2, ... ,so that there is uncertainty about the number of persons in the system. We want to have a measure of this uncertainty and we want to discuss how this uncertainty varies with λ, μ and t.

The most important and widely used measure of uncertainty is given by Shannon's measure [8],

$$S(\lambda, \mu, t) = -\sum_{n=0}^{\infty} p_n(t) \ln p_n(t) \tag{6}$$

Another important measure is Havrda-Charvat [2] entropy of second order, viz.

$$H(\lambda, \mu, t) = 1 - \sum_{n=0}^{\infty} p_n^2(t) \tag{7}$$

In the present discussion we find these measures and discuss their variations with λ, μ and t.

22.2 VARIATION OF ENTROPY IN THE STEADY-STATE-QUEUING PROCESS

In the steady case if λ and μ denote the arrival and service rates of a queuing process, then we have

$$S(\lambda, \mu) = -\sum_{n=0}^{\infty} p_n \ln p_n = -\sum_{n=0}^{\infty} [(1-\rho)\rho^n] \ln [(1-\rho)\rho^n]$$

$$= -(1-\rho)\ln(1-\rho) \sum_{n=0}^{\infty} \rho^n - (1-\rho)\ln\rho \sum_{n=0}^{\infty} n\rho^n$$

$$= -\ln(1-\rho) - \frac{\rho}{1-\rho}\ln\rho = \frac{-\rho\ln\rho - (1-\rho)\ln(1-\rho)}{1-\rho}, \tag{8}$$

so that

$$\frac{dS}{d\rho} = -\frac{\ln\rho}{(1-\rho)^2} > 0 \tag{9}$$

Thus, in the steady state queuing case, the uncertainty measure increases monotonically from 0 to ∞ as ρ increases from zero to unity. Thus, in this case as the arrival rate increases relatively to service rate, uncertainty increases. Of course, there is no steady state when $\rho \geq 1$.

22.3 VARIATION OF ENTROPY IN THE NON-STEADY BIRTH-DEATH PROCESS CASE WHEN $\lambda = \mu$, $n_0 = 1$

In this case, (3) gives

$$\sum_{n=0}^{\infty} p_n(t) s^n = \frac{\lambda t}{1+\lambda t}\left(1 - \frac{\lambda t - 1}{\lambda t} s\right)\left(1 - \frac{\lambda t}{1+\lambda t} s\right)^{-1}$$

$$= \frac{\lambda t}{1+\lambda t}\left(1-\frac{\lambda t-1}{\lambda t}s\right)\sum_{n=0}^{\infty}\left(\frac{\lambda t}{1+\lambda t}\right)^n s^n, \tag{10}$$

so that,

$$p_n(t) = \frac{\lambda t}{1+\lambda t}\left[\frac{\lambda t}{1+\lambda t}\right]^n - \left(\frac{\lambda t-1}{\lambda t}\right)\left(\frac{\lambda t\cdot}{1+\lambda t}\right)^{n-1}$$

$$= \frac{(\lambda t)^{n-1}}{(1+\lambda t)^{n+1}}, \quad n \geq 1 \tag{11}$$

Also,

$$p_0(t) = \frac{\lambda t}{1+\lambda t} \tag{12}$$

so that,

$$S(\lambda, t) = -p_0(t)\ln p_0(t) - \sum_{n=1}^{\infty} p_n(t)\ln p_n(t)$$

$$= \frac{2[(1+\lambda t)\ \ln(1+\lambda t) - \lambda t(\ln \lambda t)]}{1+\lambda t} \tag{13}$$

$$\frac{d}{d(\lambda t)}S(\lambda, t) = -\frac{2\ln \lambda t}{(1+\lambda t)^2} \tag{14}$$

Thus, uncertainty increases so long as $\lambda t < 1$ and decreases after this and maximum entropy occurs when $\lambda t = 1$, and

$$\max S(\lambda, t) = 2\ln 2 \tag{15}$$

When $t = 0$, the uncertainty is zero and when $t \to \infty$

$$\text{Lt}_{t\to\infty}\frac{2[(1+\lambda t)\ln(1+\lambda t) - \lambda t\ln \lambda t)]}{1+\lambda t}$$

$$= \text{Lt}_{x\to\infty}\frac{2[(1+x)\ln(1+x) - x\ln x]}{1+x} = \text{Lt}_{x\to\infty}\frac{2\ln(1+x)/x}{1} = 0 \tag{16}$$

Thus in this case uncertainty starts with zero value at $t = 0$ and ends with zero value as $t \to \infty$ and in between it attains its maximum value 2 ln 2 when $t = 1/\lambda$.

22.4 VARIATION OF ENTROPY IN THE NON-STEADY BIRTH-DEATH PROCESS CASE WHEN $\lambda = \mu$ AND $n_0 > 1$.

In this case, (3) gives

$$\sum_{n=0}^{\infty} p_n(t)s^n = \left(\frac{\lambda t}{1+\lambda t}\right)^{n_0}\left(1-\frac{\lambda t-1}{\lambda t}s\right)^{n_0}\left(1-\frac{\lambda t}{1+\lambda t}s\right)^{-n_0}$$

$$= \left(\frac{\lambda t}{1+\lambda t}\right)^{n_0}\left[\sum_{r=0}^{n_0} {}^{n_0}C_r\left(\frac{1-\lambda t}{\lambda t}\right)^r s^r\right]\left[1+\sum_{k=1}^{\infty}\frac{n_0(n_0+1)\ldots(n_0+k-1)}{k!}\left(\frac{\lambda t}{1+\lambda t}\right)^k s^k\right],$$

so that,

$$p_n(t) = \left(\frac{\lambda t}{1+\lambda t}\right)^{n_0} \sum_{r=1}^{n} \frac{n_0(n_0+n-r-1).......(n_0-r+1)}{r!(n_0-r)!(n-r)!}\left(\frac{1-\lambda t}{\lambda t}\right)^r \left(\frac{\lambda t}{1+\lambda t}\right)^{n-r} \tag{17}$$

This is a rather complicated expression and it will not be easy to find $-\sum_{n=0}^{\infty} p_n(t)\ln(p_n(t))$ in a closed form.

However, when $t = 0$

$$p_n(0) = 1 \text{ when } n = n_0, p_n(0) = 0 \text{ when } n \neq n_0, \tag{18}$$

so that entropy at $t = 0$ is zero. Also form (3),

$$p_0(t) = \left(\frac{\lambda t}{1+\lambda t}\right)^{n_0} \tag{19}$$

$$\text{so that} \quad \text{Lt}_{t\to\infty} p_0(t) = 1 \tag{20}$$

Thus as $t \to \infty$, $p_n(t) \to 1$ and all other probabilities tend to zero so that entropy tends to zero as $t \to \infty$.

Also entropy is always ≥ 0 and as such it is expected to be maximum for a finite value of t given by,

$$\lambda t = g(n_0), \tag{21}$$

where, $g(1) = 1$. We are not able to find a closed form expression for $g(n_0)$, but we know its value when $n_0 = 1$. We can also say that $t \propto 1/\lambda$. However, let us see whether we can find Vajda's measure of entropy in this case

Now,

$$\sum_{n=0}^{\infty} p_n(t)s^n = \left[\frac{\lambda t - (\lambda t - 1)}{1+\lambda t - \lambda ts}\right]^{n_0} \tag{22}$$

so that,

$$\sum_{n=0}^{\infty} p_n(t)s^{-n} = \left[\frac{\lambda ts - (\lambda t - 1)}{(1+\lambda t)s - \lambda t}\right]^{n_0} \tag{23}$$

Multiplying the two equations and equating the terms independent of s we get

$$\sum_{n=0}^{\infty} p_n^2(t) = \left(\frac{\lambda t - 1}{\lambda t + 1}\right)^{n_0} \tag{24}$$

These results will be valid only when $\lambda t \geq 1$. In this case

$$H(t) = 1 - \sum_{n=0}^{\infty} p_n^2(t) = 1 - 1\left(\frac{\lambda t - 1}{\lambda t + 1}\right)^{n_0} \tag{25}$$

so that,

$$\frac{d}{d(\lambda t)} \quad (H(t)) = -n_0\left(\frac{\lambda t - 1}{\lambda t + 1}\right)^{n_0 - 1} \frac{2}{(\lambda t + 1)^2} \tag{26}$$

so that, $H'(t) = 0$ when $\lambda t = 1$ and when $\lambda t > 1$, $H'(t) < 1$ and entropy is always decreasing when $t > 1/\lambda$.

This suggests, but does not prove rigorously that even when $n_0 \neq 1$, the maximum uncertainty occurs when $\lambda t = 1$. Thus the statement can be regarded as a conjecture requiring a rigorous proof.

22.5 VARIATION OF ENTROPY IN THE NON-STEADY CASE WHEN $\lambda \neq \mu$

From (3),

$$\sum_{n=0}^{\infty} p_n(t)s^n = \left[\frac{(\lambda - \mu x)s - \mu(x-1)}{\lambda(1-x)s + (\lambda x - \mu)}\right]^{n_0} \tag{27}$$

and,

$$\sum_{n=0}^{\infty} p_n(t)s^{-n} = \left[\frac{(\lambda - \mu x) - \mu(x-1)s}{\lambda(1-x) + (\lambda x - \mu)s}\right]^{n_0} \tag{28}$$

Both series may not be convergent for the same value of s (other than $s = 1$) for each value of t. However, assuming that such a common value exists and equating the terms independent of s in the product of both sides, we get

$$\sum_{n=0}^{\infty} p_n^2(t) = \left[\frac{\mu(\lambda - \mu x)(x-1)}{\lambda(1-x)(\lambda x - \mu)}\right]^{n_0} = \left[\frac{\mu(\lambda - \mu x)}{\lambda(\mu - \lambda x)}\right]^{n_0}$$

$$= \left[\frac{(\mu/\lambda)x - 1}{(\lambda/\mu)x - 1}\right]^{n_0}, \tag{29}$$

so that,

$$H(t) = 1 - \left[\frac{(\mu/\lambda)x - 1}{(\lambda/\mu)x - 1}\right]^{n_0} \tag{30}$$

Now,

$$p_0(t) = \left[\frac{\mu(x-1)}{\lambda x - \mu}\right]^{n_0} = \left[\frac{\mu e^{(\lambda-\mu)t} - \mu}{\lambda e^{(\lambda-\mu)t} - \mu}\right]^{n_0} \tag{31}$$

(*i*) If $\lambda < \mu$,

$$\operatorname*{Lt}_{t \to \infty} p_0(t) = 1 \tag{32}$$

and

$$\operatorname*{Lt}_{t \to \infty} p_n(t) = 0, \quad n > 0 \tag{33}$$

so that as $t \to \infty$, the uncertainty tends to zero. The initial uncertainty is also zero. As such the uncertainty increases from zero to a maximum value and again decreases to zero as $t \to \infty$.

$$H'(t) = \frac{d}{dt}(H(t)) = -n_0\left[\frac{(\mu/\lambda)x - 1}{(\lambda/\mu)x - 1}\right]^{n_0-1} \frac{((\lambda/\mu)x - 1)\mu/\lambda - ((\mu/\lambda)x - 1)\lambda/\mu}{((\lambda/\mu)x - 1)^2}\frac{dx}{dt}$$

$$= -n_0\left[\frac{(\mu/\lambda)x - 1}{(\lambda/\mu)x - 1}\right]^{n_0-1} \frac{(\lambda - \mu)^2(\lambda + \mu)}{\mu\lambda((\lambda/\mu)x - 1)^2}x, \tag{34}$$

so that $H(t)$ is always decreasing, whenever it is valid

Since when $\lambda < \mu$,

$$\lambda/\mu e^{(\lambda-\mu)t} - 1 < 0 \tag{35}$$

$H(t)$ must be valid when

$$\mu/\lambda e^{(\lambda-\mu)t} < 1 \quad \text{or} \quad e^{(\lambda-\mu)t} < \lambda/\mu \tag{36}$$

If

$$e^{(\lambda-\mu)t^*} = \lambda/\mu, \tag{37}$$

then the expression for $H(t)$ is valid when $t \geq t^*$. At $t = t^*$, the value of $H(t)$ is unity and it goes on decreasing when $t > t^*$. It appears therefore that when $\lambda < \mu$, the maximum value of $H(t)$ arises at $t = t^*$.

(*ii*) When $\lambda > \mu$, $\lambda/\mu\, e^{(\lambda-\mu)t} > 1$ and therefore $H(t)$ is valid when

$$\mu/\lambda e^{(\lambda-\mu)t} \geq 1 \quad \text{or} \quad e^{(\lambda-\mu)t} \geq \lambda/\mu \quad \text{or} \quad t \geq t^* \tag{38}$$

and after that $dH/dt < 0$ and $H(t)$ is decreasing. Its maximum value is again unity, but as $t \to \infty$, it does not approach zero unlike the case when $\lambda < \mu$. This is obvious from the fact that,

$$\underset{t\to\infty}{\text{Lt}}\, p_0(t) = (\mu/\lambda)^{n_0} < 1 \tag{39}$$

Also from (31),

$$\underset{t\to\infty}{\text{Lt}}\, H(t) = \underset{x\to\infty}{\text{Lt}}\, 1 - \left[\frac{(\mu/\lambda)x - 1}{(\lambda/\mu)x - 1}\right]^{n_0} = 1 - \left(\frac{\mu^2}{\lambda^2}\right)^{n_0} \tag{40}$$

Thus when $\lambda < \mu$, $H(t)$ increases from 0 to 1 as t increases from 0 to t^* and then decreases to 0 when $t \to \infty$. However, when $\lambda > \mu$, $H(t)$ increases from 0 to 1 as t increases from 0 to t^* and then decreases to $1 - (\mu^2/\lambda^2)^{n_0}$ as $t \to \infty$.

The proof of the result is not rigorous because we do not have the expression for $H(t)$ when $t < t^*$.

It is easily seen that t^* does not change when λ and μ are interchanged.

We give two illustrations of the above results.

(*i*) Suppose in a queue, 5 persons arrive per minute and 6 persons lèave per minute, then

$$t^* = \frac{1}{\lambda-\mu}\ln\lambda/\mu = \frac{1}{5-6}\ln\frac{5}{6} = \ln 6/5 = .1823 \text{ minutes} \tag{41}$$

so that maximum uncertainty occurs ·1823 minutes after we observe the queue size. If on the other hand 6 persons arrive per minute and 5 persons leave per minute, we get

$$t^* = \frac{1}{\lambda-\mu}\ln\frac{\lambda}{\mu} = \frac{1}{6-5}\ln 6/5 = \ln 6/5 = .1823 \text{ minutes} \tag{42}$$

which is the same as the earlier value of t^*.

(*ii*) Suppose in a population 2.3 persons are born per hundred persons per year and 1.5 persons die per hundred per year, then the maximum uncertainty occurs,

$$t^* = \frac{1}{2.3-1.5}\ln\frac{2.3}{1.5} = \frac{1}{.8}\ln\frac{23}{15} = .8175 \text{ years} \tag{43}$$

after we know with certainty the size of the population.

22.6 BIRTH-DEATH-IMMIGRATION PROCESS

If immigration rate is ν and $n_0 = 0$, then

$$\phi(s,t) = \left[\frac{\lambda-\mu}{\lambda e^{(\lambda-\mu)t}-\mu}\right]^{v/\lambda}\left[1-\frac{\lambda s(e^{(\lambda-\mu)t}-1)}{\lambda e^{(\lambda-\mu)t}-\mu}\right]^{-v\lambda}, \quad \lambda \neq \mu \tag{44}$$

$$= (1+\lambda t)^{v/\lambda}\left(1-\frac{\lambda t s}{1+\lambda t}\right)^{-v/\lambda} \tag{45}$$

Now $\sum_{n=1}^{v} p_n^2(t)$ is the term independent of s in the product $\phi(s,t)$, $\phi\left(\frac{1}{s},t\right)$, so that

$$\sum_{n=0}^{\infty} p_n^2(t) = \left(\frac{\lambda-\mu}{\lambda e^{(\lambda-\mu)t}-\mu}\right)^{\frac{2v}{\lambda}}\left(1+\frac{\lambda^2(e^{(\lambda-\mu)t}-1)^2}{(\lambda e^{(\lambda-\mu)t}-\mu)^2}\right)^{-v/\lambda}$$

$$= \left[\frac{2\lambda^2 e^{2(\lambda-\mu)t}-2\lambda(\lambda+\mu)e^{(\lambda-\mu)t}+\lambda^2+\mu^2}{\lambda^2-2\lambda\mu+\mu^2}\right]^{-v/\lambda} \quad ,\lambda \neq \mu \tag{46}$$

$$= (1+\lambda t)^{2v/\lambda}\left[1+\frac{\lambda^2 t^2}{(1+\lambda t)^2}\right]^{-v/\lambda} \quad ,\lambda = \mu \tag{47}$$

so that

$$H(t) = 1-\left(\frac{\lambda^2-2\lambda\mu+\mu^2}{2\lambda^2 e^{2(\lambda-\mu)t}-2\lambda(\lambda-\mu)e^{(\lambda-\mu)t}+\lambda^2+\mu^2}\right)^{v/\lambda} \quad ,\lambda \neq \mu \tag{48}$$

$$= 1-[1+2\lambda t+2\lambda^2 t^2]^{-v/\lambda} \quad ,\lambda = \mu \tag{49}$$

Also $p(0)=0, H(\infty)=1$ if $\lambda \geq \mu$ and $H(\infty) = 1-\left(\frac{\lambda^2+\mu^2-2\lambda\mu}{\lambda^2+\mu^2}\right)^{\frac{1}{\lambda}}$. If $\lambda < \mu$.

When $\lambda = \mu$, $H(t)$ increases monotonocally from 0 to 1. Again

$$\frac{d}{dt}[2\lambda^2 e^{2(\lambda-\mu)t}-2\lambda(\lambda+\mu)e^{(\lambda-\mu)t}+\lambda^2+\mu^2] \tag{50}$$

$$= \mu\lambda^2(\lambda-\mu)e^{(\lambda-\mu)t}[e^{(\lambda-\mu)t}-(\lambda+\mu)/2\lambda] \tag{51}$$

If $\lambda > \mu, e^{(\lambda-\mu)t} > 1, (\lambda+\mu)/2\mu < 1$ and if $\lambda < \mu, e^{(\lambda-\mu)} < 1, (\lambda+\mu)/2\lambda > 1$ (52)
In neither case can the derivative vanish. However, when $\lambda > \mu$, $H(t)$ increases monotonocally from 0 to 1 and when $\lambda < \mu$, it increases monotonocally from 0 to the value given by (50). From (48) and (49), at a given time if λ and μ are fixed, uncertainty increases as the immigration rate increases.

22.7 COMPARISON OF ENTROPY AND VARIANCE

It is sometimes suggested that we can use variance of a probability distribution as a measure of its uncertainty. This suggestion is made because very often these show similar trends. A good comparison between the measures is made in Kapur and Kapur [6].

We examine the relationship between these two measures in the context of birth and death processes.

(*i*) In the example of the queueing process discussed in Section 2, variance,

$$V = \frac{\rho}{(1-\rho)^2} \tag{53}$$

so that,

$$\frac{dV}{d\rho} = \frac{1+\rho}{(1-\rho)^3} > 0, \tag{54}$$

so that variance increases with ρ as ρ increases from 0 to unity.
This is similar to the behaviour of the entropy in this case.
(*ii*) In the case of birth-death processes discussed in Sections 3 and 4,

$$\text{Variance } = \sigma^2(t) = 2n_0\lambda t,$$

the variance increases upto infinity, while entropy increases upto $\lambda t = 1$ and then decreases. The behaviours in these cases are quite different.
(*iii*) In the case of example discussed in Section 5,

$$\sigma^2(t) = n_0\left(\frac{\lambda+\mu}{\lambda-\mu}\right)e^{(\lambda-\mu)t}[e^{(\lambda-\mu)t} - 1] \tag{55}$$

If $\lambda > \mu$, variance increases monotonically with t, but if $\lambda < \mu$, variance first increases and then decreases to zero. The behaviours in both cases are different from the corresponding behaviours of entropy.

22.8 USE OF DIFFERENT MEASURES OF ENTROPY

In our earlier discussion, we have used the measures of entropy of Shannon and Vajdas and we got the same results. We expect in general that the variation of entropy should show the same behaviour for all measures of entropy we can use. We illustrate this below by considering the simple examples of a pure birth process with birth rate λ. Here

$$p_n(t) = e^{-\lambda t}(1-\lambda t)^{n-1}, \qquad n = 1,2,3\ldots\ldots \tag{56}$$

Shannon's measure of entropy is given by

$$S(t) = -\sum_{n=1}^{\infty} p_n(t)\ \ln\ p_n(t)$$

$$= -\sum_{n=1}^{\infty} p_n(t)(-\lambda t + (n+1)\ln(1-e^{-\lambda t}))$$

$$= \lambda t \sum_{n=1}^{\infty} p_n(t) - \ln(1-e^{-\lambda t})\sum_{n=0}^{\infty} e^{-\lambda t} n(1-\lambda t)^n$$

$$= \lambda t - (e^{\lambda t} - 1)\ln(1-e^{-\lambda t})$$

$$= \frac{\lambda t e^{-\lambda t} - (1-e^{-\lambda t})\ln(1-e^{-\lambda t})}{e^{-\lambda t}} = \frac{-x\ \ln x - (1-x)\ln(1-x)}{x}; \quad x = e^{-\lambda t} \tag{57}$$

$$\frac{d}{dt}S(t) = \frac{x(-\ln x + \ln(1-x)) + (x\ \ln x + (1-x)\ln(1-x))}{x^2}\frac{dx}{dt}$$

$$= \ln\frac{(1-x)}{x^2}(-\lambda e^{\lambda t}) = \frac{-\lambda\ln(1-x)}{x} > 0 \tag{58}$$

Thus, $S(t)$ increases monotonically as t increases
When $t=0,\ x=1,\ S(0)=0$

When $t \to \infty, x=0, S(\infty) = \underset{x \to 0}{\text{Lt}} \dfrac{-x \ln x - (1-x)\ln(1-x)}{x}$

$$= \underset{x \to 0}{\text{Lt}} \frac{-\ln x + \ln(1+x)}{1} = \underset{x \to 0}{\text{Lt}} \ln \frac{1-x}{x} = \infty \tag{59}$$

$S(t)$ increases monotonically from 0 to ∞,

$$S''(t) = \frac{\lambda^2}{x}\left(-\frac{x}{1-x} - \ln(1-x)\right) = \frac{\lambda^2}{x(1-x)}[-x-(1-x)\ln(1-x)]$$

$$= \frac{\lambda^2}{y(1-y)}(y-1-y\ln y); \qquad y = 1-x \tag{60}$$

so that $S(t)$ is a concave function of both λt and t.

We now consider Havrda-Charvat [3] measure of entropy which includes Vajdas square entropy as a special case. For the present case,

$$H(t) = \frac{1}{1-\alpha}\left[\sum_{n=1}^{\infty} [e^{-\lambda t}(1-e^{-\lambda t})^{n-1}]^{\alpha} - 1\right] \tag{61}$$

$$= \frac{1}{1-\alpha}\left[\frac{e^{-\lambda \alpha t}}{1-(1-e^{-\lambda t})^{\alpha}} - 1\right] = \frac{1}{1-\alpha}\left[\frac{x^{\alpha}}{1-(1-x)^{\alpha}} - 1\right]$$

$$H'(t) = \frac{1}{1-\alpha} \frac{[1-(1-x)^{\alpha}]\alpha x^{\alpha-1} - \alpha x^{\alpha}(1-x)^{\alpha-1}}{[1-(1-x)^{\alpha}]^2}(-\lambda x)$$

$$= \frac{-\lambda \alpha x^{\alpha}}{1-\alpha} \frac{1-(1-x)^{\alpha-1}}{[1-(1-x)^{\alpha}]^2} \tag{62}$$

$$\text{Now } \alpha > 1 \Rightarrow (1-x)^{\alpha-1} < 1 \Rightarrow 1-(1-x)^{\alpha-1} > 0 \Rightarrow \frac{1-(1-x)^{\alpha-1}}{1-\alpha} < 0 \tag{63}$$

$$\alpha < 1 \Rightarrow (1-x)^{\alpha-1} > 1 \Rightarrow 1-(1-x)^{\alpha-1} < 0 \Rightarrow \frac{1-(1-x)^{\alpha-1}}{1-\alpha} < 0 \tag{64}$$

In both cases $H'(t) > 0$ so that $H(t)$ also increases monotonically with t.

$$\text{Now } H(0) = 0, H(\infty) = \underset{x \to \infty}{\text{Lt}} \frac{1}{1-\alpha}\left[\frac{x^{\alpha}}{1-(1-x)^{\alpha}} - 1\right] = \frac{1}{\alpha-1} \quad \text{if } \alpha > 1$$

$$= 0 \qquad \text{if } \alpha < 1 \tag{65}$$

Again Renyi's entropy,

$$R(t) = \frac{1}{1-\alpha} \ln \sum_{n=0}^{\infty} p_n^{\alpha} = \frac{1}{1-\alpha} \ln[(1-\alpha)H(t)+1] \tag{66}$$

If $\alpha < 1$, then as t increases $(1-\alpha)H(t)+1$ increases, $R(t)$ increases.
If $\alpha > 1$, then as t increases $(1-\alpha)H(t)+1$ decreases, $R(t)$ decreases.

Thus, in both cases whether $\alpha < 1$ or $\alpha > 1$, Renyi's entropy increases monotonically with t.

We have found that in the present case, all the measures of entropy lead to the same result that entropy invariably increases monotonically with time.

22.9 SUMMARY OF RESULTS

(*i*) In the queuing process with arrival and service rates λ and μ in the steady state uncertainty increases as $\rho = \lambda / \mu$ increases from 0 to 1.

In the simple birth-death process with birth and death rates λ and μ respectively:-

(*ii*) In the non-steady state when $\lambda = \mu$, $n = 1$, uncertainty increases from time $t = 0$ till time $1/\lambda$ and then decreases from the maximum value to 0 as $t \to \infty$.

(*iii*) It is highly plausible that the result (*ii*) is true even when $n_0 > 1$

(*iv*) In the steady case when $\lambda \neq \mu$, the maximum uncertainty occurs at time $\frac{1}{\lambda - \mu} \ln \lambda / \mu$

(*v*) If in addition, there is immigration at a rate ν and $n_0 = 0$, the uncertainty increases monotonically from initial value 0 as t increases.

(*vi*) In the last case, uncertainty increases with ν.

(*vii*) In general, different measures of entropy give same result about variation of entropy of a given stochastic process with time.

22.10 CONCLUDING REMARKS

Since in a stochastic process, probability distribution changes with time, it is obvious that the entropy or uncertainty of a probability distribution also changes with time. It is interesting and useful to know how the uncertainty changes with time.

Here we discuss some simple stochastic processes in which we have been able to find:

$$S(t) = -\sum_{n=0}^{\infty} p_n(t) \ln p_n(t)$$

$$\text{or, } H(t) = \frac{1}{1-\alpha}\left[\sum_{n=0}^{\infty} p_n^{\alpha}(t) - 1\right]$$

$$\text{or, } R(t) = \frac{1}{1-\alpha} \ln \sum_{n=0}^{\infty} p_n^{\alpha}(t) \tag{67}$$

in a closed form and discuss their variations with time. We have found two typical behaviour. Either uncertainty increases monotonically with time throughout or it first increases and then decreases.

However, we may not be so fortunate in case of complex stochastic processes. In fact even in a simple stochastic process this may not be possibie. Thus in the simple Poisson process,

$$p_n(t) = e^{-\lambda t}\frac{(\lambda t)^n}{n!}, \quad n = 0, 1, 2, \ldots\ldots \tag{68}$$

and,

$$S(t) = -\sum_{n=0}^{\infty} p_n(t)(-\lambda t + n\ln(\lambda t) - \ln n!)$$

$$= \lambda t - \lambda t \ln \lambda t + e^{-\lambda t}\sum_{n=0}^{\infty}\frac{(\lambda t)^n}{n!}\ln n! \tag{69}$$

It is not easy to express the last term in a simple form. Similarly,

$$H(t) = \frac{1}{1-\alpha}\left[\sum_{n=0}^{\infty} e^{-\alpha\lambda t}\frac{(\lambda t)^{n\alpha}}{(n!)^{\alpha}} - 1\right], \quad \alpha \neq 1 \tag{70}$$

and again it is not easy to sum the infinite series. However, some results can always be obtained numerically.

REFERENCES

1. M. Bartlett (1959), An introduction to stochastic processes, Cambridge University Press.
2. W. Feller (1957), Introduction to Probability Theory Vol. I, John Wiley, New York.
3. J.H. Havrda and F. Charvat (1967), Quantification Methods of Classificatory Processes, Concept of Structural α Entropy, Kybernetica, 3, 30-35.
4. J.N. Kapur (1985), "Mathematical Models in Biology and Medicine", Affiliated East West Press, New Delhi.
5. J. N. Kapur and H.C. Saxena (1991), "Mathematical Statistics", S. Chand & Co, New Delhi.
6. J.N. Kapur and S. Kapur (1992), "Some relations between entropy & variance", Metron 48, 113, 130
7. J.M. Medhi (1982), "Stochastic Processes," Wiley Eastern, New Delhi.
8. C.E. Shannon (1948), A Mathematical Theory of Communication, Univ. of Illinois Press Urbana, Chicago.
9. S.K. Srinivasan and K.M. Mehta (1976), " Stochastic Processes", Tata Mc-Graw Hill, New Delhi.

Part B

Entropy Optimisation Principles

23

ON MINIMUM ENTROPY PROBABILITY DISTRIBUTIONS

[We find the minimum entropy probability distribution for a random variate which takes the values 1, 2, . . . , n and for which the moment $E(g(i))$ is prescribed where $g(\cdot)$ is a monotonic increasing function. It is shown that the minimum value $S_{\min}$ of the entropy is a piece-wise concave function which vanishes when $E(g(x)) = g(k)$, where k is a positive integer and which has a local maximum in between every pair of consecutive positive integers. It is further shown that if $g(i) = i$, then all maximum values of $S_{\min}$ are equal. Also the maximum value of $S_{\min}$ increases if $g(x)$ is a convex function of x and it decreases if $g(x)$ is a concave function of x. The cases when $g(i) = i$ or i^2 are specially discussed.]

23.1 INTRODUCTION

Let
$$P = (p_1, p_2, ..., p_n), p_i \geq 0 \sum_{i=1}^{n} p_i = 1 \tag{1}$$

be a probability distribution. Shannon [5] gave the measure of entropy,

$$S(P) = -\sum_{i=1}^{n} p_i \ln p_i \tag{2}$$

as the measure of uncertainty associated with the distribution. Jaynes [1] suggested that if the only information available about the distribution is

$$\sum_{i=1}^{n} p_i = 1, \sum p_i g_{ri} = a_r \quad r = 1, 2, ..., m; m+1 < n \tag{3}$$

then out of all these distributions which satisfy (3), the most unbiased distribution is that which maximizes (2) subject to (3).

This maximum entropy principle has had tremendous applications in Science and Engineering (Kapur & Kesavan [4], Kapur [2]). Watanabe [6] suggested that we should also find the minimum entropy distribution subject to (3); in fact, he characterized pattern recognition as a quest for minimum entropy.

Kapur and Kesavan [3] have suggested that while $S(P)$ measures the uncertainty associated with probability distribution P, the measure of uncertainty that remains after information (3) is made available is measured by $S_{\max}-S_{\min}$. If more independent constraints are available, $S_{\max}$ decreases, $S_{\min}$ increases, $S_{\max}-S_{\min}$ decreases and the remaining uncertainty becomes smaller and smaller. When $S_{\max}$ becomes equal to $S_{\min}$, all the relevant constraints have been determined, the pattern is completely known and no further information is required.

Thus investigation of minimum entropy probability distribution is almost as important as the investigation of maximum entropy probability distribution. These are duals of each other. While maximum entropy gives the most objective, most random, most unbiased, most uniform distribution subject to the given constraints; the minimum entropy probability distribution gives the least objective, least random, least unbiased, least uniform distribution. The true distribution lies somewhere in between. Whatever be the additional constraints imposed, entropy cannot go below S_{min}.

Inspite of this importance, minimum entropy probability distributions have not been investigated very much. One possible reason is that all the entropy functions are concave and minimization of concave functions is much more different than maximizing them.

In the present discussion, we investigate the problem of minimizing measure [1] or any other concave measure of entropy subject to the constraints,

$$\sum_{i=1}^{n} p_i = 1, \sum_{i=1}^{n} p_i g\ (i) = g\ (m), \tag{4}$$

where, $g\ (i)$ is a monotonic increasing function of i. When,

(*i*) $g\ (i) = i$; this means that the mean of the distribution is prescribed.

(*ii*) $g\ (i) = i^2$, the second moment about origin is prescribed.

(*iii*) $g\ (i) = i^r$, the rth order moment about the origin is prescribed.

(*iv*) $g\ (i) = \log i$, the geometric mean of the distribution is prescribed.

We shall find analytical expression for the minimum entropy probability distribution as well as the value of S_{min}. We shall show that S_{min} is a piece-wise increasing-decreasing function of $g\ (m)$ and we shall investigate the maximum value of S_{min}.

We shall investigate specially the case when $E\ (i)$ or $E\ (i^2)$ is prescribed. We shall also discuss briefly the case when both these are prescribed and when the random variate is continuous.

23.2 THE MINIMUM ENTROPY PROBABILITY DISTRIBUTION

Since there are two constraints, the minimum entropy probability distribution will have two non-zero components. We have to find which ones these are and what their values are.

For entropy to be minimum, these probabilities should be as unequal as possible subject to the probability distribution having the prescribed mean.

If

$$g\ (m) = g\ (1) \text{ or } g\ (2) \text{ or } g\ (3) \text{ or...or } g\ (n) \tag{5}$$

the minimum entropy probability distribution is a degenerate distribution and the minimum entropy is obviously zero. If m lies between k and $k + 1$ where k is 1, 2, . . ., or $n - 1$, no degenerate distribution satisfies constraints and as such the minimum entropy probability distribution is non-degenerate and $S_{min} > 0$.

One of the non-zero probabilities has to correspond to an integer $< m$ and the other has to correspond to an integer $> m$ and these two have to be as far apart as possible.

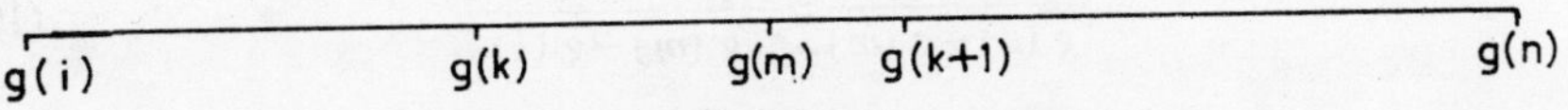

g(i) g(k) g(m) g(k+1) g(n)

Fig. 23.1

If m is very near k, then p_k can be large and $1 - p_k$ can be small and the largest value of p_k will arise if we take the other non-zero probability to be p_n.

In this case, we get

$$p_k + p_n = 1 \quad p_k g(k) + p_n g(n) = g(m),$$

so that

$$p_n(g(n) - g(m)) = p_k(g(m) - g(k))$$

or,

$$\frac{p_n}{g(m) - g(k)} = \frac{p_k}{g(n) - g(m)} = \frac{1}{g(n) - g(k)}, \tag{6}$$

so that the probability distribution is,

$$\frac{g(n) - g(m)}{g(n) - g(k)}, \quad \frac{g(m) - g(k)}{g(n) - g(k)} \tag{7}$$

When $m = k$, the probabilities are maximally unequal and $S_{\min} = 0$. As m increases beyond k, then these probabilities become more and more equal and the value of $S_{\min}$ increases. However, in this increase, as m comes close to $k + 1$, p_{k+1} can be large and $1 - p_{k+1}$ can be small and we can get the largest value of p_{k+1}, if we take the other probabilities as p_1, so that we get the non-zero probabilities as (p_1, p_{k+1}) giving

$$p_1(g(m) - g(1)) = p_{k+1}(g(k+1) - g(m))$$

or,

$$\frac{p_1}{g(k+1) - g(m)} = \frac{p_{k+1}}{g(m) - g(1)} = \frac{1}{g(k+1) - g(1)}$$

$$p_1 = \frac{g(k+1) - g(m)}{g(k+1) - g(1)}, \quad p_{k+1} = \frac{g(m) - g(1)}{g(k+1) - g(1)} \tag{8}$$

When $m = k + 1$, $p_1 = 0, p_{k+1} = 1$ and the two probabilities are maximally unequal. As m decreases, these two probabilities approach equality and $S_{\min}$ increases. Thus $S_{\min}$ increases as m increases from k towards $k + 1$ and again as m decreases from $k + 1$ towards k and there will be a value of m for which these two values of $S_{\min}$ will coincide.

This will happen when the two probability distributions coincide, i.e. when the two larger probabilities coincide, i.e., when p_k for the first distribution coincides with p_{k+1} for the second distribution, i.e., when

$$\frac{g(n) - g(m)}{g(n) - g(k)} = \frac{g(m) - g(1)}{g(k+1) - g(1)} = \frac{g(n) - g(1)}{g(n) + g(k+1) - g(1) - g(k)} \tag{9}$$

This also gives,

$$\frac{g(m)-g(k)}{g(n)-g(m)}=\frac{g(k+1)-g(m)}{g(m)-g(1)} \tag{10}$$

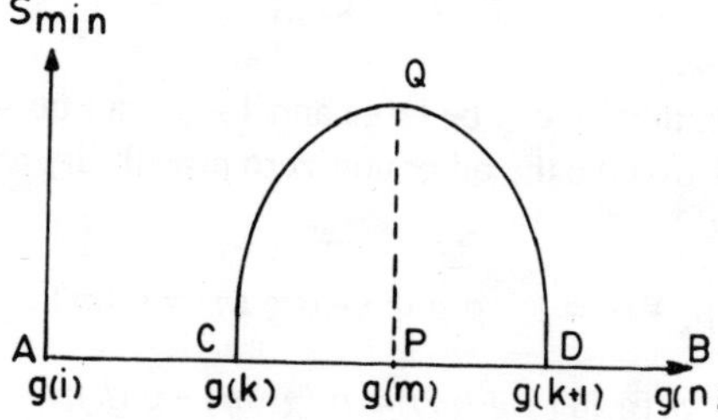

Fig. 23.2 Typical $S_{min} - g(m)$ curve in the interval $g(k)$, $g(k+1)$.

or,
$$\frac{PC}{PD}=\frac{PB}{PA} \tag{11}$$

Thus, S_{min} is increasing-decreasing between 1 and 2, 2 and 3, . . . and $(n-1)$ and n and is thus piece-wise increasing-decreasing. As a whole S_{min} is a piece-wise increasing-decreasing function of $g(m)$.

We are now interested in,

(*i*) the maximum values of S_{min}
(*ii*) the location of this maximum value,
(*iii*) the nature of S_{min} curve at points between C and D (Figure 23.1)

We investigate these in the next three sections.

23.3 THE MAXIMUM OF S_{min}

The probability distribution for which S_{min} is maximum is given by

$$\frac{g(n)-g(m)}{g(n)-g(k)}, \frac{g(m)-g(k)}{g(n)-g(k)}$$

or,
$$\frac{g(n)-g(1)}{g(n)-g(1)+g(k+1)-g(k)}, \frac{g(k+1)-g(k)}{g(n)-g(1)+g(k+1)-g(k)}$$

or
$$p=\frac{A}{A+\phi(k)}, 1-p=\frac{\phi(k)}{A+\phi(k)} \tag{12}$$

where,
$$A=g(n)-g(1), \phi(k)=g(k+1)-g(k) \tag{13}$$

Its entropy is,

$$\psi(k)=-p\log p-(1-p)\log(1-p) \tag{14}$$

so that
$$\psi'(k)=\ln\frac{1-p}{p}\frac{dp}{dk}$$

$$=\left(\ln\frac{g(k+1)-g(k)}{g(n)-g(1)}\right)\left[\frac{-A}{(A+\phi(k))^2}\right]$$

$$\times[g'(k+1)-g'(k)] \tag{15}$$

Since $g(\cdot)$ is a monotonic increasing function, the first factor is negative, and the second factor is also negative.

$$\therefore \quad \psi'(k) > 0 \text{ if } g'(k+1) > g'(k) \tag{16}$$

Case I: $g(k)$ is a convex function. In this case $g'(k)$ is an increasing function of k, $\psi'(k) > 0$, entropy increases as k increases so that the maximum value of S_{min} is the interval $(k, k+1)$ goes on increasing and the global maximum value of S_{min} occurs in the interval $(n-1\ n)$, when

$$p = \frac{g(n)-g(1)}{g(n)-g(1)+g(n)-g(n-1)} \tag{17}$$

and so,

$$(S_{min})_{max} = -\frac{g(n)-g(1)}{2g(n)-g(1)-g(n-1)}\ln\frac{g(n)-g(1)}{2g(n)-g(1)-g(n-1)}$$
$$-\frac{g(n)-g(n-1)}{2g(n)-g(1)-g(n-1)}\ln\frac{g(n)-g(n-1)}{2g(n)-g(1)-g(n-1)} \tag{18}$$

This includes the special cases when

$$g(i) = i^2, i^3, i^4, i^5, \ldots \tag{19}$$

Case II: $g(k)$ is concave function. In this case $\psi'(k) < 0$ so that $(S_{min})_{max}$ in the interval $(k, k+1)$ goes on decreasing as k increases and as such,

$$(S_{min})_{max} = -\frac{g(n)-g(1)}{g(n)+g(2)-2g(1)}\ln\frac{g(n)-g(1)}{g(n)+g(2)-2g(1)} \tag{20}$$
$$-\frac{g(2)-g(1)}{g(n)+g(2)-2g(1)}\ln\frac{g(2)-g(1)}{g(n)+g(2)-2g(1)}$$

This includes the special case when geometrical mean is prescribed.
Case III: $g(k) = ck$ or $g(i) = ci$. In this case $\psi'(k) = 0$. As such $(S_{min})_{max}$ is same in all intervals.

$$(S_{min})_{max} = -\frac{n-1}{n}\ln\frac{n-1}{n} - \frac{1}{n}\ln\frac{1}{n} \tag{21}$$

This includes the case when the arithmetic mean is prescribed. Thus, when A.M. is prescribed $(S_{min})_{max}$ remain constant and when G.M. is prescribed $(S_{min})_{max}$ goes on decreasing with k.

23.4 LOCATIONS OF $(S_{min})_{max}$

When k is small, $PB > PA$ and so from (11), $PC > PD$. As such in the earlier intervals, the maximum value of S_{min} will occur near the end of the intervals, while is the later intervals, this will occur near the beginning of the intervals. In the central interval, corresponding to $k = \frac{n}{2}$ or $\frac{n+1}{2}$, it may occur at the middle of the interval.

This is illustrated in Figure 23.2, 23.3 & 23.4.

23.5 THE NATURE OF DOUBLE POINTS

Between C and Q, $\frac{dS_{\min}}{dg(m)} > 0$, and between Q and D this derivative is negative.

Points like C, D are double points on the curve since at each of these points, there are two branches of the curve and there are two tangents. These points may be nodes, since at C, the tangent to the branch CQD has positive slope and the tangent to the earlier branch at C has negative slope. Let us see what these slopes are:

When m is very near to k, the probability distribution is $(p, 1-p)$,

where from (16) $$p = \frac{g(n) - g(m)}{g(n) - g(k)},\ 1 - p = \frac{g(m) - g(k)}{g(n) - g(k)} \tag{22}$$

and the entropy is $$S_{\min} = -p \ln p - (1-p)\ln(1-p) \tag{23}$$

so that $$\frac{dS_{\min}}{dg(m)} = \ln\frac{1-p}{p}\frac{dp}{dg(m)} = \left(\ln\frac{1-p}{p}\right)\left(\frac{-1}{g(n) - g(k)}\right) \tag{24}$$

As $m \to k$, $p \to 1$ so that the slope approaches $+\infty$.

Similarly from (8), when m is near $(k+1)$, the probability distribution is

$$p = \frac{g(k+1) - g(m)}{g(k+1) - g(1)},\ 1 - p = \frac{g(m) - g(1)}{g(k+1) - g(1)} \tag{25}$$

so that $$\frac{dS_{\min}}{dg(m)} = \ln\frac{1-p}{p}\left(-\frac{1}{g(k+1) - g(1)}\right) \tag{26}$$

As $m \to k+1$, $p \to 0$ and the slope at this point approaches $-\infty$. Thus, we find that each of the points C, D etc. is not a node but is a cusp with two vertical coincident tangents at these points.

23.6 SOME SPECIAL CASES

Case (i): $g(i) = i$

In this case the constraints are $\sum_{i=1}^{n} p_1 = 1$, $\sum_{i=1}^{n} ip_i = m$, the maximum value of $S_{\min}$ is the same in every interval and is given by (21).

In this case (9) gives

$$\frac{m-1}{k} = \frac{n-1}{n} \tag{27}$$

or

$$m = k + 1 - k/n \tag{28}$$

so that the maximum of $S_{\min}$ occurs at

$$2 - \frac{1}{n}, 3 - \frac{2}{n}, 4 - \frac{3}{n}, \ldots, n - \frac{n-1}{n} \tag{29}$$

$S_{\min} - m$ curves, are as in Figures 23.3 and 23.4 for $n = 6$ and 7, respectively.

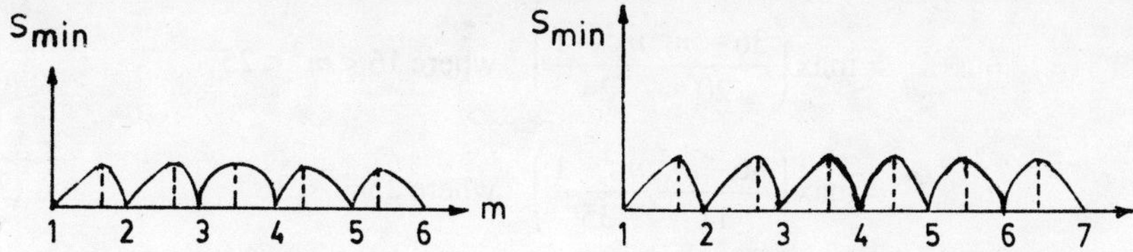

Fig. 23.3 $S_{\min} - m$ curve when $n = 6$.

Fig. 23.4 $S_{\min} - m$ curve when $n = 7$.

From (21)

$$\frac{d\,(S_{\min})_{\max}}{dn} = -\left(1+\ln\left(1-\frac{1}{n}\right)\right)\frac{1}{n^2} - \left(1+\ln\frac{1}{n}\right)\left(-\frac{1}{n^2}\right)$$

$$= \frac{1}{n^2}\left(\ln\frac{1}{n} - \ln\left(1-\frac{1}{n}\right)\right) = \frac{1}{n^2}\ln\frac{1}{n-1} < 0 \tag{30}$$

In fact, we have the Table 23.1

Table 23.1. Maximum value of $S_{\min}$ for various values of n when mean is prescribed.

n:	2	3	6	10	20	∞
$(S_{\min})_{\max}$:	.693	.6307	.4516	.325	.150	0

so that the ordinates of the peaks of the $(s_{\min})$ curve go on decreasing as n increases.

Case (ii): $g\,(i) = i^2$

The constraints are $$\sum_{i=1}^{n} p_i = 1,\ \sum_{i=1}^{n} i^2 p_i = m^2 \tag{31}$$

This gives the minimum entropy,

$$S_{\min} = -p\ln p - (1-p)\ln(1-p) \tag{32}$$

where, $$p = \max\left(\frac{n^2-m^2}{n^2-k^2},\ \frac{m^2-1^2}{(k+1)^2-1^2}\right),\quad k < m < k+1 \tag{33}$$

and $$p = 1 \qquad ,m = k \text{ or } k+1$$

$$k = 1, 2, 3, \ldots, n-1$$

Thus, let $n = 6$

$$p = \max\left(\frac{36-m^2}{35}, \frac{m^2-1}{3}\right) \quad \text{where } 1 < m^2 < 4$$

$$= \max\left(\frac{36-m^2}{32}, \frac{m^2-1}{8}\right) \quad \text{where } 4 < m^2 < 9$$

$$= \max\left(\frac{36-m^2}{27}, \frac{m^2-1}{15}\right) \quad \text{where } 9 < m^2 < 16$$

$$= \max\left(\frac{36-m^2}{20}, \frac{m^2-1}{24}\right) \quad \text{where } 16 < m^2 < 25$$

$$= \max\left(\frac{36-m^2}{11}, \frac{m^2-1}{35}\right) \quad \text{where } 25m^2 < 36 \tag{34}$$

The maximum value of S_{min} occurs when

$$\frac{36-m^2}{36-k^2} = \frac{m^2-1}{(k+1)^2-1} = \frac{35}{35-k^2+(k+1)^2} \tag{35}$$

or
$$(35-k^2+(k+1)^2)m^2 = 35((k+1)^2 - 35 + 35 - k^2 + (k+1)^2)$$

$$= 36\,(k+1)^2 - k^2$$

$$m^2 = \frac{36\,(k+1)^2 - k^2}{35-k^2+(k+1)^2} \tag{36}$$

Table 23.2. The value of m² for which S_{min} is maximum when second moment about origin is prescribed.

k	1	2	3	4	5
m^2	3.763	8.0	13.5	26.09	27.63

Also from (12), the probability distribution for a maximum of S_{max} is

$$\left(\frac{35}{36+2k}, \frac{2k+1}{36+2k}\right) \tag{37}$$

In the five intervals these are

$$\left(\frac{35}{38}, \frac{3}{38}\right), \left(\frac{35}{40}, \frac{5}{40}\right), \left(\frac{35}{42}, \frac{7}{42}\right), \left(\frac{35}{44}, \frac{9}{44}\right), \left(\frac{35}{46}, \frac{11}{46}\right) \tag{38}$$

and their entropies are:

.276, .377, .464, .576, .1550

The graph of S_{min} against m^2 is given in Figure 23.5.

Case (iii) : $g\,(i) = \ln i$

The prescribed moments are,

$$\sum_{i=1}^{n} p_i = 1, \sum_{i=1}^{n} p_i \ln i = \ln m, \tag{39}$$

so that the geometric mean is prescribed.

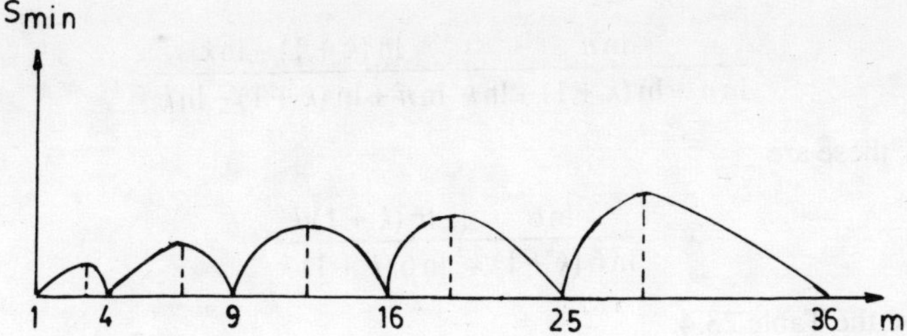

Fig. 23.5 Graph of S_{min} against m^2, where $E(i^2) = m^2$

$$S_{min} = -p \ln p - (1-p)\ln(1-p)$$

where,
$$p = \max\left(\frac{\ln n/m}{\ln n/k}, \frac{\ln m}{\ln (k+1)}\right) \tag{40}$$

so that for $n = 6$

$$\begin{aligned} p &= \max\left(\frac{\ln 6/m}{\ln 6/1}, \frac{\ln m}{\ln 2}\right) && 0 < \ln m < .693 \\ &= \max\left(\frac{\ln 6/n}{\ln 6/2}, \frac{\ln n}{\ln 3}\right) && .693 < \ln m < 1.098 \\ &= \max\left(\frac{\ln 6/m}{\ln 6/3}, \frac{\ln m}{\ln 4}\right) && 1.098 < \ln m < 1.386 \\ &= \max\left(\frac{\ln 6/m}{\ln 6/4}, \frac{\ln m}{\ln 5}\right) && 1.386 < \ln m < 1.609 \\ &= \max\left(\frac{\ln 6/m}{\ln 6/5}, \frac{\ln m}{\ln 6}\right) && 1.609 < \ln m < 1.792 \end{aligned} \tag{41}$$

The maximum value of S_{min} occurs when

$$\frac{\ln n - lm}{\ln n - \ln k} = \frac{\ln m}{\ln\ (k+1)} = \frac{\ln n}{\ln n - \ln k + \ln\ (k+1)} \tag{42}$$

or
$$\ln m = \frac{\ln n\ \ln\ (k+1)}{\ln n - \ln k + \ln\ (k+1)} \tag{43}$$

for $n = 6$, we get the table 3

Table 23.3. Value of ln m for which S_{min} is maximum in different intervals when E (ln, x) = ln m

k	1	2	3	4	5
ln m	.4998	.8959	1.1831	1.4312	1.6263

Also from (12), the probabilities are

$$\frac{\ln n}{\ln n + \ln (k+1) - \ln k}, \frac{\ln (k+1) - \ln k}{\ln n + \ln (k+1) - \ln k} \tag{44}$$

For $n = 6$, these are

$$\frac{\ln 6}{\ln 6 (k+1)/k}, \frac{\ln (k+1)/k}{\ln 6 (k+1)/k} \tag{45}$$

These give the Table 23.4.

Table 23.4. Value of larger probability for which S_{min} is maximum in different intervals when $E(\ln x) = \ln m$.

k	1	2	3	4	5
p	.72106	.81546	.86165	.88925	.92078
S_{min}	.63195	.47791	.40195	.39800	.15325

The graph is as in Figure 23.6.

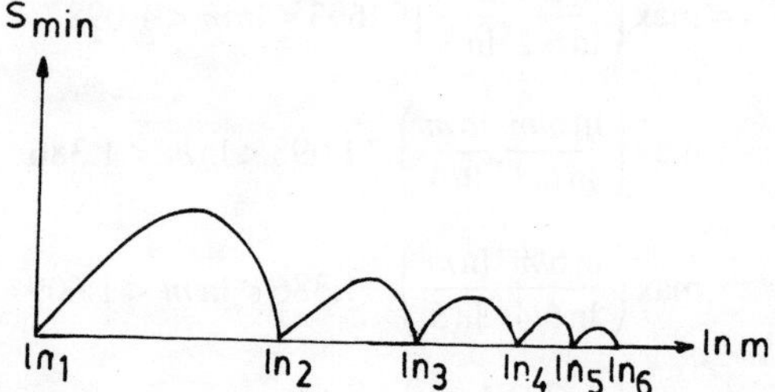

Fig. 23.6. S_{min}-ln m curve when geometrical mean is prescribed.

In the first three interval, maximum value occurs near the ends of the intervals and in the last two near the beginning.

Case (iv): ∴ Maxwell-Boltzmann Distribution

$$g(i) = \varepsilon_i = i^{\text{th}} \text{energy level.}$$

We can rename the energy levels so that $g(i)$ is an increasing function but it need not be a concave or convex function, e.g. we may have

i	1	2	3	4	5	6
ε_i	1	3	6	7	11	12

The maximum value of S_{min} arises when

$$\frac{\varepsilon_m - \varepsilon_1}{\varepsilon_{k+1} - \varepsilon_1} = \frac{\varepsilon_n - \varepsilon_1}{\varepsilon_n + \varepsilon_{k+1} - \varepsilon_1 - \varepsilon_k} \tag{46}$$

$$\text{when,} \quad k = 1 \quad \frac{\varepsilon_m - 1}{3 - 1} = \frac{12 - 1}{12 + 3 - 1 - 1} \text{ or } \varepsilon_m = 1 + \frac{22}{13} = \frac{35}{13} = 2.69$$

$$k = 2 \quad \frac{\varepsilon_m - 1}{6 - 1} = \frac{12 - 1}{12 + 6 - 1 - 3} \text{ or } \varepsilon_m = 1 + \frac{55}{14} = \frac{69}{14} = 4.93$$

$$k = 3 \quad \frac{\varepsilon_m - 1}{6} = \frac{12 - 1}{11 + 1} \Rightarrow \varepsilon_m = 1 + \frac{66}{12} = \frac{13}{2} = 6.5$$

$$k = 4 \quad \frac{\varepsilon_m - 1}{10} = \frac{11}{11 + 4} \Rightarrow \varepsilon_m = 1 + \frac{110}{15} = \frac{25}{3} = 8.35$$

$$k = 5 \quad \frac{\varepsilon_m - 1}{11} = \frac{11}{11 + 1} \Rightarrow \varepsilon_m = 1 + \frac{121}{12} = \frac{133}{12} = 11.01 \tag{47}$$

The probabilities are $(p, 1 - p)$ where p is given in Table 23.5

Table 5. Typical values for larger probabilities for given MB distribution

k	1	2	3	4	5
p	$\frac{11}{13}$	$\frac{11}{14}$	$\frac{11}{12}$	$\frac{11}{15}$	$\frac{11}{12}$
$S_{\min}$	.42936	.51947	.28683	.57992	.28683

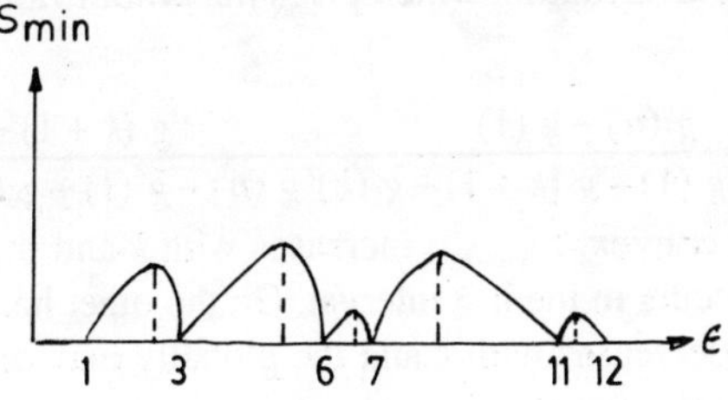

Fig. 23.7

23.7 USE OF OTHER MEASURES OF ENTROPY

It may be noted that in finding minimum entropy probability distributions in the present case, we do not make use of any measure of entropy. Also having found the minimizing probability distribution, we use it to find $S_{\min}$ and there we need the measure of entropy. As such, the minimum entropy probability distributions are independent of the measure of entropy used, since these were obtained from the considerations that the two non-zero probability should be as unequal as possible.

However, whatever measure of entropy we use, the graphs will be similar. This is unlike the case of maximum entropy probability distributions which essentially depend on the measure of entropy used and we have to know the measure of entropy before we can find the maximum entropy probability distribution.

23.8 SUMMARY OF RESULTS

Let the random variate take values 1, 2, . . . , n and let the constraints be,

$$\sum_{i=1}^{n} p_i = 1, \sum_{i=1}^{n} p_i g_r(i) = g(m) \tag{48}$$

where, $g(i)$ is a continuous monotonic increasing function of m then,

(*i*) S_{min} is a piece-wise increasing-decreasing function of $g(m)$ which vanishes where $m = k$, where k is a positive integer. When $k \le m \le k+1$. S_{min} is an increasing-decreasing function which attains its maximum value when,

$$\frac{g(m)-g(k)}{g(k+1)-g(m)} = \frac{g(n)-g(m)}{g(m)-g(1)} \tag{49}$$

(*ii*) There are cusps when $m = 1, 2, \ldots, n$ and at each point there are two coincident vertical tangents.

(*iii*) The probability distribution for calculating S_{min} for any value of m between k and $k+1$ is given by $(p, 1-p)$, where

$$p = \max\left(\frac{g(n)-g(m)}{g(n)-g(k)}, \frac{g(m)-g(1)}{g(k+1)-g(k)}\right) \tag{50}$$

(*iv*) S_{min} is maximum for that value of m is the interval $(k, k+1)$ for which

$$\frac{g(n)-g(m)}{g(n)-g(k)} = \frac{g(m)-g(1)}{g(k+1)-g(k)} \tag{51}$$

(*v*) The probability distribution which gives maximum value of S_{min} in $(k, k+1)$ is

$$\frac{g(n)-g(1)}{g(n)-g(1)+g(k+1)-g(k)}, \frac{g(k+1)-g(k)}{g(n)-g(1)+g(k+1)-g(k)} \tag{52}$$

(*vi*) When $g(x)$ is convex. $(S_{min})_{max}$ increases with k and the globally maximum value of S_{min} occurs in the last interval. On the other hand when $g(x)$ is concave, $(S_{min})_{max}$ decreases with k and the globally maximum value of $(S_{min})_{max}$ occurs in the first interval.

(*vii*) When $g(x)$ is a linear function of x, $(S_{min})_{max}$ is independent of k, i.e., it is the same as all interval, but its value depends on n and decreases as n increases and is symmetry about central value.

(*viii*) In all cases, in the earlier intervals the maximum value of S_{min} occurs near the end of the interval, which in later intervals it occurs near the middle of the intervals.

(*ix*) The probability distribution for which S_{min} is maximum does not depend on the measure of entropy used, but value of S_{min} depends on the measure.

(*x*) S_{min} gives the minimum uncertainty subject to the given constraints. In particular if $S_{min} \neq 0$, the system cannot be deterministic. Of course, the converse is not true i.e., if $S_{min} = 0$, the system may or may not be deterministic.

(*xi*) For any value of m, $S_{max} - S_{min}$ measures uncertainty remaining after the given information is used. If at any step $S_{max} - S_{min} = 0$, thus there is no uncertainty remaining, and the probability distribution is completely known.

(*xii*) In the case when $g(k)$ is monotonic decreasing; the first factor on the R.H.S. of (15) is still negative, but the second factor is now positive, so that

$$\psi'(k) \gtrless 0 \quad \text{according as} \quad g'(k+1) \lessgtr g'(k)$$

Combining this result with the earlier results, we get (*a*) $g(k)$ increasing convex (concave) $\Rightarrow$ $(S_{min})_{max}$ increasing (decreasing) (*b*) $g(k)$ decreasing convex (concave) $\Rightarrow$ $(S_{min})_{max}$ decreasing (increasing). In particular, if $g(i) = 1/i$ this function is decreasing and convex, so that if the harmonic mean is prescribed $(s_{min})_{max}$ decreases with k.

Thus when or second moment is prescribed, $(S_{min})_{max}$ increases, but when geometric or harmonic mean is prescribed, $(S_{min})_{max}$ decreases with k.

23.9 THE CASE WHEN MORE THAN TWO CONSTRAINTS ARE PRESCRIBED

We have discussed exhaustively the cases when the variate is discrete and one moment in addition to the natural constraint is prescribed. In this case our results are even more elegant than those for the maximum entropy probability distributions.

However, difficulties arise when two or more constraints are prescribed in addition to the natural constraint. Thus consider the simple case when $n = 6$ and

$$E(i) = 2.5, \qquad E(i^2) = 17$$

In this case, three are these non-zero probabilities. The only triplets we have to consider are those which include p_5 or p_6 or both p_5 and p_6. and include, p_1 or p_2 or both p_1 and p_2

Thus, we consider the following:

$$\begin{aligned}&(p_1, p_2, p_5), (p_1, p_2, p_6), (p_1, p_5, p_6), (p_2, p_3, p_5), (p_2, p_3, p_6), (p_2, p_5, p_6)\\&(p_2, p_4, p_5), (p_1, p_4, p_6), (p_1, p_3, p_5), (p_2, p_3, p_6), (p_2, p_5, p_6)\\&(p_2, p_4, p_6), (p_2, p_4, p_6), (p_2, p_5, p_6)\end{aligned} \tag{53}$$

and find which of these gives smallest entropy.

In general, given that $E(i) = a$, $E(i^2) = b^2$

where
$$1 \le a \le 6, 1 \le b^2 \le 36, b^2 \ge a^2, \tag{54}$$

we can in principle find the minimum entropy probability distributions and S_{min}. When n is small, it should not be difficult to try out all the possibilities and a convenient computational programme can be easily developed.

However, it will be worthwhile to develop analytical results for this case in the same way in which we developed these for the simple case of one constraint in addition to the natural constraint.

23.10 CASE WHEN THE VARIATE IS CONTINUOUS

Let the random variate x vary from a to b and let the constraints be

$$\int_a^b f(x)\,dx = 1, \int_a^b g(x) f(x) dx = g(m)\; a < m < b \tag{55}$$

then it is easily seen that the minimum entropy probability distribution which has necessarily to be the most concentrated distributions has its density function,

$$f(x) = \delta(x - m) \tag{56}$$

where δ denotes Dirac's delta function.
If however we are given the constraints

$$\int_a^b f(x)dx = 1, \int_a^b g_1(x) f(x)dx = g_1(m), \int_a^b g_2(x) f(x)\,dx = g_2(n),$$

$$a < m, n < b \tag{57}$$

we shall obviously require two delta functions, i.e., we require two probability masses p_1, p_2 and c and d so that

$$p_1 g_1(c) + p_2 g_1(d) = g_1(m)$$

$$p_1 g_2(c) + p_2 g_2(c) = g_2(n)$$

$$p_1 + p_2 = 1 \tag{58}$$

In principle, we can eliminate d between the first two equations, solve for p_1, p_2 in terms of c and choose c so that p_1, p_2 are as unequal as possible and then solve for c and d to finally get the function,

$$f(x) = p_1 \delta(x - c) + p_2 \delta(x - d) \tag{59}$$

We have to see that c and d lie in the interval $[a, b]$ and $p_1, p_2 \geq 0$

It is obvious that while each special problem may be solvable, it may not be easy to obtain a general solution.

23.11 MAXIMUM CROSS-ENTROPY PROBABILITY-DISTRIBUTIONS

Consider a six-faced dice for which the *a priori* probability distribution is $(q_1, q_2, \ldots, q_0)$ and the mean is prescribed as m, then we have to minimize

$$\sum_{i=1}^{n} p_i \ln \frac{p_i}{q_i} \tag{60}$$

subject to

$$\sum_{i=1}^{n} p_i = 1, \qquad \sum_{i=1}^{n} i p_i = m \tag{61}$$

Since there are two constraints, in the maximum cross-entropy probability distribution, only two probabilities will be non-zero, one corresponding to an index $< m$ and the other corresponding to an index $> m$. Thus we have to choose from the following pairs:

$1 < m < 2$ $\quad (p_1, p_2), (p_1, p_3), (p_1, p_4), (p_1, p_5), (p_1, p_6).$
$2 < m < 3$ $\quad (p_1, p_3), (p_1, p_4), (p_1, p_5), (p_1, p_6); (p_2, p_3), (p_2, p_4), (p_2, p_5), (p_2, p_6)$

$$\begin{array}{ll} 3<m<4 & (p_1, p_4), (p_1, p_5), (p_1, p_6), (p_2, p_4), (p_2, p_5), (p_2, p_6), (p_3, p_5), (p_3, p_6), (p_5, p_6) \\ 4<m<5 & (p_1, p_5), (p_1, p_6), (p_2, p_5), (p_2, p_6), (p_3, p_5), (p_3, p_6), (p_4, p_5), (p_5, p_6) \\ 5<m<6 & (p_1, p_6), (p_2, p_6), (p_3, p_6), (p_4, p_6), (p_5, p_6) \end{array} \quad (62)$$

Given any value of m, we can choise class of pairs from (62), and then calculate (60) for all 5 or 8 or 9 feasible pairs of that class and choose the pair for which the cross-entropy is maximum.

Here, the maximizing cross-entropy probability distribution will depend on the measure of cross-entropy used. This is more complicated than is the case of minimum entropy probability distributions since the final distribution depends also on Q.

Moreover, even when m is an integer, the maximum cross-entropy need not occur at one degenerate distribution. Thus, if $m = 2$, the degenerate distribution is (0, 1, 0, 0, 0, 0) and the value of the cross-entropy here is $-\ln q_2$. While the value of the cross-entropy at (1/2, 0, 1/2, 0, 0, 0) is,

$$\frac{1}{2}\ln\frac{1/2}{q_1}+\frac{1}{2}\ln\frac{1/2}{q_3}=-\ln 2-\frac{1}{2}\ln q_1-\frac{1}{2}\ln q_3=-\ln 2-\ln\sqrt{q_1q_3}$$

$$=-\ln 2\sqrt{q_1q_3} \quad (63)$$

and this will be > the cross-entropy at the single degenerate distribution, if

$$-\ln 2\sqrt{q_1q_3}>-\ln q_2 \text{ or } 2\sqrt{q_1q_3}<q_2 \quad (64)$$

which can be satisfied, e.g. if $Q = .1, .4, .1, .1, .2, .1$. Here, we have also to consider pairs (p_1, p_3), (p_1, p_4), (p_1, p_5), (p_1, p_6), (0, 1) and find which gives maximum cross-entropy.

23.12 CONCLUDING REMARKS

We have given a complete solution for the discrete-variate case for the problems of finding the minimum entropy probability distributions when in addition to the natural constraints, one more constraint is the prescribed, viz. the expected value of a monotonic increasing function of the random variable is given. Our solution is as simple and elegant as for the maximum entropy probability distribution.

For the discrete variate case, when more than one additional constraint is prescribed, the solution can be found, though it may require some computer time. For the continuous variate case, every problem may have to be solved separately and further research is required.

REFERENCES

1. E.T. Jaynes (1957), "Information Theory and Statistical Mechanics," Physics Reviews Vol. 106, page 620–33.
2. J.N. Kapur (1989), "Maximum Entropy Models in Science and Engineering", Wiley Eastern, New Delhi & John Wiley, New York.
3. J.N. Kapur and H.K. Kesavan (1986), "Generalised maximum entropy principle with applications", Sanford Educational Press, University of Waterloo.
4. J.N. Kapur and H.K. Kesavan (1992), "Entropy Optimization Principles and their Applications", Academic Press, New York.

5. C.E. Shannon (1948), "A Mathematical Theory of Communication" Bell systems Tech J. Vol. 27, pp. 379–423, 623–659.
6. S. Watanabe (1981), "Pattern recognition as a quest for maximum entropy" Pattern Recognition Vol. 13, pp. 381–387.

24

THE MOST FEASIBLE PROBABILITY DISTRIBUTION

[The concept of the most feasible probability distribution has been introduced and such a distribution has been obtained in some cases. A method of handling inequality constraints for finding maximum entropy probability distribution is given. The formulae for finding most feasible and maximum entropy probability distributions are compaired and it is shown that in both cases the maximum value of entropy is a concave function of the parameters. A method for dividing the uncountable set of feasible solutions of a set of inequalities to a countable set of families of solutions is also given.]

24.1 INTRODUCTION

Let $p_0, p_1, p_2, \ldots, p_n$ be the proportions of persons living in n + *1 residential* colonies, zeroeth colony being the Central Business Districts itself. Let $c_1, c_2, \ldots, c_n$ be the costs of travel from the n colonies to C.B.D. and let ĉ be the maximum average travel budget provided per person, so that we have

$$p_0 + p_1 + p_2 + \ldots + p_n = 1 \tag{1}$$

$$p_1 c_1 + p_2 c_2 + \ldots + p_n c_n \le \bar{c} \tag{2}$$

$$p_0 \ge 0, p_1 \ge 0, p_2 \ge 0, \ldots, p_n \ge 0 \tag{3}$$

These may be an infinity of feasible solution of (1), (2) and (3). One method of getting a unique most unbiased estimate for them is given by maximizing Shannon's measure

$$-\sum_{i=0}^{n} p_i \ln p_i \tag{4}$$

subject (1), (2) and (3)

In the present paper, our object is more modest. We want to find in some sense an optimal feasible solution of the set of inequalities

$$p_1 + p_2 + \ldots + p_n \le 1 \tag{5}$$

$$p_i c_1 + p_2 c_2 + \ldots + p_n c_n \le \bar{c} \tag{6}$$

$$p_1 \ge 0, \ldots, p_n \ge 0 \tag{7}$$

For this purpose we maximize the objective function.

$$S = -\sum_{i=1}^{n} p_i \ln p_i - (1 - p_1 - p_2 - \ldots - p_n)\ln(1 - p_1 - p_2 - \ldots - p_n)$$

$$- (\hat{c} - p_1 c_1 - p_2 c_2 - \ldots - p_n c_n)\ln(\overline{c} - p_1 c_1 - p_2 c_2 - \ldots - p_n c_n) \tag{8}$$

This function has the following properties

(*i*) It is a continuous function for all feasible probability distributions if we define $0 \ln 0 = 0$

(*ii*) Since

$$\frac{\partial S}{\partial p_i} = -(1 + \ln p_i) + (1 + \ln(1 - p_1 - p_2 - \ldots - p_n)$$

$$+ c_i[1 + \ln(\hat{c} - c_1 p_1 - c_2 p_2 - \ldots - c_n p_n)] \tag{9}$$

$$\frac{\partial^2 S}{\partial p_i^2} = -\frac{1}{p_i} - \frac{1}{1 - p_1 - \ldots - p_n} - \frac{c_i^2}{\hat{c} - c_1 p_1 - \ldots - c_n p_n} \tag{10}$$

$$\frac{\partial^2 S}{\partial p_i \partial p_j} = -\frac{1}{1 - p_1 - p_2 - \ldots - p_n} - \frac{c_i c_j}{\hat{c} - c_1 p_1 - \ldots - c_n p_n} \tag{11}$$

$$\sum_{i=1}^{n} \sum_{j=1}^{n} \frac{\partial^2 S}{\partial p_i \partial p_j} u_i u_j = -\sum_{i=1}^{n} \frac{u_i^2}{p_i} - \frac{\left(\sum_{i=1}^{n} u_i\right)^2}{1 - p_i - p_2 - \ldots - p_n}$$

$$- \frac{\left(\sum_{i=1}^{n} c_i u_i\right)^2}{\hat{c} - c_1 p_1 - c_2 p_2 - \ldots - c_n p_n} \tag{12}$$

This is a negative definite quadratic form, so that S is a concave function of p_1, $p_2, \ldots, p_n$

(*iii*) S is defined for only feasible probability distributions. Its partial derivatives are also defined for only feasible probability distribution. It will therefore be maximum for a feasible probability distribution.

(*iv*) The distributions we are considering are incomplete probability distributions.

(*v*) Since S is a concave function, there will be a unique probability distribution giving the globally maximum value of S.
We shall define this maximizing probability distribution as the most feasible probability distribution function satisfying the inequality constraints (5), (6), (7).

24.2 THE MOST FEASIBLE PROBABILITY DISTRIBUTION

(1) We first consider the simple case when there is no constraint on budget, so that we have to maximize

$$S = -\sum_{i=1}^{n} p_i \ln p_i - \sum_{i=1}^{n} (1 - p_1 - p_2 - \ldots - p_n)\ln(l - p_i - p_2 - \ldots - p_n) \tag{13}$$

This gives $$\frac{p_1}{1}=\frac{p_2}{1}=\ldots=\frac{p_n}{1}=\frac{1-p_1-p_2-\ldots-p_n}{1}=\frac{1}{n+1},$$

so that $$p_i=\frac{1}{n+1}, \quad i=1,2,\ldots,n. \tag{14}$$

This is expected, since in this case, the C.B.D. and the others n colonies will share the population equally. It may be noted that if we maximise $-\sum_{i=1}^{n} p_i \ln p_i$ subject to $\sum_{i=1}^{n} p_i = 1$, we get each $p_i = 1/n$, but the most feasible probability distribution subject to $\sum_{i=1}^{n} p_i \leq 1$ is given by $p_i = \frac{1}{n+1}$, so that for the distribution the constraint is a strict inequality and the probability distribution is incomplete. Similarly the most feasible probability distribution subject to $\sum_{i=1}^{n} p_i \leq c$ is given by $p_i = \frac{c}{n+1}$

(*i*) Now we consider the maximization (8). Using (9), we get

$$\frac{p_i}{1-p_1-p_2-\ldots-p_n}=[e(\hat{c}-c_1p_1-c_2p_2-\ldots-c_np_n]^{c_i}=d^{c_i}(\text{say}) \tag{15}$$

or $$\frac{p_1}{d^{c_1}}=\frac{p_2}{d^{c_2}}=\ldots=\frac{p_n}{d^{c_n}}=\frac{1-p_1-p_2-\ldots-p_n}{1}=\frac{p_1+p_2+\ldots+p_n}{d^{c_1}+d^{c_2}+\ldots d^{c_n}} \tag{16}$$

or $$p_i=\frac{1}{1+d^{c_1}+d^{c_2}+\ldots+d^{c_n}}, \tag{17}$$

so that $$1-\sum_{i=1}^{n} p_i=\frac{1}{1+d^{c_1}+d^{c_2}+\ldots+d^{c_n}} \tag{18}$$

Also $$\hat{c}-\sum_{i=1}^{n} c_ip_i=\frac{d}{e} \tag{19}$$

The equation to determine d is

$$\frac{d}{e}=\hat{c}-\frac{\sum_{i=1}^{n} c_id^{c_i}}{1+\sum_{i=1}^{n} d^{c_i}} \tag{20}$$

Differentiating

$$\frac{1}{e}=\frac{d\hat{c}}{dd}-\frac{\left(1+\sum_{i=1}^{n} d^{c_i}\right)\sum_{i=1}^{n} c_i^2d^{c_i}-\left(\sum_{i=1}^{n} c_id^{c_i}\right)^2}{d\left(1+\sum_{i=1}^{n} d^{c_i}\right)^2} \tag{21}$$

By using Cauchy-Sehwartz in equality

$$\frac{d\hat{c}}{dd} - 1/e > 0 \quad \text{or} \quad \frac{d\hat{c}}{dd} > \frac{1}{e} > 0 \quad \text{and} \quad \frac{dd}{d\hat{c}} > 0\,, \tag{22}$$

assuming $c_1, c_2, \ldots, c_n$ are not all equal, so that $\hat{c}$ is a monotonic increasing function of d and d is a monotonic increasing function of $\hat{c}$.

If $d = 0$, from (19), $p_1 = p_2 = p_3 = \ldots = p_n = 0$ and from (15) $\hat{c} = 0$ and as d increases, $\hat{c}$ increases. As $d \to \infty$, from (17) $p_1, p_2, \ldots, p_{n-1} \to 0$ and $p_n \to 1$, $\hat{c}$ also approaches infinity. As such as $\hat{c}$ varies from 0 to ∞, d also goes from 0 to ∞ and for every value of $\hat{c}$ there is a unique value of d.

From (20) if $d = 1$

$$\frac{1}{e} = \hat{c} - \frac{\sum_{i=1}^{n} c_i}{n+1} \quad \text{or} \quad \hat{c} - \frac{\sum_{2}^{n} c_i}{n+1} - \frac{1}{e} = \hat{c}^* \text{ (say)} \tag{23}$$

Now $\hat{c} \gtreqless \hat{c}^*$ according $d \gtreqless 1$ and the probability increases, remains constant or decreases according to $\hat{c} > \hat{c}^*$

Now

$$S_{\max} = -\sum_{i=1}^{n} p_i \ln \frac{d^{c_i}}{1 + d^{c_1} + \ldots + d^{c_n}} - \frac{1}{1 + d^{c_1} + \ldots + d^{c_n}} \ln \frac{1}{1 + d^{c_1} + \ldots + d^{c_n}}$$

$$-\frac{d}{e} \ln d/e$$

$$= -\left(\hat{c} - \frac{d}{e}\right)\ln d + \left(\frac{\sum_{i=1}^{n} d^{c_i}}{1 + d^{c_1} + \ldots + d^{c_n}} + \frac{1}{1 + d^{c_1} + d^{c_2} + \ldots + d^{c_n}}\right)$$

$$\ln\left(1 + d^{c_1} + d^{c_2} + \ldots + d^{c_n}\right) - \frac{d}{e} \ln d/e$$

$$= -\hat{c} \ln d + \frac{d}{e} + \ln\left(1 + d^{c_1} + d^{c_2} + \ldots + d^{c_n}\right), \tag{24}$$

so that

$$\frac{dS_{\max}}{d\hat{c}} = -\ln d + \frac{\sum_{i=1}^{n} c_i d^{c_i - 1}}{1 + \sum_{i=1}^{n} d^{c_i}} \frac{dd}{d\hat{c}} - \frac{\hat{c}}{d} \frac{dd}{d\hat{c}} + \frac{1}{c} \frac{dd}{d\hat{c}}$$

$$= -\ln d + \frac{1}{d}\left(\hat{c} - \frac{d}{e}\right)\frac{dd}{d\hat{c}} - \frac{c^*}{d} \frac{dd}{dc^*} + \frac{1}{\hat{c}} \frac{dd}{d\hat{c}}$$

$$= -\ln d \tag{25}$$

$$\frac{d^2 S_{\max}}{d\hat{c}^2} = -\frac{1}{d} \frac{dd}{d\hat{c}} < 0, \tag{26}$$

since $\frac{dd}{d\hat{c}} > 0$, so that $S_{\max}$ is a concave function of $\hat{c}$.

$$\frac{dS_{\max}}{dd} = -\ln d \,\frac{dd}{d\hat{c}}, \tag{27}$$

so that $S_{\max}$ increases with d when $d < 1$ and decreases with d when $d > 1$ and $S_{\max}$ is maximum when $d = 1$. The maximum value of $S_{\max}$ is

$$\frac{1}{e} + \ln(n+1) \tag{28}$$

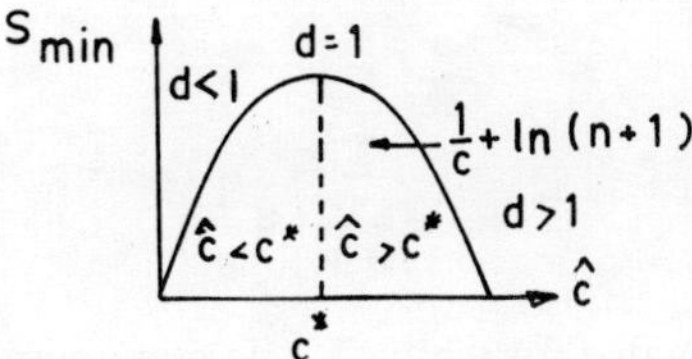

Fig. 24.1

It may be noted that

$$S = -\sum_{i=1}^{n} p_i \ln p_i - (1 - p_1 - p_2 - \dots - p_n)\ln(1 - p_1 - p_2 - \dots - p_n)$$

$$-(\hat{c} - c_1 p_1 - c_2 p_2 - \dots - c_n p_n)\ln(\hat{c} - c_1 p_1 - c_2 p_2 - \dots - c_n p_n) \tag{29}$$

The first sum is > 0, the second sum > 0, and the third term can be negative if $\hat{c}$ is large. Thus for large value of $\hat{c}$, S can be negative and $S_{\max}$ can also be negative.

24.3 COMPARISON WITH MAXIMUM ENTROPY SOLUTION

If we maximize

$$-\sum_{i=0}^{n} p_i \ln p_i \tag{30}$$

subject to $\sum_{i=0}^{n} p_i = 1,\ \sum_{i=0}^{n} p_i c_i = \hat{c},\ c_0 = 0,$ (31)

we get $p_i = \frac{b^{c_1}}{1 + b^{c_1} + b^{c_2} + \dots + b^{c_n}},\ i = 0, 1, 2, \dots, n,$ (32)

where $\frac{c_1 b^{c_1} + c_2 b^{c_2} + \dots + c_n b^{c_n}}{1 + b^{c_1} + b^{c_2} + \dots + b^{c_n}} = \hat{c},$ (33)

which differs from (20) by

$$\frac{d}{e}+\frac{c_1d^{c_1}+c_2d^{c_2}+...+c_nd^{c_n}}{1+d^{c_1}+d^{c_2}+...+d^{c_n}}=\hat{c} \tag{34}$$

Now $\hat{c}$ is a monotonic increasing function of d. As such from (33) and (34)

$$d<b, \tag{35}$$

so that the common ratio for the GP of the most feasible distribution is less than the common ratio of the GP of the most unbiased distribution. For maximum entropy distribution also $S_{\max}$ is a concave function of $\hat{c}$, but its maximum value occurs when $\hat{c}=\frac{\sum_{i=1}^{n} c_i}{n+1}$, while for the most feasible probability distribution, $S_{\max}$ is maximum when from (23)

$$\hat{c}=\frac{\sum_{i=1}^{n} c_i}{n+1}+\frac{1}{e}, \tag{36}$$

which is larger than the value for which $S_{\max}$ is maximum for the maximum entropy distribution

Also while the entropy measure for maximum entropy distribution is always non-negative, the S-function for the most feasible probability distribution can be negative also.

The differences are not surprising since the problems are different and the objective functions are different.

24.4 THE GENERAL CASE

Here we have a number of inequality constraints

$$\sum_{i=1}^{n} p_i g_{ri} \le a_r, r=1,2,3,...,m \tag{37}$$

Our objective function becomes

$$S=-\sum_{i=1}^{n} p_i \ln p_i-(1-p_1-p_2-...-p_n)\ln(1-p_1-p_2-...-p_n)$$
$$-\sum_{r=1}^{m}\left(a_r-\sum_{i}^{n} p_i g_{ri}\right)\ln\left(a_r-\sum_{i=1}^{n} p_i g_{ri}\right) \tag{38}$$

It can be easily shown that this a concave function of $p_1, p_2, \ldots, p_n$ and its maximum value is attained when

$$\frac{p_i}{1-p_1-p_2-...-p_n}=\prod_{r=1}^{m}\left(a_r-\sum_{i=1}^{n} p_i g_{ri}\right)^{g_{ri}}, i=1,2,...,n. \tag{39}$$

There are the n equalities to solve for the n maximizing probabilities $p_1, p_2, \ldots, p_n$.

In fact $\frac{\partial S}{\partial p_i}=0$ gives

$$-\ln p_i + \ln(1 - p_1 - p_2 - \ldots - p_n) + \sum_{r=1}^{m} g_{ri}\left(1 + \ln\left(a_r - \sum_{i=1}^{n} p_i g_{ri}\right)\right) = 0 \tag{40}$$

If D_0 is the defecit of $\sum_{i=1}^{n} p_i$ from unity

and D_r is the defecit of $\sum_{i=1}^{n} p_i g_{ri}$ from a_r

and G_i is the gain of p_i over zero, and get

$$-\ln G_1 G_2 \ldots G_n + n \ln D_0 + \sum_{r=1}^{m}\sum_{i=1}^{n} g_{ri} + \sum_{r=1}^{m}\sum_{i=1}^{n} g_{ri} \ln D_r = 0 \tag{41}$$

or

$$\frac{D_0^n D_1^{\sum_{i=1}^{n} g_{1i}} \ldots D_m^{\sum_{i=1}^{n} g_{mi}}}{G_1 G_2 \ldots G_n} = C^{-\sum_{r=1}^{m}\sum_{i=1}^{n} g_{ri}} \tag{42}$$

From (30)

$$\frac{p_i}{1 - p_1 - p_2 - \ldots - p_n} = \prod_{r=1}^{m} d_r^{g_{ri}}, i = 1, 2, \ldots, n. \tag{43}$$

where

$$\frac{d_r}{e} = a_r - \sum_{i=1}^{n} p_i g_{ri}, r = 1, 2, \ldots, m. \tag{44}$$

so that

$$\frac{p_1}{d_1^{g_{11}} d_2^{g_{21}} \ldots d_m^{g_{m1}}} = \frac{p_2}{d_1^{g_{12}} d_2^{g_{22}} \ldots d_m^{g_{m2}}} = \ldots = \frac{p_n}{d_1^{g_{1n}} d_2^{g_{2n}} \ldots d_m^{g_{mn}}}$$

$$= \frac{1 - p_1 - p_2 - \ldots - p_n}{1} = \frac{1}{1 + \sum_{i=1}^{n} d_1^{g_{1i}} d_2^{g_{2i}} \ldots d_m^{g_{mi}}} \tag{45}$$

so that

$$p_i = \frac{d_1^{g_{1i}} d_2^{g_{2i}} \ldots d_m^{g_{mi}}}{1 + \sum_{i=1}^{n} d_1^{g_{1i}} d_2^{g_{2i}} \ldots d_m^{g_{mi}}}, i = 1, 2, \ldots, n \tag{46}$$

Substituting in (44)

$$\frac{d_r}{e} = a_r - \frac{\sum_{i=1}^{n} g_{ri}\, d_1^{g_{1i}} d_2^{g_{2i}} \ldots d_m^{g_{mi}}}{1 + \sum_{i=1}^{n} d_1^{g_{1i}} d_2^{g_{2i}} \ldots d_m^{g_{mi}}} r = 1, 2, \ldots, m. \tag{47}$$

(47) gives m equations to determine $d_1, d_2, \ldots, d_m$. On knowing these, (46) will determine $p_1, p_2, \ldots, p_n$. Now

$$S_{\max} = -\sum_{i=1}^{n} p_i \ln p_i - (1 - p_1 - p_2 - \ldots - p_n) \ln(1 - p_1 - p_2 - \ldots - p_n)$$

$$-\sum_{r=1}^{m}(a_r - p_1 g_{1r} - \ldots - p_n g_{nr}) \tag{48}$$

$$= -\sum_{i=1}^{n} p_i \sum_{r=1}^{m} g_{ri} \ln d_r - \left(\sum_{i=1}^{n} p_i\right) \ln\left(1 + \sum_{i=1}^{n} d_1^{g_{1i}} \ldots d_m^{g_{mi}}\right)$$

$$-\frac{1}{1+\sum_{i=1}^{n} d_1^{g_{1i}} \ldots d_m^{g_{mi}}} \ln \frac{1}{1+\sum_{i=1}^{n} d_1^{g_{1i}} \ldots d_m^{g_{mi}}} - \sum_{r=1}^{m} \frac{d_r}{e} \ln \frac{d_r}{e}$$

$$= -\sum_{r=1}^{m}\left(a_r - \frac{d_r}{e}\right) \ln d_r + \ln\left(1 + \sum_{i=1}^{n} d_1^{g_{1r}} d_2^{g_{2r}} \ldots d_m^{g_{mr}}\right)$$

$$-\sum_{r=1}^{m} \frac{d_r}{c} \ln d_r + \sum_{r=1}^{m} \frac{d_r}{e}$$

$$= -\sum_{r=1}^{m} a_r \ln d_r + \frac{1}{e} \sum_{r=1}^{m} d_r + \ln\left(1 + \sum_{i=1}^{n} d_1^{g_{1i}} \ldots d_m^{g_{mi}}\right) \tag{49}$$

Now
$$\frac{\partial S_{\max}}{\partial a_s} = -\ln d_s + \sum_{r=1}^{m} \left[-\frac{a_r}{d_r} + \frac{1}{e} + \frac{\frac{1}{dr} \sum_{i=1}^{n} g_{ri} d_1^{g_{i1}} \ldots d_m^{g_{mi}}}{1 + \sum_{i=1}^{n} d_1^{g_{i1}} \ldots d_m^{g_{mi}}} \right] \frac{\partial d_r}{\partial a_s} \tag{50}$$

and using (47)

$$\frac{\partial S_{\max}}{\partial a_s} = -\ln d_s, \quad s = 1, 2, \ldots, m. \tag{51}$$

$$\frac{\partial^2 S_{\max}}{\partial a_s^2} = -\frac{1}{d_s} \frac{\partial d_s}{\partial a_s}, \frac{\partial^2 S_{\max}}{\partial a_r \partial a_s} = \frac{-1}{d_s} \frac{\partial d_s}{\partial d_r} \tag{52}$$

Here Hessian matrix of the second order partial derivatives of $S_{\max}$ with respect to $a_1, a_2, \ldots, a_m$ will be negative definite if the Jocobian

$$\frac{\partial (\ln d_1, \ln d_2, \ldots, \ln d_m)}{\partial (a_1, a_2, \ldots, a_m)} \tag{53}$$

is positive i.e., if the Jacobian

$$\frac{\partial (a_1, a_2, \ldots, a_m)}{\partial (\ln d_1, \ln d_2, \ldots, \ln d_m)} \tag{54}$$

is positive or i.e., if the determinant

$$\begin{vmatrix} d_1 \frac{\partial a_1}{\partial d_1} & d_1 \frac{\partial a_2}{\partial d_1} & \ldots & d_1 \frac{\partial a_m}{\partial d_1} \\ d_2 \frac{\partial a_1}{\partial d_2} & d_2 \frac{\partial a_2}{\partial d_2} & \ldots & d_2 \frac{\partial a_m}{\partial d_2} \\ \ldots & \ldots & \ldots & \ldots \\ d_m \frac{\partial a_1}{\partial d_m} & d_m \frac{\partial a_2}{\partial d_m} & \ldots & \partial_m \frac{\partial a_m}{\partial d_m} \end{vmatrix} \tag{55}$$

is positive. Now from (47)

$$a_r = \frac{d_r}{e} + \frac{\sum_i^n g_{ri} d_1^{g_{i1}} d_2^{g_{i2}} \ldots d_m^{g_{im}}}{1 + \sum_{i=1}^n d_1^{g_{i1}} d_2^{g_{i2}} \ldots d_m^{g_{im}}} \tag{56}$$

$$\frac{\partial a_r}{\partial d_r} = \frac{1}{e} + \frac{\left(1 + \sum_{i=1}^n d_1^{g_{i1}} d_2^{g_{i2}} \ldots d_m^{g_{im}}\right)\left(\sum_{i=1}^n g_{ri}^2 d_1^{g_{i1}} \ldots d_m^{g_{mi}}\right) - \left(\sum_{i=1}^n g_{ri} d_1^{g_{i1}} d_2^{g_{i2}} \ldots d_m^{g_{mi}}\right)^2}{d_r \left[1 + \sum_{i=1}^n d_1^{g_{i1}} d_2^{g_{i2}} \ldots d_m^{g_{im}}\right]^2} \tag{57}$$

$$\frac{\partial a_r}{\partial d_s} = \frac{\left(1 + \sum_{i=1}^n d_1^{g_{i1}} d_2^{g_{i2}} \ldots d_m^{g_{im}}\right)\left(\sum_{i=1}^n g_{si} d_1^{g_{i1}} \ldots d_m^{g_{im}}\right) - \left(\sum g_{si} d_1^{g_{1i}} \ldots d_m^{g_{mi}}\right)\left(\sum_{i=1}^n g_{si} d_1^{g_{1i}} \ldots d_m^{g_{mi}}\right)}{\partial_s \left(1 + \sum_{i=1}^n d_1^{g_{i1}} d_2^{g_{i2}} \ldots d_m^{g_{im}}\right)^2} \tag{58}$$

Now let
$$p_0 = \frac{1}{\left(1 + \sum_{i=0}^n d_1^{g_{i1}} d_2^{g_{i2}} \ldots d_m^{g_{im}}\right)}, \quad g_{in} = 0 \tag{59}$$

then

$$\frac{\partial a_r}{\partial d_r} = \frac{1}{e} + \frac{1}{d_r}\left(\sum_{i=0}^n g_{ri}^2 p_i - \left(\sum_{i=0}^n g_{ri} p_i\right)^2\right) \tag{60}$$

$$\frac{\partial a_r}{\partial d_s} = \frac{1}{ds}\left(\sum_{i=0}^n g_{ri} g_{si} p_i - \sum_{i=0}^n g_{ri} p_i \sum_{i=0}^n g_{si} p_i\right) \tag{61}$$

we can write
$$\frac{\partial a_r}{\partial d_m} = \frac{1}{e} + \frac{1}{d_r} \sigma r^2, \frac{\partial a_r}{\partial d_s} = \frac{1}{d_s} \rho_{rs} \sigma_r \sigma_s \tag{62}$$

σ_r^2 is the variance of the variate which takes values $g_{r0}, g_{r1}, \ldots, g_{rn}$ with probabilities $p_0, p_1, \ldots, p_n$ and ρ_{rs} is the correlation coefficient between the random variates which take values $g_{r0}, g_{r1}, \ldots, g_{rn}$ and $g_{s0}, g_{s1}, g_{s2}, \ldots, g_{sn}$ with probabilities $p_0, p_1, \ldots, p_n$ so that from (62)

$$\frac{\partial a_r}{\partial (\ln d_r)} = \frac{d_r}{e} + \sigma_r^2, \frac{\partial a_r}{\partial (\ln d_s)} = \rho_{rs} \sigma_r \sigma_s \tag{63}$$

so that
$$\frac{\partial(a_1, a_2, \ldots, a_m)}{\partial(\ln d_1, \ln d_2, \ldots, \ln d_m)} = \begin{vmatrix} \frac{d_1}{e} + \sigma_1^2 & 0 + \rho_{12}\sigma_1\sigma_2 & \ldots & 0 + \rho_{1m}\sigma_1\sigma_m \\ 0 + \rho_{21}\sigma_1\sigma_2 & \frac{d_2}{e} + \sigma_2^2 & \ldots & 0 + \rho_{2m}\sigma_2\sigma_m \\ \ldots & \ldots & \ldots & \ldots \\ 0 + \rho_{m1}\sigma_1\sigma_m & 0 + \rho_{m2}\sigma_2\sigma_m & \ldots & \frac{d_m}{e} + \sigma_m^2 \end{vmatrix} \tag{64}$$

$$= D + \sum_{r=1}^{m} \frac{d_r}{e} D_r + \sum_{r=1}^{m} \sum_{s=1}^{m} \frac{d_r d_s}{e^2} Dr_s$$

$$+ \sum_{l=1}^{m} \sum_{s=1}^{m} \sum_{r=1}^{m} \frac{d_r d_s d_\ell}{c^3} D_{rsl} + \ldots + \frac{d_1 d_2 \ldots d_m}{e^m} \tag{65}$$

where
$$D = \begin{vmatrix} \sigma_1^2 & \rho_{12}\sigma_1\sigma_2 & \ldots & \rho_{1n}\sigma_1\sigma_n \\ \rho_{21}\sigma_2\sigma_1 & \sigma_2^2 & \ldots & \rho_{2n}\sigma_2\sigma_n \\ \ldots & \ldots & \ldots & \ldots \\ \rho_{1n}\sigma_1\sigma_n & \rho_{2n}\sigma_2\sigma_n & \ldots & \sigma_n^2 \end{vmatrix} \tag{66}$$

and D_i is a determinant obtained from D by deleting r^{th} row and column, D_{rs} is determinant obtained from D by omitting r^{th} & s^{th} rows and columns from D and so on. Since all these determinants are positive, Jocabiain is also positive and Hessian matrix of partial derivatives of $S_{\max}$ w.r.t. $a_1, a_2, \ldots, a_m$ –ve definite so that we conclude that $S_{\max}$ is a concave function of $a_1, a_2, \ldots, a_m$.

24.5 COMPARISON WITH JAYNE'S FORMALISM FOR FINDING MAXIMUM ENTROPY PROBABILITY DISTRIBUTION

The above formalism for finding the most feasible probability distribution is quite parallel to Jaynes formalism for finding the maximum entropy probability distribution. Here $\ln d_1, \ln d_2, \ln d_3, \ldots, \ln d_n$ play the role of Lagrange multipliers and we have to find first, these m (« n) entities in order to solve for the n maximizing probabilities. We also find

$$\frac{\Delta S_{\max}}{\Delta a_r} = -\ln d_r \text{ corresponding to } \frac{\Delta S_{\max}}{\Delta a_r} = \lambda_r \tag{67}$$

in Jaynes formalism, so that $-\ln d_r$ gives the change in $S_{\max}$ due to a small unit change is a_r. If $d_r < 1$ then $S_{\max}$ increases as d_r decreases. If $d_r > 1$, then $S_{\max}$ decreases as d_r increases. Again as there, here, $S_{\max}$ is a concave function of $a_1, a_2, \ldots, a_n$.

24.6 THE MAXIMUM-ENTROPY PROBABILITY DISTRIBUTION SUBJECT TO INEQUALITY CONSTRAINTS

Let $P = (p_1, p_2, \ldots, p_n)$ be a complete probability distribution subject to the non negativity constraints.

$$p_1 \geq 0, p_2 \geq 0, \ldots, p_n \geq 0. \tag{68}$$

$$a_r \leq \sum_{i=1}^{n} p_i g_{ri} \leq b_r, r = 1, 2, \ldots, m. \tag{69}$$

and the equality constraints

$$p_1 + p_2 + \ldots + p_n = 1 \tag{70}$$

Thus we can find a maximum entropy probability distribution by maximizing the entropy measure

$$S' = -\sum_{i=1}^{n} p_i \ln p_i - \sum_{r=1}^{m}\left(b_r - \sum_{i=1}^{n} p_i g_{ri}\right)\ln\left(b_r - \sum_{i=1}^{n} p_i g_{ri}\right)$$

$$-\sum_{r=1}^{m}\left(\sum_{i=1}^{n} p_i g_{ri} - a_r\right)\ln\left(\sum_{i=1}^{n} p_i g_{ri} - a_r\right) \tag{71}$$

subject to (70), the maximizing probabilities will automatically satisfy the inequality constraints (68) and (69). The entropy S' is obviously a concave function of p_1, $p_2, \ldots, p_n$ and its local maximum will be the global maximum. Maximizing it we get

$$-1 - \ln p_i + \sum_{r=1}^{m} g_{ri}\left[1 + \ln\left(-\sum_{i=1}^{n} p_i g_{ri} + b_r\right)\right] - \sum_{r=1}^{n} g_{ri}\left(1 + \ln\left(\sum_{i=1}^{n} p_i g_{ri} - a_r\right)\right) - \lambda = 0$$

or

$$p_i = c\prod_{r=1}^{m}\frac{\left(b_r - \sum_{i=1}^{n} p_i g_{ri}\right)^{g_{ri}}}{\left(\sum_{i=1}^{m} p_i g_{ri} - a_r\right)^{g_{ri}}} = \frac{D_1^{g_{1i}} D_2^{g_{2i}} \ldots D_m^{g_{mi}}}{\sum_{i=1}^{n} D_1^{g_{1i}} D_2^{g_{2i}} \ldots D_m^{g_{mi}}}, \tag{72}$$

where $D_1, D_2, \ldots, D_m$ are obtained by solving the m equations

$$D_r\left[\sum_{i=1}^{n}(g_{ri} - a_r) D_1^{g_{1i}} D_2^{g_{2i}} \ldots D_m^{g_{mi}}\right] = \left[\sum_{i=1}^{n}(b_r - g_{ri}) D_1^{g_{1i}} D_2^{g_{2i}} \ldots D_m^{g_{mi}}\right],$$

$$r = 1, 2, \ldots, m \tag{73}$$

24.7 COMPARISON WITH MOST FEASIBLE PROBABILITY DISTRIBUTION

If we replace the equality constraint (70) by the inequality constraint

$$p_1 + p_2 + \ldots + p_n \le 1, \tag{74}$$

the most feasible probability distribution is given by

$$p_i = \frac{D_i^{g_{1i}} D_2^{g_{2i}} \ldots D_m^{g_{mi}}}{1 + \sum_{i=0}^{n} D_1^{g_{1i}} D_2^{g_{2i}} \ldots D_m^{g_{mi}}} \tag{75}$$

where $D_1, D_2, \ldots, D_m$ are determined by m equation

or

$$D_r\left[\sum_{i=1}^{n}(g_{ri} - a_r) D_1^{g_{1i}} \ldots D_m^{g_{mi}} - a_r\right] = \left[b_r + \sum_{i=1}^{n}(b_r - g_{ri}) D_1^{g_{1i}} \ldots D_m^{g_{mi}}\right] \tag{76}$$

Comparison of (72) and (73) with (75) and (76) give the comparison between maximum entropy and most-feasible probability distributions.

24.8 ALTERNATIVE MAXIMUM-ENTROPY PROBABILITY DISTRIBUTION

In the usual maximum entropy method we maximize

$$S = -\sum_{i=1}^{n} p_i \ln p_i \tag{77}$$

subject to

$$\sum_{i=1}^{n} p_i = 1, \sum_{i=1}^{n} p_i g_{ri} = c_r, \tag{78}$$

so that

$$p_i = \exp(-\lambda_0 - \lambda_1 g_{i1} - \ldots - \lambda_m g_{mi}), \tag{79}$$

where $\lambda_0, \lambda_1, \lambda_2, \ldots, \lambda_m$ are determined by using

$$\sum_{i=1}^{n} \exp(-\lambda_0 - \lambda_1 g_{1i} - \ldots - \lambda_m g_{mi}) = 1 \tag{80}$$

$$\sum_{i=1}^{n} g_{ri} \exp(-\lambda_0 - \lambda_1 g_{1i} - \ldots - \lambda_m g_{mi}) = c_r, r = 1, 2, \ldots, m \tag{81}$$

then

$$S_{\max} = \lambda_0 + \lambda_1 c_1 + \lambda_2 c_2 + \ldots + \lambda_m c_m \tag{82}$$

$S_{\max}$ comes out to be a concave function of $c_1, c_2, \ldots, c_m$. We find the maximum value of $S_{\max}$

(83)

when

$$a_r \le c_r \le b_r, \quad r = 1, 2, \ldots, m. \tag{84}$$

Suppose it is maximum for $c_{10}, c_{20}, \ldots, c_{m0}$, then we find the maximum entropy probability distribution by replacing $c_1, c_2, \ldots, c_n$ in (28) by $c_{10}, c_{20}, \ldots, c_{m0}$.

This will of course be different from the most feasible probability distribution as well as from the maximum entropy distribution found in the last section.

24.9 OTHER FEASIBLE PROBABILITY DISTRIBUTIONS

All solutions obtained by maximizing

$$\begin{aligned} S_{\lambda\mu r} = & -\sum_{i=1}^{n} p_i \ln p_i - \lambda(1 - p_1 - \ldots - p_n) \ln(1 - p_1 - \ldots - p_n) \\ & - \sum_{r=1}^{m} v_r (b_r - g_{r1} p_1 - \ldots - g_{rn} p_n) \\ & - \sum_{r=1}^{m} \mu_r (g_{r1} p_1 + g_{r2} p_2 + \ldots + g_{rn} p_n - a_r) \end{aligned} \tag{85}$$

will satisfy the inequality constraints

$$p_1 \ge 0, p_2 \ge 0, \ldots, p_n \ge 0, p_1 + p_2 + \ldots + p_n \le 1$$

$$a_r \le \sum_{i=1}^{m} p_i g_{ri} \le b_r, r = 1, 2, \ldots, m. \tag{86}$$

and will give different feasible situations of (85) for different feasible values of λ, $\mu_1, \ldots, \mu_m, v_1, v_2, \ldots, v_m$. This can be one method of solving inequalities like (85).

In the simple case of the one constraint only, we get

$$\frac{p_1}{1}=\frac{p_2}{1}=\dots=\frac{p_n}{1}=\frac{(1-p_1-p_2-\dots-p_n)^\lambda}{1}=\frac{p_1+p_2+\dots+p_n}{n} \tag{87}$$

so if

$$p_1+p_2+\dots+p_n=x$$

we get

$$1-x=\left(\frac{x}{n}\right)^{1/\lambda} \tag{88}$$

When $\lambda=0, x=1$, and when $\lambda=1, x=\dfrac{n}{n+1}$

Also

$$x=\frac{n}{n+k}\Rightarrow\frac{k}{n+k}=\left(\frac{1}{n+k}\right)^{\frac{1}{\lambda}}\Rightarrow\frac{1}{\lambda}=\frac{\ln(k/(n+k))}{\ln(1/n+k)} \tag{89}$$

For $n=1$ we have

k	1	2	10	20	100
$1/\lambda$	1	1.39	4.32	8.38	49.3

Thus for every feasible solution, there is a value of λ and for every value of λ there is a family of feasible probability distributions lying in the region

$$p_1+p_2+p_3+\dots+p_n=x\le 1 \tag{90}$$

In the general case for every feasible solution of (86) there is a unique set of values $(\lambda;\mu_1,\mu_2,\dots,\mu_m,\nu_1,\nu_2,\dots,\nu_m)$ and for every set of values of $(\lambda,\mu_1,\mu_2,\dots,\mu_m,\nu_1,\nu_2,\dots,\nu_m)$ there is a set of feasible solutions. Thus the non-countable set of feasible solutions can be changed into a countable class of feasible solutions.

24.10 DUALITY THEORY FOR MOST FEASIBLE PROBABILITY DISTRIBUTIONS

Let

$$\ln d_r=\lambda_r, d_r=\exp(\lambda_r) \tag{91}$$

and

$$Z=\sum_{r=1}^{m}\frac{e^{\lambda_r}}{e}+\ln\left(1+\sum_{i=1}^{n}\exp(g_{i1}\lambda_1+g_{i2}\lambda_2+\dots+g_{im}\lambda_m)\right)-\sum_{r=1}^{m}a_r\lambda_r \tag{92}$$

then

$$\frac{\partial Z}{\partial\lambda r}=\frac{e^{\lambda r}}{e}+\frac{\sum_{i=1}^{n}g_{ri}e^{g_{1i}\lambda_1+g_{2i}\lambda_2+\dots+g_{mi}\lambda_m}}{1+\sum_{i=1}^{n}e^{g_{1i}\lambda_1+g_{i2}\lambda_2+\dots+g_{im}\lambda_m}}-a_r \tag{93}$$

$$\frac{\partial^2 Z}{\partial\lambda_r^2}=\frac{e^{\lambda r}}{e}+\frac{\left(1+\sum_{i=1}^{n}e^{g_{i1}\lambda_1+\dots+g_{im}\lambda_m}\right)\left(\sum_{i=1}^{n}g_{ri}^2e^{g_{1i}\lambda_1+\dots+g_{ni}\lambda_m}\right)-\left(\sum_{i=1}^{n}g_{ri}e^{g_{1i}d\lambda_1+\dots+g_{mi}\lambda_m}\right)^2}{\left(1+\sum_{i=1}^{n}e^{g_{i1}\lambda_1+g_{i2}\lambda_2+\dots+g_{im}\lambda_m}\right)^2} \tag{94}$$

$$\frac{\partial^2 Z}{\partial\lambda_r\partial\lambda_s}=\frac{\left(1+\sum_{i=1}^{n} e^{g_{i1}\lambda_1+\ldots+g_{im}\lambda_{\in}}\right)\left(\sum_{r=1}^{n} g_{ri}g_{si}e^{g_{1i}\lambda_1+\ldots+g_{mi}\lambda_m}\right)-\left(\sum_{i}^{n}=1 g_{ri}e^{\sum_{i=1}^{m} g_{ri}\lambda_r}\right)\left(\sum_{r=1}^{n} g_{ri}e^{\sum_{r=1}^{m} g_{ri}\lambda_r}\right)}{\left(1+\sum_{i=1}^{n} e^{g_{i1}\lambda_1+g_{i2}\lambda_2+\ldots+g_{12}\lambda_m}\right)^2} \tag{95}$$

It is easily seen that $\sum_{s=1}^{m}\sum_{r=1}^{m}\frac{\partial^2 z}{\partial\lambda_r\partial\lambda_s}u_r u_s$ is a positive definite quadratic function so that z is a convex function of $\lambda_1, \lambda_2, \ldots, \lambda_m$ whose minimum value occurs when

$$a_r=\frac{e^{\lambda_r}}{e}+\frac{\sum_{i=1}^{n} g_{ri}e^{g_{1i}\lambda_1+g_{2i}\lambda_2+\ldots+g_{mi}\lambda_m}}{1+\sum_{i=1}^{n} e^{g_{1i}\lambda_1+g_{2i}\lambda_2+\ldots+g_{mi}\lambda_m}}, \quad r=1,2,\ldots,m, \tag{96}$$

so that

$$a_r=\frac{dr}{e}+\frac{\sum_{i=1}^{n} g_{ri}d_1^{g_{i1}}d_2^{g_{i2}}\ldots d_m^{g_{im}}}{1+\sum_{i=1}^{n} d_1^{g_{i1}}d_2^{g_{i2}}\ldots d_m^{g_{im}}}, r=1,2,\ldots,m. \tag{97}$$

and the minimum value of Z is given by

$$\frac{1}{e}\sum_{r=1}^{m}\ln d_r+\ln\left(1+\sum_{i}^{n} d_1^{g_{i1}}d_2^{g_{i2}}\ldots d_m^{g_{im}}\right)-\sum_{r=1}^{m} a_r\ln d_r \tag{98}$$

We see (97) is the same as (47) and (98) is the same as (49), so that the maximum value of S = minimum value of Z.

Thus we have two problems

Primal Problem: Maximize,

$$S=-\sum_{i=1}^{n} p_i\ln p_i-(1-p_1-p_2-\ldots-p_n)\ln(1-p_1-p_2-\ldots-p_n)$$

$$-\sum_{r=1}^{m}(a_r-p_1g_{r1}-p_2g_{r2}-\ldots-p_ng_{rn})\ln(a_r-p_1g_{r1}-\ldots-p_ng_{rn}) \tag{101}$$

Dual Problem: Minimize,

$$Z=\frac{1}{e}\sum_{r=1}^{m} e^{\lambda_r}+\ln\left(1+\sum_{i=1}^{m} e^{g_{i1}\lambda_1+g_{i2}\lambda_2+\ldots+g_{im}\lambda_m}\right)-\sum_{r=1}^{m} a_r\lambda_r \tag{102}$$

and we find that the maximum value of S = minimum of Z.

S is a function of n values and Z is a function of a much smaller number m values.

25

THE MOST LIKELY PROBABILITY DISTRIBUTION

[It is shown that by maximizing Shannon's measure of entropy or minimising Kullback-Leibler's measure of cross-entropy subject to given constraints, we may not always get the most likely probability distribution. However, there will in general always be a measure of entropy or cross-entropy by maximizing or minimizing which we can get the most likely probability distribution. Some new measures are derived using this consideration and some new derivations are given for existing measures.]

25.1 INTRODUCTION

In a recent paper (Kapur [7]) the following results were established by using Stirling's asymptotic formula for $N!$

(*i*) Let q_i $(i = 1, 2, \ldots, n;\ \sum_{i=1}^{n} q_i = 1)$ be the probability of a ball thrown at random

folling in the ith box, so that the probability of the n boxes receiving $Np_1, Np_2, \ldots, Np_n$ balls when N balls are thrown at random in the boxes is given by

$$\overline{P} = \frac{N!}{Np_1!Np_2!\ldots Np_n!} q_1^{Np_1} q_2^{Np_2} \ldots q_n^{Np_n}, \tag{1}$$

then if N is large

$$\overline{P} \cong -N \sum_{i=1}^{n} p_i \ln\frac{p_i}{q_i} = -\sum_{i=1}^{n} Np_i \ln\frac{Np_i}{Nq_i} \tag{2}$$

(*ii*) Let N_i balls be thrown in the box according to Pisson distribution with mean N_iq_i and let the balls be thrown independently in the boxes so that

$$\overline{P} = \prod_{i=1}^{n} e^{-N_iq_i} \frac{(N_iq_i)^{N_ip_i}}{(N_iP_i)!}, \tag{3}$$

then when $N_1, N_2, \ldots, N_n$ are large.

$$\ln \overline{P} \cong -\sum_{i=1}^{n} N_i \left[p_i \ln\frac{p_i}{q_i} - p_i + q_i \right] \tag{4}$$

(*iii*) Let N_i balls be thrown in the ith box according to the binomial distribution with probability q_i of success and let the balls be thrown independently in the boxes, so that

$$\overline{P} = \prod_{i=1}^{n} {}^{N_i}C_{N_i P_i} \, q_i^{N_i P_i} (1-q_i)^{N_i - N_i P_i}, \tag{5}$$

then when $N_1, N_2, \ldots, N_n$ are large

$$\ln \overline{P} \cong -\sum_{i=1}^{n} N_i \left[p_i \ln \frac{p_i}{q_i} + (1-p_i) \ln \frac{1-p_i}{1-q_i} \right]$$

$$= -\sum_{i=1}^{n} \left[N_i p_i \ln \frac{N_i p_i}{N_i q_i} + (N_i - N_i p_i) \ln \frac{N_i - N_i p_i}{N_i - N_i q_i} \right] \tag{6}$$

(*iv*) Let
$$q_i = \frac{Q_i}{1+Q_i}, 1 - q_i = \frac{1}{1+Q_i}, \tag{7}$$

so that the probability that we have to throw exactly $N_i + N_i p_i$ balls to get $N_i p_i$ balls in the ith box, i.e. in such a way that the ratio of balls in the ith box to the total number of balls thrown is $p_i / (1 + p_i)$, then

$$\overline{P} = \prod_{i=1}^{n} {}^{N_i + N_i p_i - 1}C_{N_i p_i - 1} \left(\frac{Q_i}{1+Q_i} \right)^{N_i p_i - 1} \left(\frac{1}{1+Q_i} \right)^{N_i} \frac{Q_i}{1+Q_i}$$

$$= \prod_{i=1}^{n} \frac{(N_i + N_i P_i - 1)!}{N_i!(N_i P_i - 1)!} \left(\frac{Q_i}{1+Q_i} \right)^{N_i P_i} \left(\frac{1}{1+Q_i} \right)^{N_i} \tag{8}$$

and,

$$\ln \overline{P} \cong -\sum_{i=1}^{n} N_i \left[P_i \ln \frac{P_i}{q_i} - (1+P_i) \ln \frac{1+P_i}{1+Q_i} \right]$$

$$= -\sum_{i=1}^{n} \left[N_i P \ln \frac{N_i P_i}{N_i Q_i} - (N_i + N_i P_i) \ln \frac{N_i + N_i P_i}{N_i + N_i Q_i} \right] \tag{9}$$

In Section 2, we make some comments on these results pointing out their importance and significance in the context of obtaining most likely probability distributions and the relationship of these distributions with the maximum entropy and minimum cross-entropy probability distributions. In Sections 3 and 4 we shall obtain some more results of the same type and discuss their significance.

25.2 COMMENTS ON THE IMPORTANCE AND SIGNIFICANCE OF THE RESULTS

(*i*) From (2) as $\overline{P} \le 1$, $\ln \overline{P} \le 0$,

$$D_i(p:Q)=\sum_{i=1}^{n} p_i \ln\frac{p_i}{q_i}\geq 0 \tag{10}$$

As $N\to\infty$, Np_i approaches asymptotically Nq_i and $\overline{P}$ approaches unity. The deficit of $\overline{P}$ from unity or of $\ln \overline{P}$ from zero gives a measure of the divergences of Np_i's from Nq_i's as N times the directed divergence of P from Q.

Given any information about P, that distribution will be most likely which satisfies the given information, but otherwise makes $\overline{P}$ maximum or makes $D_1\,(P:Q)$ minimum.

However, minimizing $D_1\;(P:Q)$ subject to given information in what is required by Kullback's [11] principle of minimum discrimination information or of minimum directed divergence or of minimum cross-entropy. According to it P gives that distribution which out of all distributions consistent with the given information, it 'closest' to the given *a priori* distribution Q. P is then called the minimum cross-entropy distribution.

Our above discussion shows that the distribution which minimizes Kullback-Leibler [12] measure (10) subject to given, information is also the most likely distribution subject to the same information. Conceptually being 'closest to Q' and being 'most likely' are two different concepts, but both lead to the same final distribution.

(*ii*) Further, if each $q_i = 1/n$ so that the ball is *a priori* equally likely to fall in any box, then the probability of the boxes receiving $Np_1, Np_2, \ldots, Np_n$ balls is given asymptotically by

$$\ln\overline{P}\cong -N\sum_{i=1}^{n} p_i \ln np_i = -N\left[\sum_{i=1}^{n} p_i \ln p_i + \ln n\right] \tag{11}$$

so that $\overline{P}$ is maximum subject to any given information when Shannon's [14] measure of entropy,

$$H_1(P) = -\sum_{i=1}^{n} p_i \ln p_i \tag{12}$$

is maximum subject to the given information.

However, this is what is required by Jayne's [4] maximum entropy principle to obtain the most unbiased, most objective, most random, most uniform distribution subject to the given information.

Thus, we find that in this case the most objective distribution is also the most likely distribution.

(*iii*) When the underlying distribution is the Poisson distribution, the measure of directed divergence is

$$D_2(P:Q) = \sum_{i=1}^{n}\frac{N_i}{N}\left[p_i \ln\frac{p_i}{q_i} - p_i + q_i\right] \tag{13}$$

where,

$$N = \sum_{i=1}^{n} N_i \tag{14}$$

If all the N_i's are equal, it becomes

$$D_3(P:Q) = \sum_{i=1}^{n}\left[p_i \ln\frac{p_i}{q_i} - p_i + q_i\right] \tag{15}$$

In general (13) may be written as,

$$D_2(P:Q) = \sum_{i=1}^{n} w_i \left(p_i \ln\frac{p_i}{q_i} - p_i + q_i\right) \tag{16}$$

and may be called a measure of weighted directed divergence.

This is more satisfactory than the measure,

$$D_4(P:Q) = \sum_{i=1}^{n} w_i p_i \ln\frac{p_i}{q_i} \tag{17}$$

proposed by Taneja and Tuteja [16] since while $D_2(P:Q)$ is always ≥ 0 $D_4(P:Q)$ can also be negative.

If each $q_i = \frac{1}{n}$, (16) shows that we have to maximize

$$H_2(P) = \sum_{i=1}^{n} w_i \left(p_i - p_i \ln p_i - p_i \ln n - \frac{1}{n}\right) + k \tag{18}$$

where k as a constant.

We however want a measure of entropy to be always ≥ 0 and vanish only for degenerate distributions. This suggests that Guiasu's [2] measure of weighted entropy,

$$H_3(P) = -\sum_{i=1}^{n} w_i p_i \ln p_i \tag{19}$$

can also be used.

Thus, when the underlying distribution is Poisson, we should minimize $D_2(P:Q)$ or $D_3(P:Q)$ rather then $D_1(P:Q)$, In case no information is available about *a priori* distribution, or the weights, we should take these to uniform.

It may be noted that if $\mu_1, \mu_2 \ldots, \mu_n$ are the means of the Poisson distribution, then

$$q_1 + q_2 + \ldots + q_n = \frac{\mu_1}{N_1} + \frac{\mu_2}{N_2} + \ldots + \frac{\mu_n}{N_n} \tag{20}$$

and this is not necessarily unity. Similarly, $p_1, p_2, \ldots, p_n$ are n different proportions and their sum is not necessarily unity. As such, measures (10) and (15) are essentially different. Measure (10) is applicable when $\sum_{i=1}^{n} p_i = \sum_{i=1}^{n} q_i = 1$, while measure (15) is applicable when their sums can be different from unity.

We may be given other constraints on p_i's, e.g. suppose each ball of the ith category is marked i and has a mass m_i and we may be given the total of all the numbers or total of all the masses received in the n boxes, so that we have the constraints

$$\sum_{i=1}^{n} N_i i p_i = A \tag{21}$$

$$\sum_{i=1}^{n} N_i m_i p_i = B \tag{22}$$

and, we will have to minimise (13) subject to (21) and (22) to get the most likely probability distribution. The use of Kullback–Leibler measure in this case is not warranted.

Now suppose $\mu_1, \mu_2, \ldots, \mu_n$ are the means and $Np_1, Np_2, \ldots, Np_n$ are the number of balls received in the boxes where $\sum_{i=1}^{n} p_i = 1$. We can then define

$$q_i = \mu_i \mid N, \tag{23}$$

so that both $p_1, p_2 \ldots p_n$ and $q_1, q_2, \ldots q_n$ are probability distribution and in this case (15) would reduce to (10).

(*iv*) To illustrate the use of (6), consider the situation where n players play a game of chance. Their *a priori* chances of winning in a single trial are $q_1, q_2, \ldots, q_n$. For each win a player gets 1 dollar and for each loss, he loses one dollar. The players play $N_1, N_2, \ldots, N_n$ games, respectively. At the end of the play, the total loss of all players is c dollars. Estimate the proportion of times each player won.
Here the constraint is

$$(N_1 p_1 - N_1(1-p_1)) + (N_2 p_2 - N_2(1-p_2)) + \ldots + (N_n p_n - N_n(1-p_n)) = -c \tag{24}$$

or,
$$2(N_1 p_1 + N_2 p_2 + \ldots + N_n p_n) = N_1 + N_2 + \ldots + N_n - c \tag{25}$$

Minimising (6) subject to (24) we get

$$N_i \ln \frac{p_i}{1-p_i}\frac{1-q_i}{q_i} = \text{const.}\, N_i$$

$$\text{or,} \quad \frac{p_i}{1-p_i} = k\frac{q_i}{1-q_i} \quad \text{or} \quad p_i = \frac{1}{\frac{kq_i}{1-q_i}} + 1 = \frac{1-q_i}{kq_i} + 1 = \frac{1+\overline{k-1}q_i}{kq_i},$$

where k is determined from

$$2\left[\sum_{i=1}^{n} \frac{N_i(1+\overline{k-1}q_i)}{kq_i}\right] = \sum_{i=1}^{n} N_i - c \tag{26}$$

(*v*) By using the present combinatorial approach, we have get not only the measures due to Shannon and Kullback-Leibler, we have also got at least three new measures each of weighted directed divergence weighted entropy and entropy, and we shall get more in the next two sections. Thus the most likely probability distribution approach may be used to generate new measures of entropy and directed divergence from 'physical considerations'.

25.3 SOME IMPORTANT MEASURES OF ENTROPY

(*a*) Suppose N balls are to be distributed among n boxes in such a way that the ith box does not contain less then Na_i balls and not more then Nb_i balls ($0 \le a_i < b_i \le 1$), then the number of ways in which the ith box receives the Np_i balls is given by

$$W = \frac{[N(b_1 + b_2 + \dots + b_n)]!}{(N(a_1 + a_2 + \dots + a_n)! \prod_{i=1}^{n} (Np_i - Na_i)!(Nb_i - Np_i)!} \quad . \tag{27}$$

This would obviously be possible if

$$A = a_1 + a_2 + \dots + a_n \le 1; B = b_1 + b_2 + \dots + b_n \ge 1 \tag{28}$$

Using Stirling's formula, we get

$$W \cong \frac{\sqrt{2\pi}\, e^{-N(b_1 + \dots + b_n)} [N(b_1 + b_2 + \dots + b_n)]^{N(b_1 + b_2 + \dots + b_n) + \frac{1}{2}}}{\sqrt{2\pi}\, e^{-N(a_1 + a_2 + \dots + a_n)} [N(a_1 + a_2 + \dots + a_n)]^{N(a_1 + a_2 + \dots + a_n) + \frac{1}{2}}} \times$$

$$\frac{1}{\prod_{i=1}^{n} \sqrt{2\pi}\, e^{-(Np_i - Na_i)} (Np_i - Na_i)^{Np_i - Na_i + \frac{1}{2}} \sqrt{2\pi}\, e^{-(Nb_i - Np_i)} (Nb_i - Np_i)^{Nb_i - Np_i + \frac{1}{2}}} \tag{29}$$

so that,

$$\begin{aligned} \ln W \cong \text{const.} &+ N(b_1 + b_2 + \dots b_n) \ln(b_1 + b_2 + \dots + b_n) \\ &- N(a_1 + a_2 + \dots + a_n) \ln(a_1 + a_2 + \dots + a_n) \\ &- N \sum_{i=1}^{n} (p_i - a_i) \ln(p_i - a_i) - N \sum_{i=1}^{N} (b_i - p_i) \ln(b_i - p_i) \end{aligned} \tag{30}$$

To get the most likely distribution, we have to choose $p_1, p_2, \dots, p_n$ to maximize W, or

$$\text{Constant} - N \sum_{i=1}^{n} (p_i - a_i) \ln(p_i - a_i) - N \sum_{i=1}^{N} (b_i - p_i) \ln\ (b_i - p_i)$$

or,

$$- \sum_{i=1}^{n} (p_i - a_i) \ln\ (p_i - a_i) - \sum_{i=1}^{n} (b_i - p_i) \ln (b_i - p_i) \tag{31}$$

since the constant and multiplier do not matter for maximization purposes, As special cases, we have

(*i*) when $a_i = 0, b_i = 1,$ we have to maximize

$$- \sum_{i=1}^{n} p_i \ln p_i - \sum_{i=1}^{n} (1 - p_i) \ln (1 - p_i) \tag{32}$$

(*ii*) $a_i = 0, b_i = \frac{1}{c}$, we have to maximize

$$-\sum_{i=1}^{n} p_i \ln p_i - \frac{1}{C}\sum_{i=1}^{n}(1-cp_i)\ln(1-cp_i) \quad (33)$$

(*b*) We now consider directly in terms of the number of balls in the n boxes, rather than in terms of probabilities.
Suppose we want that the number of balls in the ith box should be between l_i and m_i, then the number of ways of getting $n_1, n_2, \ldots, n_n$ balls in the n boxes is

$$W = \frac{(m_1+m_2+\ldots+m_n)!}{(l_1+l_2+\ldots+l_n)!\prod_{i=1}^{n}(n_i-l_i)!(m_i-n_i)!} \quad (34)$$

giving,

$$\ln W \cong -\sum_{i=1}^{n}(n_i-l_i)\ln(n_i-l_i) - \sum_{i=1}^{n}(m_i-n_i)\ln(m_i-n_i) \quad (35)$$

To get the most likely distribution we have to maximize it subject to the constraints on n_i's that may be given to us.

Suppose the constraints are on the total number of balls and on the total masses of balls (or total number of particles and total energy of the particles) and these are

$$n_1+n_2+\ldots+n_n = N$$

$$n_1\varepsilon_1+n_2\varepsilon_2+\ldots+n_n\varepsilon_n = N\hat{\varepsilon} \quad (36)$$

Maximizing we get,

$$\frac{m_i-n_i}{n_i-l_i} = e^{-\lambda-\mu\varepsilon_i}, \quad i = 1,2,\ldots,n$$

or
$$m_i + l_i e^{-\lambda-\mu\varepsilon_i} = n_i\left(e^{-\lambda-\mu\varepsilon_i}+1\right)$$

or,
$$n_i = \frac{m_i + l_i e^{-\lambda-\mu\varepsilon_i}}{e^{-\lambda-\mu\varepsilon_i}+1} \quad (37)$$

(*i*) If each $l_i = 0$, we get

$$n_i = \frac{m_i}{e^{-\lambda-\mu\varepsilon_i}+1} \quad (38)$$

which is an intermediate statistics distribution.

(*ii*) if each $l_i = 0, m_i = 1$, we get

$$n_i = \frac{1}{e^{-\lambda-\mu\varepsilon_i}+1} \quad (39)$$

which is Fermi-Dirac distribution.

25.4 ANOTHER CLASS OF MEASURES OF ENTROPY

We have already obtained Shannon's, Kapur's, Fermi-Dirac's, Bose-Einstein's, Intermediate Statistics and some other measures of entropy and weighted entropy. We have also obtained Kullback-Leibler's Skilling's [15], Kapur's and some other measures of directed divergence for getting most likely distributions. These can also be used for obtaining maximum entropy and minimum cross-entropy distributions for generalised entropy optimization principles (Kapur and Kesavan [9, 10]). All these measures have been obtained from combinatorial arguments which in some sense, are more satisfying than axiomatic arguments and appear to be give a physical meaning to measures of entropy and cross-entropy.

We have not so far obtained Burg's [1] and Havrda-Charvat's [2] (or Renyi's) [13] measures from these considerations.

Now consider $\overline{P}$ given by (2). Here the probabilities $q_1, q_2, \ldots, q_n$ were supposed to be given and $p_1, p_2, \ldots, p_n$ have to be estimated. Now consider, the inverse problem. We are given $p_1, p_2, \ldots, p_n$ and we have to estimate $q_1, q_2, \ldots, q_n$ for maximizing $\overline{P}$. In other words, we have to find $q_1, q_2, \ldots, q_n$ which would maximize the probabilities of getting given $p_1, p_2, \ldots, p_n$ subject to any information that may be given about $q_1, q_2, \ldots, q_n$. This would require choosing $q_1, q_2, \ldots, q_n$ to maximize

$$-\sum_{i=1}^{n} p_i \ln q_i \tag{40}$$

If $p_1, p_2, \ldots, p_n$ are equal, this requires us to maximize

$$\sum_{i=1}^{n} \ln q_i \tag{41}$$

which is Burg's entropy for distribution $q_1, q_2, \ldots, q_n$. Thus Burg's entropy is that entropy by maximizing which subject to given information, we can find the values of $q_1, q_2, \ldots, q_n$ for which $\overline{P}$ is maximum when $p_1 = p_2 = \ldots = p_n = 1/n$, i.e. for which the probability of receiving the same number of balls in all boxes is maximum.

The measure (40) may be called Burg's weighted measure of entropy, the weights being proportional to $p_1, p_2, \ldots, p_n$ and its maximization will give us the values of $q_1, q_2, \ldots, q_n$ for which the probability of getting balls in the ratio $p_1, p_2, \ldots, p_n$ in the boxes is maximized.

We can similarly get other measures of entropy for solving other inverse problem. Thus from the measure (6) we get the measure of entropy,

$$-\sum_{i=1}^{n} [p_i \ln q_i + (1-p_i) \ln (1-q_i)] \tag{42}$$

for the distribution Q for given $p_1, p_2, \ldots, p_n$. This may be called Fermi–Dirac–Burg entropy measure. Similarly,

$$-\sum_{i=1}^{n} [p_i \ln q_i - (1+p_i) \ln (1+q_i)]$$

can be called a Bose–Einstein–Burg measure of entropy.

REFERENCES

1. J.P. Burg (1972), ''The relationship between maximum entropy spectra and maximum likelihood spectra'' in Modern Spectral Analysis, ed. D.G. Childers, pp. 130–131.
2. S. Guiasu (1971), ''Weighted Entropy'', Reports on Math. Physics, vol. 2, pp. 165–179.
3. J.H. Havrda and F. Charvat (1967), ''Quantification Methods of Classification Processes: Concept of Structural α Entropy'', Kybernatica, vol. 3, pp. 30–35.
4. E.T. Jaynes (1957), '' Information theory and statistical mechanics,'' Physical Reviews, vol. 106, pp. 620–630.
5. J.N. Kapur (1972), ''Measures of uncertainty, mathematical programming and physics,'' Jour. Ind. Soc. Agri. Stat., vol. 24 pp. 47–66.
6. J.N. Kapur (1986), '' Four families of measures of entropy'', Ind. Jour. Pure and App. Maths. vol. 17, no. 4, pp. 429–449.
7. J.N. Kapur (1989), '' Monkeys and Entropies'', Bull. Math. Ass. India 21, 39–54, 1990.
8. J.N. Kapur (1990), ''Maximum Entropy Models in Science and Engineering'', Wiley Eastern, New Delhi and John Wiley, New York
9. J.N. Kapur and H.K. Kesavan (1989), '' Generalised Maximum Entropy Principle with Applications, Sandford Educational Press, University of Waterloo, Canada.
10. J.N. Kapur and H.K. Kesavan (1992), ''Entropy Optimization Principles and their Applications,'' Academic Press, New York.
11. S. Kullback, (1959), ''Information Theory and Statistics'', John Wiley, New York.
12. S. Kullback, and R.A. Leibler (1991), ''On Information and Sufficiency,'' Ann. Math Stat. vol. 22, pp. 79–86.
13. A. Renyi (1961), ''On measures of entropy and information,'' Proc. 4th Berkeley Symp. Maths. Stat. Prob. vol. 1, pp. 547–561.
14. C.E. Shannon (1948), ''A Mathematical Theory of Communication'', Bell System Tech & vol. 23, pp. 379–423, 623–655.
15. J.S. Skilling (1989), ''Quantified maximum entropy'', 9th Max. Ent. Proceedings edited by P.F. Fougere, pp. 341–350.
16. H.C. Taneja and R.K. Tuteja (1984), ''Characterization of a qualitative quantitative measure of relative information'' Information sc., vol. 30, pp. 1–6.

26

DUALS OF ENTROPY OPTIMISATION PROBLEMS

[Dual problems have been formulated for a number of primal problems concerned with maximization of various measures of entropy or minimisation of various measures of cross-entropy. In every case, it is shown that the maximum value of entropy (or minimum value of cross-entropy) is equal to the minimum (or maximum) value of the objective function of the dual problem. The recovery of the primal problem, when the dual problem is given, is also discussed.]

26.1 INTRODUCTION

When an entropy functional

$$I(f) = \int_a^b \phi(f(x))dx, \tag{1}$$

where $\phi(\cdot)$ is a concave function, is maximized subject to some constraints,

$$\int_a^b f(x)g_r(x) = a_r, \quad r = 1, 2, ..., m \tag{2}$$

by using Lagrange's method, we get

$$\phi'(f(x)) = \sum_{r=1}^{m} \lambda_r g_r(x), \tag{3}$$

where, λ_r's are determined by using the m equations,

$$\int_a^b \phi'^{-1}\left[\sum_{r=1}^{m} \lambda_r g_r(x)\right] g_r(x)dx = a_r, r = 1, 2, ..., m \tag{4}$$

These λ_r's can also be determined by finding the unconstrained minimum of a suitable function $Z(\lambda) = Z(\lambda_1, \lambda_2, \ldots, \lambda_m)$.

The primal problem is concerned with constrained maximization of $I(f)$ subject to constraints (2). The choice is to be made out of the possibly non-countable infinity of functions $f(x)$ satisfying (2).

The dual problem is concerned with unconstrained minimisation of the function $Z(\lambda)$. The choice is to be made out of the countable infinity of m-tuples $(\lambda_1, \lambda_2, \ldots, \lambda_m)$ of real numbers.

One object of the present discussion is to construct appropriate functions $Z(\lambda)$ for various measures of entropy and to show that in every case, the maximum value of $I(f)$ is equal to the minimum value of $Z(\lambda)$.

Another object is to show that a similar programme can be carried out for constrained minimisation of various measures of cross-entropy. In fact, we shall in every case construct appropriate objective function of Lagrange multipliers $\lambda_1, \lambda_2, \ldots, \lambda_m$ so that the unconstrained maximum value of this function is equal to the constrained minimum of the cross-entropy function.

The measures of entropy and cross-entropy we shall consider are due to Shannon [5], Burg [1], Havrda-Charvat [2], Kapur [3] and Kullback-Leibler [4].

26.2 SHANNON MEASURE OF ENTROPY

Primal Problem I.

Maximize
$$I(f) = -\int_a^b (f(x)\ln f(x) - f(x))dx \tag{5}$$

subject to (2).

Dual Problem I.

Minimize
$$Z(\lambda) = \sum_{r=1}^{m} a_r\lambda_r - \int_a^b \exp\left(\sum_{r=1}^{m} \lambda_r g_r(x)\right)dx \tag{6}$$

It is easily verified that $I(f)$ is the definite integral of a concave function of f and $Z(\lambda)$ is a convex function of $\lambda_1, \lambda_2, \ldots, \lambda_m$.

Maximizing (5) subject to (2), we get

$$f(x) = \exp\left(\sum_{r=1}^{m} \lambda_r g_r(x)\right), \tag{7}$$

where, λ_r's are determined by using

$$\int_a^b g_r(x)\exp\left(\sum_{r=1}^{m} \lambda_r g_r(x)\right)dx = a_r, r = 1,2,\ldots,m \tag{8}$$

and then the maximum value of $I(f)$ is given by

$$\int_a^b \left(\sum_{r=1}^{m} \lambda_r g_r(x)\right)\exp\left(\sum_{r=1}^{m} \lambda_r g_r(x)\right)dx - \int_a^b \exp\left(\sum_{r=1}^{m} \lambda_r g_r(x)\right)dx \tag{9}$$

Minimizing Z for variations in $\lambda_1, \lambda_2, \ldots, \lambda_m$, we get

$$a_r - \int_a^b g_r(x)\exp\left(\sum_{r=1}^{m} \lambda_r g_r(x)\right)dx = 0 \tag{10}$$

which is the same as (8) and then the minimum value of $Z(\lambda)$ is

$$\int_a^b \left(\sum_{r=1}^{m} \lambda_r g_r(x)\right)\exp\left(\sum_{r=1}^{m} \lambda_r g_r(x)\right)dx - \int_a^b \exp\left(\sum_{r=1}^{m} \lambda_r g_r(x)\right)dx \tag{11}$$

which is the same as (9). The value of λ_r's in (9) and (11) are the same and expressions are the same and as such $\max_f I(f) = \min_\lambda Z(\lambda)$.

26.3 BURG'S MEASURE OF ENTROPY

Burg's measure of entropy is given by

$$I(f) = \int_a^b \ln f(x)dx \tag{12}$$

We consider its modified version,

$$I(f) = \int_a^b \ln(1 + cf(x))dx \tag{13}$$

Primal Problem II Maximize entropy measure (13) subject to (2).

Dual Problem II Minimize

$$Z(\lambda) = \sum_{r=1}^{m} \lambda_r a_r - \int_a^b \ln\left(\sum_{r=1}^{m} \lambda_r g_r(x)\right)dx$$

$$+ \int_a^b \frac{\sum_{r=1}^{m} \lambda_r g_r(x)}{c} + (b-a)\ln c^{-(b-a)} \tag{14}$$

Again it is easy to verify that $\ln f$ is a concave function of f and $Z(\lambda)$ is a convex function of $\lambda_1, \lambda_2, \ldots, \lambda_m$.

Maximizing (13) subject to (2) we get

$$f(x) = \frac{1}{\sum_{r=1}^{m} \lambda_r g_r(x)} - \frac{1}{c}, \tag{15}$$

where λ_r's are determined from

$$\int_a^b \frac{g_r(x)}{\sum_{r=1}^{m} \lambda_r g_r(x)} dx - \frac{1}{c} \int_a^b g_r(x)dx = a_r, \quad r = 1, 2, \ldots m \tag{16}$$

On using (15) the maximum value of $I(f)$ is

$$\int_a^b \ln \frac{c}{\sum_{r=1}^{m} \lambda_r g_r(x)} dx = (b-a)\ln c - \int_a^b \ln\left(\sum_{r=1}^{m} \lambda_r g_r(x)\right)dx \tag{17}$$

Now, differentiating (14) with respect to λ_r and putting the derivative equal to zero, we get

$$a_r - \int_a^b \frac{g_r(x)}{\sum_{r=1}^m \lambda_r g_r(x)} dx + \frac{1}{c}\int_a^b g_r(x)\,dx = 0, \quad r = 1, 2, \ldots, m \tag{18}$$

which is same as (16).

Again the minimum value of $Z(\lambda)$ is

$$\int_a^b \frac{\sum_{r=1}^m \lambda_r g_r(x)}{\sum_{r=1}^m \lambda_r g_r(x)} dx - \frac{1}{c}\int_a^b \sum_{r=1}^m \lambda_r g_r(x) dx - \int_a^b \ln\left(\sum_{r=1}^m \lambda_r g_r(x)\right) dx$$

$$+ \frac{1}{c}\int_a^b \sum_{r=1}^m \lambda_r g_r(x) dx + (b-a)\ln c - (b-a), \tag{19}$$

which is easily seen to be same as (17), so that again max $I(f)$ = min $Z(\lambda)$.

Now considering the measure,

$$\int_a^b \ln \frac{1 + cf(x)}{1 + c} dx \tag{20}$$

and letting $c \to \infty$, we get the special case of Burg's entropy measure.

Primal Problem III. Maximize entropy measure (12) subject to (2).

Dual Problem III. Minimize,

$$Z(\lambda) = \sum_{r=1}^m a_r \lambda_r - (b-a) - \int_a^b \ln\left(\sum_{r=1}^m \lambda_r g_r(x)\right) dx \tag{21}$$

26.4 HAVRDA–CHARVAT MEASURE OF ENTROPY

Primal Problem IV. Maximize

$$\frac{1}{1-\alpha}\int_a^b f^\alpha(x) dx, \quad \alpha \neq 1, \alpha > 0 \tag{22}$$

subject to (2),

Dual Problem IV. Minimize,

$$Z(\lambda) = \frac{\alpha}{1-\alpha}\sum_{r=1}^m \lambda_r a_r + \int_a^b \left(\sum_{r=1}^m \lambda_r g_r(x)\right)^{\frac{\alpha}{\alpha-1}} dx \tag{23}$$

It is easily varified that $(1-\alpha)^{-1} f^\alpha$ is a concave function of f and $Z(\lambda)$ is a convex function of $\lambda_1, \lambda_2, \ldots, \lambda_m$.

Maximizing (22) subject to (2), we get

$$f(x) = \left[\sum_{r=1}^m \lambda_r g_r(x)\right]^{\frac{1}{\alpha-1}} \tag{24}$$

where, λ_r's are determined from

$$\int_a^b g_r(x)\left(\sum_{r=1}^m \lambda_r g_r(x)\right)^{\frac{1}{\alpha-1}} dx = a_r \tag{25}$$

The maximum value of $I(f)$ is

$$\frac{1}{1-\alpha}\int_a^b \left(\sum_{r=1}^m \lambda_r g_r(x)\right)^{\frac{\alpha}{\alpha-1}} dx \tag{26}$$

Now, $Z(\lambda)$ is minimum when

$$\frac{\alpha}{1-\alpha} a_r + \frac{\alpha}{\alpha-1}\int_a^b \left(\sum_{r=1}^m \lambda_r g_r(x)\right)^{\frac{1}{\alpha-1}} g_r(x) = 0 \tag{27}$$

which is the same as (25) and the minimum value of $Z(\lambda)$ is

$$\frac{\alpha}{1-\alpha}\int_a^b \left(\sum_{r=1}^m \lambda_r g_r(x)\right)^{\frac{\alpha}{\alpha-1}} dx + \int_a^b \left(\sum_{r=1}^m \lambda_r g_r(x)\right)^{\frac{\alpha}{\alpha-1}} dx$$

$$= \frac{1}{1-\alpha}\int_a^b \left(\sum_{r=1}^m \lambda_r g_r(x)\right)^{\frac{\alpha}{\alpha-1}} dx \tag{28}$$

which is the same as (26) so that $\max_f I(f) = \min_\lambda Z(\lambda)$.

26.5 KAPUR'S MEASURE OF ENTROPY

Primal Problem V. Maximize

$$I(f) = -\int_a^b f(x)\ln f(x)\,dx$$

$$+\frac{1}{c}\int_a^b (1 + cf(x))\ln(1 + cf(x))dx \tag{29}$$

subject to (2).

Dual Problem V. Minimise,

$$Z(\lambda) = \sum_{r=1}^m \lambda_r a_r - \frac{1}{c}\int_a^b \ln\left(e^{\sum_{r=1}^m \lambda_r g_r(x)} - c\right)$$

$$+\frac{1}{c}\int_a^b \sum_{r=1}^m \lambda_r g_r(x)\,dx \tag{30}$$

It is easily verified that $-f\ln f+\frac{1}{c}(1+cf)\ln(1+cf)$ is a concave function of f and $Z(\lambda)$ is a convex function of $\lambda_1, \lambda_2, \ldots, \lambda_m$.

Maximizing (29) subject to (2), we get

$$f(x)=\frac{1}{\exp\left(\sum_{r=1}^{m}\lambda_r g_r(x)\right)-c} \tag{31}$$

where λ_r's are determined from

$$\int_a^b \frac{g_r(x)}{\exp\left(\sum_{r=1}^{m}\lambda_r g_r(x)\right)-c}\,dx=a_r,\quad r=1,2,\ldots,m \tag{32}$$

or
$$\frac{1}{c}\int_a^b \frac{g_r(x)\exp\left(\sum_{r=1}^{m}\lambda_r g_r(x)\right)}{\exp\left(\sum_{r=1}^{m}\lambda_r g_r(x)\right)-c}\,dx-\frac{1}{c}\int_a^b g_r(x)dx=a_r \tag{33}$$

The maximum value of $I(f)$

$$=\frac{1}{c}\int_a^b \frac{\left(\sum_{r=1}^{m}\lambda_r g_r(x)\right)\exp\left(\sum_{r=1}^{m}\lambda_r g_r(x)\right)}{\exp\left(\sum_{r=1}^{m}\lambda_r g_r(x)\right)-c}-\frac{1}{c}\int_a^b \ln\left(\exp\left(\sum_{r=1}^{m}\lambda_r g_r(x)-c\right)dx\right. \tag{34}$$

The maximum value of $Z(\lambda)$ arises when

$$a_r-\frac{1}{c}\int_a^b \frac{g_r(x)\exp\left(\sum_{r=1}^{m}\lambda_r g_r(x)\right)}{\exp\left(\sum_{r=1}^{m}\lambda_r g_r(x)\right)-c}\,dx+\frac{1}{c}\int_a^b g_r(x)=0, \tag{35}$$

which is the same as (32) and the minimum value of $Z(\lambda)$ comes out to be the same as (34) so that $\max_f I(f)=\min_\lambda Z(\lambda)$.

26.6 KULLBACK-LEIBLER'S MEASURE OF CROSS-ENTROPY

Primal Problem VI. Minimize the measure,

$$D(f:g)=\int_a^b\left(f(x)\ln\frac{f(x)}{g(x)}-f(x)+g(x)\right)dx \tag{36}$$

subject to (2).

Dual Problem VI. Maximize,

$$Z(\lambda)=\sum_{r=1}^{m}\lambda_r a_r-\int_a^b g(x)\exp\left(\sum_{r=1}^{m}\lambda_r g_r(x)\right)dx+\int_a^b g(x)dx \tag{37}$$

It is easily verified that $f\ln\dfrac{f}{g}-f+g$ is a convex function of f and $Z(\lambda)$ is a concave function of $\lambda_1, \lambda_2, \ldots, \lambda_m$.

Minimizing $D(f:g)$ subject to (2), we get

$$f(x)=g(x)\exp\left(\sum_{r=1}^{m}\lambda_r g_r(x)\right) \tag{38}$$

where λ_r's are determined by using,

$$\int_a^b g(x)g_r(x)\exp\left(\sum_{r=1}^{m}\lambda_r g_r(x)\right)dx=a_r, \qquad r=1,2,\ldots,m \tag{39}$$

The minimum value of $D(f:g)$ is

$$\int_a^b g(x)\exp\left(\sum_{r=1}^{m}\lambda_r g_r(x)\right)\left(\sum_{r=1}^{m}\lambda_r g_r(x)\right)dx$$

$$-\int_a^b g(x)\exp\left(\sum_{r=1}^{m}\lambda_r g_r(x)\right)dx \tag{40}$$

The maximum value of $Z(\lambda)$ arises when

$$a_r-\int_a^b g(x)\,g_r(x)\exp\left(\sum_{r=1}^{m}\lambda_r g_r(x)\right)dx=0 \tag{41}$$

which is the same as (39). Also the maximum value of $Z(\lambda)$ is

$$\int_a^b g(x)\sum_{r=1}^{m}\lambda_r g_r(x))\exp\left(\sum_{r=1}^{m}\lambda_r g_r(x)\right)dx$$

$$-\int_a^b g(x)\exp\left(\sum_{r=1}^{m}\lambda_r g_r(x)\right)dx \tag{42}$$

which is same as (40), so that

$$\min_f \quad D(f:g)=\max_\lambda Z(\lambda) \tag{43}$$

26.7 HAVRDA–CHARVAT MEASURE OF CROSS-ENTROPY

Primal Problem VII. Minimize,

$$D\,(f{:}g)=\frac{1}{\alpha-1}\int_a^b f^{\alpha}(x)g^{1-\alpha}(x)dx,\quad \alpha\neq 1 \tag{44}$$

subject to (2).

Dual Problem VII. Maximize,

$$Z\,(\lambda)=\frac{\alpha}{\alpha-1}\sum_{r=1}^{m}\lambda_r a_r-\int_a^b g\,(x)\left(\sum_{r=1}^{m}\lambda_r g_r\,(x)\right)^{\frac{\alpha}{\alpha-1}}dx \tag{45}$$

It is easily verified that $(\alpha-1)^{-1}f^{\alpha}g^{1-\alpha}$ is a convex function of f and $Z\,(\lambda)$ is a concave function of $\lambda_1, \lambda_2, \ldots, \lambda_m$.

Minimizing $D\,(f:g)$ subject to (2), we get

$$\frac{f(x)}{g\,(x)}=\left(\sum_{r=1}^{m}\lambda_r g_r\,(x)\right)^{\frac{1}{\alpha-1}} \tag{46}$$

where λ_r's are determined by using

$$\int_a^b g(x)g_r(x)\left(\sum_{r=1}^{m}\lambda_r g_r(x)\right)^{\frac{1}{\alpha-1}}dx=a_r,\quad r=1,2,\ldots,m \tag{47}$$

The minimum value of $D\,(f:g)$ is

$$\frac{1}{\alpha-1}\int_a^b g(x)\left(\sum_{r=1}^{m}\lambda_r g_r\,(x)\right)^{\frac{\alpha}{\alpha-1}}dx \tag{48}$$

$Z\,(\lambda)$ is maximum when

$$a_r-\int_a^b g(x)g_r(x)\quad\left(\sum_{r=1}^{m}\lambda_r g_r(x)\right)^{\frac{1}{\alpha-1}}dx=0,\quad r=1,2,\ldots,m \tag{49}$$

which is the same as (47). Also the minimum value of $Z\,(\lambda)$ is

$$\frac{\alpha}{\alpha-1}[\int_a^b g(x)\left(\sum_{r=1}^{m}\lambda_r g_r(x)\right)^{\frac{\alpha}{\alpha-1}}dx-\int_a^b g(x)\,(\sum_{r=1}^{m}\lambda_r g_r(x)^{\frac{\alpha}{\alpha-1}}dx$$

$$=\frac{1}{\alpha-1}\int_a^b g(x)\left[\sum_{r=1}^{m}\lambda_r g_r(x)\right]^{\frac{\alpha}{\alpha-1}}dx \tag{50}$$

which is the same as (48) so that

$$\min_f D\,(f{:}g)=\max_\lambda Z\,(\lambda) \tag{51}$$

26.8 GENERALISED BURG MEASURE OF CROSS–ENTROPY

Primal Problem VIII. Minimize,

$$\int_a^b \left(\frac{1+cf(x)}{1+cg(x)} - \ln \frac{1+cf(x)}{1+cg(x)} - 1 \right) dx \tag{52}$$

subject to (2).

Dual Problem VIII. Maximize,

$$Z(\lambda) = \sum_{r=1}^{m} \lambda_r a_r + \frac{1}{c} \int_a^b \sum_{r=1}^{m} \lambda_r g_r(x) + \int_a^b \ln \left(\frac{c}{1+cg(x)} - \sum_{r=1}^{m} \lambda_r g_r(x) \right) dx$$

$$- \int_a^b \ln \frac{c}{1+cg(x)} dx \tag{53}$$

It can easily be verified that

$$\frac{1+cf}{1+cg} - \ln \frac{1+cf}{1+cg} - 1$$

is a convex function of f and $Z(\lambda)$ is a concave function of $\lambda_1, \lambda_2, \ldots, \lambda_m$.

Minimizing (52) subject to (2) we get

$$f(x) = \frac{1}{\frac{c}{1+cg(x)} - \sum_{r=1}^{m} \lambda_r g_r(x)} - \frac{1}{c}, \tag{54}$$

where λ_r's are determined from

$$\int_a^b \frac{g_r(x)}{\frac{c}{1+cg(x)} - \sum_{r=1}^{m} \lambda_r g_r(x)} dx - \int_a^b \frac{g_r(x)}{c} dx = a_r \tag{55}$$

From(54)

$$\frac{1+cf(x)}{1+cg(x)} = \frac{1}{1 - \frac{1+cg(x)}{c} \sum_{r=1}^{m} \lambda_r g_r(x)},$$

so that the minimum value of (52) is

$$\int_a^b \left[\frac{1}{1 - \frac{1+cg(x)}{c} \sum_{r=1}^{m} \lambda_r g_r(x)} + \ln \left(1 - \frac{1+cg(x)}{c} \sum_{r=1}^{m} \lambda_r g_r(x) \right) - 1 \right] dx \tag{56}$$

The maximum value of (53) is given by

$$a_r + \frac{1}{c} \int_a^b g_r(x) dx - \int_a^b \frac{g_r(x)}{\frac{c}{1+cg(x)} \sum_{r=1}^{m} \lambda_r g_r(x)} dx = 0 \tag{57}$$

which is the same as (55) and the maximum value of (53) is given by

$$\int_a^b \frac{\sum_{r=1}^m \lambda_r g_r(x)}{\frac{c}{1+cg(x)} - \sum_{r=1}^m \lambda_r g_r(x)} dx - \frac{1}{c}\int_a^b \sum_{r=1}^m \lambda_r g_r(x) dx + \frac{1}{c}\int_a^b \sum_{r=1}^m \lambda_r g_r(x) dx$$

$$-\int_a^b \ln\left(\frac{c}{1+cg(x)} - \sum_{r=1}^m \lambda_r g_r(x)\right) dx + \int_a^b \ln \frac{c}{1+cg(x)} dx \tag{58}$$

which is the same as (56) so that

$$\min_f \quad D(f{:}g) = \max_\lambda Z(\lambda) \tag{59}$$

26.9 KAPUR'S MEASURE OF CROSS–ENTROPY

Primal Problem IX. Minimise,

$$D(f{:}g) = \int_a^b f(x) \ln \frac{f(x)}{g(x)} dx - \frac{1}{c}\int_a^b (1+cf(x) \ln \frac{1+cf(x)}{1+cg(x)} \tag{60}$$

subject to (2)

Dual Problem IX. Maximize,

$$Z(\lambda) = \sum_{r=1}^m \lambda_r a_r - \int_a^b \ln\left(1 + cg(x) - cg(x) e^{\sum_{r=1}^m \lambda_r g_r(x)}\right) dx \tag{61}$$

It can easily be verified that $f \ln \frac{f}{g} - \frac{1}{c}(1+c) \ln \frac{1+cf}{1+cg}$ is a convex function of f and Z is a concave function of $\lambda_1, \lambda_2, \ldots, \lambda_m$.

Minimizing $D(f : g)$ subject to (2) we get

$$\ln \frac{f(x)/g(x)}{(1+cf(x))/(1+cg(x))} = \sum_{r=1}^m \lambda_r g_r(x) \tag{62}$$

or,

$$f(x) = \frac{g(x)}{(1+cg(x)) \exp\left(-\sum_{r=1}^m \lambda_r g_r(x)\right) - cg(x)} \tag{63}$$

where λ_r's are determined by using,

$$\int_a^b \frac{g(x) g_r(x) \exp\left(\sum_{r=1}^m \lambda_r g_r(x)\right) dx}{(1+cg(x)) - cg(x) \exp\left(\sum_{r=1}^m \lambda_r g_r(x)\right)} = a_r \tag{64}$$

Differentiating $Z(\lambda)$ with respect to λ_r we get (64) and the maximum value of $Z(\lambda)$ is

$$\int_a^b \frac{g(x)\sum_{r=1}^m \lambda_r g_r(x)\exp\left(\sum_{r=1}^m \lambda_r g_r(x)\right)dx}{(1+cg(x))-cg(x)\exp\left(\sum_{r=1}^m \lambda_r g_r(x)\right)}$$

$$-\int_a^b\left(\ln(1+cg(x))-cg(x)\exp\sum_{r=1}^m \lambda_r g_r(x)\right)dx \tag{65}$$

It can be shown that the minimum value of (60) is also given by (65). Hence minimum value of $D(f:g)$ = maximum value of $Z(\lambda)$.

26.10 RECOVERING THE PRIMAL FROM THE DUAL FOR ENTROPY OPTIMIZATION PROBLEMS

Let the Primal problem be,

Maximize $$\int_a^b \Phi(f(x))dx \tag{66}$$

where $\Phi(.)$ is a concave function, subject to (2). This gives

$$\Phi'(f(x)) = \sum_{r=1}^m \lambda_r g_r(x) \tag{67}$$

or, $$f(x) = \Phi'^{-1}\left(\sum_{r=1}^m \lambda_r g_r(x)\right) \tag{68}$$

assuming that Φ' has a unique inverse function, then λ_r's are determined by using

$$\int_a^b g_r(x)\Phi'^{-1}\left(\sum_{r=1}^m \lambda_r g_r(x)\right)dx = a_r \tag{69}$$

Let $$\Phi'^{-1} = F' \tag{70}$$

Then (69) gives,

$$\int_a^b g_r(x)F'\left(\sum_{r=1}^m \lambda_r g_r(x)\right)dx = a_r \tag{71}$$

We define $Z(\lambda)$ by

$$Z(\lambda) = \sum_{r=1}^m \lambda_r a_r - \int_a^b F\left(\sum_{r=1}^m \lambda_r g_r(x)\right)dx \tag{72}$$

Comparing (72) with the given function $Z(\lambda)$ is the dual problem we can find $F(\cdot)$ and thus we can find F' and therefore Φ'^{-1} and therefore Φ' and therefore Φ and therefore $I(f)$ and thus we can recover the primal problem.

Illustrations:

(*i*) For dual problem *I*,

$$Z(\lambda) = \sum_{r=1}^{m} a_r \lambda_r - \int_a^b \exp\left(\sum_{r=1}^{m} \lambda_r g_r(x)\right) dx \tag{73}$$

Comparing (72) and (73),

$$F(x) = e^x$$

$$\Rightarrow F'(x) = e^x \Rightarrow \Phi'^{-1}(x) = e^x \Rightarrow \Phi'(x) = \ln x$$

$$\Rightarrow \Phi(x) = x \ln x - x$$

$$\Rightarrow \Phi(f(x)) = f(x) \ln f(x) - f(x)$$

$$\Rightarrow I(f) = -\int_a^b (f(x) \ln f(x) - f(x)) dx \tag{74}$$

since we want $I(f)$ to be a concave funcitonal. This is the is expression in primal problem *I*.

(*ii*) For dual problem *II*,

$$Z(\lambda) = \sum_{r=1}^{m} \lambda_r a_r + \int_a^b \frac{1}{C} \sum_{r=1}^{m} \lambda_r g_r(x) - \ln \sum_{r=1}^{m} \lambda_r g_r(x)) dx \tag{75}$$

Comparing (72) and (75),

$$F(x) = \ln x - \frac{1}{c} x$$

$$\Rightarrow F'(x) = \frac{1}{x} - c \Rightarrow \Phi'^{-1}(x) = \frac{1}{x} - \frac{1}{c} \Rightarrow \Phi'(x) = \frac{1}{x + \frac{1}{c}}$$

$$\Rightarrow \Phi(x) = \ln\left(x + \frac{1}{c}\right)$$

$$\Rightarrow \Phi(f(x)) = \ln\left(f(x) + \frac{1}{c}\right) = \ln(1 + cF(x)) - \ln c$$

$$\Rightarrow I(f) = \int_a^b \ln(1 + (f(x)) dx - \int_a^b \ln c \, dx \tag{76}$$

which differs from (12) only by a constant, which does not affect maximization, but which can be so chosen that the maximum value of $I(f)$ is equal to the minimum value of $Z(\lambda)$.

(*iii*) For dual problem *IV*, we take,

$$Z(\lambda) = \sum_{r=1}^{m} \lambda_r a_r + \frac{1-\alpha}{\alpha} \int_a^b \left(\sum_{r=1}^{m} \lambda_r g_r(x) \right)^{\frac{\alpha}{\alpha-1}} dx \tag{77}$$

Comparing with (72),

$$F(x) = \frac{\alpha - 1}{\alpha} x^{\frac{\alpha}{\alpha-1}}$$

$$\Rightarrow F'(x) = x^{\frac{1}{\alpha-1}} \Rightarrow \Phi'^{-1}(x) = x^{\frac{1}{\alpha-1}} \Rightarrow \Phi'(x) = x^{\alpha-1} \Rightarrow \Phi(x) = \frac{x^\alpha}{\alpha} \Rightarrow I(f)$$

$$= \int_a^b \frac{(f(x))^\alpha}{\alpha} dx \tag{78}$$

In Primal Problem IV, we used,

$$I(f) = \frac{1}{1-\alpha} \int_a^b f^\alpha(x) dx \tag{79}$$

to ensure $\frac{1}{1-\alpha} f^\alpha$ is a concave function of f.

(*iv*) Comparing (72) and (30) for dual problem V, we get

$$F(x) = -\frac{1}{c}x + \frac{1}{c}\ln(e^x - c)$$

$$\Rightarrow F'(x) = \frac{-1}{c} + \frac{1}{c}\frac{e^x}{e^x - c} = \frac{1}{c}\frac{c}{(e^x - c)} = \frac{1}{e^x - c}$$

$$\Rightarrow \Phi'^{-1}(x) = \frac{1}{e^x - c} \Rightarrow \Phi'(x) = \ln\frac{1+cx}{x} \Rightarrow \Phi(x) = \frac{1}{c}(1+cx)\ln(1+cx) - x\ln x$$

$$\Rightarrow I(f) = -\int_a^b f(x)\ln(x)dx + \frac{1}{c}\int_a^b (1+cx)\ln(1+cx)dx \tag{80}$$

which is the same as the function for primal problem V.

In all these problems we find a certain flexibility since we can add a constraint to $I(f)$ or multiply it by a constant to ensure that

(*a*) Maximum value of $I(f)$ = minimum value of $Z(\lambda)$,

(*b*) $\Phi(f)$ is a concave function of f.

In recovering the primal problem, we have also to recover the constraints. This is easily done from (72), since the coefficients of λ_r give both $g_r(x)$ and a_r, and the constraints and given by,

$$\int_a^b f(x) g_r(x) dx = a_r, \qquad r = 1, 2, \ldots, m \tag{81}$$

26.11 RECOVERING THE PRIMAL PROBLEM IN CROSS-ENTROPY MINIMISATION PROBLEM

If we minimise

$$\int_a^b g(x)\Phi\left(\frac{f(x)}{g(x)}\right)dx\,, \tag{82}$$

where $\Phi(\cdot)$ is a convex function subject to (2), we get

$$\Phi'\left(\frac{f(x)}{g(x)}\right) = \sum_{r=1}^{m} \lambda_r g_r(x)$$

or,

$$f(x) = g(x)\Phi'^{-1}\left(\sum_{r=1}^{m} \lambda_r g_r(x)\right), \tag{83}$$

where, λ_r's are determined from

$$\int_a^b g(x) g_r(x) \Phi'^{-1}\left(\sum_{r=1}^{m} \lambda_r g_r(x)\right) dx = a_r \tag{84}$$

Let, $\quad \Phi'^{-1} = F'$ (85)

$$Z(\lambda) = \sum_{r=1}^{m} a_r \lambda_r - \int_a^b g_r(x) F\left(\sum_{r=1}^{m} \lambda_r g_r(x)\right) dx \tag{86}$$

Comparing (86) with $Z(\lambda)$ obtained in Dual Problems *VI, VII, VIII, IX*, we get the function f in each case and then we get in succession F', $\Phi'^{-1}\Phi'$. $\Phi(x)$, $\Phi(f(x))$ and then $D(f : g)$.

In this way we are able to recover the primal problem and the constraints in each case.

If we make use of the fact that the function $D(f : g)$ should be a convex function of f and the minimum value of $D(f : g)$ should be equal to the maximum value of $Z(\lambda)$, we get the measure uniquely.

26.12 CONCLUDING REMARKS

(*i*) We have carried out the entropy optimization problem for continuous variate distributions. The same treatment will hold if we have a discrete variate probability distribution $p_1, p_2, \ldots, p_n$. In the primal problem, we have to find n probabilities and in the dual problem we have to find m λ_r's and $m << n$, so that the problem in the dual space should be easier than that in the primal space.

(*ii*) We can always take one of the constriants,

$$\int_a^b f(x)dx = 1 \quad \text{or} \int_a^b f(x)g_0(x)dx = 1, \quad g_0(x) = 1$$

Our discussion will hold even if we do not take this constraint.
As such the discussion of duality holds for a wider class of functions than probability density functions.

(*iii*) It is well known that in Shannon's entropy case, the maximum value of $I(f)$ is a concave function of $a_1, a_2, \ldots, a_m$. Now $Z(\lambda)$ is a convex function of $\lambda_1, \lambda_2, \ldots, \lambda_m$, but those values of λ's which make $Z(\lambda)$ a minimum are functions of $a_1, a_2, \ldots, a_m$ and if we substitute these we get the minimum value of $Z(\lambda)$ as a function of $a_1, a_2, \ldots, a_n$. As such this minimum value of $Z(\lambda)$ is a concave function of $a_1, a_2, \ldots, a_n$.

(*iv*) Similarly, min $D(f : g)$ is a convex function of $a_1, a_2, \ldots, a_m$ and therefore in the case of cross–entropy, the maxinum value of $Z(\lambda)$ in all cases is a convex function of $\lambda_1, \lambda_2, \ldots, \lambda_m$

REFERENCES

1. J.P. Burg (1972), " The relationship between maximum entropy spectra and maximum likelihood spectra", in Modern Spectral Analysis, ed. D.G. Childrers, pp. 130–131.
2. J.H. Havrda and F. Charvat (1967), "Quantification methods of classification processes : Concept of Structural α Entropy, " Kybernatica, Vol. 3, pp. 30–35.
3. J.N. Kapur (1986), "Four families of measures of entropy", Ind Journ. Pure and Applied Maths., vol. 17, no. 4, pp. 429–449.
4. S. Kullback and R.A. Leibler (1951), "On information and sufficiency" Ann. Math. Stat., vol. 22, pp. 79–86.
5. C.E. Shannon (1948), "A Mathematical Theory of Communication", Bell System Tech. J. Vol. 27, pp. 379–423, 623–659.

27

MAXIMUM-ENTROPY PROBABILITY DISTRIBUTION WHEN MEAN OF A RANDOM VARIATE IS PRESCRIBED AND BURG'S ENTROPY MEASURE IS USED

[It is shown that whether the variate is discrete or continuous, maximization of Burg's measure of entropy subject to only natural and mean constraints, by using Lagrange's method, will always lead to positive probabilities or a positive density function. It is also shown that the measure of cross-entropy of a probability distribution P from Q corresponding to Burg's measure of entropy is given by Kullback-Leibler measure $D(Q : P)$ and when this is minimized subject to mean being prescribed, the minimising probabilities and the minimising density function are always positive.

27.1 INTRODUCTION

Let the only information about a probability distribution $p_1, p_2, \ldots, p_n$ be in terms of the given value of its mean, i.e. we know that

$$\sum_{i=1}^{n} p_i = 1, \ \sum_{i=1}^{n} i p_i = m; \quad p_1 \geq 0, p_2 \geq 0, \ldots, p_n \geq 0 \tag{1}$$

These constraints are not sufficient to determine $p_1, p_2, \ldots, p_n$ uniquely and as such we appeal to Jayne's [2] maximum-entropy principle and maximize Shannon's [7] measures of entropy,

$$S(P) = -\sum_{i=1}^{n} p_i \ln p_i \tag{2}$$

subject to (1) to get

$$p_i = \exp(-\mu i) / \sum_{i=1}^{n} \exp(-\mu i), \tag{3}$$

where μ is determined from the equation

$$\sum_{i=1}^{n} i \exp(-\mu i) / \sum_{i=1}^{n} \exp(-\mu i) = m \tag{4}$$

This is the distribution known as Maxwell-Boltzmann distribution which arises in statistical mechanics, in distribution of population in a city and in marketing [3, 4, 5].

We want to find the corresponding probability distribution when we use Burg's [1] measure of entropy

$$B(P) = \sum_{i=1}^{n} \ln p_i \tag{5}$$

instead of Shannon's measure (2).

Burg's entropy measure is always negative, but this does not matter in entropy maximization, where we have to find a probability distribution with greater entropy than all others satisfying the same constraints, and it does not matter if all the entropies are negative.

We can use (5) only if p_i's are positive. We shall show that all the probabilities obtained by using (5) are in fact positive. In fact this result is the main contribution of this discussion. In section 2, we obtain the maximum-entropy probability distribution using Burg's measure (5) and show that for this MEPD, all the probabilities are positive. We also compare this MEPD with the MEPD obtained by using Shannon's measure. In section 3, we extend the results to the continuous-variate case. In section 4, we give a heuristic argument why the results should hold for general moment constraints. In section 5, we introduce Burg's cross-entropy measure and show that the probabilities obtained by using minimum cross-entropy principle will all be positive.

27.2 BURG'S MAXENT PROBABILITY DISTRIBUTION

Maximizing (5) subject to the mean constraint and the natural constraint in (1), we get

$$1/p_i = \lambda + \mu i, i = 1, 2, \ldots, n, \tag{6}$$

where λ and μ are obtained by solving the equations

$$\sum_{i=1}^{n} \frac{1}{\lambda + \mu i} = 1; \sum_{i=1}^{n} \frac{i}{\lambda + \mu i} = m \tag{7}$$

Equations (6), (7) do not necessarily imply that all the probabilities in (6) have to be positive. The result has to be proved and we proceed to give this proof.

Multiplying equations (7) by λ and μ and adding, we get

$$\sum_{i=1}^{n} \frac{\lambda + \mu i}{\lambda + \mu i} = \lambda + \mu m \quad \text{or} \quad \lambda + \mu m = n, \tag{8}$$

so that from (7)

$$\sum_{i=1}^{n} \frac{1}{n - \mu(m - i)} = 1 \tag{9}$$

One obvious solution of this equation is, $\mu = 0$, but this will give $\lambda = n$ and this will satisfy the second equation of (7) only if $m = (n + 1)/2$.

We are, therefore, interested in finding a non-zero solution of equation (9) in μ. This equation is,

$$\frac{1}{n - \mu(m-1)} + \frac{1}{n - \mu(m-2)} + \ldots + \frac{1}{n - \mu(m-n)} = 1 \tag{10}$$

When simplified, this is an equation of nth degree in μ, one of whose roots is zero. Its non-zero roots will be obtained by solving an equation of $(n-1)^{\text{th}}$ degree in μ.

We start by proving the following lemma:

Lemma: None of the roots of the equation (10) can be complex; in other words, all the roots are real.

Proof: If possible let,

$$\mu = \alpha \pm \beta j, \beta \neq 0, j = \sqrt{-1} \tag{11}$$

be a pair of complex conjugate roots of (10), then

$$\frac{1}{n-(\alpha+j\beta)(m-1)}+\frac{1}{n-(\alpha+j\beta)(m-2)}+\ldots+\frac{1}{n-(\alpha+j\beta)(m-n)}=1 \tag{12}$$

$$\frac{1}{n-(\alpha-j\beta)(m-1)}+\frac{1}{n-(\alpha-j\beta)(m-2)}+\ldots+\frac{1}{n-(\alpha-j\beta)(m-n)}=1 \tag{13}$$

Subtracting (13) from (12), we get

$$\frac{2j\beta(m-1)}{(n-\alpha(m-1))^2+\beta^2}+\frac{2j\beta(m-2)}{(n-\alpha(m-2))^2+\beta^2}+\ldots+\frac{2j\beta(m-n)}{(n-\alpha(n-m))^2+\beta^2}=0 \tag{14}$$

This gives $\beta = 0$, or

$$\frac{m-1}{(n-\alpha(m-1))^2+\beta^2}+\frac{m-2}{(n-\alpha(m-2))^2+\beta^2}+\ldots+\frac{m-n}{(n-\alpha(n-m))^2+\beta^2}=0 \tag{15}$$

The second possibility can easily be ruled out. However, we are not doing it here because in the succeeding discussion, we shall find the actual locations of the n real roots. Let,

$$f(\mu) \equiv \frac{1}{n-\mu(m-1)}+\frac{1}{n-\mu(m-2)}+\ldots+\frac{1}{n-\mu(m-n)}-1 \tag{16}$$

This function is discontinuous at the following points

$$-\frac{n}{n-m}, -\frac{n}{n-(m-1)}, -\ldots -\frac{n}{1-l}, \frac{n}{l}, \ldots, \frac{n}{m-2}, \frac{n}{m-1} \tag{17}$$

where $m = k + l$, $k = [m]$ and l is a positive fraction.

In fact at each of these points, $f(\mu)$ approaches infinity; to be more precise the function approaches $+\infty$ if μ approaches a point from one side and it approaches $-\infty$, if it approaches the point from other side (Figure 27.1).

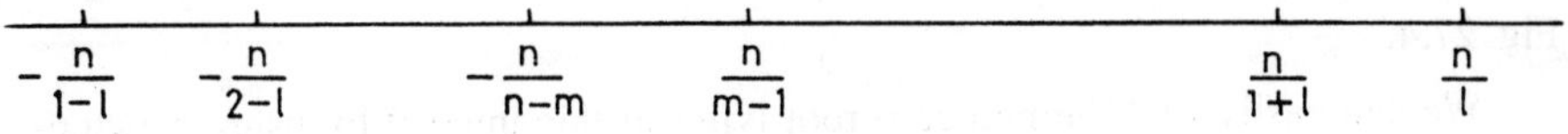

$-\infty\ \infty$ $\quad -\infty\ \infty$ $\quad -\infty\ \infty$ $\quad \infty\ -\infty$ $\quad \infty\cdot\infty$ $\quad \infty\text{-}\infty$

$-\frac{n}{1-l}$ $\quad -\frac{n}{2-l}$ $\quad -\frac{n}{n-m}$ $\quad \frac{n}{m-1}$ $\quad \frac{n}{1+l}$ $\quad \frac{n}{l}$

Fig. 27.1

The graph of this function is given in the Figure (27.2).

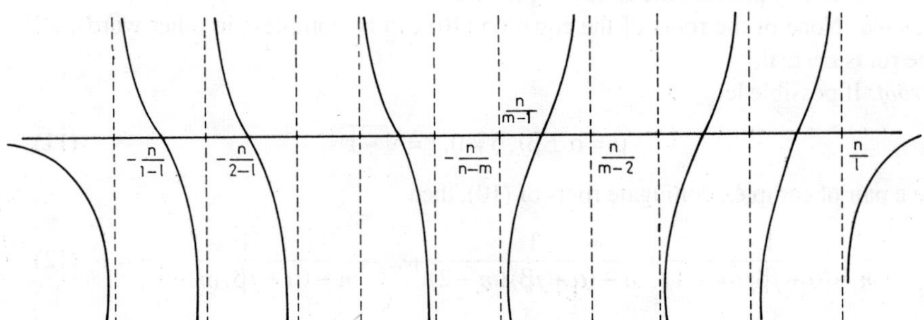

Fig. 27.2

In drawing Fig. 27.2, we use the results of Fig. 27.1 and the results,

$$\operatorname*{Lt}_{\mu\to\infty} f(\mu) = -1, \quad \operatorname*{Lt}_{\mu\to-\infty} f(\mu) = -1 \tag{18}$$

It appears that

(*i*) $k-1$ roots of (10) lie in the intervals

$$\left(\frac{n}{m-1}, \frac{n}{m-2}\right), \left(\frac{n}{m-2}, \frac{n}{m-3}\right), \ldots, \left(\frac{n}{m-k+1}, \frac{n}{m-k}\right) \tag{19}$$

(*ii*) $n-k-1$ roots lie in the intervals

$$\left(-\frac{n}{1-l}, -\frac{n}{2-l}\right), \left(-\frac{n}{2-l}, -\frac{n}{3-l}\right), \ldots, \left(-\frac{n}{n-k-1-l}, -\frac{n}{n-k-l}\right) \tag{20}$$

so that $n-2$ roots lie in these intervals.

The remaining two roots lie in the interval

$$\left(\frac{n}{n-m}, \frac{n}{m-1}\right) \tag{21}$$

One of these roots is zero and the other can be positive or negative.

The shape of graph in the interval $\left(-\frac{n}{n-m}, \frac{n}{m-1}\right)$ can be as in Fig. 27.3 or Fig. 27.4.

We can easily find the non-zero root lying in this interval by using Newton-Raphson method.

In fact, it is only for this root that all the probabilities

$$\frac{1}{n-\mu(m-1)}, \frac{1}{n-\mu(m-2)}, \ldots, \frac{1}{n-\mu(m-n)} \tag{22}$$

will be positive. For other roots, some of the probabilities will become negative. Thus though there are n pair of values of Lagrange's multipliers λ and μ, it is only one pair which is meaningful.

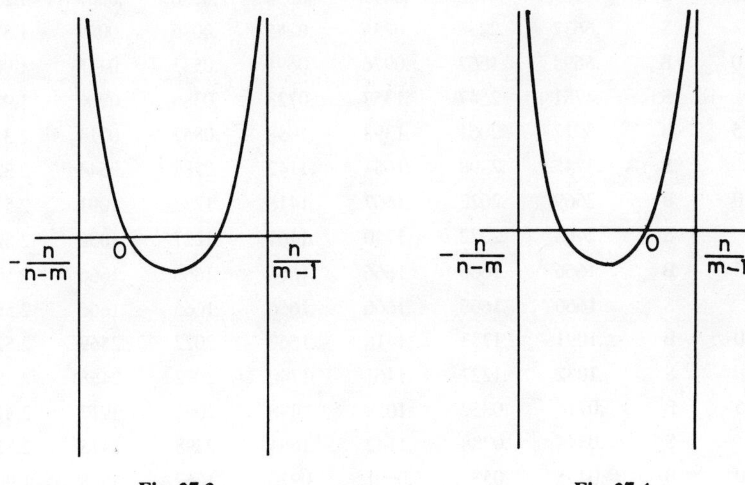

Fig. 27.3 **Fig. 27.4**

Now expanding $f(\mu)$ is powers of μ, we get

$$f(\mu) = \frac{\mu}{n}\left(m - \frac{n+1}{2}\right) + 0(\mu^2) \tag{23}$$

Thus if $m < (n + 1)/2$, then $f(\mu) < 0$ when $\mu > 0$ and small. This corresponds to Figure 27.3, so that in this case the root lying between $-n/(n - m)$ and $n/(m - 1)$ is positive and in this case the probabilities form a decreasing harmonic progression. Similarly, if $m > (n + 1)/2$, $f(\mu) > 0$ when $\mu > 0$ and small. This corresponds to Figure 27.4, so that in this case the root is negative and the probabilities form an increasing harmonic progression. Of course when $m = (n + 1)/2$, the MEPD is the uniform distribution.

This is what is expected.

For both Burg's and Shannon's entropy measures, the Table 27.1 gives the MEPD's for $n = 6$ and for various values of m. [TABLE 27.1]

It may be noted that the value of maximum entropy for either Shannon's or Burg's measure is the same when mean $= m$ or $7 - m$. As shown in Figure 27.5, in both cases S_{max} is a concave function of m which is symmetric about $m = 3.5$ where the entropy is maximum.

Figure 27.6 can be explained by the consideration that both sets of probabilities add upto unity and have the same mean but Burg's measure MEPD follows a harmonic progression and Shannon's measure MEPD follows a geometric distribution and in general the geometrical progression increases or decreases faster than the harmonic progression.

Table 27.1

		p_1	p_2	p_3	p_4	p_5	p_6	$S = -\sum_{i=1}^{6} p_i \log_2 p_i$
$m = 1.5$	B	.7769	.0933	.0497	.0338	.0256	.0207	1.234
	S	.6637	.2238	.0755	.0255	.0086	.0029	1.375
$m = 2.0$	B	.5698	.1667	.0976	.0691	.0533	.0435	1.909
	S	.4781	.2547	.1357	.0723	.0386	.0206	1.973
$m = 2.5$	B	.3917	.2061	.1398	.1058	.0852	.0714	2.313
	S	.3745	.2398	.1654	.1142	.0788	.0544	2.328
$m = 3.0$	B	.2569	.2022	.1667	.1418	.1233	.1091	2.522
	S	.2468	.2072	.1740	.1461	.1227	.1032	2.523
$m = 3.5$	B	.1666	.1666	.1666	.1666	.1666	.1666	2.585
	S	.1666	.1666	.1666	.1666	.1666	.1666	2.585
$m = 4.0$	B	.1091	.1233	.1418	.1667	.2022	.2569	2.522
	S	.1032	.1227	.1461	.1740	.2072	.2468	2.523
$m = 4.5$	B	.0714	.0852	.1058	.1398	.2061	.3917	2.313
	S	.0544	.0788	.1142	.1654	.2398	.3475	2.328
$m = 5.0$	B	.0435	.0533	.0691	.0976	.1667	.5698	1.909
	S	.0206	.0386	.0723	.1357	.2547	.4781	1.973
$m = 5.5$	B	.0207	.0256	.0338	.0497	.0933	.7769	1.234
	S	.0029	.0086	.0255	.0755	.2238	.6637	1.375

27.3 THE CONTINUOUS-VARIATE CASE

By maximizing $\int_a^b \ln f(x)dx$ (24)

subject to,

$$\int_a^b f(x)dx = 1, \int_a^b xf(x)dm = m, \tag{25}$$

we get ,

$$f(x) = \frac{1}{\lambda + \mu x}, \tag{26}$$

where,

$$\int_a^b \frac{1}{\lambda + \mu x} dx = 1, \int_a^b \frac{x}{\lambda + \mu x} dx = m, \tag{27}$$

so that,

$$\int_a^b \frac{\lambda + \mu x}{\lambda + \mu x} dx = \lambda + \mu m = b - a \tag{28}$$

and,

$$f(x) = \frac{1}{b - a - \mu m + \mu x}, \tag{29}$$

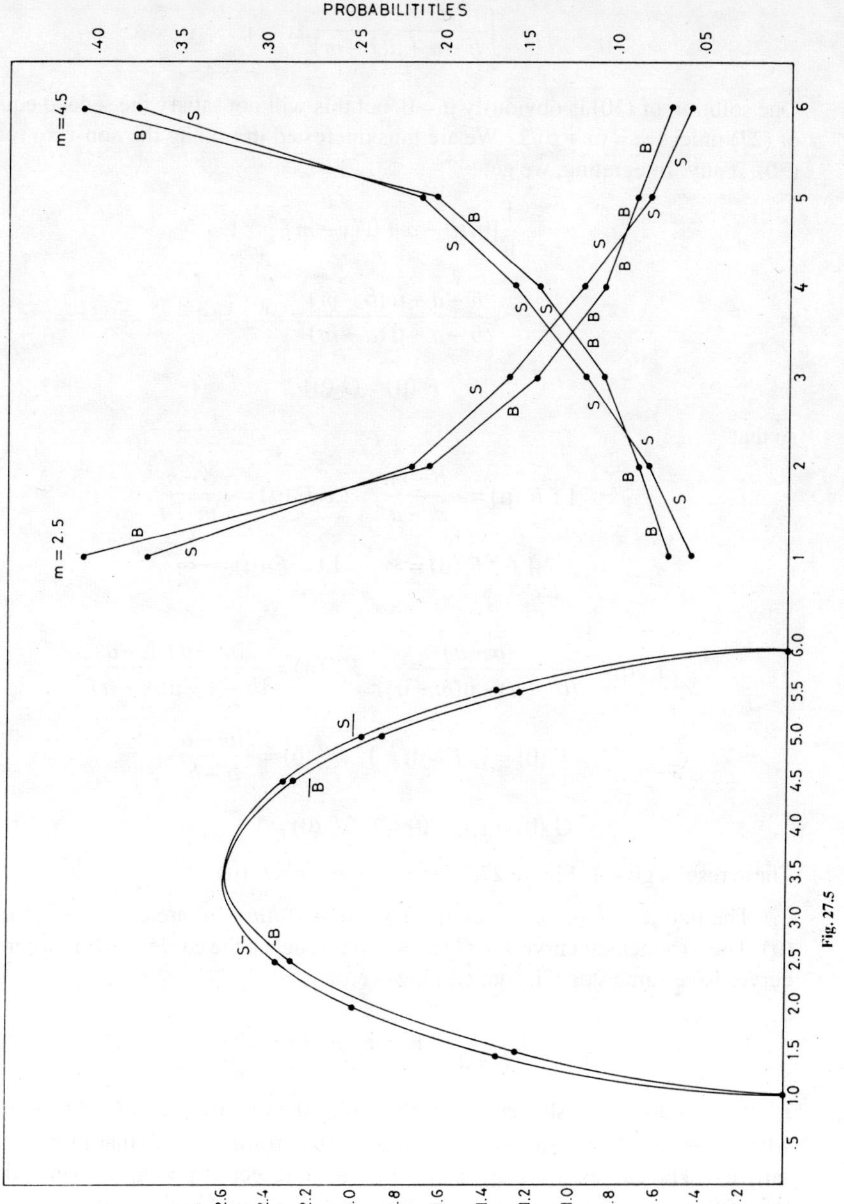
PROBABILITITLES
.40
.35
.30
.25
.20
.15
.10
.05
m = 4.5
m = 2.5
B
S
1
2
3
4
5
6
2.6
2.4
2.2
2.0
1.8
1.6
1.4
1.2
1.0
.8
.6
.4
.2
.5
1.0
1.5
2.0
2.5
3.0
3.5
4.0
4.5
5.0
5.5
6.0

Fig. 27.5

where μ is determined from the equation,

$$\int_a^b \frac{1}{b-a+\mu(x-m)}dx = 1 \tag{30}$$

One solution of (30) is obviously $\mu = 0$, but this will not satisfy the second equation of (27) unless $m = (a+b)/2$. We are thus interested in finding the non-zero roots of (30), if any. Integrating, we get

$$\frac{1}{\mu}[\ln (b-a+\mu (x-m)]_a^b = 1 \tag{31}$$

or,

$$\frac{b-a+\mu(b-m)}{b-a+\mu(a-m)} = e^{\mu} \tag{32}$$

or,

$$F(\mu) = G(\mu) \tag{33}$$

so that

$$\operatorname*{Lt}_{\mu\to\infty} F(\mu) = -\frac{b-m}{m-a}, \quad \operatorname*{Lt}_{\mu\to-\infty} F(\mu) = -\frac{b-m}{m-a} \tag{34}$$

$$\operatorname*{Lt}_{\mu\to\frac{b-a}{m-a}-0} F(\mu) = \infty, \quad \operatorname*{Lt}_{\mu\to\frac{b-a}{m-a}+0} F(\mu) = -\infty \tag{35}$$

$$F'(\mu) = \frac{(b-a)^2}{((b-a)-\mu(m-a))^2}, F''(\mu) = \frac{2(m-a)(b-a)^2}{[b-a-\mu(m-a)]^3} \tag{36}$$

$$F(0) = 1,\ F'(0) = 1,\ F''(0) = \frac{2(m-a)}{b-a} \tag{37}$$

$$G(0) = 1, G'(0) = 1, G''(0) = 1 \tag{38}$$

These results give us Figure 27.7 for the curve, $y = F(\mu)$

The line $\mu = (b-a)/m - a)$ and $Y = -(b-a)/(m-a)$ are asymptotic to $Y = F(\mu)$. The exponential curve $Y = G(\mu) = \exp \mu$ touches the curve at (0, 1) where both curves have same slope, i.e. unity. However, if

$$2\frac{m-a}{b-a} < 1 \quad \text{or} \quad m < (a+b)/2 \tag{39}$$

$F'(\mu)$ increases at a slower rate at this point than $G'(\mu)$. Also $G(\mu) \to \infty$ only when $\mu \to \infty$, while $f(\mu) \to \infty$ when $\mu \to (b-a)/(m-a)$ so that in case (39) is satisfied. The curves, $y = F(\mu)$ and $y = G(\mu)$ intersect at a point for which $0 < \mu < (b-a)/(m-a)$ and the equation (32) has a positive root. Similarly, if

$$m > (a+b)/2, \tag{40}$$

then the two curves intersect at a point for which μ is negative. Thus (32) has a positive root $< (b-a)/(m-a)$ if $m < (a+b)/2$ and a negative root $> -(b-a)/(b-m)$ if $m > (a+b)/2$. In either case $\mu = 0$ is also a root, but since this will not satisfy the second equation of (27), it is not relevant. However, when

$$m = (a+b)/2 \tag{41}$$

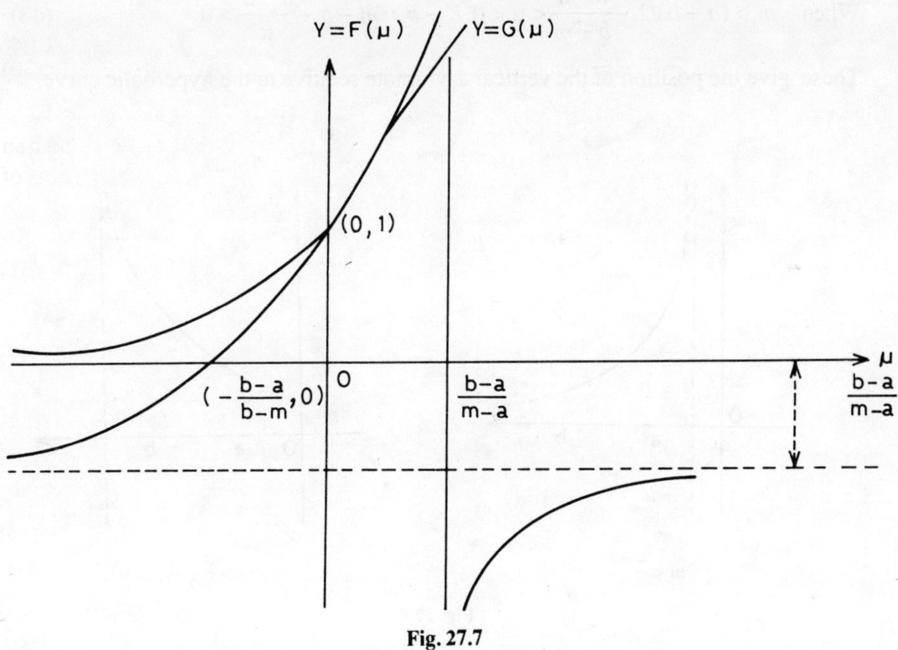

Fig. 27.7

$\mu = 0$ is the only root and it satisfies the second equation of (27) also.

Thus for Burg's measure, maximum entropy density function is constant and the MEPD distribution is uniform if $m = (a + b)/2$.

If $m \gtrless (a + b)/2$, the curve, $y = f(x)$ is a rectangular hyperbola given by

$$y = \frac{1}{\lambda + \mu x} \tag{42}$$

This has two asymptotes, $y = 0$ and

$$x = -\frac{\lambda}{\mu} = -\frac{b - a - \mu m}{\mu} = m - \frac{b - a}{\mu} = K \quad \text{(say)} \tag{43}$$

Also $$f(a) = [\lambda + \mu a]^{-1} = [b - a - \mu (m - a)]^{-1} \tag{44}$$

$$f(b) = [\lambda + \mu b]^{-1} = [b - a + \mu (b - m)]^{-1} \tag{45}$$

If $m < (a + b)/2$, $0 < \mu < (b - a)/(m - a)$ and both $f(a)$ and $f(b)$ are > 0. If $m > (a + b)/2$, $-(b - a)/(b - m) < \mu < 0$ and again both $f(a)$ and $f(b)$ are > 0.

Also, $$f(x) = (\lambda + \mu x)^{-1} = (b - a + \mu(x - m))^{-1} \tag{46}$$

and this is positive in both cases when $a \le x \le b$.

Now from (43),

when $$m < (a + b)/2, 0 < \mu < (b - a)/(m - a), a - K = a - m + \frac{b - a}{\mu} < 0 \tag{47}$$

When $m > (a+b)/2, -\frac{b-a}{b-m} < \mu < 0, K - b = m - b + \frac{b-a}{\mu} > 0$ (48)

These give the position of the vertical asymptote relative to the hyperbolic curve.

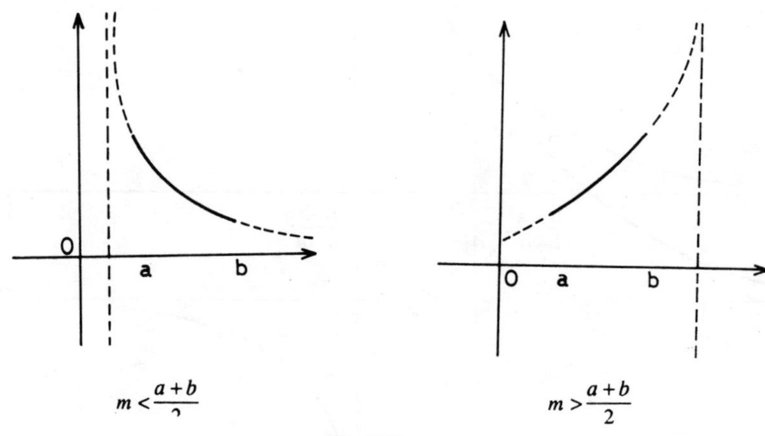

Fig. 27.8

The density function curves are shown in Figure 27.8. From (44) and (45), it is easily seen that

$$f(a) - \frac{1}{b-a} = \frac{1}{b-a-\mu(m-a)} - \frac{1}{b-a} = \frac{\mu(m-a)}{(b-a)(b-a-\mu(m-a))} \tag{49}$$

$$f(b) - \frac{1}{b-a} = \frac{1}{(b-a)+\mu(b-m)} - \frac{1}{b-a} = \frac{\mu(m-b)}{(b-a)(b-a+\mu(b-m))} \tag{50}$$

so that if $\quad 0 < \mu < (b-a)/(m-a), \quad f(a) > \frac{1}{b-a}, \quad f(b) < \frac{1}{b-a}$ (51)

and, if $\quad -\frac{b-a}{m-b} < \mu < 0, \quad f(a) < \frac{1}{b-a}, \quad f(b) > \frac{1}{b-a}$ (52)

These results are also illustrated in Figure 27.8 which also shows Shannon's MEPD's in the two cases.

The results of this section are quite analogous to those of the last section for the discrete variate case.

27.4 DISCUSSION OF THE GENERAL CASE

We have seen above that when Burg's entropy measure is maximized subject to the mean being prescribed, the maximum entropy probabilities and probability density function are all positive. We are naturally interested in knowing whether this result will continue to hold when other moments are prescribed.

If the constraints are,

$$\sum_{i=1}^{n} p_i = 1, \sum_{i=1}^{n} p_i g_r(x_i) = a_r, r = 1, 2, \ldots, m \tag{53}$$

then maximizing (5) subject to (53), we get

$$1/p_i = \lambda_0 + \lambda_1 g_{1i} + \ldots + \lambda_m g_{mi}, i = 1, 2, \ldots, n \tag{54}$$

Now, p_i's are continuous functions of $a_1, a_2, \ldots, a_m$ and if some p_i's have to become negative for some values of the moments, then p_i's will have to becomes zero for some values of the moments. If for any set of values $a_{10}, a_{20}, \ldots, a_{m0}$, one p_i become zero, then some of λ_i will have to be infinite and in that case all p_i's will have to be zero and this is not possible.

As such even in the general case, we do not expect p_i's to be negative. This is however not a rigorous proof of the results.

The same argument will not be valid, for example, for the measure

$$1 - \sum_{i=1}^{n} p_i^2 \tag{55}$$

since, this will give instead of (54)

$$p_i = \lambda_0 + \lambda_1 g_{1i} + \ldots + \lambda_m g_{mi} \tag{56}$$

and in this case p_i's can vary continuously from positive to negative values. This is the reason why in using the mean square error algorithm in image processing, the non-negativity constraints have to be introduced explicitely.

27.5 BURG'S MEASURE OF CROSS-ENTROPY: MINIMUM CROSS ENTROPY PROBABILITY DISTRIBUTION

We have shown elsewhere [4] that if $\phi(x)$ is a convex function for which $\phi(1) = 0$ and if $-\sum_{i=1}^{n} \phi(p_i)$ is a measure of entropy, then $\sum_{i=1}^{n} q_i \phi(p_i/q_i)$ can be used as a measure of cross-entropy of P from Q. In our case,

$$\phi(p_i) = -\ln p_i, \phi(1) = 0, \tag{57}$$

so that the measure of cross-entropy is

$$D_B(P:Q) = \sum_{i=1}^{n} q_i \left(-\ln \frac{p_i}{q_i}\right) = \sum_{i=1}^{n} q_i \ln \frac{q_i}{p_i} = D_K(Q:P), \tag{58}$$

where $D_K(P: Q)$ is Kullback-Leibler's [6] measure of cross-entropy, so that Burg's entropy does not lead to any essentially new measure of cross-entropy, though it shows that Kullback-Liebler measure $D_K(Q: P)$ is as good a measure of cross-entropy as $D_K(P: Q)$.

In fact if U is the uniform distribution

$$D_K(P:U) = \sum_{i=1}^{n} p_i \ln \frac{p_i}{1/n} = \ln n - \left(-\sum_{i=1}^{n} p_i \ln p_i\right) = \ln n - S(P) \tag{59}$$

$$D_K(U:P) = \sum_{i=1}^{n} \frac{1}{n} \ln \frac{1/n}{p_i} = -\ln n - \left(\sum_{i=1}^{n} \ln p_i \right) = -\ln n - B\,(P) \tag{60}$$

so that wile the first cross-entropy is associated with Shannon's measure of entropy, the second cross-entropy is associated with Burg's measure of entropy.

Now minimizing, $\sum_{i=1}^{n} q_i \ln (q_i/p_i)$ subject to constraint

$$\sum_{i=1}^{n} p_i = 1, \sum_{i=1}^{n} i p_i = m, \tag{61}$$

we get

$$p_i = \frac{q_i}{\lambda + \mu i}\,, \tag{62}$$

where λ and μ are determined by using

$$\sum_{i=1}^{n} \frac{q_i}{\lambda + \mu i} = 1, \sum_{i=1}^{n} \frac{i q_i}{\lambda + \mu i} = m, \tag{63}$$

so that,

$$\lambda + \mu m = \sum_{i=1}^{n} \frac{(\lambda + \mu i) q_i}{\lambda + \mu i} = 1 \tag{64}$$

and,

$$p_i = \frac{q_i}{1 - \mu(m - i)} \tag{65}$$

The equation (5) is modified to

$$\frac{q_1}{1 - \mu(m-1)} + \frac{q_2}{1 - \mu(m-2)} + \ldots + \frac{q_n}{1 - \mu(m-n)} = 1, \tag{66}$$

all of whose roots are real and one root is zero. The function $f(\mu)$ of equation (19) now becomes

$$\bar{f}(\mu) \equiv \frac{q_1}{1 - \mu(m-1)} + \frac{q_2}{1 - \mu(m-2)} + \ldots + \frac{q_n}{1 - \mu(m-n)} - 1 \tag{67}$$

This will have k roots in the intervals.,

$$\left(\frac{1}{m-1}, \frac{1}{m-2} \right), \left(\frac{1}{m-2}, \frac{1}{m-3} \right), \ldots, \left(\frac{1}{m-n+1}, \frac{1}{m-n} \right) \tag{68}$$

and $n - k - 2$ roots in the interval,

$$\left(-\frac{1}{1-l}, \frac{1}{2-l} \right), \left(-\frac{1}{2-l}, \frac{1}{3-l} \right) \ldots \left(-\frac{1}{n-k-1-l}, -\frac{1}{n-k-l} \right) \tag{69}$$

and two roots in the interval,

$$\left(-\frac{1}{n-m}, \frac{1}{m-1} \right) \tag{70}$$

Again the non-zero root will be >0 if $m<(n+1)/2$ and will be <0 if $m<(n+1)/2$. The value of this root will of course depend on $(q_1, q_2, \ldots, q_n)$. The probabilities will still be all positive, but unlike the case of uniform *a priori* probability distribution, there is no guarantee that the minimising probabilities will always decrease or always increase.

A similar discussion applies to the continuous-variate case when

$$f(x)=\frac{g(x)}{\lambda+\mu x}, \tag{71}$$

where, $g(x)$ is the *a priori* density function and $\mu \lesseqgtr 0$ according as $m \lesseqgtr (a+b)/2$, and $f(x)$ need not necessarily be a monotonic increasing or decreasing function of x.

REFERENCES

1. J.P. Burg (1972), "The relationship between maximum entropy spectra and maximum likelihood method" in Modern Spectral Analysis ed. D.G. Childers, 130–131, M.S.A.
2. E.T. Jaynes (1957), "Information Theory and Statistical Mechanics", Physical Reviews, 108, 620–630.
3. J.N. Kapur (1983) "Twenty-five years of maximum-entropy" Jour. Math. Phys. Sci. 17 (2), 103–156.
4. J.N. Kapur (1989) "Maximum Entropy Models of Science and Engineering", Wiley Eastern; New Delhi.
5. J.N. Kapur and H.K. Kesavan (1987), "Generalised Maximum Entropy Principle", Sandford Educational Press, Waterloo, Canada.
6. S. Kullback and R.A. Leibler (1951), "On information and sufficiency" Ann. Math. Stat. vol. 22, 79–86.
7. C.E. Shannon (1948) "A Mathematical Theory of Communication" Bell. Syst. Tech. J. vol. 27, 379–423, 623–659.

28

SOME THEOREMS CONCERNING MAXIMUM ENTROPY AND MINIMUM CROSS-ENTROPY PROBABILITY DISTRIBUTIONS

[Some theorems are proved concerning the increasing or decreasing nature of probabilities obtained when some measures of entropy (cross-entropy) are maximized (minimized) subject to moment constraints.]

28.1 INTRODUCTION

Let a factory arrange to get samples of its product examined in batches of size n. The number of defectives in a sample can be $0, 1, 2, \ldots, n$. Let p_i be the probability of there being i defectives in the sample ($i = 0, 1, 2, \ldots, n$) and let the expected value of the number of defectives be known as m so that we have the following information about the probabilities.

$$p_0 + p_1 + p_2 + \ldots. + p_n = 1; \quad p_0 \geq 0, p_1 \geq 0, p_2 \geq 0 \ldots, p_n \geq 0 \tag{1}$$

$$\sum_{i=0}^{n} i\, p_i = m \tag{2}$$

We can have an infinity of probability distributions consistent with (1) and (2). According to the maximum–entropy principle (*MEP*) of Jaynes [3], we should choose that probability distribution, out of all those satisfying (1) and (2) which maximizes Shannon's [1, 2] measure of entropy,

$$S(P) = -\sum_{i=0}^{n} p_i \ln p_i \tag{3}$$

Using Lagrange's method, we get

$$p_i = B y^i; \quad i = 0, 1, 2, \ldots n, \tag{4}$$

where B and y are determined by using

$$B \sum_{i=0}^{n} y^i = 1, \quad B \sum_{i=0}^{n} i y^i = m \tag{5}$$

From this, it can be shown (Ferdinand [2] , Kapur [8]) that

$$y \gtreqless 1 \quad \text{according as} \quad m \gtreqless \frac{1}{2} n \tag{6}$$

The object of the present paper is to generalise this result in four directions:

(*a*) when the *a priori* probability distribution is not necessarily uniform, but is any given probability distribution,

$$Q = (q_0, q_1, q_2, \ldots, q_n) \tag{7}$$

(*b*) when instead of the maximum entropy probability distribution, we consider the more general minimum cross-entropy probability distribution (Kullbuck [10]).

(*c*) when instead of the mean being prescribed, a more general moment (which includes the mean as a special case) is prescribed.

(*d*) when the measure of entropy used is not necessarily the same as Shannon's and when the measure of cross-entropy used is not necessarily the corresponding Kullback-Liebler's [11],

$$D(P:Q) = \sum_{i=0}^{n} p_i \ln \frac{p_i}{q_i} \tag{8}$$

In section 2, we consider the minimum cross-entropy probability distribution obtained when (8) is minimized and in section 3, we consider some special cases.

In sections 4 and 5, we consider maximum entropy probability distribution for the measures of entropy,

$$K(P) = -\sum_{i=0}^{n} p_i \ln p_i - \frac{1}{a} \sum_{i=0}^{n} (1 - a p_i) \ln (1 - a p_i),\ a > 0 \tag{9}$$

and,

$$B(P) = \sum_{i=0}^{n} \ln p_i \tag{10}$$

due to Kapur [7] and Burg [1], respectively.

In every case, we shall show that p_i (or p_i / q_i as the case may be) increases or decreases with i according as the prescribed value of the moment $\gtrless$ the value of the same moment for the uniform distribution (or the *a priori* distribution).

28.2 THE FIRST THEOREM

If $p_0, p_1, \ldots, p_n$ are the probabilities obtained by minimizing Kullback-Leibler's measure of cross entropy (8) subject to the natural constraints (1) and the moment constraint

$$\sum_{i=0}^{n} p_i\, g(i) = M, \tag{11}$$

where $g(i)$ is a positive integral-valued increasing function of i for which $g(0) = 0$, then p_i/q_i is an increasing, constant, decreasing function of i according as $M \gtreqless M_0$ where M_0 is defined by

$$\sum_{i=0}^{n} q_i g(i) = M_0 \tag{12}$$

Proof: Minimizing (8) subject to (1) and (11), we get

$$p_i = q_i\, A x^{g(i)}, \quad i = 0, 1, 2, \ldots, n, \tag{13}$$

where A and x are obtained by solving the equations,

$$A \sum_{i=0}^{n} q_i x^{g(i)} = 1, \quad A \sum_{i=0}^{n} q_i\, g(i) x^{g(i)} = M, \tag{14}$$

so that x can be obtained by solving

$$\sum_{i=0}^{n} q_i g(i) x^{g(i)} / \sum_{i=0}^{n} q_i x^{g(i)} = M \tag{15}$$

or

$$\sum_{i=0}^{n} q_i (g(i) - M) x^{g(i)} = 0 \tag{16}$$

since, $g(0) = 0$ and

$$g(0) < g(1) < g(2) < \ldots < g(n) \tag{17}$$

and all these values are positive integers, (16) is a polynomial equation.

Moreover from (11), M is the weighted mean of $g(0), g(1), \ldots, g(n)$ with weights $p_0, p_1, \ldots, p_n$, so that M has to be between $g(0)$ and $g(n)$.

As such, the sequence,

$$g(0) - M, g(1) - M, g(2) - M, \ldots, g(n) - M \tag{18}$$

will start with a negative value, will go on increasing with i and will finally end with a positive value. There will be one stage and only one stage when this member will change from a negative value to a positive value. Thus the polynomial

$$f(x) = q_0(g(0) - M)x^{g(0)} + q_1(g(1) - M)x^{g(1)} + \ldots + q_n(g(n) - M)x^{g(n)} \tag{19}$$

has only one change of sign as x goes from 0 to ∞, so that by Desecrates rule of signs, $f(x) = 0$ has only one positive real root. The graph of $f(x)$ will be as in Figure 28.1, where x_0 denotes the positive real root of $f(x) = 0$. Now,

$$f(1) = \sum_{i=0}^{n} q_i (g(i) - M) = M_0 - M \tag{20}$$

so that $f(1) \gtreqless 0$ according as $M \lesseqgtr M_0$.

From Figure 28.1, it is easily seen that $x_0 \geq$ or ≤ 1 according as $f(1) \leq$ or ≥ 0, i.e. according as $M \geq$ or $\leq M_0$ (21)

Now,

$$\frac{p_i}{q_i} = A x^i \tag{22}$$

so that $\frac{p_i}{q_i}$ increases (remains constant, decreases) with i according as $x_0 \gtreqless 1$, i.e. according as $M \gtreqless M_0$.

This was the result to be proved.

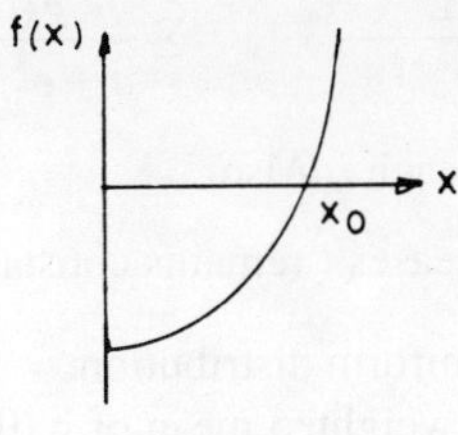

Fig. 28.1

28.3 SOME SPECIAL CASES

(*i*) If $g(i) = i^k$. i.e. if the value of the kth moment is prescribed, the conditions of the theorem are satisfied and it follows that in this case p_i/q_i increases (remains constant, decreases) with i according as the moment value $M > (=, <) M_0$, the value of the kth order moment for the *a priori* distribution.

(*ii*) If $g(i) = i^k$ and $q_i = 1/n$, so that Q is the uniform distribution and the minimum cross-entropy distribution becomes the maximum entropy distribution. In this case p_i increases (remains constant, decreases) according as

$$M > (=,<)\frac{1}{n}(1^k + 2^k + \ldots n^k) \tag{23}$$

(*iii*) If $g(i) = i$ and $q_i = 1/n$, we get the result proved in Ferdinand [2] and Kapur [8] when the probabilities are in geometrical progression. In fact, the probabilities will be in geometric progression whenever $g(0), g(1) \ldots, g(n)$ is an increasing arithmetic progression.

(*iv*) The result will continue to hold if

$$g(i) = a_1 i + a_2 i^2 + \ldots + a_k^{i^k}, a_1, a_2, \ldots a_k > 0 \tag{24}$$

(*v*) The result may not hold if the variance is prescribed, since in that case $g(i) = (i - \bar{i})^2$ and this will not increase as i increases.

28.4 THE SECOND THEOREM

If $p_0, p_1, \ldots, p_n$ are the probabilities obtained by maximizing entropy measure (9) subject to constraints (1) and (11), then p_i increases (remains constant, decreases) with i according as $M > (=, <) M_0$.

Proof: Maximizing (9) subject to (1) and (11) by using Lagrange's method, we get

$$\frac{p_i}{1 - a p_i} = e^{-\lambda - \mu g(i)} \tag{25}$$

or,

$$p_i = \frac{1}{a + e^{\lambda + \mu g(i)}} \tag{26}$$

when λ and μ are obtained by using

$$\sum_{i=0}^{n} \frac{1}{a+e^{\lambda+\mu g(i)}} = 1, \quad \sum_{i=0}^{n} \frac{g(i)}{a+e^{\lambda+\mu g(i)}} = M \tag{27}$$

It is easily seen that $p_i > 0$ for each i. Also

$$\mu < (=, >) 0 \Rightarrow p_i \text{ increases (remains constant, decreases) with } i$$

If $\mu = 0$, (26) gives the uniform distribution.

Now from (11), M is the weighted mean of g (0), g (1), . . . , g (n), the weights being $p_0, p_1, \ldots, p_n$, respectively. If p_i is an increasing function of i, then the weighted mean will be more than the weighted mean obtained when the weights are equal. As such

$$\mu < 0 \Rightarrow p_i \text{ is increasing function of } i \Rightarrow M > M_0$$

and, $$\mu > 0 \Rightarrow p_i \text{ is decreasing function of } i \Rightarrow M < M_0$$

This proves that for the maximum entropy probability distribution when measure (9) is used, the probability will always be increasing (constant, decreasing) with i according as the prescribed value of the moment $>$ ($=$, $<$) the value of the corresponding moment for the uniform distribution.

If $a \to 0$, the measure (9) reduces to Shannon's measure and we get a generalisation of the earlier result of Ferdinand [2] and Kapur [8] to the case when the moment prescribed is not necessarily the mean, but may be algebraic moment of any order or expected value of an increasing function and in particular it may be the geometric mean.

28.5 THE THIRD THEOREM

The probabilities $p_0, p_1, \ldots, p_n$ obtained by maximizing Burg's entropy measure (10) subject to (1) and (11) are increasing (constant, decreasing) with i according as $M > (=, <) M_0$.

Proof: Maximizing (10) subject to (1) and (2), we get

$$\frac{1}{p_i} = \lambda + \mu g(i) \quad \text{or} \quad p_i = \frac{1}{\lambda + \mu g(i)}, \tag{28}$$

when λ and μ are determined by using

$$\sum_{i=0}^{n} \frac{1}{\lambda + \mu g(i)} = 1, \quad \sum_{i=0}^{n} \frac{g(i)}{\lambda + \mu g(i)} = M, \tag{29}$$

so that $$\lambda + \mu M = n + 1 \tag{30}$$

and μ is determined from the equation

$$g(\mu) = \sum_{i=0}^{n} \frac{1}{n + 1 + \mu(g(i) - M)} = 0 \tag{31}$$

It can be shown (Kapur [9]) that $g(\mu) = 0$ has only one real root apart from $\mu = 0$ which will make all probabilities positive. If this root is negative , probabilities are increasing and $M > M_0$ and if this root is positive the probabilities are decreasing and $M < M_0$. The result of the theorem follows.

It can also be seen that $p_i > 0$ for all permissible values of M. When $M = M_0$, the distribution is uniform and p_i's are positive. As M changes continuously, so also do p_i's. If p_i has ever to become negative, the p_i has to pass through a zero value so that for that value of M, $1 / p_i$ will become infinite and so λ and μ will have to be zero and thus all probabilities will have to be infinite which is not possible. In fact when $\mu = 0, \lambda = n + 1$.

Thus in the case of Burg's entropy, probabilities cannot be negative.

In fact in this paper, we have only considered those measures of entropy or cross–entropy which lead automatically to positive values of probabilities, and in these cases we have so far found that p_i's or p_i / q_i's have always increased or decreased with i according as prescribed value of the moment of the distribution is > or < the corresponding value for the uniform or *a priori* distribution. In the next section, we consider a measure of cross–entropy which automatically leads to positive probabilities when it is minimised subject to linear constraints, but in which p_i's do not necessarily always increase or decrease with i.

28.6 THE FOURTH THEOREM

The probability distribution obtained by minimizing the measure of cross entropy [6]

$$K(P:Q) = \sum_{i=0}^{n} p_i \ln\frac{p_i}{q_i} + \frac{1}{a}\sum_{i=0}^{n}(1 - ap_i)\ln\frac{(1 - ap_i)}{1 - aq_i}, a < 1 \tag{32}$$

subject to (1) and (11) is such that each

$p_i > 0$ and $p_i(1 - q_i)/q_i(1 - p_i)$

always increases with i or always decreases with i or remains equal to unity.

Proof: Minimising (32) subject to (1) and (11), we get

$$\frac{p_i}{1 - p_i} = \frac{q_i}{1 - q_i} e^{-\lambda - \mu g(i)}, \tag{33}$$

where λ and μ are determined by using

$$\sum_{i=0}^{n} \frac{1}{\frac{1-q_i}{q_i} e^{\lambda + \mu g(i)} + 1} = 1, \quad \sum_{i=0}^{n} \frac{g(i)}{\frac{1-q_i}{q_i} e^{\lambda + \mu g(i)} + 1} = M \tag{34}$$

If $\mu = 0$, then $\lambda = 0$ and $p_i (1-q_i) / q_i (1-p_i) = 1$ for each i.
If $\mu < 0$, $p_i (1-q_i) / q_i (1-p_i)$ increases with i.
If $\mu > 0$, $p_i (1-q_i) / q_i (1-p_i)$ decreases with i.

However, in the last two cases p_i / q_i may sometimes increase and sometimes decrease with i.

If $q_i = 1/n$, we get the maximum entropy probability distribution and $p_i/(1-p_i)$ will always increases with i or always decrease with i or remain constant with value $1/(n-1)$ for all i. Now,

$$\frac{p_i}{1-p_i} = k_i \quad \Rightarrow \quad p_i = \frac{k_i}{1+k_i} = \frac{1}{\frac{1}{k_i}+1}, \tag{35}$$

so that if k_i always increases or decreases with i, then p_i always increases or decreases with i. This is what we found in Section 4.

In the general case,

$$\frac{p_i}{1-p_i} = k_i \frac{q_i}{1-q_i} \tag{36}$$

(*i*) If k_i and q_i both increase with i, p_i also increases with i.

(*ii*) If both k_i and q_i decrease with i, p_i also decreases with i.

(*iii*) If k_i increases and q_i decreases with i, then p_i will increase or decrease with i according as $k_i\, q_i/(1-q_i)$ increases or decreases.

REFERENCES

1. J.P. Burg (1972), "The Relationship between Maximum Entropy Spectra and Maximum Likelihood Spectra", In Modern Spectral Analysis, ed D.G. Childers, pp. 130–131, M.S.A.
2. A.E. Ferdinand (1974), "A Theory of General Complexity, " Int. J. Gen. Syst., vol. 1, pp. 19–33.
3. E.T. Jaynes (1957), "Information Theory and Statistical Mechanics", Physical Reviews, vol. 106, pp. 620–630.
4. J.N. Kapur (1972), "Measures of Uncertainty, Mathematical Programming and Physics" Jour. Ind. Soc. Agri. Stat. vol. 24, pp. 47–66.
5. J.N. Kapur (1983), "On Maximum–entropy complexity measures", Int. Jour. of General Systems, vol. 9, pp. 95–102.
6. J.N. Kapur (1984), "A Comparative Assessment of various measures of directed divergence" Adv. Mang. Stud. 3, 1–10.
7. J.N. Kapur (1986), "Four families of measures of entropy", Ind Jour. Pure and Applied Maths. vol. 17 no. 4, pp. 429–449.
8. J.N. Kapur (1990) , "Maximum –Entropy Models in Science and Engineering, Wiley Eastern New Delhi & John Wiley, New York.
9. J.N. Kapur (1989), "Which is in better entropy $\sum_{i=0}^{n} \ln p_i$ or $-\sum_{i=0}^{n} p_i \ln p_i$? MSTS Res. Report.
10. S. Kullback (1959), "Information Theory and Statistics", John Wiley, New York.
11. S. Kullback and R.A. Leibler (1951), "On Information and Sufficiency", Ann. Math. Stat., vol. 22, pp. 79–88.
12. C.E. Shannon (1948), " A Mathematical Theory of Communication", Bell System Tech. J., vol. 27, pp. 379–423, 623–659.

29

ON CONCAVITY OF MAXIMUM ENTROPY WHEN BURG'S ENTROPY IS MAXIMISED SUBJECT TO ITS MEAN BEING PRESCRIBED

[It is proved that if Burg's measure of entropy is maximized subject to its mean m being prescribed then the maximum entropy S_{max} will be a concave function of m].

29.1 INTRODUCTION

Given a number of constraints,

$$\sum_{i=1}^{n} p_i = 1,\ \sum_{i=1}^{n} p_i g_{ri} = a_r, r = 1,2,\ldots,m, m+1 \ll n, p_1 \geq 0,\ldots,p_n \geq 0 \tag{1}$$

there may be an infinity of probability distributions $(p_1, p_2, \ldots, p_n)$ satisfying (1).

If there is no other information, which one of these should we choose? Jaynes [2] gave his principle of maximum entropy according to which we should choose that probability distribution, out of all these, which maximizes Shannon's [5] measure of entropy,

$$S(P) = -\sum_{i=1}^{n} p_i \ln p_i \tag{2}$$

This ensures that the distribution is most unbiased, most objective, most random and most spread-out. The maximum entropy distribution obtained is

$$p_i = \exp(-\lambda_0 - \lambda_1 g_{1i} - \ldots. - \lambda_m g_{mi}), i = 1,2,\ldots,n, \tag{3}$$

where $\lambda_0, \lambda_1, \ldots, \lambda_m$ are determined as functions of $a_1, a_2, \ldots, a_m$ by using (1). Then the maximum entropy S_{max} is

$$S_{max} = \lambda_0 + \lambda_1 a_1 + \lambda_2 a_2 + \ldots + \lambda_m a_m \tag{4}$$

Kapur [3] showed that this is always a concave function of $a_1, a_2, \ldots, a_m$. We also note that all the probabilities given by (3) are always positive. We naturally want to know whether there is another measure of entropy or spread-out or uncertainty, other than (2), which when maximized, subject to (1) gives positive probabilities and for which S_{max} is possibly a concave function of the parameters.

Kapur [4] showed that Burg's [1] measure of entropy, which has been very successfully used in spectral analysis does always give positive probabilities. The object of the present paper is to see whether in this case S_{max} is a concave function of the parameters. We consider the special case when the moment prescribed is mean and show that at least in this case $S_{\max}$ is a concave function of m. We also show that $S_{\max}$ will be concave even if we prescribe $E(g(x))$ instead of $E(x)$ where $g(x)$ is a monotonic increasing function of x.

29.2 MAXIMUM ENTROPY PROBABILITY DISTRIBUTION

Maximizing Burg's measure of entropy,

$$\sum_{i=1}^{n} \ln p_i \tag{5}$$

subject to,

$$\sum_{i=1}^{n} p_i = 1, \sum_{i=1}^{n} i p_i = m, \tag{6}$$

we get,

$$p_i = \frac{1}{\lambda + \mu i}, \tag{7}$$

where,

$$\sum_{i=1}^{n} \frac{1}{\lambda + \mu i} = 1, \sum_{i=1}^{n} \frac{i}{\lambda + \mu i} = m \tag{8}$$

so that,

$$\lambda + \mu m = \sum_{i=1}^{n} \frac{\lambda + \mu i}{\lambda + \mu i} = n \tag{9}$$

so that,

$$p_i = \frac{1}{n - \mu m + \mu i}, i = 1, 2, \ldots, n, \tag{10}$$

where μ is determined as a function of m from

$$\sum_{i=1}^{n} \frac{1}{n - \mu m + \mu i} = 1 \tag{11}$$

The probabilities are all positive because if any one of these is negative, the corresponding denominator will have to vanish somewhere giving rise to an infinite probability. From (5) and (1)

$$S_{\max} = -\sum_{i=1}^{n} \ln(n - \mu m + \mu i) \tag{12}$$

$$\frac{dS_{\max}}{dm} = -\sum_{i=1}^{n} \frac{1}{n - \mu m + \mu i}\left[(i - m)\frac{d\mu}{dm} - \mu\right]$$

$$= -\frac{d\mu}{dm}\sum_{i=1}^{n} p_i (i - m) + \mu \sum_{i=1}^{n} p_i \tag{13}$$

Using (6)

$$\frac{dS_{max}}{dm} = \mu, \tag{14}$$

so that,

$$\frac{d^2S_{max}}{dm^2} = \frac{d\mu}{dm} \tag{15}$$

However from (11),

$$\sum_{i=1}^{n} \frac{1}{(n - \mu m + \mu i)^2}\left[(i-m)\frac{d\mu}{dm} - \mu\right] = 0$$

so,

$$\frac{d\mu}{dm} = \frac{\mu \sum_{i=1}^{n} p_i^2}{\sum_{i=1}^{n} p_i^2 (i-m)} \tag{16}$$

From (15) and (16) S_{max} will be a concave function of m if either,

$$(i)\ \mu > 0, \sum_{i=1}^{n} p_i^2(i-m) < 0 \quad \text{or} \quad (ii)\ \mu < 0, \sum_{i=1}^{n} p_i^2(i-m) > 0 \tag{17}$$

Of course when $\mu = 0$, $d^2S_{max}/dm^2 = 0$

Now from (7), the probabilities are in *H.P*, which is decreasing if $\mu > 0$ and is increasing if $\mu < 0$.

If $\mu = 0$, all the probabilities are equal, each probability is $1/n$ and from (6) $m = (n+1)/2$.

If $\mu > 0$, probabilities are decreasing and $m < \frac{n+1}{2}$.

If $\mu < 0$, probabilities are increasing and $m > \frac{n+1}{2}$.

Let us consider two cases,

Case (*i*): $m < \frac{n+1}{2}$.

In this case $\mu > 0$, probabilities are decreasing and,

$$\sum_{i=0}^{n} p_i^2(i-m) = \sum_{i=1}^{n} i p_i^2 - \sum_{i=1}^{n} p_i^2 \sum_{i=1}^{n} i p_i$$

$$= \sum_{i=0}^{n} p_i^2 \left[\frac{\sum_{i=1}^{n} i p_i^2}{\sum_{i=1}^{n} p_i^2} - \frac{\sum_{i=1}^{n} i p_i}{\sum_{i=1}^{n} p_i}\right] \tag{18}$$

The sign of the R.H.S. depends in the difference between two weighted averages of $1, 2, \ldots, n$.

In the first weighted average, weights are proportional to $p_1^2, p_2^2, \ldots, p_n^2$ and in the second weighted average, weights are proportional to $p_1, p_2, \ldots, p_n$.

In this case, the probabilities are decreasing, in the first case, the weights are decreasing faster than in the second case, and as such the mean will relatively be less in the first case so that in case (*i*) the sign of (18) is negative.

Case (*ii*) : Here $m > \frac{n+1}{2}$, $\mu < 0$, probabilities are decreasing and the sign of (18) is positive, so that again $d\mu/dm$ is negative.

Thus, in both cases $\frac{d^2 S_{\max}}{dm^2} < 0$, and $S_{\max}$ is a concave function of m.

29.3 GRAPH OF $S_{\max}$

When $m = 1$, there is only probability distribution $(1, 0, \ldots, 0)$ and

$$(S_{\max})_{m=1} = -\infty \tag{19}$$

Similarly, when $m = n$ there is only one probability distribution $(0, 0, \ldots, 1)$ and

$$(S_{\max})_{m=n} = -\infty \tag{20}$$

When $m = (n+1)/2$, the maximum entropy distribution is the uniform distribution $(1/n, 1/n, \ldots, 1/n)$ and,

$$(S_{\max})_{m=\frac{(n+1)}{2}} = -n \ln n \tag{21}$$

When m lies between 1 and $(n+1)/2$, $\mu > 0$, $d S_{\max}/dm > 0$, $S_{\max}$ is increasing.
When m lies between $(n+1)/2$ and n, $\mu < 0$, $d S_{\max}/dm < 0$, $S_{\max}$ is decreasing.
When $m = (n+1)/2$, $\mu = 0$, $d S_{\max}/dm = 0$.
As such the graph of $S_{\max}$ is as given in Figure 29.1.

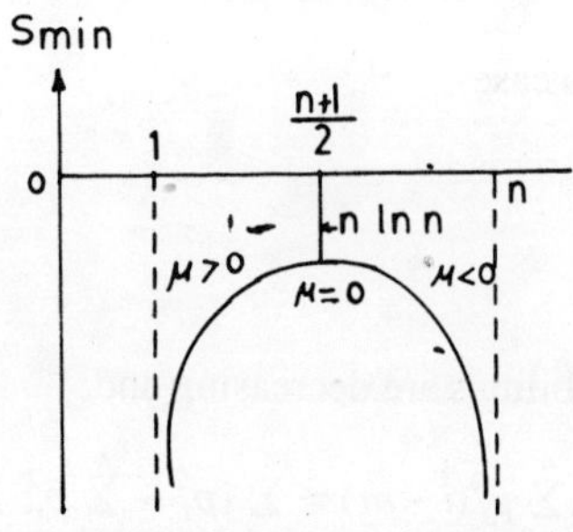

Fig. 29.1

29.4 REMARKS

(*i*) Burg's entropy is always negative, but this does not matter in entropy maximization. Here we are interested in the probability distribution which maximizes the entropy, rather than the value of this entropy and thus probability distribution has also its components positive.

(*ii*) We have nevertheless found S_{max} and discussed its concavity. The reason is that this can enable us to handle inequality constraints of the type,

$$m_1 \leq \sum_{i=1}^{n} i p_i \leq m_2 \tag{22}$$

If m_1 and m_2 are both $< \frac{n+1}{2}$, S_{max} occurs when $m = m_2$; if they are both greater than $(n + 1)/2$, S_{max} occurs when $m = m_1$, and if $m_1 < \frac{n+1}{2}$ and $m_2 > \frac{n+1}{2}$, it occurs when $m = \frac{n+1}{2}$.

(*iii*) If we plot p_i against i, we shall get rectangular hyperbolic curves (Figure 29.2)

$$y = \frac{1}{n - \mu m + \mu x} \tag{23}$$

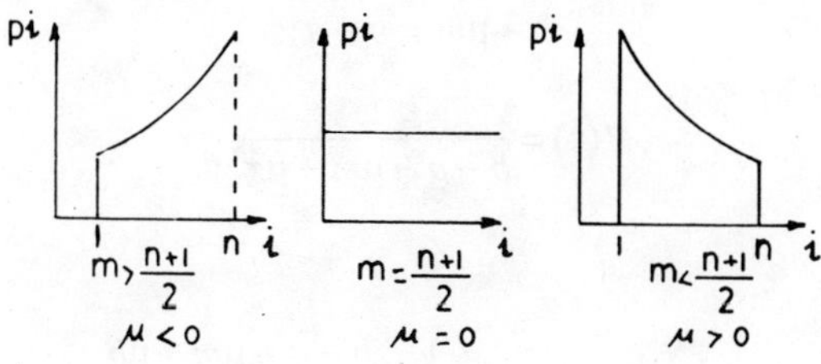

Fig. 29.2

(*iv*) If instead of prescribing the value of $E(i)$, we prescribe $E(g(i))$ where $g(\cdot)$ is any monotonic increasing function, then our discussion will apply with necessary changes and S_{max} will still be a concave function of m.
Instead of (22), we shall have

$$y = \frac{1}{n - \mu m + \mu g(x)}$$

$$\frac{dS_{max}}{dm} = \mu$$

$$\frac{d^2 S_{max}}{dm^2} = \sum_{i=1}^{n} p_i^2 \left[\sum_{i=1}^{n} \frac{g(i) p_i^2}{\sum_{i=1}^{n} p_i^2} - \frac{\sum_{i=1}^{n} g(i) p_i}{\sum_{i=1}^{n} p_i} \right] \tag{24}$$

(v) If in particular, $g(i) = \varepsilon_i$ we shall have the modified Maxwell Boltzmann distribution,

$$p_i = \frac{1}{n - \mu\hat{\varepsilon} + \mu\varepsilon_i}, \quad i = 1, 2, \ldots n \tag{25}$$

29.5 CONTINUOUS-VARIATE CASE

In this case maximizing

$$\int_a^b \ln f(x) dx$$

subject to
$$\int_a^b f(x) dx = 1, \quad \int_a^b x f(x) dx = m, \tag{26}$$

we get,
$$f(x) = \frac{1}{\lambda + \mu x} \tag{27}$$

where
$$\int_a^b \frac{1}{\lambda + \mu x} dx = 1, \quad \int_a^b \frac{x}{\lambda + \mu x} dx = m \tag{28}$$

giving,
$$\lambda + \mu m = b - a \tag{29}$$

and,
$$f(x) = \frac{1}{b - a - \mu m + \mu x}, \tag{30}$$

so that

$$\int_a^b \frac{1}{b - a - \mu m + \mu x} dx = 1, \quad \frac{1}{\mu} \ln \frac{b - a - \mu m + \mu b}{b - a - \mu m + \mu a} = 1$$

$$\Rightarrow \quad \frac{b - a - \mu m + \mu b}{b - a - \mu m + \mu a} = \frac{e^\mu}{1}$$

$$\Rightarrow \quad \frac{\mu(b-a)}{b - a - \mu m + \mu a} = \frac{e^\mu - 1}{1} \Rightarrow b - a - \mu m + \mu a = \frac{\mu(b-a)}{e^\mu - 1}$$

$$\Rightarrow \quad \frac{b-a}{\mu} - m + a = \frac{b-a}{e^\mu - 1} \Rightarrow m - a = (b-a)\left(\frac{1}{\mu} - \frac{1}{e^\mu - 1}\right) \tag{31}$$

Thus m is expressed explicitly as a function of μ and $\mu = -\infty$ gives $m = b$,

$$\mu = 0 \text{ gives } m = \frac{b+a}{2}, \mu = \infty \quad \text{gives} \quad m = a \tag{32}$$

Also from (31)

$$dm/d\mu = (b-a)\left(-\frac{1}{\mu^2} + \frac{e^\mu}{(e^\mu - 1)^2}\right) = \frac{(b-a)g(\mu)}{\mu^2(e^\mu - 1)^2} \tag{33}$$

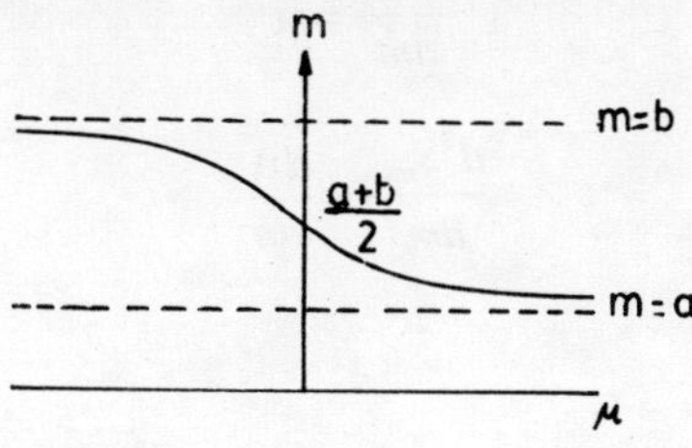

Fig. 29.3

where,

$$g(\mu) = \mu^2 e^{\mu} - (e^{\mu} - 1)^2, g'(\mu) = e^{\mu}(\mu^2 + 2\mu + 2 - 2e^{\mu}) = e^{\mu} h(\mu) \quad (34)$$

where,

$$h(\mu) = \mu^2 + 2\mu + 2 - 2e^{\mu}, h'(\mu) = 2(\mu + 1 - e^{\mu}), h''(\mu) = 2(1 - e^{\mu}) \quad (35)$$

So that, $h''(\mu) \overset{\leq}{\underset{>}{=}} 0$ according as $\mu \overset{\geq}{\underset{<}{=}} 0$ and $h'(\mu)$ is decreasing when $\mu > 0$ and increasing when $\mu < 0$ and $h'(\mu) = 0$ only if $\mu = 0$.

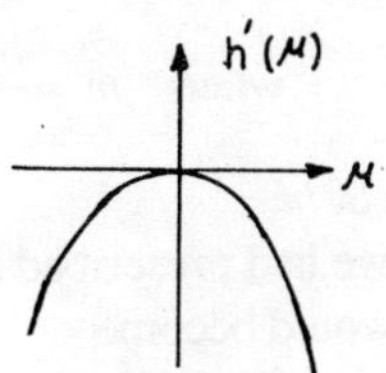

Fig. 29.4

so that $h'(\mu) \leq 0$ and $h'(\mu) = 0$ only if $\mu = 0$.
Also,

$$h(\mu) = 0 \quad \text{so that } h'(\mu) \gtrless 0 \quad \text{according as } \mu \lessgtr 0$$

$$g'(\mu) \gtrless 0 \text{ according as } \mu \lessgtr 0. \quad \text{Also } g(\mu) = 0 \quad (36)$$

so that, $g(\mu) \leq 0$ and $g(\mu) = 0$ iff $\mu = 0$.
From (33) and (36),

$$\frac{dm}{d\mu} < 0 \quad \text{and} \quad \frac{d\mu}{dm} < 0 \quad \text{when} \quad \mu \neq 0 \quad (37)$$

Now, using (25) and (30)

$$S_{\max} = -\int_a^b \ln(b - a - \mu m + \mu x)dx$$

$$\frac{dS_{\max}}{dm} = -\frac{d\mu}{dm}\int_a^b \frac{1}{b - a - \mu m + \mu x}(x - m)dx + \int_a^b \frac{\mu}{b - a - \mu m + \mu x}dx \quad (38)$$

Using (28) and (38),

$$\frac{dS_{\max}}{dm} = \mu \tag{39}$$

$$\frac{d^2S_{\max}}{dm^2} = \frac{d\mu}{dm} \tag{40}$$

but from (37)

$$\frac{d\mu}{dm} \le 0 \quad \text{when} \quad \mu \ne 0 \tag{41}$$

When

$$\mu = 0, \underset{\mu \to 0}{Lt}\left(\frac{e^{\mu}}{(e^{\mu}-1)^2} - \frac{1}{\mu^2}\right) = \underset{\mu \to 0}{Lt}\left(\frac{1+\mu+\ldots}{\mu^2} - \frac{1}{\mu^2}\right) = \infty \tag{42}$$

From (33) and (42),

$$\frac{d\mu}{dm} = 0 \qquad \text{when} \quad \mu = 0 \tag{43}$$

so that,

$$\frac{d^2S_{\max}}{dm^2} < 0 \qquad \text{when} \quad m \ne \frac{b+a}{2}$$

$$\frac{d^2S_{\max}}{dm^2} = 0 \qquad \text{when} \quad m = \frac{b+a}{2} \tag{44}$$

and $S_{\max}$ is again a concave function of m.

If instead of prescribing $E(x)$, we had prescribed $E(g(x))$ where $g(x)$ is a monotonic increasing function of x, (38) would become

$$\frac{dS_{\max}}{dm} = -\frac{d\mu}{dx}\int_a^b \frac{1}{b-a-\mu m+\mu g(x)}(g(x)-m)\,dx + \int_a^b \frac{\mu}{b-a-\mu m+\mu g(x)}dx$$

$$= \mu \tag{45}$$

However, it may not be possible to express m explicitly as a function of μ. We would get

$$\int_a^b \frac{1}{b-a-\mu m+\mu g(x)}dx = 1 \tag{46}$$

giving,

$$\frac{d\mu}{dm}\int_a^b \frac{1}{(b-a-\mu m+\mu g(x))^2}(g(x)-m)dx = \mu \int_a^b \frac{1}{(b-a-\mu m+\mu g(x))^2}dx$$

or

$$\frac{d\mu}{dm} = \frac{\mu\int_a^b f^2(x)dx}{\int_a^b (g(x)-m)f^2(x)dx} \tag{47}$$

and, we shall have to use an argument similar to that used in the discrete case to show that $d\mu/dm < 0$ and S_{max} is a concave function of m.

REFERENCES

1. J.P. Burg, (1972), ''The relationship between Maximum Entropy Spectra and Maximum Likelihood Spectra,'' in Modern Spectral Analysis, ed. D.G. Childers, pp. 130–131.
2. E.T. Jaynes, (1973), ''Information Theory and Statistical Physics,'' 1962 Brandies lectures in Math Physics, Vol. 3, pp. 181–216.
3. J.N. Kapur, (1988) *Maximum Entropy Models in Science and Engineering*, Wiley Eastern, New Delhi.
4. J.N. Kapur (1989), ''Maximum Entropy Probability Distribution when mean of random variate is prescribed and Burg's entropy measure is used'' MSTS Res. Rep. 476.
5. C.E. Shannon, (1948), ''A Mathematical Theory of Communication,'' Bell System Tech. J., Vol. 27, pp. 379–423, 623–659.

30

THE PRINCIPLE OF MINIMUM WEIGHTED INFORMATION

[The principle of minimum information of Lind and Solana has been extended to give the principle of minimum weighted information.]

30.1 INTRODUCTION

Let a random variate X vary over the interval $[a, b]$. On the basis of our information, experience or theory, we may have reason to believe that the probability density function for X is $g(x, \Theta)$ where the parameter Θ may be a scalar or an m-dimensional vector. Now to estimate Θ, we draw a random sample $x_1, x_2, \ldots, x_n$ from the population. We can assume, without loss of generality, that (Fig 30.1)

$$a = x_0 \leq x_1 \leq x_2 < x_3 \leq \ldots \leq x_n \leq x_{n+1} = b \tag{1}$$

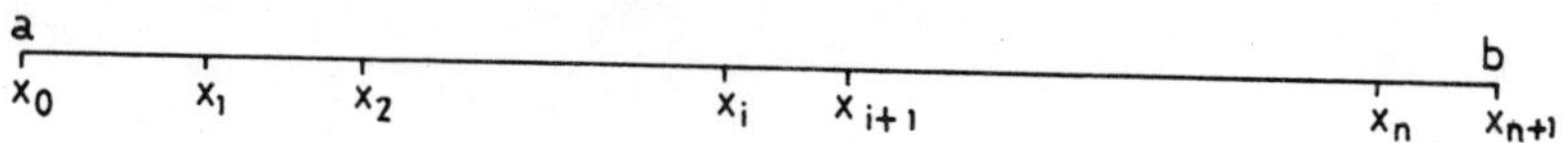

Fig. 30.1

For the estimation of Θ, Pearson [15] gave his method of moments according to which we calculate first m algebraic moments of the sample $x_1, x_2, \ldots, x_n$ and equate these to the first m algebraic moments of the population with density function $g(x, \Theta)$. Later Fisher [1] suggested that Θ should be estimated so as to maximize the likelihood function,

$$L = g(x_1, \Theta)\, g(x_2, \Theta) \ldots g(x_n, \Theta) \tag{2}$$

The two methods gave in general quite different results and there was a fierce controversy about it. Recently Kapur [4] showed that if in Pearson's method of moments, instead of using the m algebraic moments, we use the m characterising moments of the probability distribution, i.e. we use those moments for which the given probability distribution is a maximum-entropy probability distribution, then Fisher's and Pearson's modified methods give the same results. Kapur [4, 6] also derived the principle of maximum likelihood by using Jaynes [3] maximum entropy principle and Kullback's [10] principle of minimum cross entropy.

Both these methods are based on principles which are stated as axioms. Recently, Lind [12] and Lind and Solana [13] have given a third principle, the so-called principle of minimum information for estimating Θ. We discuss this principle in the next section. This is based on the concept of directed divergence or cross-entropy and specifically this uses Kullback and Leibler's [11] measure of directed divergence to implement this principle.

Kapur [7] discussed how other measures of directed divergence can be used to implement this principle.

If all the outcomes or all values of x do not have equal importance, we use the concept of weighted entropy, introduced first by Guiasu [2]. Taneja and Tuteja [17] tried to extend this concept to get the concept of weighted directed divergence. However, their measure had a weakness, viz. that it could be negative, while we want a measure of directed divergence to be always ≥ 0. Kapur [9] gave a number of correct measures of weighted directed divergence which are always ≥ 0. We use one of these in section 3 to enunciate a principle of minimum weighted information which is the main contribution of this paper.

30.2 PRINCIPLE OF MINIMUM INFORMATION

Suppose on the basis of intuition, theory or experience, we have reason to believe that the probability density function for X is $g(x, \Theta)$ where Θ is unknown. To estimate Θ, we draw a random sample $x_1, x_2, \ldots, x_n$ and rearrange its members in increasing order so that (1) is satisfied.

We now use Laplace's principle of insufficient reason to deduce that

$$P(x_i \leq X \leq x_{i+1}) = \frac{1}{n+1}; \quad i = 0, 1, 2, \ldots, n \tag{3}$$

since we have no reason to assume that these $(n + 1)$ probabilities are unequal. If our *a priori* probability density function is correct, we should have

$$\int_{x_i}^{x_{i+1}} g(x, \Theta) dx = \frac{1}{n+1}, i = 0, \ldots, n \tag{4}$$

There is likely to be no value of Θ which will make all the $(n + 1)$ equations (4) correct. As such, we are forced to modify our intuition in the light of the random sample and the constraints, but we would like to modify it as little as possible. Suppose the modified density function is $f(x, \Theta)$, then

$$\int_{x_i}^{x_{i+1}} f(x, \Theta) dx = \frac{1}{n+1}, i = 0, 1, 2, \ldots, n \tag{5}$$

are satisfied, since these are the given constraints and we would like $f(x, \Theta)$ to be as close to $g(x, \Theta)$ as possible.

To achieve this objective, we choose $f(x, \Theta)$ to minimize

$$\int_{x_0}^{x_{n+1}} f(x, \Theta) \ln \frac{f(x, \Theta)}{g(x, \Theta)} dx \tag{6}$$

subject to (5) being satisfied. Using Lagrange's method, this gives

$$\frac{f(x, \Theta)}{g(x, \Theta)} = k_i, \quad x_i \leq x \leq x_{i+1} \tag{7}$$

Geometrically this means that in each sub-interval $(x_i, x_{i+1}), f(x,\Theta)$ is proportional to $g(x,\Theta)$, but the constants of proportionality can be different for different sub-intervals and are chosen so that the area under $f(x,\Theta)$ in each sub-interval is $1/(n+1)$ (Fig. 30.2)

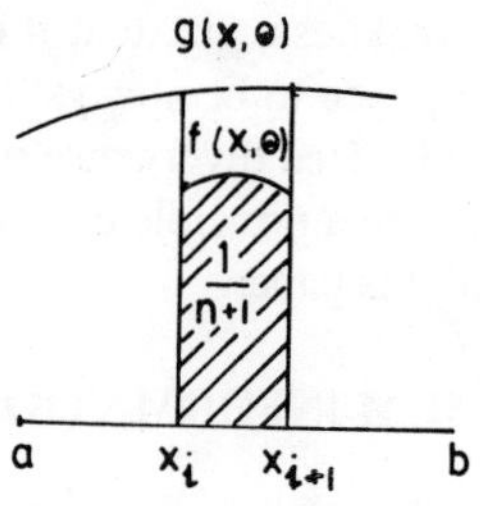

Fig. 30.2

In this discussion, Θ is fixed and for a given Θ, we have chosen $f(x,\Theta)$. Thus, Θ is still at our disposal and we can choose it so that

$$I = \int_{x_0}^{x_{n+1}} f(x,\Theta) \ln \frac{f(x,\Theta)}{g(x,\Theta)} dx = \sum_{i=0}^{n} \int_{x_i}^{x_{i+1}} f(x,\Theta) \ln \frac{f(x,\Theta)}{g(x,\Theta)} dx \tag{8}$$

is minimum. Now using (6) and (7),

$$I = \sum_{i=0}^{n} \int_{x_i}^{x_{i+1}} k_i g(x,\Theta) \ln k_i dx$$

$$= \sum_{i=0}^{n} k_i \ln k_i \int_{x_i}^{x_{i+1}} g(x,\Theta) dx$$

$$= \sum_{i=0}^{n} k_i \ln k_i \frac{1}{k_i} \int_{x_i}^{x_{i+1}} f(x,\Theta) dx$$

$$= \frac{1}{n+1} \sum_{i=0}^{n} \ln k_i \tag{9}$$

Now from (6) and (7)

$$\int_{x_i}^{x_{i+1}} g(x,\Theta) dx = \frac{1}{k_i} \int_{x_i}^{x_{i+1}} f(x,\Theta) dx = \frac{1}{k_i(n+1)}$$

or,

$$\frac{1}{k_i} = (n+1) \int_{x_i}^{x_{i+1}} g(x,\Theta) dx \tag{10}$$

From (9) and (10),

$$I = -\frac{1}{n+1}\sum_{i=0}^{n} \ln (n+1) \int_{x_i}^{x_{i+1}} g(x,\Theta)dx, \tag{11}$$

so that we have to choose Θ so as to minimise

$$J = \sum_{i=0}^{n} \ln \int_{x_i}^{x_{i+1}} g(x,\Theta)dx \tag{12}$$

Thus the principle of minimum information requires us to choose Θ so as to minimise J or to maximize

$$K = \sum_{i=0}^{n} \ln \int_{x_i}^{x_{i+1}} g(x,\Theta)dx \tag{13}$$

This can be compared with the principle of maximum likelihood which requires us to maximize

$$\ln L = \sum_{i=1}^{n} \ln g(x_i,\Theta) \tag{14}$$

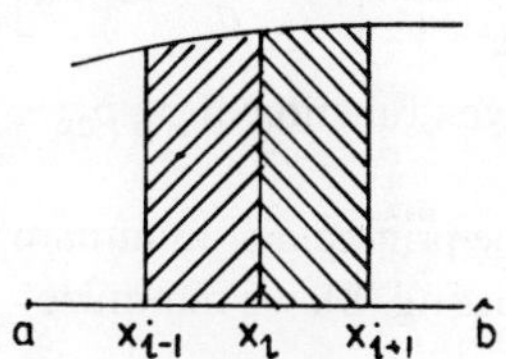

Fig. 30.3

Thus in the method of maximum likelihood, we choose Θ so as to maximize the product of the n ordinates of the probability density curve at $x_1, x_2, \ldots, x_n$ while in principle of minimum information, we choose Θ to maximize the product of the $(n+1)$ areas under the probability density curve (Fig. 30.3). In the former case, we multiply probability densities, while in the latter case, we multiply probabilities themselves.

30.3 THE PRINCIPLE OF MINIMUM WEIGHTED INFORMATION

For a discrete random variate which takes values $x_1, x_2, \ldots, x_n$ with probabilities $p_1, p_2, \ldots, p_n$ and for which the weights of the n outcomes are $w_1, w_2, \ldots, w_n$, Guiasu [2] gave the measure of weighted entropy as

$$-\sum_{i=1}^{n} w_i p_i \ln p_i, \tag{15}$$

which is a concave non–negative function of $p_1, p_2, \ldots, p_n$ and which reduces to Shannon's [16] measure of entropy,

$$-\sum_{i=1}^{n} p_i \ln p_i \tag{16}$$

when all the weights are equal.

For a continuous random variate for which the density function is $g(x, \Theta)$ and the weight function is $w(x)$, the corresponding measure is

$$-\int w(x) g(x,\Theta) \ln g(x,\Theta) dx \tag{17}$$

This is also a concave function. Corresponding to (15), Taneja and Tuteja [17] suggested the measure of weighted directed divergence of probability distribution P from the probability distribution Q as

$$\sum_{i=1}^{n} w_i p_i \ln \frac{p_i}{q_i} \tag{18}$$

While this reduces to Kullback-Leibler's [11] measure of directed divergence, when all the weights are equal and this is a convex function of $p_1, p_2, \ldots, p_n$, it can be negative for some weight functions and it cannot be used as a measure of weighted directed divergence. Kapur [9] for the first time gave the corresponding correct measure of directed divergence, viz.

$$\sum_{i=1}^{n} w_i \left[p_i \ln \frac{p_i}{q_i} - p_i + q_i \right] \tag{19}$$

which is always ≥ 0, is a convex function of $p_1, p_2, \ldots, p_n$ and which vanishes iff $p_i = q_i$ for each i.

As such, according to the principle of minimum weighted directed divergence, to obtain Θ, instead of minimizing (8), we minimize

$$I_W = \int_{x_0}^{x_{n+1}} w(x) \left[f(x\Theta) \ln \frac{f(x,\Theta)}{g(x,\Theta)} - f(x,\Theta) + g(x,\Theta) \right] dx$$

$$= \sum_{i=0}^{n} \int_{x_i}^{x_{i+1}} w(x) \left[k_i g(x,\Theta) \ln \frac{1}{k_i} - k_i g(x,\Theta) + g(x,\Theta) \right] dx$$

$$= \sum_{i=0}^{n} \left(k_i \ln \frac{1}{k_i} - k_i + 1 \right) \int_{x_i}^{x_{i+1}} w(x) g(x,\Theta) dx$$

$$= \sum_{i=0}^{n} \left(k_i \ln \frac{1}{k_i} - k_i + 1 \right) l_i, \tag{20}$$

where,

$$l_i = \int_{x_i}^{x_{i+1}} w(x) g(x,\Theta) dx \tag{21}$$

In the special case when $w(x)$ is a constant, so that using (10)

$$l_i = C \int_{x_i}^{x_{i+1}} g(x, \Theta)dx = C / k_i(n+1) \tag{22}$$

so that we have to minimise

$$\sum_{i=0}^{n} \left(k_i \ln \frac{1}{k_i} - k_i + 1 \right) \frac{1}{k_i} \tag{23}$$

However, in this case using (7)

$$\int_{x_0}^{x_{n+1}} g(x, \Theta)dx = \int_{x_0}^{x_{n+1}} f(x, \Theta)dx$$

or,

$$\sum_{i=0}^{n} \int_{x_i}^{x_{i+1}} g(x, \Theta)dx = \sum_{i=0}^{n} \int_{x_i}^{x_{i+1}} f(x, \Theta)dx$$

$$= \sum_{i=0}^{n} k_i \int_{x_i}^{x_{i+1}} g(x, \Theta)dx,$$

so that using (8)

$$\sum_{i=0}^{n} \left(\frac{1}{k_i} - 1 \right) = 0 \tag{24}$$

From (23) and (24), we have to minimise $-\sum_{i=0}^{n} \ln k_i$ or maximize $\sum_{i=0}^{n} \ln k_i$, which is the result we got earlier in the case when $w(x)$ is constant.

In the general case we have to maximize

$$\sum_{i=0}^{n} (k_i \ln k_i + k_i - 1) l_i \tag{25}$$

where,

$$k_i = \frac{1}{n+1} \frac{1}{\int_{x_i}^{x_{i+1}} g(x, \Theta)dx} \tag{26}$$

and,

$$l_i = \int_{x_i}^{x_{i+1}} w(x) g(x, \Theta)dx \tag{27}$$

Thus the implementation of the principle of minimum weighted information is in two stages.

In the first stage, we keep Θ fixed and choose $f(x, \Theta)$ in such a way that $f(x, \Theta)$ is as close as possible to the *a priori* density $g(x, \Theta)$ subject to its satisfying the fractile constraints, viz. that for $f(x, \Theta)$, the probability of the random variate lying in each sub interval is $1/(n + 1)$.

In the second stage, the functional form of $f(x, \Theta)$ is known and we choose Θ so as to minimise the weighted directed divergence of $f(x, \Theta)$ from $g(x, \Theta)$.

As such this principle may also be called fractiles constrained minimum weighted cross–entropy principle.

The main difference from Kullback's principle of *MINXENT* lies in that here the constraints are not in the form of prescribing the values of certain moments, but are in the form of prescribing of some fractiles and also we use here a weight function.

30.4 CONCLUDING REMARKS

In both stages of FC MIN W XENT (Fractile constrained Minimum Weighted Cross Entropy) or *MIN W INF* (Minimum Weighted Information) we have to minimise a directed divergence, but the principle leaves us free to choose any appropriate measure of directed divergence [5].

To make the distinction clear, we call the principle discussed above in which we use Kullback-Liebler measure of directed divergence as Lind-Solana principle of minimum information and we call the principle in which we use any arbitrary appropriate measure of directed divergence as the generalised principle of minimum information.

We have two corresponding principles, viz. the principle of least weighted information and generalised principle of least weighted information.

The implications of the generalised principles would be discussed separately later.

REFERENCES

1. R.A. Fisher, (1921), "On the Mathematical Foundations of Theoretical Statistics," Phil Trans Roy. Soc. Vol. 222(A), pp. 309–368.
2. S. Guiasu (1971), "Weighted Entropy", Reports on Math. Physics, vol. 2, pp. 165–179.
3. E.T. Jaynes (1957), "Information Theory and Statistical Mechanics", Physical Reviews, vol. 106, pp. 620–630.
4. J.N. Kapur (1984), "The Role of Maximum-Entropy and Minimum Discrimination Information Principles in Statistics", Jour. Ind. Soc. Agri. Stat., vol. 36 no. 3, pp. 12–55.
5. J.N. Kapur (1984), "A comparative assessment of various measures of directed divergences", Advances in Management studies, vol. 3, pp. 1–16.
6. J.N. Kapur (1988), "Maximum-Entropy Models in Science and Engineering", Wiley Eastern, New Delhi John Wiley, New York.
7. J.N. Kapur (1990), "On Lind's Principle of Minimum Information", MSTS Res. Rep # 488.
8. J.N. Kapur and H.K. Kesavan (1992), "Entropy Optimization Principles and their Applications," Academic Press, New York.
9. J.N. Kapur (1991), "Corrected Measures of directed divergence", M.S.T.S. Res. Rep. # 557. Chapter11 of the present book
10. S. Kullback (1959), "Information Theory and Statistics", John Wiley, New York.
11. S. Kullback and R.A. Liebler (1951), "On Information and Sufficiency", Ann. Math. Stat., vol. 22, pp. 79–86.
12. N.C. Lind (1989), "Information Theory and Estimation of Random Variable," Paper No. 15, Institute for Risk Research, University of Waterloo.
13. N.C. Lind and V. Solana (1989), "Cross–Entropy Estimation of Random Variables with Fractile Constraints", Paper No. 11, Institute of Risk Analysis, University of Waterloo.

14. N.C. Lind and V. Solana (1990), ''Fractile Constraints cross-entropy, Civil Engineering Systems, 7(2).
15. K. Pearson (1920), ''The Fundamental Problems of Practical Statistics'', Biometrica vol. 13, 1–16.
16. C.E. Shannon (1948), ''A Mathematical Theory of Communication'', Bell System Tech. J. vol. 27, pp. 379–423, 623–659.
17. H.C. Taneja and R.K. Tuteja (1984), ''Characterisation of Qualitative–Quantitative Measure of Relative Information'', Information Sciences vol. 33, pp. 1–6.

31

A MAXIMUM WEIGHTED ENTROPY PRINCIPLE

[The weights $w_1, w_2, ..., w_n$ in Guiasu's measure of weighted entropy are so determined that its maximum subject to $\sum_{i=1}^{n} p_i = 1$ arises when $p_i = q_i$ for each i, where $q_1, q_2, ..., q_n$ is a specified *a priori* probability distribution for which each $q_i < 1 / e$. Principle of maximum weighted entropy is enunciated and is shown to be equivalent to the principle of minimum cross-entropy when we use a new measure of cross-entropy derived here. The results are extended to the case when we use a measure of weighted entropy corresponding to Havrda and Charvat's measure of unweighted entropy.]

31.1 INTRODUCTION

For a given probability distribution $P = (p_1, p_2, ..., p_n)$ Shannon [14] defined his measure of uncertainty or entropy or information as

$$-\sum_{i=1}^{n} p_i \ln p_i \tag{1}$$

Jaynes [4] later suggested that when we are given only the constraints,

$$\sum_{i=1}^{n} p_i = 1, \quad \sum_{i=1}^{n} p_i\, g_{ri} = a_r, \quad r = 1, 2, ---, m, \quad m << n, \tag{2}$$

we should choose $p_1, p_2, ..., p_n$, so as to maximize (1) subject to (2). This is known as Jayne's maximum-entropy principle and has a tremendous variety of applications in science and engineering [6]. It is easy to implement because (1) is a concave function and the constraints (2) are all linear.

Later Guiasu [1, 2] generalised (1) to define his measure of weighted entropy,

$$-\sum_{i=1}^{n} w_i\, p_i \ln p_i, \tag{3}$$

where $w_1, w_2, ..., w_n$ are certain given positive numbers called weights or utilities of the n outcomes.

According to the principle of weighted maximum entropy, for given weights $w_1, w_2, ..., w_n$, we should choose that probability distribution which maximizes the concave function (3) subject to (2).

However, it is not specified how the weights $w_1, w_2, ..., w_n$ are to be determined except that these mesasure the utilities or usefulness or importance of outcomes and these are subjective or qualitative.

Earlier Kullback and Leibler [13] and Kullback [12] had talked of *a priori* probability distribution $Q = (q_1, q_2, ..., q_n)$ of the n outcomes and Kullback and Leibler had given a measure of the 'directed distance' or 'directed divergence' or 'discrimination information' or 'cross-entropy' of P from Q. The probability distribution Q is also to be determined through intuition and experience, is subjective and qualitative.

However, the concepts of 'weighted entropy' and 'cross-entropy' have been developed quite independently and so far no attempt has been made to determine a relationship between them, inspite of the fact that conceptually there is a good deal common between the concepts of 'weights' and '*a priori* probabilities.'

The obejct of the present paper is to investigate this relationship and to show that we can define a principle of maximum weighted entropy and this principle is equivalent to the principle of minimum cross-entropy provided we use a measure of cross-entropy which we also derive in the presnt discussion.

In this approach, we use the principles of generalised maximum entropy and generalised minimum cross-entropy [8, 9, 10, 11] which give us the flexibility to use any suitable measure of entropy or cross-entropy and release us from the restriction of using only Shannon's or Kullback-Leibler's measures.

Incidentally this approach enables us to derive new measures of cross-entropy, not given earlier, and enables us to discuss afresh the connection between cross-entropy and difference of entropies.

As a first step in section 2, we determine weights $w_1, w_2, ..., w_n$ so that when (3) is maximized subject to the natural constraint,

$$\sum_{i=1}^{n} p_i = 1, \tag{4}$$

the maximum arises when $p_i = q_i$ where $Q = (q_1, q_2, ..., q_n)$ is a specified *a priori* probability distribution and obtain a measure of weighted entropy in terms of $q_1, q_2, ..., q_n$. In section 3, we enunciate the principle of maximum weighted entropy for this measure of weighted entropy and in section 4, we show that this principle is equivalent to a principle of minimum cross-entropy when a suitable measure of cross-entropy is used. In section 5, we extend the results to the case when we use measure of weighted entropy corresponding to Havrda and Charvat [3] and other measure of entropy. In section 6, we give some concluding remarks.

31.2 DETERMINATION OF $w_1, w_2, ..., w_n$ AND OF THE WEIGHTED MEASURE OF ENTROPY INTERMS OF $q_1, q_2, ..., q_n$

Maximizing (3) subject to (4) we get,

$$\frac{1+\ln p_i}{\frac{1}{w_1}} = \frac{1+\ln p_2}{\frac{1}{w_2}} = \ldots = \frac{1+\ln p_n}{\frac{1}{w_n}} = k \quad \text{(say)}, \tag{5}$$

so that if the maximum occur when $p_i = q_i$ for each i, we get

$$1+\ln q_i = \frac{k}{w_i}, \quad i = 1, 2, \ldots, n \tag{6}$$

We shall assume that each $q_i < 1/e$ so that
$\ln q_i < -1$ or $1 + \ln q_i < 0$ or $k < 0$ or $k' > 0$ when $k' = -k$, and

$$1 + \ln q_i = -\frac{k'}{w_i}, \quad k' > 0 \tag{7}$$

If some of the q_i's are $> 1/e$, then for those, k will be > 0 while for others it will be < 0. Since this is not possible, it implies that the maximum of (3) cannot occur at Q. We are of course assuming that $n \geq 3$.

As such our discussion applies to only those *a priori* probability distribution for which each $q_i < 1/e$. This is not a serious restriction if n is not small.

Now from (7)

$$w_i = -\frac{k'}{1 + \ln q_i}, \tag{8}$$

and the measure for weighted entropy is

$$\sum_{i=1}^{n} \frac{p_i \ln p_i}{1 + \ln q_i}, \tag{9}$$

since we can always neglect a positive multiplier.

Thus (9) gives a measure of weighted entropy which when maximized subject to the natural constraint (4) has its maximum when $p_i = q_i$ for each i provided $q_i < 1/e$ for each i.

The maximum value of weighted entropy is

$$\sum_{i=1}^{n} \frac{q_i \ln q_i}{1 + \ln q_i} = \sum_{i=1}^{n} \frac{q_i(1 + \ln q_i - 1)}{1 + \ln q_i}$$

$$= 1 - \sum_{i=1}^{n} \frac{q_i}{1 + \ln q_i} \tag{10}$$

This is of course positive since both $\ln q_i$ and $1 + \ln q_i$ are negative.

31.3 THE PRINCIPLE OF MAXIMUM WEIGHTED ENTROPY

According to the principle if the weights $w_1, w_2, \ldots, w_n$ are known and the only other information available about $p_1, p_2, \ldots, p_n$ is given by is (2), then we choose $p_1, p_2, \ldots, p_n$ so as to maximize (3) subject to (2).

If instead of the weights $w_1, w_2, \ldots, w_n$, we know the *a priori* probabilities $q_1, q_2, \ldots, q_n$, then according to the principle of maximum weighted entropy, we choose $p_1, p_2, \ldots, p_n$ to maximize (9) subject to (2). This gives

$$\frac{1 + \ln p_i}{1 + \ln q_i} = \lambda_0 + \sum_{r=1}^{m} \lambda_r g_{ri}, \quad i = 1, 2, \ldots, n, \tag{11}$$

where $\lambda_0, \lambda_1, \ldots, \lambda_m$ are obtained by substituting the values of $p_1, p_2, \ldots, p_n$ obtained from (11) in (2).

31.4 THE PRINCIPLE OF MINIMUM CROSS-ENTROPY

According to Kullback's [12] principle of minimum cross-entropy, we should choose $p_1, p_2, \ldots, p_n$ so as to minimise Kullback-Leibler's [13] measure,

$$\sum_{i=1}^{n} p_i \ln \frac{p_i}{q_i} \tag{12}$$

subject to (2) to get

$$1 + \ln \frac{p_i}{q_i} = \lambda_0 + \sum_{r=1}^{m} \lambda_r g_{ri}, \tag{13}$$

which is different from (11).

We now ask whether there exists a measure of cross-entropy other than (12) whose minimization subject to (2) will give us (11). Let such a measure be

$$\sum_{i=1}^{n} \frac{f(p_i) - f(q_i)}{g(q_i)} \tag{14}$$

This vanishes when $p_i = q_i$ for each i and this will be a convex function of p_1, $p_2, \ldots, p_n$ if $f''(p_i) / g(q_i) > 0$.

Now (14) will lead to (11) if

$$\frac{f'(p_i)}{g(q_i)} = \frac{1 + \ln p_i}{1 + \ln q_i}, \tag{15}$$

so that we choose

$$f(p_i) = p_i \ln p_i \quad g(q_i) = 1 + \ln q_i \tag{16}$$

and our measure cab be

$$\sum_{i=1}^{n} \frac{p_i \ln p_i - q_i \ln q_i}{1 + \ln q_i} \tag{17}$$

However, since $1 + \ln q_i < 0$, this is a concave function of $p_1, p_2, \ldots, p_n$. As such we choose the measure,

$$\sum_{i=1}^{n} \frac{q_i \ln q_i - p_i \ln p_i}{1 + \ln q_i} \tag{18}$$

This is a convex function whose minimum value zero occurs when $p_i = q_i$ for each i and as such this function is always ≥ 0. Thus (18) represents a valid measure of cross-entropy which is such that its minimisation subject to (2) gives us the same results as are given by maximization of weighted entropy (10) subject to (2).

In other words the principle of maximum weighted entropy when measure of entropy used is (10) is equivalent to the principle of minimum cross-entropy when the measure of cross-entropy is given by (18).

31.5 WEIGHTED ENTROPY CORRESPONDING TO HAVRDA-CHARVAT AND KAPUR'S MEASURES

Havrda and Charvats [3] weighted entropy measure is given by

$$\sum_{i=1}^{n} \frac{w_i (p_i^{\alpha} - p_i)}{1 - \alpha}, \quad \alpha \neq 1 \tag{19}$$

This is maximum subject to (4) when $p_i = q_i$ if

$$\frac{\alpha q_i^{\alpha-1} - 1}{\frac{1}{w_i}} = k, \quad i = 1, 2, ---, n \tag{20}$$

so that our measure of weighted entropy is

$$\frac{1}{(\alpha-1)} \sum_{i=1}^{n} \frac{p_i^{\alpha} - p_i}{\alpha q_i^{\alpha-1} - 1}; \quad q_i < (\alpha)^{\frac{1}{1-\alpha}} \quad \text{for each } i \tag{21}$$

The upper bounds for q_i for various values of α are given by

α	1/10	1/4	1/2	3/4	1	2	3	5	∞
$(q_i)_{v\cdot B}$	.77426	.15744	.25000	.31641	.36788	.50000	.54754	.66874	1.0000

Thus we can use the measure (21) for most *a priori* probability distributions unless α is very small.

Alternatively, we can always choose an appropriate value for α for every non-degenerate *a priori* probability distribution.

The corresponding measure of cross-entropy, by minimising which we get the same probability distribution as we get by maximizing the weighted entropy is given by

$$\frac{1}{(\alpha-1)} \sum_{i=1}^{n} \frac{q_i^{\alpha} - p_i^{\alpha} - q_i + p_i}{\alpha q_i^{\alpha-1} - 1} \quad \text{or} \quad \frac{1}{(\alpha-1)} \sum_{i=1}^{n} \frac{q_i - q_i^{\alpha} + p_i^{\alpha} - p_i}{1 - \alpha q_i^{\alpha-1}} \tag{22}$$

which is easily seen to be a valid measure of cross-entropy since it is a convex function with minimum value zero when $q_i = p_i$ for each i.

Similarly, the measure of weighted entropy corresponding to Kapur's [5] measure of entropy is

$$-\sum w_i \left[p_i \ln p_i - \frac{1}{a}(1 + a\, p_i) \ln (1 + a\, p_i) + \frac{1}{a}(1 + a) \ln (1 + a)\, p_i \right] \tag{23}$$

and the measure in terms of $q_1, q_2, \ldots, q_n$ is

$$\sum_{i=1}^{n} \frac{p_i \ln p_i - \frac{1}{a}(1 + a\, p_i) \ln (1 + a\, p_i) + \frac{1}{a}(1 + a) \ln (1 + a)\, p_i}{\ln q_i - \ln (1 + a\, q_i) + \frac{1}{a}(1 + a) \ln (1 + a)} \tag{24}$$

and the corresponding measure of cross-entropy is

$$\sum_{i=1}^{n} \frac{\left[q_i \ln q_i - \frac{1}{a}(1 + a\, q_i) \ln (1 + a\, q_i) + \frac{1}{a}(1 + a) \ln (1 + a)\, q_i - p_i \ln p_i + \frac{1}{a}(1 + a\, p_i) \ln (1 + a\, p_i) - \frac{1}{a}(1 + a) \ln (1 + a)\, p_i \right]}{\ln q_i - \ln (1 + a\, q_i) + \frac{1}{a}(1 + a) \ln (1 + a)} \tag{25}$$

provided,

$$\frac{q_i}{1 + a\, q_i}(1 + a)^{\frac{1+a}{a}} < 1 \qquad \text{for each } i \tag{26}$$

In general if $\sum_{i=1}^{n} \phi(p_i)$ is a measure of unweighted entropy where $\phi(p_i)$ is a positive concave function which vanishes both when $p_i = 0$ or 1, then $\sum_{i=1}^{n} w_i\, \phi(p_i)$ is the corresponding measure of weighted entropy and the corresponding measure of weighted entropy in terms of Q is

$$\sum_{i=1}^{n} \frac{\phi(p_i)}{\phi'(q_i)} \quad \text{or} \quad -\sum \frac{\phi(p_i)}{\phi'(q_i)} \tag{27}$$

according as $\phi'(q_i) >$ or < 0 for each i (28)

and the corresponding measure of cross-entropy is

$$\sum_{i=1}^{n} \frac{\phi(q_i) - \phi(p_i)}{\phi'(q_i)} \quad \text{or} \quad -\sum_{i=1}^{n} \frac{\phi(q_i) - \phi(p_i)}{\phi'(q_i)} \tag{29}$$

Maximizing (27) subject to (2) gives the same probability distribution as minimising (28) subject to (2).

For the uniform distribution (27) becomes a multiple of the measure of entropy and (29) is a monotonic decreasing function of this measure.

31.6 CONCLUDING REMARKS

(i) $\sum_{i=1}^{n} \phi(p_i) - \sum_{i=1}^{n} \phi(q_i)$, which is the difference of entropies of two probability distributions cannot be used as a measure of 'discrepancy' or 'distance' since it can be sometimes positive and sometimes negative.

$\sum_{i=1}^{n} |\phi(p_i) - \phi(q_i)| \geq 0$ and is also symmetric, but is inconvenient for use in optimization problems.

The other measures like

$$\sum_{i=1}^{n} p_i \ln \frac{p_i}{q_i}, \frac{1}{\alpha - 1}\left(\sum_{i=1}^{n} p_i^{\alpha} q_i^{1-\alpha} - 1\right), \frac{1}{\alpha - 1} \ln \sum_{i=1}^{n} p_i^{\alpha} q_i^{1-\alpha} \tag{30}$$

are always ≥ 0 and are used to measure directed divergences, but they do not correspond to differences in entropies.

Our new measure (29) is always ≥ 0 and is weighted sum of differences in entropies of individual outcomes.

(ii) Corresponding to every measure of entropy $\sum_{i=1}^{n} \phi(p_i)$, we have a measure of cross-entropy $\sum_{i=1}^{n} (\phi(q_i) - \phi(p_i))/(\phi'(q_i))$ and corresponding to every measure of cross-entropy,

$\sum_{i=1}^{n} (\phi(q_i) - \phi(p_i))/(\phi'(q_i))$, we have a measure of entropy $\sum_{i=1}^{n} \phi(p_i)$

(iii) Thus corresponding to the trigonometric measure of entropy [7]

$$\sum_{i=1}^{n} \sin \pi p_i, \quad \sum_{i=1}^{n} \cos \frac{\pi p_i}{2} - (n-1), \quad \sum_{i=1}^{n} \sin \frac{\pi p_i}{2} - 1 \tag{31}$$

we get measures of directed divergence

$$\sum_{i=1}^{n} \frac{\sin \pi q_i - \sin \pi p_i}{\cos \pi q_i} \quad \text{provided each} \quad q_i < \frac{1}{2} \tag{32}$$

$$\sum_{i=1}^{n} \frac{\cos \frac{\pi q_i}{2} - \cos \frac{\pi p_i}{2}}{\sin \pi q_i/2} \tag{33}$$

and,

$$\sum_{i=1}^{n} \frac{\sin \frac{\pi q_i}{2} - \sin \frac{\pi p_i}{2}}{\cos \frac{\pi q_i}{2}} \tag{34}$$

(iv) corresponding to our new measures of directed divergence, the measures of symmetric divergence are :

(a)
$$\sum_{i=1}^{n} \frac{\phi(q_i) - \phi(p_i)}{\phi'(q_i)} + \sum_{i=1}^{n} \frac{\phi(p_i) - \phi(q_i)}{\phi'(p_i)} = \sum_{i=1}^{n} [\phi(p_i) - \phi(q_i)] \left[\frac{1}{\phi'(p_i)} - \frac{1}{\phi'(q_i)} \right] \tag{35}$$

(b)
$$\sum_{i=1}^{n} [p_i \ln p_i - q_i \ln q_i] \left[\frac{1}{1 + \ln p_i} - \frac{1}{1 + \ln q_i} \right] = \frac{\ln q_i/p_i}{(\ln p_i e)(\ln q_i e)} \ln \frac{p_i^{p_i}}{q_i^{q_i}} \tag{36}$$

(c)
$$\sum_{i=1}^{n} q_i^{\alpha} - p_i^{\alpha} - q_i + p_i) \left(\frac{1}{\alpha q_i^{\alpha-1} - 1} - \frac{1}{\alpha p_i^{\alpha-1} - 1} \right), \quad \alpha \neq 1 \tag{37}$$

(d)
$$\sum_{i=1}^{n} [\sin \pi q_i - \sin \pi p_i] \left[\frac{1}{\cos \pi q_i} - \frac{1}{\cos \pi p_i} \right] \tag{38}$$

(e)
$$\sum_{i=1}^{n} \left[\cos \frac{\pi q_i}{2} - \cos \frac{\pi p_i}{2} \right] \left[\frac{1}{\sin \frac{\pi q_i}{2}} - \frac{1}{\sin \frac{\pi p_i}{2}} \right] \tag{39}$$

(f)
$$\sum_{i=1}^{n} \left[\sin \frac{\pi q_i}{2} - \sin \frac{\pi p_i}{2} \right] \left[\frac{1}{\cos \frac{\pi q_i}{2}} - \frac{1}{\cos \frac{\pi p_i}{2}} \right] \tag{40}$$

REFERENCES

1. S. Guiasu (1971),'' Weighted Entropy,'' Reports on Math. Physics, Vol 2, pp. 165-179.
2. S. Guiasu (1977),''Information Theory with Applications,'' Mc-Graw Hill, New York.
3. J.H. Havrda and F. Charvat, (1967),'' Quantification Methods of Classification Processes: Concept of Structural α Entropy,'' Kybernatica, Vol.3, pp. 30-35.
4. E.T. Jaynes (1957),''Information Theory and Statistical Mechanics,'' Physical Reviews, Vol. 106. pp. 620-630.
5. J.N. Kapur (1986),''Four Families of Measures of Entropy,'' Ind. Jour. Pure and app. Maths, Vol. 17, no.4, pp. 429-449.

6. J.N. Kapur (1990)," Maximum Entropy Models in Scinece and Engineering," Wiley Eastern, New Delhi.
7. J.N. Kapur and G.P. Tripathi (1989)." On Trigonometric Measure of Information Joun-Math. Phys. Scineces, Vol 24. No.1, pp 1-10.
8. J.N. Kapur and H.K. Kesavan (1987), "Generalised Maximum Entropy Principles (with Applications)" Sand Ford Educational Press, Waterloo University, Canada.
9. J.N. Kapur and H.K. Kesavan (1992), "Entropy Optimization Principles, and their Applications," Academic Press, USA.
10. H.K. Kesavan and J.N. Kapur (1989), "The Generalised Maximum Entropy Principle," IEEE, Trans, Syst.Man Cyb 19(9) / 1042-1059.
11. H.K. Kesavan and J.N. Kapur (1990)," On the families of solutions of generalised maximum entropy and minimum cross-entropy problems", Int. Journ. Gen.Systs 11, pp. 199-219.
12. S. Kullback(1959), Information Theory and Statistics, John Wiley, New York.
13. S. Kullback and R.A. Leibler (1951), On Information and Sufficiency", Ann. Math. Stat, Vol.22, pp. 79-86.
14. C.E. Shannon (1948)," A Mathematical Theory of Communication", Bell System Tech. J., Vol. 27, pp. 379-423, 623-659.

32

ON LIND AND SOLANA'S PRINCIPLE OF MINIMUM INFORMATION

[The implications of the use of different measures of cross-entropy in the application of Lind and Solana's principle of minimum information are examined.]

32.1 LIND AND SOLANA'S PRINCIPLE OF MINIMUM INFORMATION

Let $g(x, \theta)$ be a known probability density function of a continuous random variate defined over the interval $[x_0, x_{n+1}]$. However, the parameter θ is not known and to estimate it, we draw a random sample $x_1, x_2, ..., x_n$ from the population and rearrange its members, so that

$$x_0 < x_1 < x_2 < ... < x_i < x_{i+1} < ... < x_n < x_{n+1} \tag{1}$$

The usual method for estimation of θ is based on Fisher's [4] principle of maximum likelihood according to which we estimate θ by maximizing the likelihood function;

$$L \equiv g(x_1, \theta)\, g(x_2, \theta) \ldots g(x_n, \theta) \tag{2}$$

Recently, Lind and Solana [11-18] have suggested a two-stage method of estimating θ. In the first stage we assume θ to be known and choose a function $f(x, \theta)$ which satisfies

$$\int_{x_i}^{x_{i+1}} f(x,\theta)\, dx = \frac{1}{n+1}, \quad i = 0, 1, 2, \ldots, n \tag{3}$$

so that the probabilities in each of the $n + 1$ intervals defined by the n sample points are $1/(n + 1)$ each and which is such that

$$\int_{x_0}^{x_n} f(x,\theta) \ln \frac{f(x,\theta)}{g(x,\theta)}\, dx \tag{4}$$

is minimum, i.e. out of all the density functions satisfying constraints (3), we choose that function which is 'closest' to given $g(x, \theta)$ in the sense that it minimises Kullback-Leibler's [9] measure of cross-entropy of $f(x, \theta)$ from $g(x, \theta)$.

Thus, the first stage determies $f(x, \theta)$ for any given θ. The object of the second stage is to choose θ to minimize (4) for the function $f(x, \theta)$ determined by the first stage.

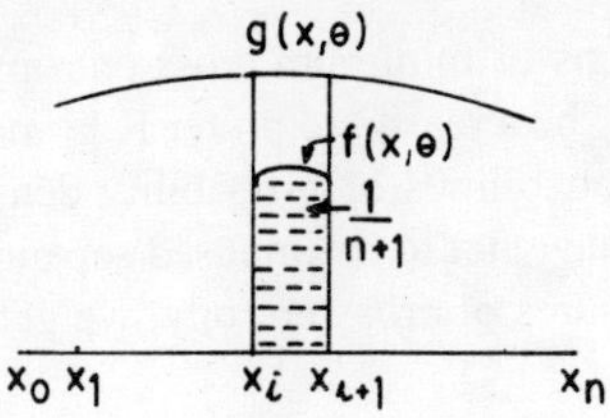

Fig. 32.1

The first stage gives,

$$f(x,\theta) = \frac{g(x,\theta)}{k_i}, \quad x_i < x \leq x_{i+1} \tag{5}$$

where, $$k_i = (n+1) \int_{x_i}^{x_{i+1}} g(x,\theta)dx \,, i = 0, 1, \ldots, n \tag{6}$$

This is illustrated in Figure 32.1.

The second stage gives that we choose θ to minimise

$$\sum_{i=0}^{n} \int_{x_i}^{x_{i+1}} \frac{g(x,\theta)}{k_i} \ln \frac{1}{k_i} dx = -\frac{1}{n+1} \sum_{i=0}^{n} \ln k_i \tag{7}$$

In other words, we choose θ by maximizing

$$\sum_{i=0}^{n} \ln \int_{x_i}^{x_{i+1}} g(x,\theta)dx \tag{8}$$

This can be compared with the principle of maximum likelihood which suggests that we choose θ to maximize

$$\sum_{i=1}^{n} \ln g(x_i, \theta) \tag{9}$$

It may be observed that in the principle of minimum information, we choose θ to maximize the product of the areas under the curve, $y = g(x, \theta)$ in the $(n+1)$ intervals determined by n sample points, while in the principle of maximum likelihood, we choose θ so as to maximize the product of the n ordinates of the probability curve at the n sample points (Fig. 32.1).

32.2 THE PROBLEM

It will be observed that in both stages, we minimize the Kullback-Leibler measure of cross-entropy (4). In the first stage, θ is kept fixed and the functional form of $f(x, \theta)$ is varied. In the second stage, the functional form determined in the first stage is fixed and θ is varied.

However, as far as the basic philosophy of the principle is concerned, there is no reason why we should use only Kullback-Leibler's measure and even if we decide to use. K.L. measure , why we should minimize the cross-entropy of f from g and not of g from f or why should we not minimize the symmetric measure of cross-entropy.

In the usual applications of minimum cross-entropy principle when there are linear constraints, there may be a reason to prefer K.L. measure, because it automatically leads to positive probabilities or probability density functions and as such non-negativity constraints have not to be imposed separately. Here we have fractile constraints and for all measures of cross-entropy, we get positive probabilities and probability density functions.

Again in that case, there is a reason to minimize the cross-entropy of distributions from a given distribution. Here in the first stage, we may talk of $f(x, \theta)$ being chosen to be close to $g(x, \theta)$ but in the second stage it is θ that is being chosen and both $f(x, \theta)$ and $g(x, \theta)$ have the same status.

One consideration however has to be seen that (4) is a convex function of θ so that its local minimum is a global minimum.

We shall show that (8) which has to be maximized is a concave function of θ and thus in this case, we need not worry about the sign of its second order derivative, but for other measures of cross-entropy, the corresponding function need not be concave. In fact, one of our objects will be to find out those measures for which the corresponding function is a concave function of θ.

32.3 DISCUSSION FOR CSISZER'S CLASS OF MEASURES OF CROSS-ENTROPY

Csiszer [3] gave the classes of measures of cross-entropy,

$$\int g(x, \theta)\, \phi\left(\frac{f(x, \theta)}{g(x, \theta)}\right) dx \quad \text{and} \quad \int f(x, \theta)\, \phi\left(\frac{g(x, \theta)}{f(x, \theta)}\right) dx, \tag{10}$$

where ϕ (.) is a twice -differentiable convex function for which $\phi(1) = 0$,

For different functions ϕ(.), (10) can represent a variety of measures of cross-entropy.

Using these for the first stage, we get

$$\phi'\left(\frac{f(x, \theta)}{g(x, \theta)}\right) = \text{const} \quad \text{or,} \quad \phi\left(\frac{g(x, \theta)}{f(x, \theta)}\right) - \frac{g(x, \theta)}{f(x, \theta)}\, \phi'\left(\frac{g(x, \theta)}{f(x, \theta)}\right) = \text{const} \tag{11}$$

Whatever be the function. ϕ(.), this gives (5) and (6), so that the first stage gives the same result for all Csiszer's measure of cross-entropy. For the second stage, we have to choose θ to minimize

$$\sum_{i=0}^{n} \int_{x_i}^{x_{i+1}} g(x, \theta)\, \phi\left(\frac{1}{k_i}\right) dx \quad \text{or} \quad \sum_{i=0}^{n} \int_{x_i}^{x_{i+1}} \frac{g(x,\theta)}{k_i}\, \phi(k_i)\, dx$$

that is,

$$\frac{1}{n+1} \sum_{i=0}^{n} k_i\, \phi\left(\frac{1}{k_i}\right) \quad \text{or} \quad \frac{1}{n+1} \sum_{i=0}^{n} \phi(k_i) \tag{12}$$

where,

$$k_i = (n+1) \int_{x_i}^{x_{i+1}} g(x, \theta)\, dx \tag{13}$$

Now let, $$F(\theta) = \sum_{i=0}^{n} \phi(k_i),$$

then, $$\frac{1}{n+1}\frac{dF}{d\theta} = \sum_{i=0}^{n} \phi'(k_i) \int_{x_i}^{x_{i+1}} \frac{\partial g}{\partial \theta} dx$$

$$\frac{1}{n+1}\frac{d^2F}{d\theta} = \sum_{i=0}^{n} \phi''(k_i)(n+1)\left[\int_{x_i}^{x_{i+1}} \frac{\partial g}{\partial \theta} dx\right]^2$$

$$+\sum_{i=0}^{n} \phi'(k_i) \int_{x_i}^{x_{i+1}} \frac{\partial^2 g}{\partial \theta^2} dx \tag{14}$$

Now since $\phi(.)$ is convex, $\phi''(k_i) > 0$ and the first term on the R.H.S > 0. Again $g(x, \theta)$ is a concave function of θ so that the second term will also be > 0 if $\phi'(k_i) < 0$, so that if $\phi'(k_i) < 0$, i.e. if $\phi(\cdot)$ is a decreasing function at k_i, $F(\theta)$ will be a convex function of θ and its local minimum will be its global minimum.

Again let $$G(\theta) = \sum_{i=0}^{n} k_i\, \phi\left(\frac{1}{k_i}\right)$$

so that, $$G'(\theta)/(n+1) = \sum_{i=0}^{n}\left[\phi\left(\frac{1}{k_i}\right) - \frac{1}{k_i}\phi'\left(\frac{1}{k_i}\right)\right] \int_{x_i}^{x_{i+1}} \partial g/\partial\theta \; dx$$

and $$\frac{G''(\theta)}{n+1} = \sum_{i=0}^{n}\left\{\left[\phi\left(\frac{1}{k_i}\right) - \frac{1}{k_i}\phi'\left(\frac{1}{k_i}\right)\right] \int_{x_i}^{x_{i+1}} \frac{\partial^2 g}{\partial \theta^2} dx\right.$$

$$\left. + \frac{1}{k_i^3}\phi''\left(\frac{1}{k_i}\right)\left[\int_{x_i}^{x_{i+1}} \frac{\partial g}{\partial \theta} dx\right]^2\right\} \tag{15}$$

The second term is > 0 and the first term will also be > 0, if

$$\phi\left(\frac{1}{k_i}\right) - \frac{1}{k}\phi'\left(\frac{1}{k_i}\right) < 0 \tag{16}$$

In this case, the local minimum will be the global minimum.

32.4 DISCUSSION OF HAVRDA & CHARVAT MEASURES OF CROSS-ENTROPY

These measures are defined by either

$$\frac{\int (f^{\alpha}(x)\, g^{1-\alpha}(x) - f(x))dx}{\alpha(\alpha-1)} \quad \text{or} \quad \frac{\int (g^{\alpha}(x)\, f^{1-\alpha}(x) - g(x))dx}{\alpha(\alpha-1)}, \; \alpha \neq 0,\; \alpha \neq 1 \tag{17}$$

It $\alpha \to 1$, these gives Kullback-Leibler measures,

$$\int f(x) \ln \frac{f(x)}{g(x)} dx \quad \text{or} \quad \int g(x) \ln \frac{g(x)}{f(x)} dx \tag{18}$$

If $\alpha \to 0$, these will give the measures,

$$\int g(x) \ln \frac{g(x)}{f(x)} dx \quad \text{or} \quad \int f(x) \ln \frac{f(x)}{g(x)} dx \tag{19}$$

which are the same as (18).
For our second stage, these suggest that we choose θ to minimize

$$\sum_{i=0}^{n} \int_{x_i}^{x_{i+1}} \int \left(\frac{g(x)}{k_i^{\alpha}} - \frac{g(x)}{k_i} \right) dx \quad \text{or} \quad \sum_{i=0}^{n} \int_{x_i}^{x_{i+1}} \int \left(\frac{g(x)}{k_i^{1-\alpha}} - \frac{g(x)}{k_i} \right) dx \tag{20}$$

in the general case, and

$$\sum_{i=0}^{n} \int_{x_i}^{x_{i+1}} \frac{g(x)}{k_i} \ln \frac{1}{k_i} dx \quad \text{or} \quad \sum_{i=0}^{n} \int_{x_i}^{x_{i+1}} g(x) \ln k_i \, dx \tag{21}$$

in the Kullback-Leibler case.
Thus we have to minimize

$$\frac{1}{\alpha(\alpha-1)} \frac{1}{n+1} \left[\sum_{i=1}^{n} k_i^{1-\alpha} - 1 \right] \quad \text{or} \quad \frac{1}{\alpha(\alpha-1)} \frac{1}{n+1} \left[\sum_{i=1}^{n} k_i^{\alpha} - 1 \right] \tag{22}$$

in the general case, and

$$-\frac{1}{n+1} \sum_{i=0}^{n} \ln k_i \quad \text{or} \quad \frac{1}{n+1} \sum_{i=0}^{n} k_i \ln k_i \tag{23}$$

in the Kullback-Leibler case.

32.5 DISCUSSION OF KULLBACK-LEIBLER CASE

In this case we have to maximize either

$$\sum_{i=0}^{n} \ln k_i \quad \text{or} \quad -\sum_{i=0}^{n} k_i \ln k_i \tag{24}$$

Now, k_i's are proportional to probabilities and

$$\sum_{i=0}^{n} k_i = n+1 \tag{25}$$

If we write $p_i = \dfrac{k_i}{n+1}$, we have to maximize

$$\sum_{i=0}^{n} \ln p_i \quad \text{or} \quad -\sum_{i=0}^{n} p_i \ln p_i \tag{26}$$

The first of these is Burg's [1] measure of entropy and the second is Shannons's [19] measure of entropy.

However, we have not to maximize these as functions of $p_1, p_2, ..., p_n$, but as functions of θ.

Now,

$$k_i = (n+1) \int_{x_i}^{x_{i+1}} g(x, \theta)\, dx$$

$$\frac{d k_i}{d\theta} = (n+1) \int_{x_i}^{x_{i+1}} \frac{\partial g}{\partial \theta}\, dx$$

$$\frac{d^2 k_i}{d\theta^2} = (n+1) \int_{x_i}^{x_{i+1}} \frac{\partial^2 g}{\partial \theta^2}\, dx, \tag{27}$$

so that $k_1, k_2, ..., k_n$ are all concave functions of θ and $\ln k_1, \ln k_2, ..., \ln k_n$ are monotonic increasing functions of $k_1, k_2, ..., k_n$ so that $\sum_{i=1}^{n} \ln k_i$ is also a concave function of θ and its local maximum is its global maximum.

However, while $-k_i \ln k_i$ is a concave function of k_i, it is not necessarily a monotonic increasing function and as such $-\sum_{i=1}^{n} k_i \ln k_i$ is not necessarily a concave function of θ. As such its maximization may not be easy and straight forward.

32.6 DISCUSSION OF THE MORE GENERAL CASE

Let,

$$P(\theta) = \frac{1}{(n+1)^2 \alpha(\alpha-1)} \left(\sum_{i=1}^{n} k_i^{1-\alpha} - 1 \right) \tag{28}$$

$$\frac{dP}{d\theta} = \frac{1}{(n+1)^2\, \alpha\, (\alpha-1)} (1-\alpha) \sum_{i=0}^{n} k_i^{-\alpha} \frac{d k_i}{d\theta}$$

$$= -\frac{1}{n+1} \frac{1}{\alpha} \sum_{i=0}^{n} k_i^{-\alpha} \int_{x_i}^{x_{i+1}} \frac{\partial g}{\partial \theta}\, dx$$

$$\frac{d^2 P}{d\theta^2} = \sum_{i=0}^{n} k_i^{-\alpha-1} \left(\int_{x_i}^{x_{i+1}} \frac{\partial g}{\partial \theta}\, dx \right)^2$$

$$-\frac{1}{n+1} \frac{1}{\alpha} \sum_{i=0}^{n} k_i^{-\alpha} \int_{x_i}^{x_{i+1}} \frac{\partial^2 g}{\partial \theta^2}\, dx \tag{29}$$

Both the terms are positive and as such $P(\theta)$ is convex function of θ if $\alpha > 0$

Let,
$$Q(\theta) = \frac{\frac{1}{n+1}\sum_{i=0}^{n} k_i^{\alpha} - 1}{\alpha(\alpha-1)} \tag{30}$$

so that,

$$\frac{dQ}{d\theta} = \frac{1}{\alpha-1}\sum_{i=0}^{n} k_i^{\alpha-1}\int_{x_i}^{x_{i+1}} \frac{\partial g}{\partial \theta}\,dx$$

$$\frac{d^2 Q}{d\theta^2} = \sum_{i=0}^{n} k_i^{\alpha-2}\left(\int_{x_i}^{x_{i+1}} \frac{\partial g}{\partial \theta}\,dx\right)^2 (n+1)$$

$$+\frac{1}{\alpha-1}\sum_{i=0}^{n} k_i^{\alpha-1}\int_{x_i}^{x_{i+1}} \frac{\partial^2 g}{\partial \theta^2}\,dx \tag{31}$$

The first term is positive and the second is also positive if $\alpha < 1$. Thus minimization of (30) for all values of $\alpha < 1$ will give a global minimum. This includes the special case $\alpha = 0$ where (30) reduces to $-\sum_{i=0}^{n} \ln k_i$.

32.7 CONCLUDING REMARKS

(i) In the first stage of application of Lind and Solana's principle of least information, we get results (5) and (6), for all Csiszer's measure of cross-entropy, whether we take cross-entropy of $f(x, \theta)$ from $g(x, \theta)$ or of $g(x, \theta)$ from $f(x, \theta)$ or we use the symmetric measure,

$$\int\left[f(x,\theta)\,\phi\left(\frac{g(x,\theta)}{f(x,\theta)}\right) + g(x,\theta)\,\phi\left(\frac{f(x,\theta)}{g(x,\theta)}\right)\right]dx \tag{32}$$

(ii) In the second stage, the estimate of θ will depend on the measure of cross-entropy used and whether we take cross-entropy of $f(x, \theta)$ from $g(x, \theta)$ or of $g(x, \theta)$ from $f(x, \theta)$ or we use the symmetric measure (32).

(iii) In fact, in the second stage, some measures may lead to concave functions of θ for minimization or convex function of θ for maximization, and thus a great deal of care will have to be used in finding the global minimum or maximum.

(iv) Havrda and Charvat's measure

$$\frac{\int g^{\alpha}(x,\theta)\,f^{1-\alpha}(x,\theta)\,dx - 1}{\alpha(\alpha-1)} \tag{33}$$

will give a convex functions of θ when $\alpha < 1$ and so this measure can be conveniently used when $\alpha < 1$. Similarly,

$$\left[\int f^{\alpha}(x)\,g^{1-\alpha}(x)\,(dx - 1)\right]/\alpha(\alpha-1) \tag{34}$$

can be used when $\alpha > 0$.

(v) The estimate $\hat{\theta}$ which will correspond to the value α for (33) will correspond to the value $1 - \alpha$ for (34).

(vi) We thus get a wide choice of estimates and the question now arises as to which one we should use. Lind and Solana have suggested the principles of monotonocity and invariance [13, 20, 21] for this purpose and these can be applied to see whether these restrict our choice, otherwise the choice has to be left to the users or decided by considerations of computational convenience.

(vii) The principle of least information approaches objectively in the estimation of random variables from sparse data. However, there are cases when objectivity is not the preferred criterion, e.g. high demands and high stresses can cause failure and high intensities of earthquake can cause great damage and as such these, have to be treated differently from low demands, low stresses and low intensities. The weights attached to different outcomes will depend on the specific applications we have in view. An application-oriented least information principle has been given by Lind [17]. A least weighted information principle has been given by Kapur [9].

(viii) Chang and Amin [2] gave the Maximum Product of Spacings [MPS] Method, seeking to maximize the geometric mean of spacings. Independently Renneby [18] observed that a good inference method ought to minimize the distance between the true distribution and the model in terms of a suitable metric. These provide similar but distinctly different rationales for estimation from that provided by PLI.

(ix) Some other papers which discuss the PLI are [6, 11, 12, 13, 14, 15, 16, 17].

REFERENCES

1. J.P. Burg (1972),'' The Relationship between Maximum Entropy Spectra and Maximum Likelihood Spectra'', In Modern Spectral Analysis, ed. D.G. Childers, pp 130-131.
2. R.C.H. Chang and, N.A.K. Amin, ''Estimating Parameters of Continuous Univariate Distributions with a shifted origin'', J.R.Statist. Soc, B, 45, no. (3), pp. 394-403.
3. I. Csiszer (1972),''A class of measures of informativity of observation channels'', Periodica Math. Hungarika vol.2, pp. 191-213.
4. R.A. Fisher (1921),''On the Mathematical Foundations of Theoretical Statistics'', Phil. Trans. Roy Soc, Vol. 222 (A), pp 309-368.
5. J.H. Havrda and F, Charvat (1967),''Quantification Methods of Classification Processes; Concepts of Structural α Entropy'', Kybernetica, Vol 3, pp. 30-35.
6. H.P. Hong, N.C. Lind and V. Solana (1989)'', Probability assignment of earthquake inter arrival times using cross-entropy'', pp 3-17 in proceedings of the 4th International Conference in Social Dynamics and Earthquake Engineering UNAM, Maxico.
7. J.N. Kapur and H.K. Kesavan (1987), Generalised Maximum Entropy Principle (with Applications). Sandford Educational Press, University of Waterloo.
8. J.N. Kapur (1990),''On Linds Principle of Minimum Information'', MSTS, Research Reports # 488.
9. J.N. Kapur (1991),''Principle of Least Weighted Information'', MSTS, Research Report # 564.
10. S. Kullback and R.A. Leibler (1951),''On Information and Sufficiency,'' Ann Math. Stat, Vol 22, pp, 79-86.
11. N.C. Lind and V. Solana (1988),'' Cross-entropy estimation of random variables with fractile constraints, Paper No 11. Institute for Risk Research, University of Waterloo.
12. N.C. Lind and V. Solana (1988)'', Cross-entropy estimation of distributions based on scarce data'', Presented at Society for Risk Analysis, Annual Meeting, Boston.

13. N.C. Lind and X. Chen (1987),"Consistent distribution parameter estimation for reliability analysis," Structural Safety, 4, pp. 141-149.
14. N.C. Lind (1989),"Information Theory and Estimation of Random Variables," Paper No 15, Institute for Risk Research, University of Waterloo.
15. N.C. Lind and V. Solana (1990),"Fractile constraints entropy estimation of distributions based on scarce data, Civil Engineering System, 7 (2).
16. N.C. Lind and H.P. Hong (1991),"The Entropy Approximation", To appear in Structural Safety:
17. N.C. Lind (1991), "Applications Oriented Least Information" Principle to appear in CSIR Journal.
18. B. Rameby (1984)," The Maximum spacing method–An estimating method related to the maximum likelihood method," Scand, J. Statist, 11, pp. 93-112.
19. C.E. Shannon (1948),"A Mathematical Theory of Communication", Bell System. Tech. J, Vol 27, pp 379-423, 623-659.
20. V. Solans and N.C. Lind (1989)," A monotonic property of distributions based on entropy fractile constraints," pp. 481-490, in Maximum Entropy and Bayesian Method ed. by J. Skilling, Kluwer Academic Publishers.
21. V. Solana (1990),"Consistency principle for data based probabilistic inference in Maximum Entropy and Bayesian Methods ed. by P. Fougere, Kluwer Academic Publishers.

33

ON LIND'S PRINCIPLE OF MINIMUM INFORMATION

[Some alternatives to Lind's Principle of minimum information are given and are compared with the original principle. The principle is also compared with Fisher's Principle of Maximum Likelihood.]

33.1 INTRODUCTION

Let $x_1, x_2, ..., x_n$ be a random sample drawn from a population with density function $f(x, \theta)$ where $x_0 \le x \le x_{n+1}$ then without loss of generality, we can assume that

$$x_1 \le x_2 \le \ldots \le x_n \tag{1}$$

We can also allow x_0 to be $-\infty$ and x_{n+1} to be $+\infty$.

Our object is to estimate θ

Lind [8] first considered the case when $x_1, x_2, ..., x_n$ are all distinct and suggested a two-stage process for estimating θ.

In the first stage, we find *a prior* density function $g(x, \theta)$, depending on θ, which, in each of the intervals $(x_0, x_1), (x_1, x_2) ..., (x_n, x_{n+1})$ is as close to $f(x, \theta)$ as possible, subject to its enclosing an area $1/(n+1)$ in each interval.

In the second stage, we find θ in such a manner that $f(x, \theta)$ is as close to g (x, θ) as possible.

He concluded that we should choose θ so as to maximize Burg's [1] measure of entropy,

$$I_B = \sum_{i=0}^{n} \ln p_i, \tag{2}$$

where,

$$p_i = \int_{x_i}^{x_{i+1}} f(x,\theta)\, dx \tag{3}$$

We show in the present paper that, the principle can as well be implemented by maximizing Shannon's [9] measure of entropy,

$$I_s = -\sum_{i=0}^{n} p_i \ln p_i \tag{4}$$

In fact it can be implemented by maximizing any other measure of entropy of the distribution $(p_1, p_2, ..., p_n)$.

We also compare the principle with the principle of maximum likelihood of Fisher [3].

33.2 THE PRINCIPLE OF MINIMUM INFORMATION: FIRST STAGE

Minimizing Kullback-Leibler [7] measure,

$$\int_{x_i}^{x_{i+1}} g(x,\theta) \ln \frac{g(x,\theta)}{f(x,\theta)} dx \tag{5}$$

subject to,

$$\int_{x_i}^{x_{i+1}} g(x,\theta)\, dx = \frac{1}{n+1} \tag{6}$$

we get,

$$1 + \ln \frac{g(x,\theta)}{f(x,\theta)} = \lambda_i$$

or,

$$g(x,\theta) = k_i f(x,\theta), \tag{7}$$

where, k_i is determined by using (6) so that

$$k_i \int_{x_i}^{x_{i+1}} f(x,\theta)\, dx = \frac{1}{n+1} \tag{8}$$

Using (3),

$$k_i = \frac{1}{(n+1)\, p_i} \tag{9}$$

or,

$$\frac{g(x,\theta)}{f(x,\theta)} = \frac{1}{(n+1)\, p_i} \tag{10}$$

This shows that in each sub interval $g(x, \theta)$ is proportional to $f(x, \theta)$; only its magnitude is so adjusted that the area under it is $1 / (n + 1)$ as compared with the area p_i under $f(x, \theta)$ (Figure 33.1)

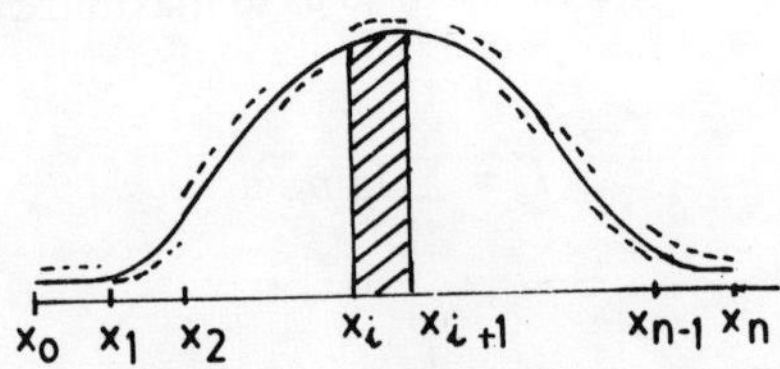

Fig. 33.1

Dotted curve gives $g(x, \theta)$
Full curve gives $f(x, \theta)$

Thus, $g(x, \theta)$ is piece-wise continuous while its cumulative distribution function, $G(x, \theta)$ is continuous.

In the above analysis, we minimized Kullback-Leibler measure $D(g(x, \theta) : f(x, \theta))$, If instead, we minimize $D(f(x, \theta) : g(x, \theta))$ i.e., we minimize

$$\int_{x_i}^{x_{i+1}} f(x,\theta) \ln \frac{f(x,\theta)}{g(x,\theta)} dx \tag{11}$$

subject to (6), we get again

$$\frac{f(x,\theta)}{g(x,\theta)} = k_i \tag{12}$$

which again gives us (10).

Thus, whether we minimize $D(f(x, \theta) : g(x, \theta))$ or we minimize $D(g(x, \theta) : f(x, \theta))$ we get the same minimizing $g(x, \theta)$. This is interesting.

What is even more interesting is that we get the same result by minimizing any valid measure of directed divergence of $f(x, \theta)$ from $g(x, \theta)$ or of $g(x, \theta)$ from $f(x, \theta)$. Let us minimize Csiszer's [2] measure

$$\int_{x_i}^{x_{i+1}} f(x,\theta)\, \phi\left(\frac{g(x,\theta)}{f(x,\theta)}\right) dx, \tag{13}$$

subject to (6). Here $\phi(x)$ is any convex twice-differentiable function of x with $\phi(1) = 0$. Minimizing (13) subject to (4), we get

$$\phi'\left(\frac{g(x,\theta)}{f(x,\theta)}\right) = \text{constant} \tag{14}$$

and, since $\phi'(y)$ is a monotonic increasing function, there is only one value of y corresponding to every value of $\phi'(y)$, we get

$$\frac{g(x,\theta)}{f(x,\theta)} = \mu_i \tag{15}$$

which gives the same result as before. Similarly, minimizing

$$\int_{x_i}^{x_{i+1}} g(x,\theta)\, \phi\left(\frac{f(x,\theta)}{g(x,\theta)}\right) dx, \tag{16}$$

subject to (6), we get

$$\phi\left(\frac{f(x,\theta)}{g(x,\theta)}\right) - \frac{f(x,\theta)}{g(x,\theta)} \quad \phi'\left(\frac{f(x,\theta)}{g(x,\theta)}\right) = \text{constant} \tag{17}$$

Now,
$$\frac{d}{dx}(\phi(\mu) - \mu\,\phi'(\mu)) = -\mu\,\phi''(\mu) < 0 \tag{18}$$

so that corresponding to every value of $\phi(\mu) - \mu\,\phi'(\mu)$, we have a unique value of μ. Thus (17) again gives (15).

The measure (13) includes a wide variety of measures of directed divergence. Thus, whatever be the measure of directed divergence we use and whether we minimize $D(f(x, \theta), g(x, \theta))$ or $D(g(x, \theta) : f(x, \theta))$ we always get (5).

33.3 THE PRINCIPLE OF MINIMUM INFORMATION: SECOND STAGE

In the first stage, θ was supposed to be known and we found $g(x, \theta)$ to be as near to $f(x, \theta)$ as possible in each interval.

In the second stage, the function $g(x, \theta)$ is known, but θ is unknown and we have to choose θ so that $D(f(x, \theta) : g(x, \theta))$ or $D(g(x, \theta) : f(x, \theta))$ is as small as possible.

However, here the value of θ depends on the measure we use.

Since $g(x, \theta)$ is given by (10), if we use Kullback-Leibler [7] measure, we have to choose θ to minimize

$$\sum_{i=0}^{n} \int_{x_i}^{x_{i+1}} f(x,\theta) \ln (n+1)p_i \, dx \quad \text{or} \quad \sum_{i=0}^{n} \int_{x_i}^{x_{i+1}} \frac{f_i(x,\theta)}{(n+1)p_i} \ln \frac{1}{(n+1)} p_i \, dx \tag{19}$$

i.e.,

$$\sum_{i=0}^{n} p_i \ln (n+1) p_i \quad \text{or} \quad \sum_{i=0}^{n} \frac{1}{n+1} \ln \frac{1}{(n+1) p_i} \tag{20}$$

i.e.,

$$\ln (n+1) + \sum_{i=0}^{n} p_i \ln p_i \quad \text{or} - \frac{1}{n+1} \ln (n+1) - \frac{1}{n+1} \sum_{i=0}^{n} \ln p_i \tag{21}$$

so that we have to choose θ to maximize

$$-\sum_{i=0}^{n} p_i \ln p_i \quad \text{or} \quad \sum_{i=0}^{n} \ln p_i \tag{22}$$

The first is Shannon's measure of entropy and the second is Burg's measure of entropy.

Lind has suggested the use of the second measure, but the first measure is equally good.

Thus, using Lind's Principle of Minimum Information along with Kullback-Leibler measure, we get the following Principle:

"Given a random sample $x_1, x_2, ..., x_n$ for a population with density function $f(x, \theta)$, $(x_o \leq x \leq x_{n+1})$, choose θ so as to maximize

$$-\sum_{i=0}^{n} p_i \ln p_i \quad \text{or} \quad \sum_{i=0}^{n} \ln p_i \tag{23}$$

where,

$$p_i = \int_{x_i}^{x_{i+1}} f(x,\theta) \, dx, \quad i = 0,1,2,\ldots,n" \tag{24}$$

There may be uncertainty about having to choose between two functions to be maximized. However, in the second stage both $f(x, \theta)$ and $g(x, \theta)$ are known and we have to choose θ. As such there is no reason to prefer one directed divergence over the other. In fact, there is every reason here to use the symmetric measure of divergence of Kullback [6] and to minimize

$$\ln (n+1) + \sum_{i=0}^{n} p_i \ln p_i - \frac{1}{n+1} \ln (n+1) - \frac{1}{n+1} \sum_{i=0}^{n} \ln p_i \tag{25}$$

so that we may choose θ to maximize

$$-\sum_{i=0}^{n} p_i \ln p_i - \frac{1}{n+1}\sum_{i=0}^{n} \ln p_i \tag{26}$$

This shows that if n is large, $-\sum_{i=0}^{n} p_i \ln p_i$ may be more important than $\sum_{i=0}^{n} \ln p_i$.

If we use (13) as a measure of directed divergence, we have to choose θ to minimize

$$\sum_{i=0}^{n} \int_{x_i}^{x_{i+1}} f(x,\theta)\, \phi\left(\frac{1}{(n+1)\,p_i}\right) dx = \sum_{i=0}^{n} p_i\, \phi\left(\frac{1}{(n+1)\,p_i}\right) \tag{27}$$

and we use (16), we have to choose θ to maximize

$$\sum_{i=0}^{n} \int_{x_i}^{x_{i+1}} \frac{f(x,\theta)}{(n+1)\,p_i}\, \phi\,((n+1)\,p_i)\, dx = \frac{1}{n+1}\sum_{i=0}^{n} \phi\,((n+1)\,p_i) \tag{28}$$

and if we use the symmetric measure of directed divergence, we have to minimize,

$$\sum_{i=0}^{n} p_i\, \phi\left(\frac{1}{(n+1)\,p_i}\right) + \frac{1}{n+1}\sum_{i=0}^{n} \phi\,(n+1)\,p_i \tag{29}$$

If we put $\phi(x) = x \ln x$, we get (22) and (26). If we put $\phi(x) = (x^{\alpha} - x)/(\alpha - 1)$, $\alpha > 0$, $\alpha \neq 1$, we get measures corresponding to Havrda-Charvat [4] measures of directed divergence which includes Kullback-Leibler [7] measure as a limiting case as $\alpha \to 1$.

33.4 COMPARISON WITH PRINCIPLE OF MAXIMUM LIKELIHOOD

The Principle of Maximum Likelihood gives up the first stage altogether and instead takes

$$g(x,\theta) = \frac{1}{n}\,\delta(x - x_1) + \frac{1}{n}\,\delta(x - x_2) + \ldots + \frac{1}{n}\,\delta(x - x_n), \tag{30}$$

where $\delta(x)$ is Dirac's delta functions defined by

$$\int_{-\infty}^{\infty} \delta(x)\, dx = 1, \quad \delta(x) = 0, \quad x \neq 0 \tag{31}$$

It may be noted that $g(x, \theta)$ is independent of $f(x, \theta)$: We minimize

$$\int_{x_0}^{x_{n+1}} g(x,\theta) \ln \frac{g(x,\theta)}{f(x,\theta)}\, dx \tag{32}$$

Since, $g(x, \theta)$ is independent of θ, this is equivalent to minimizing

$$-\int_{x_0}^{x_{n+1}} g(x,\theta) \ln f(x,\theta)\, dx$$

$$= -\frac{1}{n}\int_{x_0}^{x_{n+1}} [\delta(x-x_1)\ln f(x,\theta) + \delta(x-x_2)\ln f(x,\theta)$$

$$+ \ldots + \delta(x-x_n)\ln f(x,\theta)]\,dx$$

$$= -\frac{1}{n}[\ln f(x_1,\theta) + \ln f(x_2,\theta) + \ldots + \ln f(x_n,\theta)]$$

$$= -\frac{1}{n}L(x_1,x_2,\ldots,x_n,\theta), \tag{33}$$

where, $L(x_1, x_2 \ldots, x_n, \theta)$ is the likelihood function. Thus, θ will be obtained by maximizing the likelihood function.

Thus the Principle of Maximum Likelihood requires us to choose θ to maximize

$$L \equiv \ln f(x_1,\theta) + \ln f(x_2,\theta) + \ldots + \ln f(x_n,\theta), \tag{34}$$

while Lind's Principle of Minimum Information requires us to maximize

$$\overline{L} \equiv \ln\int_{x_0}^{x_1} f(x,\theta)\,dx + \ln\int_{x_1}^{x_2} f(x,\theta)\,dx + \ldots + \int_{x_n}^{x_{n+1}} \ln f(x,\theta) \tag{35}$$

$\overline{L}$ may be called the modified likelihood function.

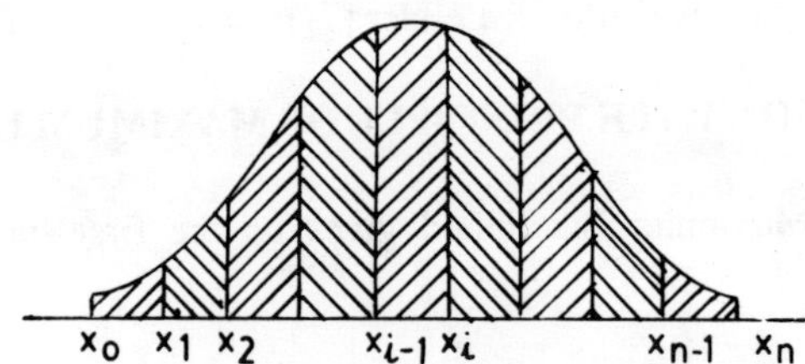

Fig. 33.2

Thus the Principle of Maximum Likelihood requires us to obtain θ so as to maximize the sum of logarithms of ordinates of the density function at $x_1, x_2, \ldots, x_n$.

On the other hand, the principle of modified maximum likelihood requires us to choose θ so as to maximize the sum of the logarithms of the $n + 1$ areas under the density function.

Now let us try to minimize the measure,

$$\int_{x_0}^{x_{n+1}} f(x,\theta)\ln\frac{f(x,\theta)}{g(x,\theta)}\,dx \tag{36}$$

Now this is defined only if $f(x, \theta)$ vanishes whenever $g(x, \theta)$ vanishes, but $g(x, \theta)$ vanishes everywhere except at n points. However we assume $f(x, \theta)$ is continuous for all values of x, for each value of θ. Thus in this case minimization of this measure does not make sense.

Now let us try to minimize the measure,

$$I = \int_{x_0}^{x_{n+1}} g(x,\theta) \ln \frac{g(x,\theta)}{f(x,\theta)} dx \tag{37}$$

and let us approximate the sample distribution by the distribution shown in Figure 33.3, so that

$$g(x,\theta) = \frac{1}{2na}, x_i - a \le x \le x_i + a, \quad i = 1,2,\ldots,n$$

$$= 0, \text{otherwise.} \tag{38}$$

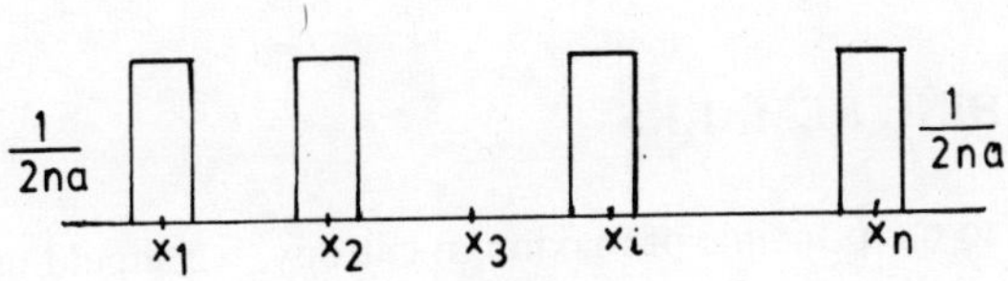

Fig. 33.3

For the function (38), (37) gives

$$I \equiv \sum_{i=1}^{n} \frac{1}{2na} \ln \frac{2a}{2na\, f(x_i,\theta)} \tag{39}$$

$$= \sum_{i=1}^{n} \frac{1}{n} \ln \frac{1}{2na} - \sum_{i=1}^{n} \frac{1}{n} \ln f(x_i,\theta) \tag{40}$$

The first sum does not contain θ and therefore for all values of a, however small these may be, minimizing I implies maximing Fisher's Likelihood function.

Thus, Fisher's maximum likelihood principle can be deduced by minimizing the directed divergence of the observed sample distribution from the given distribution (and not the other directed divergence).

Linds formulation has an edge over the *ML* formulation in the sense that while his prior distribution is piece-wise continuous, Fisher's *priori* is the sum of a number of Dirac delta functions. In other words while the cumulative distribution for Lind's *priori* is continuous, it is only piece-wise continuous in Fisher's case. Also, while in Lind's case we can take directed divergence either way, in Fisher's case, we could have taken it one way only.

However, Fisher's maximum likelihood method has two great advantages, (i) it is computationally simpler, and (ii) Its properties of consistency, efficiency and sufficiency can be easily proved.

To illustrate the computational aspect, consider the problem of estimating the mean of exponential distribution by both methods.

Now,

$$f(x) = a\, e^{-ax}, \quad f(x_1)(x_2)\ldots f(x_n) = a^n\, e^{-na\bar{x}} \tag{41}$$

$$\frac{\partial}{\partial a}(L) = \frac{n}{a} - n\,\bar{x} \quad \text{gives} \quad \hat{a} = \frac{1}{\bar{x}} \tag{42}$$

On the other hand the modified Likelihood function

$$\overline{L} = \sum_{i=0}^{n} \ln \int_{x_i}^{x_{i+1}} a\, e^{-ax}\, dx$$

$$= \sum_{i=0}^{n} \ln \left(e^{-ax_i} - e^{-ax_{i+1}} \right)$$

$$\frac{\partial \overline{L}}{\partial a} = \sum_{i=0}^{n} \frac{-x_i\, e^{-ax_i} + x_{i+1}\, e^{-ax_{i+1}}}{e^{-ax_i} - e^{-ax_{i+1}}} = 0 \tag{43}$$

From (43), we can obtain $\hat{a}$ as a function of $x_1, x_2 \ldots, x_n$, but the calculation is not going to be easy.

33.5 CONCLUDING REMARKS

(i) According to the principle of maximum entropy, we should use all the information that we have and scrupulously avoid using any information which we do not have. Thus to derive the *a priori* probability distribution, we have two pieces of information, (a) the probability in each of the $(n + 1)$ intervals is $1 / (n + 1)$, (b) the probability density of the *a posterior* distribution is $f(x, \theta)$. In order to derive the *prior* distribution. Fisher's *prior* does not make use of the second information, while Linds *prior* does.

(ii) After we have got the *a priori* distribution, the principle of minimum cross-entropy requires that we minimize the cross-entropy with respect to the *a prior* distribution subject to whatever constraints are given. Both methods use this principle.

(iii) We can use any method of non-parametric density estimation based on the basis of the random sample to estimate the density function $g(x)$ and then try to minimize,

$$\int_{x_0}^{x_n} g(x) \ln \frac{g(x)}{f(x, \theta)}\, dx$$

or maximize,

$$\int_{x_0}^{x_n} g(x) \ln f(x, \theta)\, dx$$

to get $\hat{\theta}$. In particular we can consider the maximum entropy density estimation method of Theil [10].

REFERENCES

1. J.P. Burg (1972), "The Relationship between Maximum Entropy Spectra and Maximum Likelihood Spectra," in Modern Spectral Analysis, ed. D.G. Childers, pp. 130-131.
2. I. Csiszer (1972), "A class of measures of informativity of observation channels," Periodic Math. Hungarica, Vol. 22, pp. 191-213.

3. R.A. Fisher (1921), "On the Mathematical Foundations of Theoretical Statistics," Phil. Trans. Ray. Soc, Vol. 222 (A), pp. 309-368.
4. J.H. Havrda and F. Charvat (1967), "Quantification methods of classification process : Concepts of Structural α Entropy, Kybernetica, Vol 3, pp. 30-35.
5. J.N. Kapur (1990), Maximum Entropy Models in Science and Engineering, John Wiley & Sons, New York.
6. S. Kullback (1959), Information Theory and Statistics, John Wiley, New York.
7. S. Kullback and R.A. Leibler (1951), "On Information and Sufficiency," Ann. Math. Stat., Vol. 22, pp. 79-86.
8. N. Lind (1989), "Principle of Minimum Information," I R A Report No. 11, University of Waterloo.
9. C.E. Shannon (1948), "A Mathematical Theory of Communication," Bell System Tech. J., Vol. 27, pp. 379-423, 623-659.
10 H. Theil and D.G. Fiebig, (1984), "Exploiting Continuity : Maximum Entropy Estimation of Continuous Distributions," Ballinger, Cambridge.

34

MAXIMUM-ENTROPY AND MINIMUM CROSS-ENTROPY PROBABILITY DISTRIBUTIONS WHEN THERE ARE INEQUALITY CONSTRAINTS ON PROBABILITIES

[Events-dependent entropy and cross-entropy measures are defined and are used to solve a number of entropy maximization and cross-entropy minimization problems involving inequality constraints on probabilities.]

34.1 INTRODUCTION

Let it be required to find the most unbiased, most objective and most uncertain probability distribution satisfying the natural constraints,

$$\sum_{i=1}^{n} p_i = 1, p_1 \geq 0, \ldots, p_n \geq 0, \tag{1}$$

the moment constraints,

$$\sum_{i=1}^{n} p_i \, g_r(x_i) = a_r, \quad r = 1, 2, \ldots, m \tag{2}$$

and the inequality constraints,

$$a_i \leq p_i \leq b_i, a_i \geq 0, b_i \leq 1, i = 1, 2, \ldots, n \tag{3}$$

Of course if the inequality constraints (3) are satisfied, then the inequality constraints in (1) will automatically be satisfied.

The normal method of solving this problem is to maximize Shannon's [6] measure of entropy

$$S(P) = -\sum_{i=1}^{n} p_i \ln p_i \tag{4}$$

subject to (1), (2) and (3).

Freund and Saxena [2] solved this problem in the absence to moment constraints and gave an algorithm for obtaining the *MAX ENT* distribution. Our problem is more complicated than their's and should require a more complicated algorithm.

A heuristic algorithm for solving this problem can be the following :

Maximize S subject to constraints (1) and (2) only. If all the probabilities satisfy the constraints (3), our problem is solved, otherwise let probabilities $p_{k_1}, p_{k_2}, \ldots, p_{k_m}$ be the probabilities which violate these constraints.

If $p_{kj} < a_j$, put $p_{kj} = a_j$ and if $p_{kj} > b_j$, put $p_{kj} = b_j$, then maximize,

$$-\Sigma' p_i \ln p_i, \tag{5}$$

subject to,

$$\Sigma' p_i = 1 - p_{k_1} - p_{k_2} - \ldots - p_{k_m}$$

$$\Sigma' p_i\, g_r(x_i) = a_r - \sum_{j=1}^{m} p_{kj}\, g_r(x_{kj}), \quad r = 1, 2, \ldots, m, \tag{6}$$

where Σ' denotes summation over all values of i except $k_1, k_2, \ldots, k_m$ If these satisfy the constraints, we get the solution, otherwise we repeat the process again.

This algorithm does not however guarantee that the solution obtained would be optimal.

We therefore discuss whether Shannon's measure is the appropriate measure to be used in this case. In fact, we shall propose another measure of entropy which appears to be more appropriate and which enables us to solve our problem.

We shall also propose corresponding measure of directed divergence or cross-entropy and generalise all these measures.

34.2 AN ALTERNATIVE MEASURE OF ENTROPY

Shannon's measure of entropy satisfies the axioms of continuity, symmetry, expansionability, recursivity, additivity, subadditivity and being maximum when all the probabilities are equal. However, its main advantage in entropy maximization problems over other proposed measures of entropy is not so much due to these axioms which are used to derive it, but due to the three properties which it has, though it was not primarily designed to have these properties. These properties are :

(*i*) it is strictly a concave function of $p_1, p_2, \ldots, p_n$, so that when it is maximized subject to linear moment constraints, its local maximum is its global maximum.

(*ii*) When it is maximized subject to natural constraint (1) and linear moment constraints in (2), by using Lagrange's method, its gives

$$p_i = \exp(-\lambda_0 - \lambda_1 g_1(x_i) - \lambda_2 g_2(x_i) - \ldots - \lambda_m g_m(x_i)), \tag{7}$$

so that the resulting probabilities are automatically positive and inequality constraints in (1) are automatically satisfied.

This is a very important property since it enables us not to worry about the inequality constraints in (1). In most non-linear optimization problems (and our problem belongs to this category), these inequality constraints have to be explicitly imposed and considered.

(*iii*) The maximum value of entropy $S_{\max}$ comes out to be a concave function of $a_1, a_2, \ldots, a_m$ so that in this case, inequality constraints of the type,

$$c_r \le \sum_{i=1}^{n} p_i\, g_r(x_i) \le d_r, \quad r = 1, 2, \ldots, m \tag{8}$$

can also be handled relatively easily.

In our problem, we have one natural constraint, m linear equality constraints and n inequality constraints, so that the number of constraints exceeds the number of unknowns.

The only other measure which has these three properties is Kapur's [4] measure of entropy, viz.

$$K_1(P) = -\sum_{i=1}^{n} p_i \ln p_i + \frac{1}{a}\sum_{i=1}^{n}(1+ap_i)\ln(1+ap_i) - \frac{1}{a}(1+a)\ln(1+a), -1 \le a \quad (9)$$

but its computational consequences are a little more cumbersome than those of Shannon's measure, though it has greater flexibility in applications because of the presence of the parameter a in it.

In our problem, the advantage *(ii)* above is not relevant since we are not interested in the constraints,

$$0 \le p_1 \le 1, 0 \le p_2 \le 1, \ldots, 0 \le p_n \le 1 \quad (10)$$

We are rather interested in the more restrictive constraints,

$$a_1 \le p_1 \le b_1, a_2 \le p_2 \le b_2, \ldots, a_n \le p_n \le b_n \quad (11)$$

If constraints (11) are satisfied, constraints (10) will automatically be satisfied. However, if constraints (10) are satisfied, constraints (11) need not be satisfied.

Just as Shannon's measure automatically satisfies the constraints (10), we now look for a measure which automatically satisfies the constraints (11) and which has as many of the properties of Shannon's measure as possible.

One such measure is given by

$$K_2(P) = -\sum_{i=1}^{n}(p_i - a_i)\ln(p_i - a_i) - \sum_{i=1}^{n}(b_i - p_i)\ln(b_i - p_i)$$

$$+\sum_{i=1}^{n}(b_i - a_i)\ln(b_i - a_i) \quad (12)$$

34.3 SOME PROPERTIES OF THE NEW MEASURE

(i) If we define $0 \ln 0 = 0$, $K_2(P)$ is a continuous function of $p_1, p_2, \ldots, p_n$ over the range defined by (11) and this is the range we are interested in.

(ii) It is permutationally symmetric in the sense that if the n triplets (a_i, b_i, pi), $(i = 1, 2, \ldots, n)$ are permuted among themselves, the measure does not change.

(iii) It is strictly a concave function of $p_1, p_2, \ldots, p_n$, since

$$\frac{\partial K_2}{\partial p_i} = -(1 + \ln(p_i - a_i)) + (1 + \ln(b_i - p_i)) = \ln(b_i - p_i) - \ln(p_i - a_i) \quad (13)$$

$$\frac{\partial^2 K_2}{\partial p_i^2} = -\frac{1}{b_i - p_i} - \frac{1}{p_i - a_i} < 0, \quad \frac{\partial^2 K_2}{\partial p_i \partial p_j} = 0, \quad i \ne j \quad (14)$$

This means that when it is maximized subject to linear constraints in (2), its local maximum will also be its global maximum.

(iv) Maximizing $K_2(P)$, subject to the natural constraint (1), we get,

$$\frac{b_i - p_i}{p_i - a_i} = k, \quad p_i = \frac{b_i + a_i k}{1+k}, \quad i = 1, 2, \ldots, n, \quad (14a)$$

where k is determined from the equation

$$\sum_{i=1}^{n} b_i + k \sum_{i=1}^{n} a_i = 1 + k,$$

so that,

$$k = \left(\sum_{i=1}^{n} b_i - 1 \right) / \left(1 - \sum_{i=1}^{n} a_i \right) \tag{15}$$

and,

$$p_k = \frac{b_i - a_i - b_i \sum_{i=1}^{n} a_i + a_i \sum_{i=1}^{n} b_i}{\sum_{i=1}^{n} b_i - \sum_{i=1}^{n} a_i}, \quad i = 1, \ldots, n \tag{16}$$

As a special case of each $a_i = 0$ and each $b_i = 1$, this gives

$$k = (n-1), \quad p_i = \frac{1}{n}, \quad i = 1, 2, \ldots, n, \tag{17}$$

so that in the absence of any constraints, except the natural constraint, the maximum entropy probability distribution would be the uniform distribution.

However, because of the constraints (11), the probability distribution given by (16) will not be uniform except when

$$a_1 = a_2 = \ldots = a_n = a, b_1 = b_2 \ldots = b_n = b, a \geq 0, b \leq 1$$

$$a \leq \frac{1}{n} \leq b \tag{18}$$

This does not contradict Laplace's principle of insufficient reason since that principle states that in the absence of any information, all probabilities should be equal. In our case we have the information given by the constraints (11).

In fact, we can modify Laplace's principle to the statement, ''If we have no other information about the probability distribution except that $\sum_{i=1}^{n} p_i = 1$ and $a \leq p_i \leq b$ where $a \geq 0$, $b \leq 1$, then all the probabilities should be taken as equal''.

We can even give a more general form ;

''If all the information about $p_1, p_2, \ldots, p_n$ is permutationally symmetric in $p_1, p_2, \ldots, p_n$, we should choose

$$p_1 = p_2 = \ldots = p_n = \frac{1}{n} \tag{19}$$

(v) Now the maximum value of $K_2(P)$ is

$$-\sum_{i=1}^{n} \frac{b_i - a_i}{1+k} \ln \frac{b_i - a_i}{1+k} - \sum_{i=1}^{n} \frac{(b_i - a_i)k}{1+k} \ln \frac{(b_i - a_i)k}{1+k}$$

$$+ \sum_{i=1}^{n} (b_i - a_i) \ln (b_i - a_i)$$

$$= -\frac{1}{1+k}\left[\sum_{i=1}^{n} (b_i - a_i) \ln (b_i - a_i) - (B - A) \ln (1 + k)\right]$$

$$-\frac{k}{1+k}\left[\sum_{i=1}^{n} (b_i - a_i) \ln (b_i - a_i) - (B - A) \ln \frac{1+k}{k}\right] + \sum_{i=1}^{n} (b_i - a_i) \ln (b_i - a_i)$$

$$= (B - A)\left[-\frac{1}{1+k} \ln \frac{1}{1-k} - \frac{k}{1+k} \ln \frac{k}{1+k}\right] \tag{20}$$

where,

$$A = \sum_{i=1}^{n} a_i, \quad B = \sum_{i=1}^{n} b_i, \quad k = \frac{B-1}{1-A} \tag{21}$$

$$[K_2 (P)]_{\max} = (B - A) \ln (B - A) - (1 - A) \ln (1 - A) - (B - 1) \ln (B - 1) \tag{22}$$

$$\frac{d}{dB} [K_2(P)]_{\max} = \ln \frac{B-A}{B-1}, \quad \frac{d}{dA} [K_2(P)]_{\max} = \ln \frac{1-A}{B-A} \tag{23}$$

Since $A < 1 < B$, the maximum entropy increases with B and decreases with A.

(vi) Since $K_2 (P)$ is a concave function, its minimum value occurs at a vertex of the convex region determined by (1) and (11). However, since $0 \le a_i \le b_i \le 1$, each term of $K_2 (P) \ge 0$ and as such the maximum is always ≥ 0. If the measure does not attain the value zero, we can always subtract a suitable positive number from it to make its minimum value zero.

(vii) The measure is neither additive, nor sub-additive nor recursive, but these properties do not matter in our maximization problem.

(viii) If each $a_i = 0$, $b_i = 1$, $K_2 (P)$ reduces to Fermi-Dirac [3] entropy,

$$-\sum_{i=1}^{n} p_i \ln p_i - \sum_{i=1}^{n} (1 - p_i) \ln (1 - p_i) \tag{24}$$

34.4 MEASURES OF INSET ENTROPY

It has been very often argued that measure of entropy should depend not only on probabilities $p_1, p_2, ..., p_n$ of outcomes, but these should also depend on values $x_1, x_2, ..., x_n$ of outcomes. Aczel and his coworkers [1] tried to develop a set of functions of $(p_1, p_2, ..., p_n ; x_1, x_2, ..., x_n)$ satisfying certain properties of recursivity, additivity etc. In most cases they got measures of the form

$$-\sum_{i=1}^{n} p_i \ln p_i + \sum_{i=1}^{n} p_i \phi (x_i), \tag{25}$$

which are of the form of Lagrangian used in maximizing Shannon's measure of entropy subject to linear constraints. As such these were called Lagrangian entropy measures by Kapur [5]. They do not lead to complete integration of probabilities and events.

Our measure $K_2 (P)$ depends not only on probabilities $p_1, p_2, ..., p_n$, but also depends to some extent on the outcomes. As such it is an events-dependent measure and this measure has been developed not from axiomatic considerations, but from the needs of a certain situation.

34.5 AN APPLICATION TO DISTRIBUTION OF POPULATION IN A CITY

Let proportions $p_1, p_2, ..., p_n$ of population of a city live in n colonies and let the costs of travel from these colonies to the central business district (CBD) be $c_1, c_2, ..., c_n$, respectively. The expected cost of travel is prescribed, so that we have the constraints

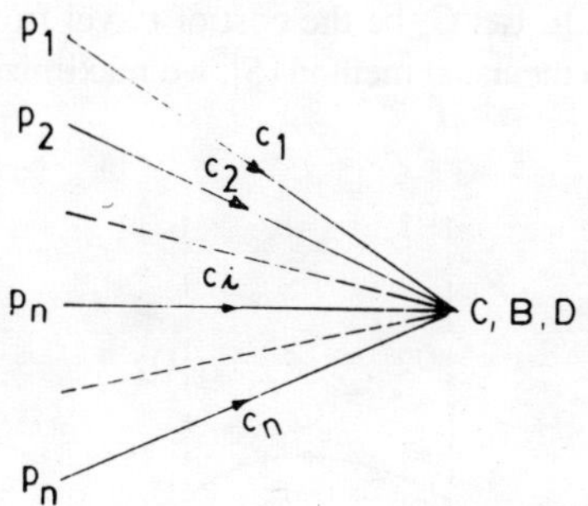

Fig. 34.1

$$\sum_{i=1}^{n} p_i = 1, \sum_{i=1}^{n} p_i c_i = \hat{c}, p_1 > 0, p_2 \geq 0 \ldots p_n \geq 0 \tag{26}$$

We want the population to be as much spreadout as possible and for this purpose, we maximize $-\sum_{i=1}^{n} p_i \ln p_i$ subject to (26) to get the Maxwell-Boltzmann distribution [5]

$$p_i = e^{-\mu c_i} / \sum_{i=1}^{n} e^{-\mu c_i}, \sum_{i=1}^{n} c_i e^{-\mu c_i} / \sum_{i=1}^{n} e^{-\mu c_i} = \hat{c} \tag{27}$$

However, this may lead to very low populations in colonies distant from the CBD and very large populations in colonies near the CBD. This may mean under-utilization of the capacity of some colonies and overcrowding in some other colonies. We may therefore prescribe some minimum and maximum proportions a_i and b_i for the ith colony and impose the additional constraints

$$a_i \leq p_i \leq b_i, i = 1, 2, ..., n \tag{28}$$

Maximizing our measure $K_2(P)$ subject to (26), we get

$$\frac{b_i - p_i}{p_i - a_i} = e^{\lambda + \mu c_i} = x_i \quad \text{(say)} \tag{29}$$

or,

$$p_i = \frac{b_i + a_i x_i}{x_i + 1}, \; p_i - a_i = \frac{b_i - a_i}{1 + x_i} > 0, \; b_i - p_i = \frac{(b_i - a_i) x_i}{1 + x_i} > 0 \tag{30}$$

This obviously satisfies the constraints (28) where now λ and μ can be determined to satisfy the constraints,

$$\sum_{i=1}^{n} \frac{b_i + a_i x_i}{x_i + 1} = 1, \quad \sum_{i=1}^{n} \frac{b_i + a_i x_i}{x_i + 1} c_i = \hat{c} \tag{31}$$

34.6 APPLICATION TO TRANSPORTATION PROBLEM

Let T_{ij} be the number of trips from the ith residential colony with population O_i to the jth office with D_j jobs in it. Let C_{ij} be the cost of travel from the ith colony to the jth office, then according to the usual method [5], we maximize

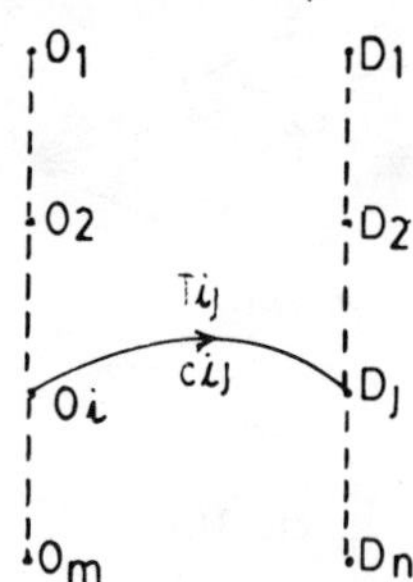

Fig. 34.2

$$-\sum_{j=1}^{n} \sum_{i=1}^{m} \frac{T_{ij}}{T} \ln \frac{T_{ij}}{T}, T = \sum_{j=1}^{n} \sum_{i=1}^{m} T_{ij} \tag{32}$$

subject to,

$$\sum_{i=1}^{m} T_{ij} = D_j, \quad \sum_{j=1}^{n} T_{ij} = 0_i, \quad \sum_{j=1}^{n} \sum_{i=1}^{m} T_{ij} C_{ij} = \overline{C} \tag{33}$$

to get,

$$T_{ij} = A_i \quad B_i \quad O_i \quad D_o \exp(-\nu C_{ij}) \tag{34}$$

These trips may however lead to under-utilization of some routes and traffic conjestions on other routes. We may therefore impose additional constraints,

$$a_{ij} \le T_{ij} \le b_{ij},$$

depending on the capacities of the routes and then maximize

$$-\sum_{j=1}^{n} \sum_{i=1}^{m} \frac{T_{ij} - a_{ij}}{T - A} \ln \frac{T_{ij} - a_{ij}}{T - A} - \sum_{j=1}^{n} \sum_{i=1}^{m} \frac{b_{ij} - T_{ij}}{B - T} \ln \frac{b_{ij} - T_{ij}}{B - T}, \tag{35}$$

subject to (32) and (33).

In practice, this constraint is automatically imposed since when the traffic on a route reaches saturation, the traffic on this route is diverted to other routes. Sometimes this information is given on radio and it leads to greater spreadout of traffic on the roads.

34.7 A NEW MEASURE OF DIRECTED DIVERGENCE

The measure of directed divergence corresponding to $K_2(P)$ is

$$K_2(P:Q) = \sum_{i=1}^{n} (p_i - a_i) \ln \frac{p_i - a_i}{q_i - a_i} + \sum_{i=1}^{n} (b_i - p_i) \ln \frac{b_i - p_i}{b_i - q_i} \tag{36}$$

where Q is an a *priori* probability distribution which satisfies the constraints (11). $K_2(P:Q)$ is a convex function of $p_1, p_2, \ldots, p_n$, as well as of $q_1, q_2, \ldots, q_n$ and attains its minimum value zero when $P = Q$.

When $Q = \overline{Q}$ is given by (16)

$$q_i = \frac{b_i - a_i - b_i A + a_i B}{B - A}, \quad i = 1, \ldots, n \tag{37}$$

we get,

$$K_2(P:\overline{Q}) = \sum_{i=1}^{n} (p_i - a_i) \ln \frac{p_i - a_i}{\frac{(b_i - a_i)(1-A)}{B-A}} + \sum_{i=1}^{n} (b_i - p_i) \ln \frac{b_i - p_i}{\frac{(b_i - a_i)(B-1)}{B-A}}$$

$$= \sum_{i=1}^{n} (p_i - a_i) \ln (p_i - a_i) + \sum_{i=1}^{n} (b_i - p_i) \ln (b_i - p_i)$$

$$- \sum_{i=1}^{n} (b_i - a_i) \ln (b_i - a_i) + (1 - A) \ln \frac{B - A}{1 - A} + (B - A) \ln \frac{B - A}{B - 1}$$

$$= K_2(\overline{Q}) - K_2(P) \tag{38}$$

Thus, $\overline{Q}$ given by (16) may be regarded as the 'uniform' distribution corresponding to our new measure, if we define the 'uniform' distribution for any measure as that distribution which maximizes this measure when there are no constraints except the natural constraints. In that case our measure of entropy of any distribution is a monotonic decreasing function of the measure of directed divergence of P from this 'uniform' distribution.

34.8 GENERALISED MEASURE OF ENTROPY AND DIRECTED DIVERGENCE

Let $\phi(x)$ be a convex twice-differentiable function for which $\phi(1) = 0$, then we define generalised measures to be used when the constraints (11) are operative, by

$$K_{2\phi}(P) = -\sum_{i=1}^{n} \phi(p_i - a_i) - \sum_{i=1}^{n} \phi(b_i - p_i) + \text{constant} \tag{39}$$

$$K_{2\phi}(P:Q) = \sum_{i=1}^{n} (q_i - a_i)\, \phi\left(\frac{p_i - a_i}{q_i - a_i}\right) + \sum_{i=1}^{n} (b_i - q_i)\, \phi\left(\frac{b_i - p_i}{b_i - q_i}\right) \tag{40}$$

The 'uniform' distribution will be defined by

$$-\phi'(p_i - a_i) + \phi'(b_i - p_i) = k, \tag{41}$$

where, k is determined by using (1). In this case $K_{2\phi}(P)$ will be a monotonic decreasing function of the directed divergence of P from the uniform distribution defined here. If $\phi(x) = x \ln x$, we get the measures given by (12) and (36). We can get other useful measures by taking

$$\phi(x) = \frac{x^\alpha - x}{\alpha - 1}, \quad \alpha \neq 1, \quad \alpha > 0 \tag{42}$$

$$\phi(x) = x \ln x - \frac{1}{a}(1+ax)\ln(1+ax),$$

$$+\frac{x}{a}(1+a)\ln(1+a), -1 < a < 1 \tag{43}$$

These give the measures,

$$K_3(P:Q) = \frac{1}{\alpha-1}\left[\sum_{i=1}^{n}(p_i - a_i)^\alpha (q_i - a_i)^{1-\alpha} - 1 + \sum_{i=1}^{n}(b_i - p_i)^\alpha (b_i - q_i)^{1-\alpha} - 1\right], \alpha \neq 1 \tag{44}$$

$$K_3(P) = \frac{1}{1-\alpha}\left[\sum_{i=1}^{n}(p_i - a_i)^\alpha - 1 + \sum_{i=1}^{n}(b_i - p_i)^\alpha - 1\right], \alpha \neq 1 \tag{45}$$

$$K_4(P) = -\sum_{i=1}^{n}(p_i - a_i)\ln(p_i - a_i) + \frac{1}{a}\sum_{i=1}^{n}(1 + a(p_i - a_i))\ln(1 + a(p_i - a_i))$$

$$-\sum_{i=1}^{n}(b_i - p_i)\ln(b_i - p_i) + \frac{1}{a}\sum_{i=1}^{n}(1 + a(b_i - p_i))\ln(1 + a(b_i - p_i))$$

$$-\left(\sum_{i=1}^{n}(b_i - a_i)\right)\frac{1}{a}(1+a)\ln(1+a) \tag{46}$$

$$K_4(P:Q) = \sum_{i=1}^{n}(p_i - a_i)\ln\frac{p_i - a_i}{q_i - a_i} + \sum_{i=1}^{n}(b_i - p_i)\ln\frac{b_i - p_i}{b_i - q_i}$$

$$-\frac{1}{a}[1 + a(p_i - a_i)]\ln\frac{(1 + a(p_i - a_i))}{(1 + a(q_i - a_i)}$$

$$+\sum_{i=1}^{n}[1 + a(b_i - p_i)]\ln\frac{1 + a(b_i - p_i)}{1 + a(b_i - q_i)} \tag{47}$$

34.9 APPLICATIONS OF NEW MEASURES OF ENTROPY AND DIRECTED DIVERGENCE

In the city population distribution model, we may have the figures for the proportions in n colonies in the previous years. Now the costs change suddenly. The population is to be redistributed but the restrictions on proportions of populations which existed in earlier years still exist. In this case, we should use the principle of minimum cross-entropy along with new measures of cross-entropy.

Similarly, in the transportation model, there may be a committment of each power station to supply electric power between certain minimum and maximum number of units to different consumers so that each T_{ij} has to be between specified limits.

Again, suppose a man has a number N of salesmen to be assigned to m stores. Each store requires a minimum and maximum number of salesmen. The incomes per salesman are $c_1, c_2, ..., c_m$ from the different stores. In this case the new measures may be useful.

Similarly, in other allocation problems where there are upper and lower bounds for each individual allocation, these measures, may be useful. Thus in allotment of children to schools, patients to hospitals, resources to different objects, electricity to different groups users etc, we do want maximum spreadout, but we do also want certain minimum and maximum targets to be met in each case.

REFERENCES

1. J.Aczel and Z. Daroczy (1978), ''A Mixed Theory of Information I'' Theoretical Computer Science, 12 (2), 149-158.
2. D. Freund and U. Saxena (1984), ''An Algorithm for a Class of Discrete Maximum Entropy Problems'' Operations Research, 32 (1), 210-215.
3. J.N. Kapur (1972), ''Measures of Uncertainty, Mathematical Programming and Physics'', Jour. Ind. Soc. Ag. Stat. 24, 47-66.
4. J.N. Kapur (1986), ''Four Families of Measures of Entropy'' Ind. Jour. Pure Appl. Math. 17 (4), 429-449.
5. J.N. Kapur (1989), Maximum Entropy Models in Science and Engineering, Wiley Eastern, New Delhi.
6. C.E. Shannon (1948), ''A Mathematical Theory of Communication'' Bell. Syst. Tech. Jour. 27, 379-425, 623-659.

35

MAXIMUM ENTROPY PROBABILITY DISTRIBUTIONS WHEN THERE ARE INEQUALITY CONSTRAINTS ON PROBABILITIES

[When there are inequality constraints of the type $0 \leq a_i \leq p_i \leq b_i \leq 1$ on probabilities, appropriate generalised measures of entropy were developed in earlier papers. It is shown here that if there are no additional constraints except the natural constraints $\sum_{i=1}^{n} p_i = 1$, then the maximum value S_{max} of entropy is a concave function of $a_1, a_2, ..., a_n, b_1, b_2, ..., b_n$. We also find the globally maximum value of entropy. If there is one additional constraint given by prescribing the value of the mean as m, when the random variate takes the values 1, 2, 3, ..., n, it is shown that even in this case, probabilities will increase or decrease with i, according as $m \gtrless (n+1)/2$].

35.1 INTRODUCTION

Let the probabilities $p_1, p_2, ..., p_n$ of a probability distribution $P = (p_1, p_2, ..., p_n)$ be subject to the inequality constraints

$$0 \leq a_i \leq p_i \leq b_i \leq 1, \quad i = 1, 2, ..., n \tag{1}$$

These are also, of course, subject to the natural constraint

$$\sum_{i=1}^{n} p_i = 1 \tag{2}$$

There may be an infinity of probability distributions satisfying constraints (1) and (2), when

$$A = \sum_{i=1}^{n} a_i \leq 1, \quad B = \sum_{i=1}^{n} b_i \geq 1 \tag{3}$$

Which of these should we choose?

If $a_i = 0$, $b_i = 1$ for each i, then Laplace had given the answer through his principle of insufficient reason that we should choose the uniform distribution,

$$v = \left(\frac{1}{n}, \frac{1}{n}, ..., \frac{1}{n}\right) \tag{4}$$

Here each p_i divides the interval [0, 1] in the same ratio $1 : n-1$ [Fig. 35.1].

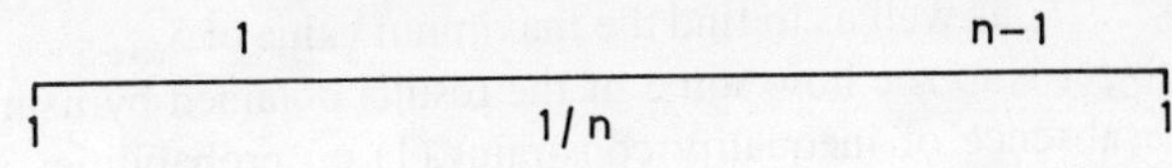

Fig. 35.1

Using this motivation, we [1] suggested that in the more general case, we should choose p_i's so that each p_i divides the interval (a_i, b_i) in the same ratio $K : 1$, i.e. we choose

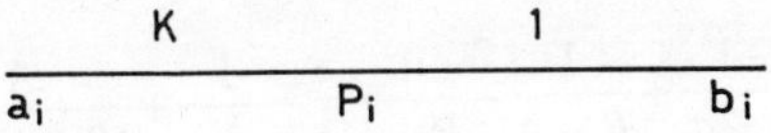

K 1

a_i P_i b_i

Fig. 35.2

$$p_i = \frac{K\, b_i + a_i}{K+1}, \quad i = 1, 2, \ldots, n \tag{5}$$

Now, $$\sum_{i=1}^{m} p_i = 1 \Rightarrow KB + A = K + 1 \Rightarrow K = \frac{1-A}{B-1} \tag{6}$$

We called this distribution given by (5) and (6) as the generalised uniform distribution since it includes the usual uniform distribution (4) as a special case when each $a_i = 0$ and each $b_i = 1$ or when each $a_i = a < 1/n$ and each $b_i = b > 1/n$.

We next used the generalised maximum entropy principle [3] to find the entropy measure $S(P)$ whose maximization subject to (2) leads to the distribution (5). We get

$$S(P) = -\sum_{i=1}^{n} (p_i - a_i) \ln (p_i - a_i) - \sum_{i=1}^{n} (b_i - p_i) \ln (b_i - p_i) \tag{7}$$

This entropy measure has its maximum value when P is given by (5), so that

$$p_i - a_i = \frac{K}{K+1}(b_i - a_i) \quad b_i - p_i = \frac{1}{K+1}(b_i - a_i) \tag{8}$$

and,

$$S_{\max} = -\sum_{i=1}^{n} \frac{K}{K+1}(b_i - a_i) \ln \frac{K}{K+1}(b_i - a_i) - \sum_{i=1}^{n} \frac{1}{K+1}(b_i - a_i) \ln \frac{(b_i - a_i)}{K+1}$$

$$= -\sum_{i=1}^{n} (b_i - a_i) \ln (b_i - a_i) + (B - A) \ln (B - A)$$

$$- (1 - A) \ln (1 - A) - (B - 1) \ln (B - 1) \tag{9}$$

Thus, S_{max} depends on $a_1, a_2, ..., a_n$ and $b_1, b_2, ..., b_n$. (10)

One object of the present paper is to show that S_{max} is a concave function of $a_1, a_2, ..., a_n, b_1, b_2, ..., b_n$ as well as to find the maximum value of S_{max}.

Another object is to see how some of the results obtained by using Shannon's measure in the absence of inequality constraint (1) on probabilities are modified when these inequality constraints are present.

35.2 CONCAVITY OF S_{max}

(*a*) S_{max} *is a concave function of* $b_1, b_2, ..., b_n$

From (9), we easily get

$$\frac{\partial S_{max}}{\partial b_i} = -1 - \ln(b_i - a_i) + \ln(B - A) - \ln(B - 1) \tag{11}$$

$$\frac{\partial^2 S_{max}}{\partial b_i^2} = \frac{-1}{b_i - a_i} + \frac{1}{B - A} - \frac{1}{B - 1} = \frac{-1}{b_i - a_i} + \frac{A - 1}{(B - 1)(B - A)}$$

$$= -u_i - v < 0 \tag{12}$$

$$\frac{\partial^2 S_{max}}{\partial b_i \, \partial b_j} = \frac{1}{B - A} - \frac{1}{B - 1} = -\frac{(1 - A)}{(B - 1)(B - A)} = -v < 0 \tag{13}$$

$$\frac{\partial^2 S_{max}}{\partial b_i^2} \frac{\partial^2 S_{max}}{\partial b_j^2} - \left(\frac{\partial^2 S_{max}}{\partial b_i \, \partial b_j} \right)^2 = (u_i + v)(u_j + v) - v^2$$

$$= u_i u_j + v(u_i + u_j) > 0 \tag{14}$$

Thus,

$$\sum_{j=1}^{n} \sum_{i=1}^{n} \frac{\partial^2 S_{max}}{\partial b_i \, \partial b_j} x_i x_j = -\sum_{i=1}^{n} (u_i + v) x_i^2 - v \sum_{j=1}^{n} \sum_{\substack{i=1 \\ i \neq j}}^{n} x_i x_j$$

$$= -\sum_{i=1}^{n} u_i x_i^2 - v \left(\sum_{i=1}^{n} x_i \right)^2 \tag{15}$$

is a negative definite quadratic form, so that S_{max} is a concave function of $b_1, b_2, ..., b_n$ when $a_1, a_2, ..., a_n$ are kept fixed.

(*b*) S_{max} *is a concave function of* $a_1, a_2, ..., a_n$

$$\frac{\partial S_{max}}{\partial a_j} = 1 + \ln(b_j - a_j) - \ln(B - A) + \ln(1 - A) \tag{16}$$

$$\frac{\partial^2 S_{max}}{\partial a_j^2} = -\frac{1}{b_j - a_j} + \frac{1}{B - A} - \frac{1}{1 - A} = -\frac{1}{b_j - a_j} - \frac{B - 1}{(1 - A)(B - A)}$$

$$= -u_j - w < 0 \tag{17}$$

$$\frac{\partial^2 S_{\max}}{\partial a_j\, \partial a_i} = \frac{1}{B-A} - \frac{1}{1-A} = -\frac{B-1}{(1-A)(B-A)} = -w < 0 \tag{18}$$

$$\sum_{j=1}^{n} \sum_{i=1}^{n} \frac{\partial^2 S_{\max}}{\partial a_i\, \partial a_j} x_i x_j = -\sum_{j=1}^{n} u_j x_j^2 - w\left(\sum_{i=1}^{n} x_j\right)^2 < 0 \tag{19}$$

so that $S_{\max}$ is a concave function of $a_1, a_2, ..., a_n$ when $b_1, b_2, ..., b_n$ are kept fixed

(c) $S_{\max}$ is a concave function of $a_1, a_2, ..., a_n$ and $b_1, b_2, ..., b_n$

$$\frac{\partial^2 s_{\max}}{\partial a_i\, \partial b_j} = \frac{\partial^2 S_{\max}}{\partial a_j\, \partial b_i} = -\frac{1}{B-A} \tag{20}$$

Using (12), (13), (17), (18) and (20) we see that the Hessian matrix of $S_{\max}$ w.r.t. $2n$ variables $a_1, a_2, ..., a_n, b_1, b_2, ..., b_n$ is negative definite, so that $S_{\max}$ is concave function of these $2n$ variables.

35.3 MAXIMUM VALUE OF $S_{\max}$

(a) When $a_1, a_2, ..., a_n$ are held constant

Using (11) and putting $\dfrac{\partial S_{\max}}{\partial b_j} = 0$, we get

$$(b_i - a_i)\, e\, (B-1) = (B-A) \Rightarrow b_i - a_i = \frac{B-A}{e\,(B-1)}$$

$$\Rightarrow B - A = \frac{n\,(B-A)}{e\,(B-1)} \Rightarrow B = 1 + n/e$$

$$\Rightarrow b_i - a_i = \frac{1}{e} + \frac{1-A}{n} \qquad \Rightarrow b_i = a_i + \frac{1}{e} + \frac{1-A}{n}$$

$$\Rightarrow B - A = \frac{n}{e} + 1 - A \tag{21}$$

and the maximum value of $S_{\max}$ is

$$= -\sum_{i=1}^{n}\left(\frac{1}{e} + \frac{1-A}{n}\right)\ln\left(\frac{1}{e} + \frac{1-A}{n}\right) + \left(\frac{n}{e} + 1 - A\right)\ln\left(\frac{n}{e} + 1 - A\right)$$

$$-(1-A)\ln(1-A) - \frac{n}{e}\ln n/e \tag{22}$$

$$= (1-A)\ln n.$$

(b) When $b_1, b_2, ..., b_n$ are kept constant

Proceeding, in the same way, $S_{\max}$ is maximum when

$$b_i - a_i = \frac{1}{e} + \frac{B-1}{n} \tag{23}$$

and the maximum value of $S_{max} = (B-1)\ln n$ (24)

(*c*) *When both* $a_1, a_2, ..., a_n$ *and* $b_1, b_2, ..., b_n$ *vary*

$$b_i - a_i = \frac{B-A}{e(B-1)} = \frac{B-1}{e(1-A)} \tag{25}$$

$$b_i - a_i = \frac{2}{e}, \quad A = 1 - \frac{n}{e}, \quad B = 1 + n/e \tag{26}$$

$$(S_{max})_{max} = -n\frac{2}{e}\ln\frac{2}{e} - \frac{2n}{e}\ln\frac{2n}{e} - \frac{n}{e}\ln\frac{n}{e} - \frac{n}{e}\ln\frac{n}{e} = \frac{2n}{e} \tag{27}$$

(*d*) *Remarks on these maximum values*

(i) If $a_1 = a_2 = ... = a_n = a, \quad b_1, = b_2 = ... = b_n = b,$ (28)

then S_{max} is maximum when

$$a = \frac{1}{n} - \frac{1}{e}, \qquad b = \frac{1}{n} + \frac{1}{e} \tag{29}$$

(ii) The above maximum value may or may not be attained since we require that $a_i \geq 0, b_i \leq 1$. Thus, in (21) b_i will exceed unity if $a_i > 1/e$ and in (23) a_i will become < 0 if $b_i < 1/e$. In such cases the maximum value of S_{max} will be attained on the boundary region which is permissible for a_i's and b_i's. In such cases the maximum value attained would be ≤ theoretical maximum value given here.

(iii)We have two maximization processes. First we keep $a_1, a_2, ..., a_n$; $b_1, b_2, ..., b_n$ fixed and consider all possible probability distributions consistant with (1) and find the maximum value of the entropy for any of these distributions. In the second maximization, we vary $a_1, a_2, ..., a_n, b_1, b_2, ..., b_n$ and find the maximum of the maximum value for various fixed values of these parameters.

(iv) Behaviour of the MEPD when an additional constraint is given :

Now we assume that in addition to inequality constraints (1) and the natural constraint (2), we have also the mean value constraint,

$$\sum_{i=1}^{n} i\, p_i = m \tag{30}$$

Now if inequality constraint (1) are not there, we can use Shannon's [4] measure of entropy. Maximizing it subject to (2) and (30) we get

$$p_i = D\, x^i, \tag{31}$$

where, D and x are determined by using

$$\sum_{i=1}^{n} D\, x^i = 1, \qquad \sum_{i=1}^{n} i\, Dx^i = m \tag{32}$$

It is well-known (Kapur [2]) that in this case the MEPD is given by the geometric distribution (32), and

$$x \gtrless 1 \text{ according as } m \gtrless \frac{n+1}{2} \tag{33}$$

i.e. the probability increases or decreases with i according as the prescribed mean $m \gtrless$ the mean of the usual uniform distribution.

Now we introduce the constraint (1) or rather a special case of it

$$a \leq p_i \leq b, \tag{34}$$

so that the generalised uniform distribution is still the usual uniform distribution and there is no change in the *MEPD* in the absence of the mean constraint.

We want to know how the *MEPD* changes due to the introduction of (34). In particular we want to know whether it remains the geometric distribution and whether the probabilities increase or decrease according as $m \gtrless (n+1)/2$. Maximizing,

$$S = -\sum_{i=1}^{n} (p_i - a)\ln(p_i - a) - \sum_{i=1}^{n} (b - p_i)\ln(b - p_i), \tag{35}$$

subject to (2) and (30) we get

$$\frac{p_i - a}{b - p_i} = E\,x^i \quad \text{or} \quad P_i = \frac{a + b\,Ex^i}{1 + E\,x^i}$$

or
$$p_i = b - \frac{(b-a)}{1+Ex^i}, \tag{36}$$

where E and x are determined by using

$$nb - (b-a)\sum_{i=1}^{n} \frac{1}{1+Ex^i} = 1, \frac{n(n+1)}{2} b - (b-a)\sum_{i=1}^{n} \frac{i}{1+Ex^i} = m \tag{37}$$

These give,

$$\frac{n+1}{2}\left[1 + (b-a)\sum_{i=1}^{n} \frac{1}{1+Ex^i}\right] = 1,\ m + (b-a)\sum_{i=1}^{n} \frac{i}{1+Ex^i} \tag{38}$$

or
$$(b-a)\left(\sum_{i=1}^{n} \frac{i - \frac{n+1}{2}}{1+Ex^i}\right) = \frac{n+1}{2} - m \tag{39}$$

Now from (37) since $p_i > 0, p_i < b, Ex^i > 0$, so that

$$x > 1 \Rightarrow 1 + Ex^i \text{ increase with } i \Rightarrow \frac{1}{1+Ex^i} \text{ decrease with } i$$

$$\Rightarrow \sum_{i=1}^{n} \frac{i}{1+Ex^i} \text{ (weighted average of } 1, 2, \ldots n \text{ with decreasing weights)}$$

$$\leq \sum_{i=1}^{n} \frac{\frac{n+1}{2}}{1+Ex^i} \text{ (weighted average of } 1, 2, \ldots, n \text{ with equal weights)}$$

$$\Rightarrow \text{L.H.S. of (39)} < 0 \tag{40}$$

$$\Rightarrow \text{R.H.S. of (39)} < 0$$

$$\Rightarrow m > \frac{n+1}{2}$$

Using (38) this mean that decreasing probabilities $\Rightarrow m < \dfrac{n+1}{2}$

Similarly, increasing probabilities $\Rightarrow m > \dfrac{n+1}{2}$

and constant probabilities $\Rightarrow m = \dfrac{n+1}{2}$

It is easily seen that the converse is also true, so that when the constraint (35) and the mean constraint (30) is present

$$m > (=, <) \frac{n+1}{2} \Rightarrow \text{increasing (constant, decreasing) probabilities} \tag{41}$$

Thus in the presence of inequality constraint (34) while the result that the probabilities are in G.P. is not true, the other result given by (41) is still valid.

35.5 USE OF ANOTHER GENERALISED MEASURE OF ENTROPY

In [1] we derived another generalised measure of entropy, viz.

$$S = -\sum_{i=1}^{n} (b_i - a_i) \tan^{-1} \sqrt{\frac{p_i - a_i}{b_i - p_i}} + \sum_{i=1}^{n} (p_i - a_i)(b_i - p_i) \tag{42}$$

which has the property that when it is maximized subject to given moment constraints, it automatically satisfies the inequality constant (1).

In the special case $a_1 = a_2 = \ldots = a_n = a$, $b_1, = b_2 = \ldots = b_n = b$, maximizing (42) subject to constraints (2) and (30) we get

$$\frac{p_i - a}{b - p_i} = (\lambda_0 + \lambda_1 i)^2 = F(1 + ci)^2, \tag{43}$$

where F and c determined by using the equations

$$\sum_{i=1}^{n} \frac{a + bF\,(1+ci)^2}{1 + a\,(1+ci)^2} = 1, \; \sum_{i=1}^{n} i\,\frac{a + bF\,(1+ci)^2}{1 + F\,(1+ci)^2} = m \tag{44}$$

$$nb - (b-a) \sum_{i=1}^{n} \frac{1}{1 + F\,(1+ci)^2} = 1$$

$$\frac{n\,(n+1)}{2} b - (b-a) \sum_{i=1}^{n} \frac{i}{1 + F\,(1+ci)^2} = m \tag{45}$$

so that,

$$\frac{n+1}{2} \left[1 + (b-a) \sum_{i=1}^{n} \frac{1}{1 + F\,(1+ci)^2} \right] = m + (b-a) \sum_{i=1}^{n} \frac{i}{1 + F\,(c+i)^2} \tag{46}$$

Arguing as in the last section, we conclude that

$$m > (=, <) \frac{n+1}{2} \Rightarrow \text{increasing (constant, decreasing) probabilities} \tag{47}$$

Thus even this measure confirms the results obtained by using the other measure.

35.6 SUMMARY OF RESULTS

(i) The most natural distribution when the only information available about the probability distribution is given by the inequality constraint (1) and the natural constraint (2) is given by

$$p_i = \frac{Kb_i + a_i}{K+1}, K = \frac{1-A}{B-1}, A = \sum_{i=1}^{n} a_i, B = \sum_{i=1}^{n} b_i \tag{48}$$

(ii) The two measures whose maximization subject to given moment constraints will automatically satisfy the inequality constraint (1) are given by

$$S = -\sum_{i=1}^{n} (p_i - a_i) \ln (p_i - a_i) - \sum_{i=1}^{n} (b_i - p_i) \ln (b_i - p_i) \tag{49}$$

and

$$S = \sum_{i=1}^{n} (b_i - a_i) \tan^{-1} \sqrt{\frac{p_i - a_i}{b_i - p_i}} + \sum_{i=1}^{n} (p_i - a_i)(b_i - p_i) \tag{50}$$

(iii) When the only constraint is $\sum_{i=1}^{n} p_i = 1$ the maximum value of S in (49) is given by

$$S_{\max} = -\sum_{i=1}^{n} (b_i - a_i) \ln (b_i - a_i) + (B - A) \ln (B - A) - (1 - A) \ln (1 - A)$$

$$- (B - 1) \ln (B - 1) \tag{51}$$

(iv) $S_{\max}$ is concave function of $a_1, a_2, ..., a_n, b_1, b_2, ..., b_n$.

(v) When $a_1, a_2, ..., a_n$ are kept constant, $S_{\max}$ is maximum when $b_i = a_i + 1/e + 1 - A/n$ provides it is < unity and then the maximum value of $S_{\max}$ is $(1 - A) \ln n$.

(vi) If $b_1, b_2, ..., b_n$ are kept constant, $S_{\max}$ is maximum when $a_i = b_i - 1/e - (b-1)/n$ provides this is > 0 and then the maximum value of $S_{\max}$ is $(B - 1) \ln n$.

(vii) The globally maximum value of $S_{\max}$ occurs when

$$a_i = \frac{1}{n} - \frac{1}{e}, b_i = \frac{1}{n} + \frac{1}{e} \text{ provides } a_i > 0, b_i < 1$$

and then the maximum value of $S_{\max}$ is $\frac{2n}{e}$.

(vii) When the conditions in (v), (vi) (vii) are violated, the maximum value occurs on the boundaries of the permissible region and are < the values given here.

(viii) If $a_1 = a_2 = ... = a_n = a$, $b_1, = b_2 = ... = b_n = b$ and if the mean is prescribed as m, then the probabilities increase or decrease with i according as $m \gtrless (n+1)/2$ and the result holds for the measure of entropy (49) and (50) and it is the same as the result obtained by using Shannon's measure in the absence of the inequality constraints.

REFERENCES

1. J.N. Kapur (1989a), Maximum Entropy and Minimum Cross-Entropy probability Distributions when there are inequality constraints on probabilities, M.S.T.S. Rep. # 475.
2. J.N. Kapur (1989), "Maximum Entropy Models in Science and Engineering," John Wiley & Sons, New York.
3 J.N. Kapur and H.K. Kesavan (1987), "Generalised Maximum Entropy Principle with Applications, Sandford Educational Press, Canada.
4. C.E. Shannon (1948) "Mathematical Theory of Communication", Bell. System Tech. J., Vol. 27, 327-423, 623-659.

36

MAXIMUM ENTROPY PROBABILITY DISTRIBUTIONS WHEN THERE ARE INEQUALITY CONSTRAINTS ON PROBABILITIES—AN ALTERNATIVE APPROACH

[A new approach is given for finding the Maximum-Entropy Probability distributions when there are inequality constraints on probabilities, in addition to the usual moment constraints.]

36.1 INTRODUCTION

The following constraints are prescribed on the probabilities $p_1, p_2, ..., p_n$ of a probability distribution $P = (p_1, p_2, ..., p_n)$

$$a_i \le p_i \le b_i, \quad i = 1, 2, \ldots n \tag{1}$$

$$\sum_{i=1}^{n} p_i = 1 \tag{2}$$

$$\sum_{i=1}^{n} p_i\, g_{ri} = a_r, \quad r = 1, 2, \ldots m \tag{3}$$

There may be an infinity of probability distributions satisfying (1), (2) and (3). Our object is to find that distribution out of these which has the maximum entropy.

Freund and Saxena [1] solved this problem by maximizing Shannon's [3,5] entropy measure,

$$-\sum_{i=1}^{n} p_i \ln p_i \tag{4}$$

subject to constraints (1) and (2) only. Even in this simple case, the algorithm they got was not very simple. The problem of maximizing (4) subject to (1), (2) and (3) has not been solved and its solution is not going to be easy.

We can get sub-optimal solutions by maximizing (4) subject to (2) and (3) and then adjusting the probabilities to satisfy (1) or maximizing (4) subject to (1) and (2) and then adjusting the probabilities to satisfy (3). Both the methods may give us solutions, which are far from optimal.

Recently Kapur [2] used the inverse maximum entropy principle of Kapur and Kesavan [4] to get the measure,

$$-\sum_{i=1}^{n} (p_i - a_i) \ln (p_i - a_i) - \sum_{i=1}^{n} (b_i - p_i) \ln (b_i - p_i) \tag{5}$$

Its maximization subject to (2) and (3) leads to a solution which automatically satisfies (1) Computationally the method is extremely simple and elegant.

However, while maximization of (4) gives us a solution which makes probabilities as equal as possible subject to the constraints, maximization of (5) gives us probabilities which make $P_1, P_2, ..., P_n$ as equal as possible, subject to the given constraints where

$$P_i = \frac{p_i - a_i}{b_i - p_i}, \quad i = 1,2,\ldots,n \tag{6}$$

The two objectives are the same only if

$$a_1 = a_2 = \ldots = a_n = a, \quad b_1 = b_2 = \ldots = b_n = b \tag{7}$$

In other words maximization of (5) will give us probabilities which will divide the intervals (a_i, b_i) in as equal ratios as possible subject to the given constraints.

In many practical problems, this is a more desirable objective than making all the probabilities equal subject to the given constraints.

However, in these problems where the objective is to make probabilities as equal as possible subject to the given constraints, inspite of the non-symmetric inequality constraints, we have to maximize (4) subject to (1), (2) and (3).

In the present paper, we suggest an alternative approach to solving the problem by modifying the problem in a different manner.

36.2 THE NEW METHOD

We assume that p_i is a random variable which lies in the interval (a_i, b_i) with density function $f_i(x)$ so that

$$\int_{a_i}^{b_i} f_i(x)\, dx = 1, \quad i = 1,2,\ldots,n \tag{8}$$

We define $\hat{p}_i$ as the expected value of this random variate so that

$$\hat{p}_i = \int_{a_i}^{b_i} x\, f_i(x)\, dx \tag{9}$$

Now the constraints (2) and (3) are replaced by

$$\sum_{i=1}^{n} \int_{a_i}^{b_i} x\, f_i(x)\, dx = 1 \tag{10}$$

$$\sum_{i=1}^{n} g_{ri} \int_{a_i}^{b_i} x\, f_i(x)\, dx = a_r, \quad r = 1,2,\ldots,m \tag{11}$$

We now seek to maximize,

$$-\sum_{i=1}^{n} \int_{a_i}^{b_i} f_i(x) \ln f_i(x)\, dx, \tag{12}$$

subject to (10) and (11) to get $f_1(x), f_2(x) \ldots, f_n(x)$ and then use (9) to get $\hat{p}_1, \hat{p}_2, \ldots, \hat{p}_n$.

In view of (8) and (9), we shall get automatically

$$a_i \le \hat{p}_i \le b_i \tag{13}$$

Even in this method we are strictly not attacking the original problem. We are using expected values of probabilities instead of probabilities themselves but instead of taking entropy of expected values, viz.

$$-\sum_{i=1}^{n} \hat{p}_i \ln \hat{p}_i \tag{14}$$

we are using the entropy measure (12).

The method aims to make the density functions as uniform as possible subject to the given constraints.

36.3 THE SOLUTION

Maximising (12) subject to (8), (10) and (11) by using Lagrange's method we get

$$-1 - \ln f_i(x) - (\mu - 1) - \lambda_0 x - \sum_{r=1}^{m} \lambda_r g_{ri} x = 0$$

or,

$$f_i(x) = C_i \exp[-x(\lambda_0 + \lambda_1 g_{1i} + \lambda_2 g_{2i} + \ldots + \lambda_m g_{mi})], \quad i = 1, 2, \ldots, n \tag{15}$$

where, the $n + m + 1$ Lagrange' multipliers are determined by using $n + m + 1$ equations (8), (10) and (11); so that

$$C_i \int_{a_i}^{b_i} \exp(-A_i x)\, dx = 1, \quad i = 1, 2, \ldots n \tag{16}$$

$$\sum_{i=1}^{n} C_i \int_{a_i}^{b_i} x \exp(-A_i x)\, dx = 1 \tag{17}$$

$$\sum_{i=1}^{n} C_i\, g_{r_i} \int_{a_i}^{b_i} x \exp(-A_i x)\, dx = a_r, \quad r = 1, 2, \ldots m, \tag{18}$$

where,

$$A_i = \lambda_0 + \lambda_1 g_{1i} + \lambda_2 g_{2i} + \ldots + \lambda_m g_{mi}, \tag{19}$$

so that

$$C_i \left[\frac{e^{-A_i a_i} - e^{-A_i b_i}}{A_i} \right] = 1, \quad i = 1, 2, \ldots, n \tag{20}$$

$$\sum_{i=1}^{n} C_i \left[\frac{a_i e^{-A_i a_i} - b_i e^{-A_i b_i}}{A_i} - \frac{e^{-A_i b_i} \cdots a_i}{A_i^2} \right] = 1 \tag{21}$$

$$\sum_{i=1}^{n} C_i\, g_{r_i}\left[\frac{a_i\, e^{-A_i a_i} - b_i e^{-A_i b_i}}{A_i} - \frac{e^{-A_i b_i} - e^{-A_i a_i}}{A_i^2}\right] = a_r, \quad r = 1,2,\ldots,m \tag{22}$$

Substituting for $-C_i$ in (21), (22) we get

$$\sum_{i=1}^{n}\left[\frac{a_i\, e^{-A_i a_i} - b_i\, e^{-A_i b_i}}{e^{-A_i a_i} - e^{-A_i b_i}} + \frac{1}{A_i}\right] = 1 \tag{23}$$

$$\sum_{i=1}^{n} g_{r_i}\left[\frac{a_i e^{-A_i a_i} - b_i\, e^{-A_i b_i}}{e^{-A_i a_i} - e^{-A_i r b_i}} + \frac{1}{A_i}\right] = a_r, \quad r = 1,2,\ldots,m \tag{24}$$

From the $(m + 1)$ equations (23) and (24), we can find $\lambda_0, \lambda_1, \ldots, \lambda_m$ and then from (20) we can find $C_1, C_2, \ldots, C_n$ and then we can find $f_1(x), f_2(x) \ldots, f_n(x)$ and $\hat{p}_1, \hat{p}_2, \ldots, \hat{p}_n$.

36.4 A SPECIAL CASE

We consider the special case considered by Freund and Saxena [1], i.e we consider only the natural constraint. In this case $A_i = \lambda_0$ and this is determined by using (23)

$$\sum_{i=1}^{n}\left[\frac{a_i\, e^{-\lambda_0 a_i} - b_i\, e^{-\lambda_0 b_i}}{e^{-\lambda_0 a_i} - e^{\lambda_0 b_i}} + \frac{1}{\lambda_0}\right] = 1 \tag{25}$$

and then C_i's are determined by using (20), so that

$$C_i = \frac{\lambda_0}{e^{-\lambda_0 a_i} - e^{-\lambda_0 b_i}} \quad i = 1,2,\ldots n \tag{26}$$

and,
$$\hat{p}_i = \frac{a_i\, e^{-\lambda_0 a_i} - b_i\, e^{-\lambda_0 b_i}}{e^{-\lambda_0 a_i} - e^{-\lambda_0 b_i}} + \frac{1}{\lambda_0}, \quad i = 1,2,\ldots,n \tag{27}$$

If
$$a_1 = a_2 \ldots = a_n = a, \quad b_1 = b_2 \ldots = b_n = b, \tag{28}$$

then $\hat{p}_i$ is independent of i and since $\sum_{i=1}^{n} \hat{p}_i = 1$ each $\hat{p}_i = \frac{1}{n}$, the distribution is the uniform distribution if $a \le \frac{1}{n} \le b$. This is expected.

36.5 A NECESSARY CONDITION FOR A SOLUTION

Since,
$$a_i \le p_i < b_i \tag{29}$$

$$\sum_{i=1}^{n} a_i \le \sum_{i=1}^{n} p_i \le \sum_{i=1}^{n} b_i \tag{30}$$

so that $\sum_{i=1}^{n} a_i$ should be less than or equal to unity and , $\sum_{i=1}^{n} b_i$ should be greater than or equal to unity.

REFERENCES

1. D. Freund and V. Saxena (1984), "An algorithm for a class of entropy problems, Ops Res. 32, 210-215.
2. J.N. Kapur (1990), "Maximum Entropy probability distributions when there are inequality constraints on probabilities and equality constraints on, moments Aligarh Journal of Statistics. 9, 28-38.
3. J.N. Kapur (1990), "*Maximum Entropy Models in Science and Engineering*, John Wiley, New York.
4. J.N. Kapur and H.K. Kesaven (1987), *Generalised Maximum Entropy Principle* (With Applications) sand ford Educational Press, Waterloo.
5. C.E. Shannon (1946), "A Mathematical Theory of Communication", Bell Sys. Tech Journal 27, 379-423, 623, 659.

37

ON MAXIMUM ENTROPY PROBABILITY DISTRIBUTIONS WHEN THE PRESCRIBED VALUES OF MOMENTS ARE FUNCTIONS OF ONE OR MORE PARAMETERS

[In many practical situations, we do not know the values of moments of a probability distribution, but we know some relations between these moments. The existence of maximum-entropy probability distribution under these conditions is discussed.]

37.1 INTRODUCTION

Let $f(x)$ be the probability density function for a continuous random variate and let it be subject to the following constraints,

$$\int_{-\infty}^{\infty} f(x)\, dx = 1 \tag{1}$$

$$\int_{-\infty}^{\infty} f(x)\, g_r(x)\, dx = a_r; \quad r = 1, 2, \ldots, m \tag{2}$$

These $(m + 1)$ equation do not determine $f(x)$ uniquely. As such to get the most unbiased estimate of $f(x)$, we maximize the Boltzmann-Shannon entropy

$$S = -\int_{-\infty}^{\infty} f(x) \ln f(x)\, dx \tag{3}$$

subject to (1) and (2) to get

$$f(x) = \exp\left(-\lambda_0 - \lambda_1 g_1(x) - \lambda_2 g_2(x) - \ldots - \lambda_m g_m(x)\right), \tag{4}$$

where the values of Lagrangian multipliers $\lambda_1, \lambda_2, \cdots, \lambda_m$ are obtained by using (1) and (2).

The problem that we consider is to discuss as to what happens when $a_1, a_2, \ldots, a_m$ are functions of one parameter θ or these are functions of k parameters $\theta_1, \theta_2, \cdots, \theta_k$ where $k < m$.

If we eliminate these parameters from (2), we get a number of possibly non-linear constraints, on $f(x)$. If we maximize the entropy (3) subject to these constraints, shall we get the same MaxEnt probability distributions? Is it possible that the MaxEnt distribution may not exist now or it may not be unique?

37.2 AN OBVIOUS EFFECT OF PARAMETERISATION OF MOMENT VALUES

The parameter, $(a_1, a_2, \cdots, a_m)$ determine a point in m-dimensional Euclidean space and the constrains determine a region in the space bounded by

$$\min_x g_r(x) \le a_r \le \max_x g_r(x), \quad r = 1, 2, \ldots, n \tag{5}$$

There are also other restrictions arising due to the nature of the functions $g_1(x)$, $g_2(x), \cdots, g_m(x)$. Thus if $g_1(x) = x$, $g_2(x) = x^2$, then we should have

$$a_2 \ge a_1^2 \tag{6}$$

Thus, $(a_1, a_2, \cdots, a_m)$ can take any position in the m-dimensional region determined by (5), (6) and similar inequalities.

However, if $a_1, a_2, \cdots, a_m$ are functions of k parameters $\theta_1, \theta_2, \cdots, \theta_k$ $(k < m)$, then the point can take any position on a curve or a surface in this region.

Thus if $m = 3$, (a_1, a_2, a_3) moves in 3-dimensional space, but if a_1, a_2, a_3 are function of one parameter, then the point moves on a curve in this space and if these are functions of two parameters, then the point moves on a surface.

37.3 SOME EXAMPLES

EXAMPLE 1 Let a random variate x vary from $-\infty$ to ∞ and let its mean and variance be prescribed as m and σ^2, then the maximum entropy distribution is $N(m, \sigma^2)$ and its entropy is $1/2 \ln 2\pi e + \ln \sigma$. Now consider the case when m and σ are given as functions of a parameter θ, then for a fixed value of θ, say θ_0, the maximum entropy distribution will be $N(m(\theta_0), \sigma^2(\theta_0))$. As θ_0 varies, mean and variance of the normal distribution will change, but the MaxEnt distribution will always be normal. However, the entropy will not depend on $m(\theta)$, but will depend on only $\sigma(\theta)$. Now we consider special cases.

(i) $m(\theta) = \theta$, $\sigma(\theta) = \theta^2$, the entropy $\to \infty$ as $\theta \to \infty$ and can be made as large as please. *In this case no maximum entropy probability distribution exists.*

(ii) If $m(\theta) = \theta^{-2}$, $\sigma(\theta) = 2/\pi \tan^{-1} \theta^2$, then $\sigma(\theta)$ is always finite and the *maximum entropy distribution, exists, and is normal. It mean approaching zero and its variance approaching unity as* $\theta \to \infty$.

(iii) If $m(\theta) = \theta$, $\sigma(\theta) = \cos^2 \theta$, then the maximum value of σ is unity and all normal distribution with mean $k\pi$ (k an integer) and variance unity are maximum entropy distribution, *Thus the maximum entropy distribution exists but is not unique, though the maximum entropy is still unique.*

(iv) If $m(\theta) = 1/(\sin^2 \theta)$, $\sigma(\theta) = \cos^2 \theta$, thus when the variance is unity mean is infinity and *MEPD* does not exist. *However we can get probability distribution with entropy as near the maximum value as we like.*

Thus when mean and variance are prescribed as functions of a parameter θ, for a random variate which varies from $-\infty$ to ∞, the maximum entropy distribution may or may not exist and when it exists it may or may not be unique.

However, when it exists, it is normal and if it is not unique, all maximum entropy distributions have to be normal with same variance and same entropy.

EXAMPLE 2 The MaxEnt distribution will be non-normal if

(i) the range of variate is not from $-\infty$ to ∞,
(ii) the moments prescribed are different from mean and variance, and
(iii) the measure of entropy used is different from Shannon's
As special cases, we consider,

(a) If $E(|x|)$ is prescribed, we get Laplace's distribution and if $E(\ln(1+x^2))$ is prescribed, we get the class of generalised Cauchy's distributions and in the second case the shape of the probability distribution curve changes with the value of $\ln(1+x^2)$ prescribed.

(b) If Havrda-Charvat measure of entropy is used, we get the distribution of the form

$$f(x) = (a + bx + cx^2)^{1/\alpha - 1}, \quad \alpha \neq 1 \tag{7}$$

and hence the form of the density curve will change with the change of parameter α also.

(c) If the range of variate is $(0, \infty)$ then if $E(x)$, $E(x^2)$ are prescribed, the distribution is not normal. It may be truncated normal or U-shaped and it may not even exist. Thus if m and σ are functions of θ, then the truncated normal distribution will exist only if

$$\sigma^2(\theta) > m^2(\theta)$$

Thus, for finite ranges or semi-infinite ranges, if mean and variance are prescribed as functions of a parameter θ then as θ varies the *MEPD* may not exist and even if it exists, it may go on changing its form and the problem of finding the distribution with the globally maximum entropy will form an interesting problem by itself.

EXAMPLE 3 Let the range be $[0, \infty)$ and let the arithmetic and geometric means be prescribed, so that

$$E(x) = m\,;\; E(\ln x) = \ln g \tag{8}$$

then the *MEPD* is

$$f(x) = \frac{a^r}{\Gamma(r)} e^{-ax} x^{r-1}, \tag{9}$$

where

$$m = \frac{r}{a}, \quad \ln\frac{g}{m} = \frac{\Gamma'(r)}{\Gamma(r)} - \ln r, \quad \sigma^2 = \frac{r}{a^2} \tag{10}$$

Now we have the Table.

r	1	3	5	7	9	11	13	15	17	20	∞
$\frac{g}{m}$	.5615	.8188	.9019	.9291	.9418	.9545	.9618	.9650	.9427	.9791	1

If m is kept fixed and g increases to m, then

$$r \to \infty \text{ and since } a = \frac{r}{m},\ a \text{ increases to } \infty, \text{ since } \sigma^2 = \frac{m}{a},\ \sigma^2 \text{ decreases to 0.}$$

If g is kept fixed and m decreases to 0, g / m increases, r increases and since $a = \frac{r}{m}$, a decreases, σ^2 decreases.

$$S = -\int_0^\infty f(x)\,[r \ln a - \ln \Gamma(r) - a\,x + (r-1)\ln x]\;dx$$

$$= \ln \Gamma(r) - r \ln a + a\,m - (r-1) \ln g$$

$$= \ln \Gamma(r) + r - \ln a - (r-1)\frac{\Gamma'(r)}{\Gamma(r)} \tag{11}$$

If $a \to \infty$, $r = 1$, the Gamma distribution becomes degenerate uniform and its entropy tends to infinity. Thus if $m \to \infty$, entropy tends to infinity.

37.4 THE CASE WHEN A FUNCTIONAL RELATION BETWEEN MOMENTS IS GIVEN

Suppose we have to maximize,

$$-\int f(x) \ln f(x)\,dx \tag{12}$$

subject to,

$$\int f(x)\,dx = 1 \tag{13}$$

and,

$$\int g_1(x)\,f(x)\,dx = \phi\left[\int g_2(x)\,f(x)\,dx\right] \tag{14}$$

then we maximize (12) subject to (13), and

$$\int g_1(x)\,f(x)\,dx = m \tag{15}$$

$$\int g_2(x)\,f(x)\,dx = \phi(m) \tag{16}$$

For every value of m, we can find the maximum entropy distribution and find its entropy S as a function of m and then proceed to find the value of m for which $S(m)$ is maximum. If the maximum value of $S(m)$ arises for $m = m_0$ and is finite, then the *MEPD* corresponding to $m = m_0$ gives the *MEPD* with the globally maximum entropy. If $S(m) \to \infty$, then the global maximum entropy distribution does not exist.

Though the constraint (14) may appear to be non-linear, it can be reduced to two linear constraints and the nature of the problem does not changes essentially.

Thus, if range is $\phi(-\infty, \infty)$, $g(x) = x$, $g_2(x) = x^2$, $\phi(m) = m^k$

we get,

$$f(x) = \frac{1}{\sqrt{2\pi}\sqrt{m^k - m^2}} \exp\left(-\frac{1}{2}\frac{(x-m)^2}{m^k - m^2}\right)$$

This distribution will exist only if $m^k > m^2$.

37.5 REMARKS

In many important problems in structures, risk analysis, turbulence theory, we can not find moments, but we can find relations between moments. In all such cases the approach given in the present paper should be useful.

38

CHARACTERISING MOMENTS OF A PROBABILITY DISTRIBUTION

[Characterising moments of a probability distribution of the exponential family are introduced in three different ways. Characterising moments for other probability distributions are also considered.]

38.1 INTRODUCTION

Let a random variate take values $x_1, x_2, \cdots, x_n$ with relative frequencies $p_1, p_2, \cdots, p_n$. Let the variate have the density function $f(x, \theta)$. We want to choose θ so that the density function $f(x, \theta)$ is as close as possible to the observed distribution.

The observed distribution has a cumulative density function, $G(x)$ so that

$$\begin{aligned} G(x) &= 0, \quad x < x_1 \\ G(x) &= p_1, \quad x_1 \le x < x_2 \\ G(x) &= p_1 + p_2, \quad x_2 \le x < x_3 \\ &\cdots\cdots\cdots\cdots\cdots\cdots \\ G(x) &= 1, \quad x \ge x_n \end{aligned} \tag{1}$$

To get the best fit, we have to choose θ to minimize Kullback-Leibler's [6] measure of directed-divergence

$$\int g(x) \ln \frac{g(x)}{f(x,\theta)}\, dx = \int g(x) \ln g(x)\, dx - \int g(x) \ln f(x,\theta)\, dx \tag{2}$$

i.e, we have to choose θ to minimize

$$-\int f(x,\theta)\, dG \tag{3}$$

This requires

$$\int \frac{1}{f}\frac{\partial f}{\partial \theta}\, dG = 0 \tag{4}$$

which gives,

$$\sum_{i=1}^{n} \left(\frac{1}{f}\frac{\partial f}{\partial \theta}\right)_{x=x_i} p_i = 0 \tag{5}$$

Thus (5) is the equation to determine θ.

38.2 ILLUSTRATIVE EXAMPLES

EXAMPLE 1

Let
$$f(x,\theta)=\theta\, e^{-x\theta}, \tag{6}$$

then (5) gives
$$\sum_{i=1}^{n}\left(\frac{1}{\theta}-x_i\right)p_i=0 \quad \text{or}\ \frac{1}{\theta}=\sum_{i=1}^{n} p_i\, x_i=\bar{x} \tag{7}$$

or
$$\frac{1}{\theta}=\text{observed mean}, \tag{8}$$

but $1/\theta$ is the population mean, so that the best estimate for θ is obtained by equating the observed sample mean to the population mean.

EXAMPLE 2

Let
$$f(x,a,r)=\frac{a^r}{\Gamma(r)}e^{-ax}x^{r-1} \tag{9}$$

$$\ln f(x,a,r)=r\ln a-\ln\Gamma(r)-ax+(r-1)\ln x$$

$$\frac{1}{f}\frac{\partial f}{\partial a}=\frac{r}{a}-x \tag{10}$$

$$\frac{1}{f}\frac{\partial f}{\partial r}=\ln a-\frac{\Gamma'(r)}{\Gamma(r)}+\ln x, \tag{11}$$

so that (5) gives
$$\sum_{i=1}^{n}\left(\frac{r}{a}-x_i\right)p_i=0;\quad \sum_{i=1}^{n}\left(\ln a-\frac{\Gamma'(r)}{\Gamma(r)}+\ln x_i\right)p_i=0 \tag{12}$$

or
$$\frac{r}{a}-A=0;\quad \ln a-\frac{\Gamma'(r)}{\Gamma(r)}+\ln G=0, \tag{13}$$

where A and G are the arithmetic and geometric means of the observed distribution. Now let us find the arithmetic and geometric mean of the Gamma distribution (9).

$$A.M.=\frac{a^r}{\Gamma(r)}\int_0^\infty x\, e^{-ax}\,x^{r-1}\,dx=\frac{a^r}{\Gamma(r)}\frac{\Gamma(r+1)}{a^{r+1}}=\frac{r}{a} \tag{14}$$

$$\ln(G.M.)=\frac{a^r}{\Gamma(r)}\int_0^\infty e^{-ax}\,x^{r-1}\ln x\, dx \tag{15}$$

but,
$$\int_0^\infty e^{-ax}\,x^{r-1}\,dx=\frac{\Gamma(r)}{a^r} \tag{16}$$

Differentiating under the sign of integration with respect to r,

$$\int_0^\infty e^{-ax} x^{r-1} \ln x \, dx = \frac{\Gamma'(r)}{a^r} - \frac{\Gamma(r) \ln a}{a^r} \tag{17}$$

From (15) and (17)

$$\ln(G.M.) = \frac{\Gamma'(r)}{\Gamma(r)} - \ln a \tag{18}$$

From (13), (14), and (18)

$$A.M. = A \; ; G.M. = G, \tag{19}$$

so that the chosest fit of the gamma distribution to the observed probability distribution is obtained by choosing a and r by equating the arithmetic and geometric means of the population to the arithmetic and geometric means of the observed distribution.

DEFINITION 1. Those moments of a probability distribution, by equating the observed values of which to their population values, we obtain the best fit to the observed distribution, are called characterising moments of the probability distribution [3,4, 5].

Thus the characterising moment of the exponential distribution is its mean and the characterising moments of the gamma distribution are its arithmetic and geometric means.

38.3 GENERAL CASE OF AN EXPONENTIAL FAMILY OF DISTRIBUTIONS

Let the density function be

$$\exp(-\lambda_0 - \lambda_1 g_1(x) - \ldots - \lambda_m g_m(x)) \tag{20}$$

where,
$$\exp(\lambda_0) = \int \exp(-\lambda_1 g_1(x) - \ldots - \lambda_m g_m(x)) \, dx \tag{21}$$

so that there are m independent parameters $\lambda_1, \lambda_2, \cdots, \lambda_m$

Let,
$$E(g_r(x)) = a_r; \quad r = 1, 2, \ldots, m \tag{22}$$

so that,

$$\int g_r(x) \exp(-\lambda_0 - \lambda_1 g_1(x) - \ldots - \lambda_m g_m(x)) \, dx = a_r \tag{23}$$

$a_1, a_2, \cdots, a_m$ are m independent population moments. Using (5),

$$\sum_{i=1}^{n} \left[\frac{-\partial \lambda_0}{\partial a_r} - \frac{\partial \lambda_1}{\partial a_r} g_1(x_i) - \ldots - \frac{\partial \lambda_m}{\partial a_r} g_m(x_i) \right] p_i = 0 \tag{24}$$

or
$$\frac{\partial \lambda_0}{\partial a_r} + \frac{\partial \lambda_1}{\partial a_r} \bar{a}_1 + \ldots. + \frac{\partial \lambda_m}{\partial a_r} \bar{a}_m = 0 \tag{25}$$

where,
$$\bar{a}_r = \sum_{i=1}^{n} p_i g_r(x_i) \tag{26}$$

Also, from (21),

$$\int \exp(-\lambda_0 - \lambda_1 g_1(x) - \ldots - \lambda_m g_n(x)) \left[\frac{\partial \lambda_0}{\partial a_r} + \ldots + \frac{\partial \lambda_m}{\partial a_r} g_m(x)\right] dx = 0 \quad (27)$$

or

$$\frac{\partial \lambda_0}{\partial a_r} + \frac{\partial \lambda_1}{\partial a_r} a_1 + \ldots + \frac{\partial \lambda_m}{\partial a_r} a_m = 0 \quad (28)$$

From (25) and (28),

$$\frac{\partial \lambda_1}{\partial a_r}(a_1 - \bar{a}_1) + \frac{\partial \lambda_2}{\partial a_r}(a_2 - \bar{a}_2) + \ldots + \frac{\partial \lambda_m}{\partial a_r}(a_m - \bar{a}_m) = 0$$

$$r = 1, 2, \ldots m \quad (29)$$

We have m homogeneous equations in $a_1 - \bar{a}_1, a_2 - \bar{a}_2, \ldots, a_m - \bar{a}_m$ and the only solution these equations will have is

$$a_1 = \bar{a}_1, \ldots, a_m = \bar{a}_m \quad (30)$$

unless,

$$\frac{\partial(\lambda_1, \lambda_2, \ldots, \lambda_m)}{\partial(a_1, a_2, \ldots, a_m)} = 0 \quad (31)$$

We are assuming all the moments are independent and therefore (31) is not satisfied and the best fit is obtained by equating observed and population values of $E(g_1(x))$, $E(g_2(x))$, ..., $E(g_m(x))$. These will then be the characterising moments of every exponential distribution (20) [2].

38.4 MAXIMUM-ENTROPY PROBABILITY DISTRIBUTION

DEFINITION 2. The characterising moments of a probability distribution are those moments such that out of all probability distribution having given values for these moments, the given distribution is the one with the maximum value for the Shannon's measure of entropy,

$$-\int f(x) \ln f(x)\, dx \quad (32)$$

Let,

$$E(g_1(x)) = a_1, \quad E(g_2(x)) = a_2, \ldots, E(g_m(x)) = a_m \quad (33)$$

so that the probability density function $f(x)$ has to satisfy the equation,

$$\int f(x)\, dx = 1; \quad \int f(x)\, g_r(x)\, dx = a_r; \quad r = 1, 2, \ldots m \quad (34)$$

there will in general, be an infinity of probability distributions satisfying (34). To get the maximum entropy distributions, we maximize (32), subject to (34) to get

$$f(x) = \exp(-\lambda_0 - \lambda_1 g_1(x) - \lambda_2 g_2(x) - \ldots - \lambda_m g_m(x)) \quad (35)$$

where $\lambda_0, \lambda_1, \cdots, \lambda_m$ are obtained by using (34).

Thus when characterising moments are given by (33), the maximum entropy probability distribution is given by (35).

38.5 DOES EVERY PROBABILITY DISTRIBUTION HAVE CHARACTERISING MOMENTS ?

If the distribution is of the type (35), it has characterising moments given by (33). Thus Cauchy distribution with

$$f(x) = \frac{1}{\pi} \frac{1}{1+x^2} \tag{36}$$

can be written as

$$f(x) = \frac{1}{\pi} \exp(-\ln(1+x^2)) \tag{37}$$

and its characterising moments is $E(\ln(1+x^2))$. However consider

$$f(x) = \frac{1}{\pi} \cdot \frac{1}{1+(x-m)^2} \tag{38}$$

If m is a known constant than the characterising moment is $E(\ln(1+(x-m)^2))$. However, if m is not known and is a parameter to be estimated, then (38) may not have a characterising moment.

38.6 USE OF OTHER MEASURES OF ENTROPY

If we use Harada-Charvat's [1] measure of entropy,

$$\frac{1}{\alpha(1-\alpha)} \left[\int f^{\alpha}(x)\, dx - 1 \right]; \quad \alpha \neq 1, \quad \alpha \neq 0 \tag{39}$$

then the maximum entropy distribution is

$$f(x) = (\lambda_0 + \lambda_1 g_1(x) + \ldots + \lambda_m(x))^{1/(\alpha-1)} \tag{40}$$

If we let $\alpha \to 0$ and put $g_1(x) = x$, $g_2(x) = x^2$, we get the measure of entropy,

$$\int \ln f(x)\, dx \tag{41}$$

and the maximum entropy probability distribution is

$$f(x) = (\lambda_0 + \lambda_1 x + \lambda_2 x^2)^{-1} = \frac{1}{\lambda_0 + \lambda_1 x + \lambda_2 x^2} \tag{42}$$

Comparising with (38), we find that if we maximize (41) subject to,

$$\int f(x)\, dx = 1, \quad \int x f(x)\, dx = A; \quad \int x^2 f(x)\, dx = B \tag{43}$$

the maximum entropy distribution will be (42) and for suitable value of A and B it may be same as (38). However, for the distribution (38), if the range of x is $(0, \infty)$, the value of A and B will be infinite. As such (38) will have characterising moments as $E(x)$, $E(x^2)$, provided the entropy measure used is given by (41) and the range of x is finite. However,

$$f(x) = \frac{1}{(\lambda_0 + \lambda_1 x + \lambda_2 x^2)^2} \tag{44}$$

has characterising moments as $E(x)$, $E(x^2)$ if we use the entropy measure as

$$\int f^{1/2}(x)\, dx \tag{45}$$

and in this case the range can be $(0, \infty)$.

Thus, a probability distribution may not have characterising moments for Shannon entropy, but may have these for a generalised measure of entropy [2] and it may not have these when x varies over any range, but it may have these when x varies over a specified range.

38.7 ANOTHER APPROACH THROUGH APPROXIMATING MAXENT DISTRIBUTIONS

A given probability distribution may be complicated or it may be only numerically specified. We may want to approximate it by a simpler analytically defined probability distribution.

One method is to find its moments $E(x)$, $E(x^2)$, ..., $E(x^r)$ and to find that distribution which has the maximum entropy out of all distributions having the same values of these r moments as the given distribution. If $H(P)$ is the entropy of the given distribution and $H(Q_r)$ is the entropy of MaxEnt distribution then it can be shown that

$$H(Q_1) \geq H(Q_2) \geq \ldots \geq H(Q_r) \geq \ldots \geq H(P) \tag{46}$$

If for some values of $r = k$, Q_k coincides with P then $E(x), \cdots, E(x^k)$ will be the characterising moments of the given distribution. If there is no such value k, then only algebraic moments cannot characterize the distribution. Thus, we get the third definition of characterising moments of a probability distribution.

DEFINITION 3. If there is a set of moments $E(g_1(x))$, $E(g_2(x))$,..., $E(g_k(x))$, so that the approximating MaxEnt distribution which has the same values for these moments, coincides with the given distribution P, then $g_1(x)$, $g_2(x)$, ..., $g_k(x)$ are called characterising moments of P.

REFERENCES

1. J.H. Havrda and F. Charvat (1967), "Quantification methods of classification processes : concepts of structural α entropy", Kybernatica, Vol. 3, pp. 30-35.
2. J.N. Kapur (1988), " On equivalence of Gauss's and maximum entropy principles of estimation of probability distributions", Ganit Sandesh, Vol. 2 (1) pp. 4-10.
3. J.N. Kapur (1989), Maximum Entropy Models in Science and Engineering, Wiley Eastern, New Delhi and John Wiley, New York.

4. J.N. Kapur and H.K. Kesavan (1987), Generalised Maximum Entropy Principle (with Applications), Sandford Educational Press, University of Waterloo, Canada
5. J.N. Kapur and H.K. Kesavan (1992), Entropy Optimization Principles with Applications, Academic Press, New York.
6. S. Kullback and R.A. Leibler (1951), "On Information and Sufficiency", Ann. Math. Stat. Vol. 22. pp. 79-86.

Part C

Application of Measures of Information and Entropy Optimization Principle

39

MAXIMUM ENTROPY ESTIMATION OF MISSING VALUES

[Given n positive numbers $x_1, x_2, \cdots, x_n$ and given that x belongs to the same set as $x_1, x_2, \cdots, x_n$ and $a \le x \le b$, different maximum-entropy estimates of the missing value x are obtained by using different methods and by using different measures of entropy. In the first method, the estimate is either a or b or x^* where x^* is an entropic mean of $x_1, x_2, \cdots, x_n$. In the second method, the estimate depends on a, b and $\sum_{i=1}^{n} x_i$. In the third method, it depends on all values $x_1, x_2, \cdots, x_n$.]

39.1 INTRODUCTION

Let $x_1, x_2, \cdots, x_n$ be n given values and let x be a missing value about which we have no other information except that it belongs to the same set to which $x_1, x_2, \cdots, x_n$ belong.

To estimate x, Kapur [3,4] proposed that we should choose x so that $x_1, x_2, \cdots, x_n$ and x are as equal among themselves as possible. For this purpose, we have to choose x to maximum a measure of equality of these $(n + 1)$ values.

One such measure is given by Shannon's [6] measure of entropy or equality, which for the probability or proportion distribution

$$\frac{x_i}{T+x}, \frac{x_2}{T+x}, \ldots, \frac{x_n}{T+x}, \frac{x}{T+x} \quad ; \quad T = \sum_{i=1}^{n} x_i \tag{1}$$

is given by

$$S(x) = -\sum_{i=1}^{n} \frac{x_i}{T+x} \ln \frac{x_i}{T+x} - \frac{x}{T+x} \ln \frac{x}{T+x} \tag{2}$$

By maximizing $S(x)$, Kapur obtained the estimate

$$X_1^* = \left(x_1^{x_1} x_2^{x_2} \ldots x_n^{x_n}\right)^{\frac{1}{T}}, \tag{3}$$

which he called the Entropic mean of $x_1, x_2, \cdots, x_n$.

If instead of using Shannon's measure, we use Burg's [1] measure of entropy

$$B(x) = \sum_{i=1}^{n} \ln \frac{x_i}{T+x} + \ln \frac{x}{T+x}, \tag{4}$$

we get the estimate

$$X_2^* = \frac{x_1 + x_2 + \ldots + x_n}{n} \tag{5}$$

which may appear to be a better estimate, than X_1^*. Similarly, if we use Gini-Simpson index or Havrda-Charvet [2] entropy of second order given by

$$H(x) = 1 - \sum_{i=1}^{n} \left(\frac{x_i}{T + x} \right)^2 - \frac{x^2}{(T + x)^2}, \tag{6}$$

we get the estimate

$$X_3^* = \sum_{i=1}^{n} x_i^2 / \sum_{i=1}^{n} x_i \tag{7}$$

The object of the present paper is to discuss how far these estimates are modified, when an additional information about x, viz it lies between given values, is available.

39.2 USE OF SHANNON ENTROPY

In this case we have to maximize (2) subject to

$$a \le x \le b \tag{8}$$

Now from (2)

$$\frac{dS}{dx} = \frac{1}{(T + x)^2} \left[\sum_{i=1}^{n} x_i \ln x_i - T \ln x \right] \tag{9}$$

$$\frac{d^2 S}{dx^2} = \frac{1}{(T + x)^3} \left[2 \left(T \ln x - \sum_{i=1}^{n} x_i \ln x_i \right) - T (T + x)/x \right] \tag{10}$$

When $x = 0, S = 0, dS/dx = \infty, \quad d^2 S/d x^2 < 0$ (11)

As x increases, $S(x)$ is a monotonic increasing concave function of x till x reaches the value X_1^* where dS/dx becomes zero, and d^2S/dx^2 is still negative. After that S begins decreasing and at some stage d^2S/dx^2 becomes zero giving rise to a point of inflexion in the $S - x$ curve and thereafter d^2S/dx^2 is positive and $S(x)$ becomes a convex function. Ultimately, as $x \to \infty$, $dS/dx \to 0$.

The graph of $S(x)$ is therefore given in Figure 39.1.

If $a < b < X_1^*$, then $S(x)$ is maximum when $x = b$ and this is our maximum entropy estimate for the missing value.

If $a < X_1^* < b$, then the maximum entropy estimate for x is X_1^*.

If $X_1^* < a < b$, then the maximum entropy estimate for x is a.

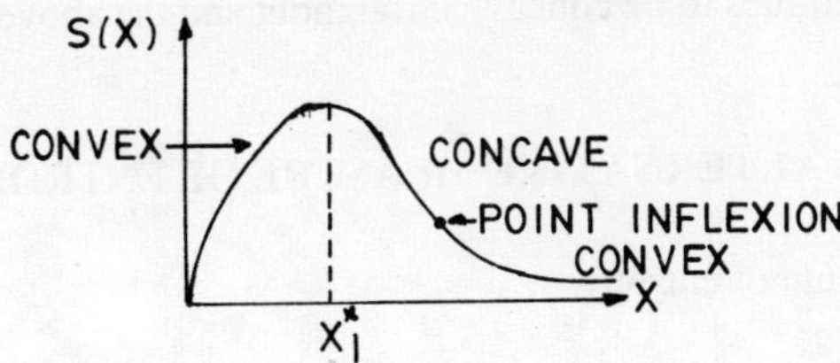

Fig. 39.1

39.3 USE OF OTHER CLASSICAL MEASURE OF ENTROPY

(a) Use of Burg's measure of entropy

From (4),

$$\frac{dB}{dx} = -\frac{n+1}{T+x} + \frac{1}{x}, \quad \frac{d^2B}{dx^2} = \frac{n+1}{(T+x)^2} - \frac{1}{x^2} \tag{12}$$

In this case, the graph is similar to that of Figure 39.1, except that $B(x)$ is negative and is as given in Figure 39.2.

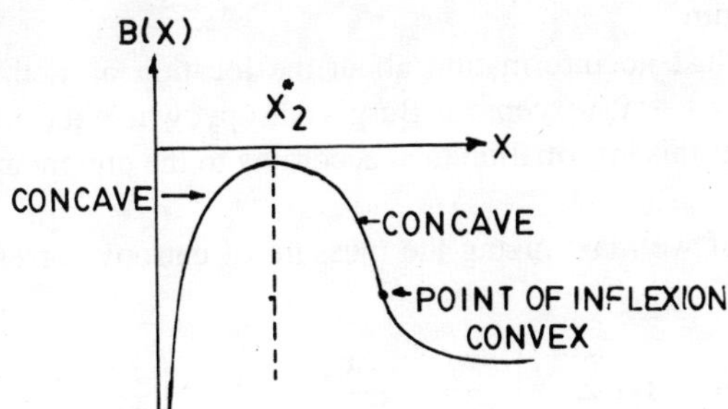

Fig. 39.2

Also,

If $a < b < X_2^*$, ME estimate for x is b
If $a < X_2^* < b$, ME estimate for x is X_2^*
If $X_2^* < a < b$, ME estimate for x is a

(b) use of Gini-Simpson Index

In this case from (6)

$$\frac{dH}{dx} = \frac{b}{(T+x)^3}\left\{\sum_{i=1}^{n} x_i^2 - T x\right\} \tag{13}$$

$$\frac{d^2 H}{d x^2} = -\frac{b}{(T+x)^4}\left\{\sum_{i=1}^{n} x_i^2 - T x\right\} - \frac{2T}{(T+x)^3} \tag{14}$$

Here $H(x)$ continues to be concave throughout and the above *results* continue to hold.

39.4 USE OF AN ALTERNATIVE MEASURE OF ENTROPY

Consider the measure of entropy,

$$\begin{aligned} K_1(x) = &-\sum_{i=1}^{n}\left(\frac{x_i}{T+a} - \frac{x_i}{T+x}\right)\ln\left(\frac{x_i}{T+a} - \frac{x_i}{T+x}\right) \\ &-\sum_{i=1}^{n}\left(\frac{x_i}{T+x} - \frac{x_i}{T+b}\right)\ln\left(\frac{x_i}{T+x} - \frac{x_i}{T+b}\right) \\ &-\left(\frac{x}{T+x} - \frac{a}{T+a}\right)\ln\left(\frac{x}{T+x} - \frac{a}{T+a}\right) - \left(\frac{b}{T+b} - \frac{x}{T+x}\right)\ln\left(\frac{b}{T+b} - \frac{x}{T+x}\right) \end{aligned} \tag{15}$$

Its maximization will automatically ensure that $a < x < b$. In fact the maximization gives

$$\frac{b-x}{x-a} = \frac{T+b}{T+a} \quad \text{or} \quad X_4^* = \frac{bT + aT + 2ab}{2T + a + b}, \tag{16}$$

so that X_4^* divides the interval [a, b] in the ratio $T + a : T + b$. The estimate of the missing value does not depend on the individual value $x_1, x_2, \cdots, x_n$, but only the value of their sum.

When we had no information about the location of x, the *ME* estimate did depend on $x_1, x_2, \cdots, x_n$ (except for Burg's entropy, where it did depend on T only), but when we get this information then according to the present approach, it depends on T, a, b only.

Similarly, if we maximizing the measure of entropy corresponding to Burg's entropy, viz.

$$\begin{aligned} K_2(x) = &\sum_{i=1}^{n}\ln\left(\frac{x_i}{T+a} - \frac{x_i}{T+x}\right) + \sum_{i=1}^{n}\ln\left(\frac{x_i}{T+x} - \frac{x_i}{T+b}\right) \\ &+ \ln\left(\frac{x}{T+x} - \frac{a}{T+a}\right) + \ln\left(\frac{b}{T+b} - \frac{x}{T+x}\right) \end{aligned} \tag{17}$$

then, X_5^* is determined by solving

$$\frac{b + a - 2x}{(x-a)(b-x)} = \frac{1}{T+x}\frac{2n}{n+1} \tag{18}$$

so that,

$$a < X_5^* < \frac{a+b}{2} < b \tag{19}$$

From (16) it also follows that,

$$a < X_5^* < \frac{a+b}{2} < b, \tag{20}$$

but while X^*_5 depends on a, b, T and n, X^*_4 depends on a, b, T only

39.5 USE OF ANOTHER PRINCIPLE

Consider the three probability distributions,

$$P: \frac{x_1}{T+x}, \frac{x_2}{T+x}, \ldots, \frac{x_n}{T+x}, \frac{x}{T+x}$$

$$A: \frac{x_1}{T+x}, \frac{x_2}{T+x}, \ldots, \frac{x_n}{T+x}, \frac{a}{T+a} \tag{21}$$

$$B: \frac{x_1}{T+x}, \frac{x_2}{T+x}, \ldots, \frac{x_n}{T+x}, \frac{b}{T+b}$$

In the absence of any other information, we have no reason to believe that the directed-divergence of A from P should be different from the directed-divergence of P from B, so that we take,

$$D\,(A:P) = D\,(P:B),$$

so that using Kullback-Leibler measure [5] of divergence

$$\sum_{i=1}^{n} \frac{x_i}{T+a} \ln \frac{T+x}{T+a} + \frac{a}{T+a} \ln \frac{T+x}{T+a} = \sum_{i=1}^{n} \frac{x_i}{T+x} \ln \frac{T+b}{T+x} + \frac{x}{T+x} \ln \frac{T+b}{T+x}$$

or $\quad \ln \dfrac{T+x}{T+a} = \ln \dfrac{T+b}{T+x} \quad$ or $\quad \dfrac{T+x}{T+a} = \dfrac{T+b}{T+x}$

or $T + x$ is the geometric mean between $T + a$ and $T + b$ and as such is less than the arithmetic mean between $T + a$ and $T + b$ so that our estimate,

$$X_6^* < \frac{a+b}{2} \tag{22}$$

In fact,

$$X_6^* = \sqrt{(T+a)\,(T+b)} - T \tag{23}$$

Again it depends on T, a, and b only.

If, however, we use Havrada-Charvat [2] measure of directed-divergence, we get

$$\sum_{i=1}^{n} \left(\frac{x_i}{T+a}\right)^{\alpha} \left(\frac{x_i}{T+x}\right)^{1-\alpha} + \left(\frac{a}{T+x}\right)^{\alpha} \left(\frac{x}{T+x}\right)^{1-\alpha} = \sum_{i=1}^{n} \left(\frac{x_i}{T+x}\right)^{\alpha} \left(\frac{x_i}{T+b}\right)^{1-\alpha} + \left(\frac{x}{T+x}\right)^{\alpha} \left(\frac{b}{T+b}\right)^{1-\alpha}$$

or

$$\frac{T + a^{\alpha} x^{1-\alpha}}{(T+a)^{\alpha}\,(T+x)^{1-\alpha}} = \frac{T + x^{\alpha} b^{1-\alpha}}{(T+x)^{\alpha}\,(T+b)^{1-\alpha}}, \tag{24}$$

so that the estimate again depends on T, a and b.

39.6 AN ALTERNATIVE APPROACH FOR ESTIMATION

Let it be known that the missing, value lies between x_k and x_{k+1}, then it is not fair to use the principle that $x_1, x_2, \cdots, x_k, x, x_{k+1}, \cdots, x_n$ should be maximally equal since here x_k and x_{k+1} have a special status.

We therefore start with the principle that

$$x-x_1, \quad x-x_2,\ldots,x-x_k, \quad x_{k+1}-x, \quad x_{k+2}-x,\ldots,x_n-x \tag{25}$$

should be maximally equal. Now let

$$A=\sum_{i=1}^{k} x_i \tag{26}$$

The corresponding probability distribution is

$$\frac{x-x_1}{(2k-n)x+T-2A},\ldots,\frac{x-x_k}{(2k-n)x+T-2A},\frac{x_{k-1}-x}{(2k-n)x+T-2A},\ldots,\frac{x_n-x}{(2k-n)x+T-2A} \tag{27}$$

Using Burg's entropy measure of equality, we choose x to maximize,

$$\sum_{i=1}^{k} \ln\frac{x-x_i}{(2k-n)x+T-2A}+\sum_{i=k+1}^{n} \ln\frac{x_i-x}{(2k-n)x+T-2A} \tag{28}$$

This gives,

$$\sum_{i=1}^{k}\frac{1}{x-x_i}-\frac{(2k-n)k}{(2k-n)x+T-2A}-\sum_{i=k+1}^{n}\frac{1}{x_i-x}-\frac{(n-k)(2k-n)}{(2k-n)x+T-2A}=0$$

or $$f(x)=\sum_{i=1}^{k}\frac{1}{x-x_i}-\sum_{i=k+1}^{n}\frac{1}{x_i-x}-\frac{n(2k-n)}{(2k-n)x+T-2A}=0 \tag{29}$$

Since, $$f(x_k)=\infty, \quad f(x_{k+1})=-\infty \tag{30}$$

$f(x)=0$ has a root lying between x_k and x_{k+1} and that is our estimate for the missing value lying between x_k and x_{k+1}

When $n=2k$, (30) simplifies to

$$\sum_{i=1}^{k}\frac{1}{x-x_i}=\sum_{i=k+1}^{2k}\frac{1}{x_i-x}, \tag{31}$$

so that in this case x is such that sum of reciprocal of deviations on the L.H.S = sum of reciprocal, of derivations on the R.H.S. If k = 1,

$$x=(x_i+x_2)/2 \tag{32}$$

as expected.

39.7 REMARKS

The estimate for the missing value depends on, (1) the available information ; (ii) the principle of inductive inference, we use.

If we are given only the values $x_1, x_2, \cdots, x_n$ and we use the principle that $x_1, x_2, \cdots, x_n$ and x should be as equal as possible, the estimate depends on the measure of equality (of entropy) we use. The estimate can be given by (3), (5), (6) or any similar measure.

If we are given the further information that x lies between x_k and x_{k+1}, we have the following alternatives :

(i) We can find the measure of entropy $S(x)$ and find for which value of x between x_k and x_{k+1}, it is maximum.

(ii) We can find x so that probability distribution based on x is equidistant from the probability distributions based on x_k and x_{k+1}.

(iii) We can choose x so that $x - x_1, x - x_2, \cdots, x - x_k, x_{k+1} - x, \cdots, x_n - x$ are maximally equal.

REFERENCES

1. J.P. Burg (1972), "The Relationship between Maximum Entropy Spectra and Maximum likelihood Spectra," in Modern Spectral Analysis, ed D.G. Childers, pp. 130-131, M.S.A.
2. J.H. Havrda and F. Charvat (1967), "Quantification Methods of Classification Processes : Concepts of Structural α Entropy," Kybernetica, Vol. 3, pp 30-35.
3. J.N. Kapur (1983),"On Maximum Entropy estimation of missing values", Nat. Acad. Science Letters, **6** (10), pp 59–65.
4. J.N. Kapur (1989), "Maximum Entropy Models in Science and Engineering, Wiley Eastern, New Delhi and John Wiley, New York.
5. S. Kullback and R.A. Leibler (1951), " On Information and Sufficiency", Ann. Math. Stat ; Volume 22, pp. 79-86.
6. C.E. Shannon (1948), " A Mathematical Theory of Communication", Bell System, Tech. J. Volume 27, pp. 379-423, 623-659.

40

ON THE USE OF MINIMAX ENTROPY PRINCIPLE IN ESTIMATION OF MISSING VALUES

[It is shown that the estimate of a missing value depends on the information available about it, as well as on the purpose for which the missing value has to be used. The rationale for the use of the principles of maximum entropy, minimum interdependence and minimax entropy are given. The implications of these principles for analysis of design of experiments are considered.]

40.1 INTRODUCTION

Estimation of missing values in experimental data is an important problem for all those experimental scientists who have to analyse data from which some values are missing for various reasons.

In design of experiments [2,3] certain techniques have been developed to carry out the analysis of variance of data when yields from some plots or some experimental units are missing. In general, the missing values are so chosen as to minimize the error sum of squares and corresponding adjustments for bias are made in the other sums of squares.

In [7, 8] we gave as method based on Jaynes [6] maximum entropy principle. If $x_1, x_2, \cdots, x_n$ are the given values and x is the missing value, it was proposed to choose x so as to maximize the entropy of the probability distribution

$$\frac{x_1}{T+x}, \frac{x_2}{T+x}, \ldots, \frac{x_n}{T+x}, \frac{x}{T+x}; \quad \sum_{i=1}^{n} x_i = T \tag{1}$$

This gave the estimate

$$\hat{x} = \left(x_1^{x_1}, x_2^{x_2}, \ldots, x_n^{x_n}\right)^{1/T} \tag{2}$$

Three objections could be raised against this estimate.

(i) $\hat{x}$ is rather unnatural. In fact $\hat{x}$ is always $\geq$ the arithmtic mean $\bar{x}$ of $x_1, x_2, \cdots, x_n$ which appears a more natural estimate from common sense considerations.

(ii) If two values are missing, the maximum entropy estimate for each is the same, viz. the one given by (2) ; this seems to be unsatisfactory.

(iii) This is not very useful for estimating missing values in design of experiments.

The first abjection can be met if we are prepared to use generalised measures of entropy [1, 11]. Thus if we use Shannon's [12] measure,

$$S(P) = -\sum_{i=1}^{n} p_i \ln p_i \tag{3}$$

for a probability distribution $P = (p_1, p_2, \cdots, p_n)$, we get the estimate (2) but if we use Burg's [1] measure,

$$B(P) = \sum_{i=1}^{n} \ln p_i \tag{4}$$

we get the estimate,

$$\hat{x} = \frac{x_1, +x_2 + \ldots + x_n}{n} \tag{5}$$

In fact according to the generalised principles of maximum entropy [10, 11] or of maximum equality, we should choose a value $\hat{x}$, which makes $\hat{x}$ as equal to $x_1, x_2, \cdots, x_n$ as possible, and for this purpose, we have to maximize a measure of equality. Now there are many measure of equality and our estimate will depend on the measure of equality we use.

The second objection can be understood by considering the information-theoretic nature of maximum entropy principles. The estimate $\hat{x}$ for the first missing value is obtained on the basis of information provided by $x_1, x_2, \cdots, x_n$. The estimate for the second missing value is obtained on the basis of the information provided by $x_1, x_2, \cdots, x_n, \hat{x}$ but this essentially is the same as the information provided by $x_1, x_2, \cdots, x_n$ and as such it is not surprising that the estimate for the second value is the same as the estimate for the first value.

In fact the information-theoretic estimate for a missing value depends on:

(a) the information available to us.

(b) the purpose for which missing value is to be used.

Thus in [9], we used the information-theoretic principle of minimum interdependence [4,13] to estimate missing values in contingency tables.

There mn-1 values x_{kl} ($k = 1,2, \cdots, m$; $l = 1, 2, \cdots, n$, $k \neq i$, $l \neq j$) were given and were to be used to estimate x_{ij}. We had not only the knowledge of the $mn-1$ values, but we also knew how these were arranged in rows and columns. We also knew that the purpose of estimation was to use the estimated value in testing for independence in the contingency tables.

Accordingly the value x_{ij} was chosen to minimize the measure of dependence,

$$D = S_1 + S_2 - S, \tag{5a}$$

where, S_1 is the entropy due to row totals, S_2 in the entropy due to column totals and S is the entropy of the cell entries.

(iii) As for the third objection, according to the principle of maximum entropy; we should use all the information given to us and scrupulously avoid using any information not given to us. In design of experiments, we are given information about the structure of the design and we should use the knowledge in estimating the missing values there. The estimate (2) is not relevant there because it does not make use of this knowledge.

However, entropy optimization principles can also be useful there. In the present paper, we show how the principle of minimax entropy can be used to get some insight into the information-provided by various designs.

40.2 THE MINIMAX ENTROPY PRINCIPLE

Let N given values be classified into n classes and let x_{ij} be the jth member of the ith class, so that

$$j = 1,2,\ldots,m_i; \quad i = 1,2,\ldots,n; \quad \sum_{i=1}^{n} m_i = N \tag{6}$$

then the entropy for this classification is

$$S = -\sum_{i=1}^{n}\sum_{j=1}^{m_i} \frac{x_{ij}}{T}\ln\frac{x_{ij}}{T}, \sum_{i=1}^{n}\sum_{j=1}^{m_i} x_{ij} = T \tag{7}$$

Let $T_{i.}$ be the total of the ith class, so that

$$\sum_{j=1}^{m_i} x_{ij} = T_{i.} \quad ; \quad \sum_{i=1}^{n} T_{i.} = T, \tag{8}$$

then we can write,

$$\begin{aligned}
S &= -\sum_{i=1}^{n}\sum_{j=1}^{m_i} \frac{T_{i.}}{T}\frac{x_{ij}}{T_{i.}} \ln\frac{x_{ij}}{T_{i.}}\frac{T_{i.}}{T} \\
&= -\sum_{i=1}^{n}\sum_{j=1}^{m_i} \frac{T_{i.}}{T}\frac{x_{ij}}{T_{i.}} \ln\frac{x_{ij}}{T_{i.}} - \sum_{i=1}^{n}\sum_{j=1}^{m_i} \frac{x_{ij}}{T_{i.}}\frac{T_{i.}}{T}\ln\frac{T_{i.}}{T} \\
&= -\sum_{i=1}^{n} \frac{T_{i.}}{T}\sum_{j=1}^{m_i}\frac{x_{ij}}{T_{i.}} \ln\frac{x_{ij}}{T_{i.}} - \sum_{i=1}^{n}\frac{T_{i.}}{T}\ln\frac{T_{i.}}{T} \\
&= \sum_{i=1}^{n}\frac{T_{i.}}{T}S_i - \sum_{i=1}^{n}\frac{T_{i.}}{T}\ln\frac{T_{i.}}{T} \\
&= S_W + S_B,
\end{aligned} \tag{9}$$

where, S_i is the entropy of the values within the i_{th} class, S_W is the weighted sum of $S_1, S_2, \cdots, S_n$ the weights being proportional to $T_{1.}, T_{2.}, \cdots, T_{n.}$, i.e to the class totals.

S_B is entropy of the class totals,
S_W is called the entropy within classes, and
S_B is called the entropy between classes.

Now we can classify the N values in n classes in a number of ways. In each way S will be the same, but S_W and S_B will be different.

We shall like the classes to be as distinguishable as possible and we shall also like the classes to be as homogeneous within themselves as possible.

Now entropy is large when the values are equal and entropy is small when the values are unequal.

As such, we shall like S_B to be as small as possible and we shall want S_W to be as large as possible.

This is the principle of minimax entropy which states that we should classify the values in such a way that the entropy between classes is as small as possible and the entropy within classes is as large as possible.

40.3 ESTIMATION OF MISSING VALUES

Let x be the missing value.

Let T be total of all known values so that the total of all values is $T + x$. Let T_1 be the total of all known values of the class to which the missing value belongs, so that $T_1 + x$ is the total of this class, then

$$S = -\Sigma' \frac{x_{ij}}{T+x} \ln \frac{x_{ij}}{T+x} - \frac{x}{T+x} \ln \frac{x}{T+x} \tag{10}$$

where Σ' denotes summation over all values of i, j except that corresponding to the missing values. Also,

$$S_B = -\Sigma'' \frac{T_{i.}}{T+x} \ln \frac{T_{i.}}{T+x} - \frac{T_1+x}{T+x} \ln \frac{T_1+x}{T+x} \tag{11}$$

where, Σ'' denotes summation over all values of i except for the class to which the missing value belongs.

Now we have to choose x

to minimize S_B/S_W

or to minimize $S_B/(S - S_B)$

or to maximize $(S - S_B)/S_B$

or to maximize $S/S_B - 1$

or to maximize S/S_B

or to minimize S_B/S

Now, $$\frac{d}{dx}\left(\frac{S_B}{S}\right) = 0 \Rightarrow S \frac{dS_B}{dx} - S_B \frac{dS}{dx} = 0$$

$$\Rightarrow \frac{dS_B/dx}{dS/dx} = S_B/S, \tag{12}$$

which gives the following equation to determine x

$$\frac{-\Sigma'' T_{i.} \ln T_{i.} + (T - T_1) \ln (T_1 + x)}{-\Sigma' x_{ij} \ln x_{ij} + T \ln x} = \frac{\ln (T_1 + x) - \ln (T + x)}{\ln x - \ln (T + x)} \tag{13}$$

If two values x, y are missing, we get

$$S = -\Sigma' \frac{x_{ij}}{T+x+y} \ln \frac{x_{ij}}{T+x+y} - \frac{x}{T+x+y} \ln \frac{x}{T+x+y} - \frac{y}{T+x+y} \ln \frac{y}{T+x+y} \tag{14}$$

$$S_B = -\Sigma'' \frac{T_{i.}}{T+x+y} \ln \frac{T_{i.}}{T+x+y} - \frac{T_1+x}{T+x+y} \ln \frac{T_1+x}{T+x+y} - \frac{T_2+y}{T+x+y} \ln \frac{T_2+y}{T+x+y} \tag{15}$$

and x and y are obtained by solving

$$\frac{\partial}{\partial x}(S_B/S) = 0; \frac{\partial}{\partial y}(S_B/S) = 0 \tag{16}$$

40.4 S_B–S_W FRONTIER

We have two goals here, viz. of minimizing S_B and of maximizing S_W and these goals can be achieved by minimizing

$$S_B - \lambda S_W, \tag{17}$$

where λ is any real number which can vary from 0 to ∞. If $\lambda = 0$, it means minimizing S_B only and if $\lambda = \infty$ it means maximizing S_W only. Other values of λ depend on the importance, we attach to the two goals of minimizing S_B and maximizing S_W.

The estimate $\hat{x}$ will be obtained by solving

$$\frac{d S_B}{d x} - \lambda \frac{d S_W}{d x} = 0 \tag{18}$$

or
$$\frac{d S_B}{d x} - \lambda \frac{d}{d x}(S - S_B) = 0 \quad \text{or} \quad \frac{d S_B}{d x}(1+\lambda) = \lambda \frac{d S}{d x} \tag{19}$$

or
$$-\Sigma''\left(1+\ln\frac{T_{i.}}{T+x}\right)\left(\frac{-T_{i.}}{(T+x)^2}\right) - \left(1+\ln\frac{T_1+x}{T+x}\right)\frac{T-T_1}{(T+x)^2}$$

$$= \frac{\lambda}{1+\lambda}\left[-\Sigma'\left(1+\ln\frac{x_{ij}}{T+x}\right)\left(\frac{-x_{ij}}{(T+x)^2}\right) + \left(1+\ln\frac{x}{T+x}\right)\frac{T}{(T+x)^2}\right]$$

or,
$$\Sigma'' T_{i.} \ln\frac{T_{i.}}{T+x} - (T - T_{i.})\ln\frac{T_{i.}+x}{T+x}$$

$$= \frac{\lambda}{1+\lambda}\left[\Sigma' x_{ij} \ln\frac{x_{ij}}{T+x} + T \ln\frac{x}{T+x}\right]$$

or,
$$\Sigma'' T_{i.} \ln T_{i.} - (T - T_{i.}) \ln (T_{i.} + x)$$

$$= \frac{\lambda}{1+\lambda}[\Sigma' x_{ij} \ln x_{ij} + T \ln x] \tag{20}$$

For every value of λ, this gives a value of x. The solution obtained in Section 3 corresponds to that value of λ for which

$$\left(\frac{d S_B}{d x}\right) / (d S / d x) = S_B/S \tag{21}$$

40.5 RANDOMISED BLOCK DESIGN

Let x_{ij} be the yield of the plot carrying the jth variety or treatment in the ith block ($i = 1, 2, \ldots, b; j = 1, 2, \ldots. t$) then

$$S_b = \text{entropy between blocks} = -\sum_{i=1}^{b} \frac{T_{i.}}{T} \ln \frac{T_{i.}}{T};\ \sum_{j=1}^{t} x_{ij} = T_{i.} \tag{22}$$

$$S_t = \text{entropy between treaments} = -\sum_{j=1}^{t} \frac{T_{.j}}{T} \ln \frac{T_{.j}}{T};\ \sum_{i=1}^{b} x_{ij} = T_{.j} \tag{23}$$

$$S = \text{total entropy} = -\sum_{j=1}^{t} \sum_{i=1}^{b} \frac{x_{ij}}{T} \ln \frac{x_{ij}}{T}, \tag{24}$$

so that,

$$S = -\sum_{j=1}^{t} \sum_{i=1}^{b} \frac{x_{ij}}{T} \ln \frac{x_{ij}/T}{(T_{i.}/T)(T_{.j}/T)} \cdot \frac{T_{i.}}{T} \cdot \frac{T_{.j}}{T}$$

$$= -\sum_{j=1}^{t} \sum_{i=1}^{b} \frac{x_{ij}}{T} \ln \frac{T_{i.}}{T} - \sum_{j=1}^{t} \sum_{i=1}^{b} \frac{x_{ij}}{T} \ln \frac{T_{.j}}{T}$$

$$= -\sum_{j=1}^{t} \sum_{i=1}^{b} \frac{x_{ij}}{T} \ln \frac{x_{ij}/T}{(T_{i.}/T).(T_{.j}/T)}$$

$$= -\sum_{i=1}^{b} \frac{T_{i.}}{T} \ln \frac{T_{i.}}{T} - \sum_{j=1}^{t} \frac{T_{.j}}{T} \ln \frac{T_{.j}}{T} - \sum_{j=1}^{t} \sum_{i=1}^{b} \frac{x_{ij}}{T} \ln \frac{x_{ij}/T}{(T_{i.}/T).(T_{.j}/T)}$$

$$= S_b + S_t - D(P:V), \tag{25}$$

where, $D(P:V)$ is the directed-divergence of the distribution x_{ij}/T from the distribution $(T_{i.}/T)\,(T_{.j}/T)$ and this is always ≥ 0 and vanishes if the blocks and treatments are independent.

Now if the blocks do not differ very much in their fertility, then the entropy S_b between blocks should be large. On the other hand if the treatments differs very much in their yield, the entropy between treatments should be small and so S_t / S_b, should be small. We can accordingly chose x to minimize S_t / S_b.

40.6 USE OF MINIMAX ENTROPY PRINCIPLE TO ESTIMATE x

We want to chose x to minimize the entropy between treatments and to maximize the entropy between blocks. We shall get a unique answer if we chose x to minimize S_t / S_b. Alternatively, we can get a family of estimates if we minimize $S_t - \lambda S_b$ for various positive real values of λ.

40.7 COMPARISON WITH USUAL ANALYSIS OF VARIANCE OF A RANDOMISED BLOCK DESIGN

In the usual analysis of variance, we use sum of squares and apparantly we do not use entropy.

However, the sum of squares are closely related to a measure of entropy other than Shannon's, viz. it is related to Havrada-Charvat entropy of second order [5].

Also entropy of a probability distribution is a monotonic decreasing function of the directed divergence of P from the uniform distribution. Moreover, the sums of squares can be expressed in terms of this directed-divergence.

The usual analysis of variance table for a RBD is :

Due to	D.F	$S.S_2$
Blocks	$b-1$	$\sum_{i=1}^{b} T_{i\cdot}^2/b - T^2/bt$
Treatments	$t-1$	$\sum_{j=1}^{t} T_{\cdot j}^2/t - T^2/bt$
Error	$(b-1)(t-1)$	$\sum_{j=1}^{t}\sum_{i=1}^{b} (x_{ij} - \hat{x}_{i.} - \hat{x}_{.j} + \hat{x}_{...})^2$
Total	$bt-1$	$\sum_{j=1}^{t}\sum_{i=1}^{b} x_{ij}^2 - T^2/bt$

Now, consider the probability distributions

$$\frac{T_{1\cdot}}{T}, \frac{T_{2\cdot}}{T}, \ldots, \frac{T_{b\cdot}}{T}$$

$$\frac{T_{\cdot 1}}{T}, \frac{T_{\cdot 2}}{T}, \ldots, \frac{T_{\cdot t}}{T}$$

$$\frac{x_{11}}{T}, \frac{x_{12}}{T}, \ldots, \frac{x_{tb}}{T} \tag{26}$$

and their Havrda-Charvat's [5] measures of directed-divergence of second order from the uniform distribution in each Case. These are:

$$\sum_{i=1}^{b} \left(\frac{T_{i\cdot}}{T}\right)^2 \left(\frac{1}{b}\right)^{-1} - 1 = \frac{bt}{T^2}\left(\sum_{i=1}^{b} \frac{T_{i\cdot}^{\ 2}}{t} - \frac{T^2}{bt}\right)$$

$$\sum_{j=1}^{t} \left(\frac{T_{\cdot j}}{T}\right)^2 \left(\frac{1}{t}\right)^{-1} - 1 = \frac{bt}{T^2}\left(\sum_{j=1}^{t} \frac{T_{\cdot j}^{\ 2}}{t} - \frac{T^2}{bt}\right)$$

$$\sum_{j=1}^{t}\sum_{i=1}^{b} \left(\frac{x_{ij}}{T}\right)^2 \left(\frac{1}{bt}\right)^{-1} - 1 = \frac{bt}{T^2}\left(\sum_{j=1}^{t}\sum_{i=1}^{b} x_{ij}^2 - \frac{T^2}{bt}\right) \tag{27}$$

Now,

$$\sum_{j=1}^{t}\sum_{i=1}^{b} (x_{ij} - \bar{x})^2 = t\sum_{i=1}^{b} (\bar{x}_{i.} - \bar{x})^2 + b\sum_{j=1}^{t} (\bar{x}_{.j} - \bar{x})^2$$

$$+ \sum_{j=1}^{t}\sum_{i=1}^{b} (x_{ij} - \bar{x}_{i.} - \bar{x}_{.j} + \bar{x})^2$$

$$= \sum_{i=1}^{b}\left(\frac{T_{i.}^2}{t} - \frac{T^2}{bt}\right) + \sum_{j=1}^{t}\left(\frac{T_{\cdot j}^{\ 2}}{b} - \frac{T^2}{bt}\right)$$

$$+ \sum_{j=1}^{t}\sum_{i=1}^{b} [x_{ij} - \bar{x}_{i.} - \bar{x}_{.j} + \bar{x}]^2 \tag{28}$$

In usual analysis of variance we have,

Total $S.S = S.S$ due to blocks + $S.S$ due to treatments + error, $S.S$

Here we have,

$D.D$ of all observations distribution from uniform distribution = $D.D$ of block totals distributions from corresponding uniform distribution + $D.D$ of treatments totals distribution from uniform distribution + error $D.D$.

In the usual analysis of variance, we choose the missing value so as to minimize the error sum of squares and adjust the sum of squares due to treatments for bias, and then finally

$$F = \frac{M.S \quad \text{for adjusted } S.S \text{ due to treaments}}{M.S \quad \text{for error}}$$

From our present point of view, we argue as follows :

We want to minimize the entropy between treatments, i.e we want to maximize the directed-divergence of the treatment distributions from the uniform distribution. For the same reason we want to maximize the error directed-divergence and as such we want to maximize the ratio of these directed-divergence or the corresponding $S.S$ and $M.S.S$. We can use an estimate of x as that value which maximizes the ratio.

40.8 AN ALTERNATIVE APPROACH TO ESTIMATION OF A MISSING VALUE IN RANDOMISED BLOCK DESIGN

Since we do not know the missing value, we cannot get a perfect solution. However, we know that there was a certain value x. We therefore calculate the F-ratio as a function of x. In fact F would be of the form

$$F = \frac{a_1 x^2 + a_2 x + a_3}{b_1 x^2 + b_2 x + b_3} \tag{29}$$

Its maximum and minimum value will occur when

$$(b_1 x^2 + b_2 x + b_3)(2a_1 x + a_2) - (a_1 x^2 + a_2 x + a_3)(2b_1 x + b_2) = 0$$

or

$$x^2 (b_2 a_1 - b_1 a_2) + x(2a_1 b_3 - 2a_3 b_1) + b_3 a_2 - a_3 b_2 = 0 \tag{30}$$

In general, one root of (30) will give $F_{\max} > F^*$ where F^* is the critical value of F for significance at a specified level of significance for the corresponding number of degrees of freedom. We can say that, whatever be the value of x, F will be significant and we shall reach definite conclusion from the experiment.

If both values of x are not positive, we can still find $F_{\max}$ and $F_{\min}$ of the range of $x > 0$ and we can still reach a conclusion of $> F^*$ or $F_{\max} < F^*$.

The principle used here is that when we have some missing information ; we should consider all possible values and if all these values lead to the same decision, we should take that decision. If same values point to one decision and other values point to another decision, then honestly we should just make *this* statement.

40.9 APPLICATION OF THIS PRINCIPLE TO OTHER DESIGN

In other design, we have a number of F values. Thus in Latin square design, we have an F_1 corresponding to rows and F_2 corresponding, to column and an F_3 corresponding to treatments. We can find maximum and minimum value for each and take our decision. These maximum and minimum value of F's may correspond to different value of x's but this does not matter. In fact, even in ordinary analysis, of variance it is proposed to minimize the total sum of squares and error sum of squares separately (and this will be for different value of x), then obtain the *S.S.* due to treatments by subtraction [2,33].

REFERENCES

1. J.P Burg : (1970), ''The Relationship between Maximum Entropy Spectra and Maximum Likelihood Spectra, in Modren Spectra Analysis'' ed D.G. Childers, pp 130-131.
2. W. Cochran and G.W Cox, ''Experimental Designs'', 2nd edition John Wiley & Sons, New York.
3. R.A. Fisher (1947), ''The Design of Experiments", Oliver and Loyd, London.
4. S. Guiasu [1977], '' Information Theory with Applications'', Mcgraw Hill.
5. J.H. Havrda and F. Charvat (1967), ''Quantification methods of classification Process : Concept of structural entropy'', Kybernetica Vol. 3, pp. 30-35.
6. E.T. Jaynes (1957), '' Information Theory and statistical Mechanics'' Physical Reviews, Vol. 106 pp 620-630.
7. J.N. Kapur (1989),''On Maximum Entropy Estimation of Missing Values'' Not Acad Sci Letter 6(2), 59-65.
8. J.N. Kapur (1989), ''Maximum Entropy Models in Science & Engineering", Wiley Eastern, New Delhi and John Wiley, New York.
9. J.N. Kapur (1991), '' Estimation of missing values, in Contingency tables by using principle of minimum interdependence'', MSTS Res. Rep.
10. J.N. Kapur and H.K. Kesavan (1987), '' Generalised Maximum Entropy Principle (with Applications)'', Sandford Educational Press, University of Waterloo.
11. J.N. Kapur and H.K. Kesvan (1992), ''Entropy Optimization Principles with Applications'', Academic Press, New York.
12. C.E. Shannon (1948),'' A Mathematical Theory of Communication'', Bell System Tech. J. Vol. 27 pp. 379-423, 623-659.
13. S. Watanabe (1969), ''Knowing and Guessing'', John Wiley & Sons, New York.

41

HIGHER ORDER MOMENTS OF DISTRIBUTIONS OF STATISTICAL MECHANICS

[The maximum-entropy principle has been earlier used to derive the Bose-Einstein, Fermi-Dirac and Intermediate Statistics distributions of statistical mechanics. In the present paper, variances and higher order moments of the distributions of number of particles in each energy level are obtained. The conditions under which the expected number of particles in an energy level increases or decreases with the energy level are also given. A similar discussion is given for the total energy in each level.]

41.1 DERIVATIONS OF THE DISTRIBUTIONS

Let there be n energy levels with energies $\varepsilon_1, \varepsilon_2, \ldots, \varepsilon_n$ Let q_i be the probability of the ith energy level being chosen and once this level is chosen, let p_{ij} be the conditional probability of there being j particles in this level. Let

$$P_{ij} = q_i \, p_{ij}, \tag{1}$$

then assuming that the number of particles in the ith energy level can vary from 0 to m_i, we get

$$\sum_{j=0}^{m_i} P_{ij} = 1, \quad i = 1, 2, \ldots, n \tag{2}$$

Also,

$$\sum_{i=1}^{n} \sum_{j=0}^{m_i} P_{ij} = \sum_{i=1}^{n} q_i \sum_{j=0}^{m_i} p_{ij} = \sum_{i=1}^{n} q_i = 1 \tag{3}$$

Let the total expected number of particles and the total expected energy be prescribed, so that

$$\sum_{i=1}^{n} \sum_{j=0}^{m_i} j \, P_{ij} = A \tag{4}$$

$$\sum_{i=1}^{n} \varepsilon_i \sum_{j=0}^{m_i} j \, P_{ij} = B \tag{5}$$

The entropy of the system is

$$-\sum_{i=1}^{n} \sum_{j=0}^{m_i} P_{ij} \ln P_{ij} = -\sum_{i=1}^{n} \sum_{j=0}^{m_i} q_i \, p_{ij} \ln q_i \, p_{ij}$$

$$= -\sum_{i=1}^{n} q_i \ln q_i - \sum_{i=1}^{n} q_i \sum_{j=0}^{m_i} p_{ij} \ln p_{ij} \tag{6}$$

Since we have no information on q_i's, we may take $q_i = 1/n$ for each i, so that we have to maximize,

$$S = -\sum_{i=1}^{n} \sum_{j=0}^{m_i} p_{ij} \ln p_{ij}, \tag{7}$$

subject to (2), (4) and (5). Using Lagrange's method, we get

$$p_{ij} = a_i\, e^{-(\lambda + \mu \varepsilon_i) j} = \frac{e^{-(\lambda+\mu\varepsilon_i)j}}{\sum_{j=0}^{m_i} e^{-(\lambda+\mu\varepsilon_i)j}} \tag{8}$$

Also the expected number of particles in the i_{th} energy level is given by

$$\bar{n}_i = \sum_{j=0}^{m_i} j\, p_{ij} = \frac{x_i + 2x_i^2 + \ldots + m_i x_i^{m_i}}{1 + x_i + x_i^2 + \ldots + x_i^{m_i}}, \tag{9}$$

where, $$x_i = e^{-(\lambda+\mu\varepsilon_i)}, \quad i = 1, 2, \ldots, n \tag{10}$$

If $m_i = \infty$, i.e if each energy level can have any number of particles from 0 to ∞, we get the Bose-Einstein distribution

$$\bar{n}_i = \frac{1}{e^{\lambda+\mu\varepsilon_i} - 1} \tag{11}$$

If $m_i = 1$, i.e if each energy level can have either 0 or 1 particles, we get the Fermi-Dirac distribution.

$$\bar{n}_i = \frac{1}{e^{\lambda+\mu\varepsilon_i} + 1} \tag{12}$$

The distribution (9) is itself called the Intermediate Statistics distribution.

The details of the above derivations are given in Forte and Sempi [1], Kapur and Kesavan [5] and Kapur [2,3,4].

41.2 VARIANCE AND HIGHER MOMENTS

For the Intermediate Statistics distribution, Variance σ_i^2 of the number of particles in the ith energy level is given by

$$\sigma_i^2 = \sum_{j=0}^{m_i} j^2 p_{ij} - \left(\sum_{j=0}^{m_i} j\, p_{ij}\right)^2$$

$$= \frac{x_i + 2^2 x_i^2 + \ldots + m_i^2 x_i^{m_i}}{1 + x_i + x_i^2 + \ldots + x_i^{m_i}} - \left(\frac{x_i + 2x_i^2 + \ldots + m_i x_i^{m_i}}{1 + x_i + \ldots + x_i^{m_i}}\right)^2 \tag{13}$$

For the Bose-Einstein distribution, this gives

$$\sigma_i^2 = \frac{x_i}{(1-x_i)^2} \tag{14}$$

Similarly, for the Fermi-Dirac distribution, we get

$$\sigma_i^2 = \frac{x_i}{(1+x_i)^2} \tag{15}$$

In both cases,

$$\frac{\text{Standared deviation}}{\text{Mean}} = \sqrt{x_i} = e^{-\frac{1}{2}(\lambda+\mu\,\varepsilon_i)} \tag{16}$$

The kth order moment about the origin is

$$\sum_{j=0}^{m_i} p_{ij}\, j^k \tag{17}$$

and this can be easily summed up for any positive integral value of k for both $m_i = \infty$ and $m_i = 1$.

41.3 VARIATION OF THE NUMBER OF PARTICLES IN EACH ENERGY LEVEL

Maximizing entropy means equalising P_{ij}'s as much as possible. In the absence of constraints, (8) would give on putting, $\lambda = 0$, $\mu = 0$,

$$p_{ij} = \frac{1}{m_i + 1}, \tag{18}$$

so that there is a uniform distribution in each energy level.

In the presence of constraints, (8) gives a Maxwell-Boltzmann distribution (Kapur [4]) for the distribution of number of particles in each energy level. In fact we get n dependent M.B distributions with means $\overline{n}_1, \overline{n}_2, \ldots, \overline{n}_n$ and variances $\sigma_1^2, \sigma_2^2, \ldots, \sigma_n^2$. The parameters of these n distribution are of course related, since these are $\lambda + \mu\,\varepsilon_1, \lambda + \mu\,\varepsilon_2, \ldots, \lambda + \mu\,\varepsilon_n$.

When the number of particles in each energy level can go upto ∞, the n distributions are geometric distributions and when each $m_i = 1$, then each distribution is a Binomial distribution.

For the Bose-Einstein distribution, λ, μ are determined by

$$\sum_{i=1}^{n} \frac{1}{e^{\lambda+\mu\,\varepsilon_i} - 1} = A, \quad \sum_{i=1}^{n} \frac{\varepsilon_i}{e^{\lambda+\mu\,\varepsilon_i} - 1} = B, \tag{19}$$

so that,

$$\frac{B}{A} = \sum_{i=1}^{n} \frac{\varepsilon_i}{e^{\lambda+\mu\,\varepsilon_i} - 1} \Big/ \sum_{i=1}^{n} \frac{1}{e^{\lambda+\mu\,\varepsilon_i} - 1} \tag{20}$$

Thus B/A is the weighted mean of $\varepsilon_1, \varepsilon_2, \ldots, \varepsilon_n$, the weights being proportional to $(e^{\lambda+\mu\,\varepsilon_i} - 1)^{-1}$.

If $\mu = 0$, weights are equal and $B/A = \bar{\varepsilon}$,
If $\mu > 0$, the weights decrease with ε_i and if we assume

$$\varepsilon_1 < \varepsilon_2 < \varepsilon_3 \ldots < \varepsilon_n \tag{21}$$

the weights decrease as we go to higher energy levels and $B/A < \bar{\varepsilon}$. If $\mu < 0$, the $B/A > \bar{\varepsilon}$, so that we get Figure 41.1

$$\begin{array}{ccc} \mu < 0 & \mu = 0, & \mu > 0 \\ \hline \frac{B}{A} > \bar{\varepsilon} & B/A = \bar{\varepsilon} & \frac{B}{A} < \bar{\varepsilon} \end{array} \tag{22}$$

Fig. 41.1

We first consider the case when $B/A < \bar{\varepsilon}$, so that we should get $\mu > 0$

Now,
$$\bar{n}_i = \frac{1}{e^{\lambda+\mu\varepsilon_i} - 1}, \quad \frac{\bar{n}_j}{\bar{n}_i} = \frac{e^{\lambda+\mu\varepsilon_i} - 1}{e^{\lambda+\mu\varepsilon_j} - 1} \tag{23}$$

so that,
$$\bar{n}_i \mathrel{\substack{\geq \\ <}} \bar{n}_j \text{ according as } \varepsilon_j \mathrel{\substack{\geq \\ <}} \varepsilon_i \tag{24}$$

Thus *higher energy levels contain relatively smaller expected number of particles when $B/A < \bar{\varepsilon}$. However, higher energy level will contain larger expected number of particles when $B/A > \bar{\varepsilon}$*

Again,
$$\bar{n}_i \varepsilon_i = \frac{\varepsilon_i}{e^{\lambda+\mu\varepsilon_i} - 1} \tag{25}$$

Now consider,

$$\frac{d}{dx}\left(\frac{x}{Ae^{\mu x} - 1}\right) = \frac{A e^{\mu x} - 1 - \mu x A e^{\mu x}}{(Ae^{\mu x} - 1)^2} = \frac{A e^{y} - 1 - A y e^{y}}{(Ae^{y} - 1)^2} \tag{26}$$

Let,
$$g(y) = A e^{y} - 1 - A y e^{y} \tag{27}$$

so that, ,
$$g'(y) = -A y e^{y} < 0 \tag{28}$$

Also, $g(0) = A - 1$ (29)

If $A \leq 1$, $g(y)$ will be negative, $x/(A e^{\mu x} - 1)$ will always decrease and $\bar{n}_i \varepsilon_i$ will always decrease with ε_i. Thus when $\lambda \leq 0$ and $B/A < \bar{\varepsilon}$, the energy contained in higher energy levels will be less than that contained in lower levels. If $\lambda > 0$ and $B/A < \bar{\varepsilon}$, the energy contained in higher energy levels will be more.

These results will be reversed if $B/A > \bar{\varepsilon}$

Thus we get the following Figures 41.2 - 41.5.

41.4 CONCLUDING REMARKS

(i) The *BE*, *FD* and *IS* distributions of statistical mechanics give us the distributions of expected number of particles in each energy level. These are no

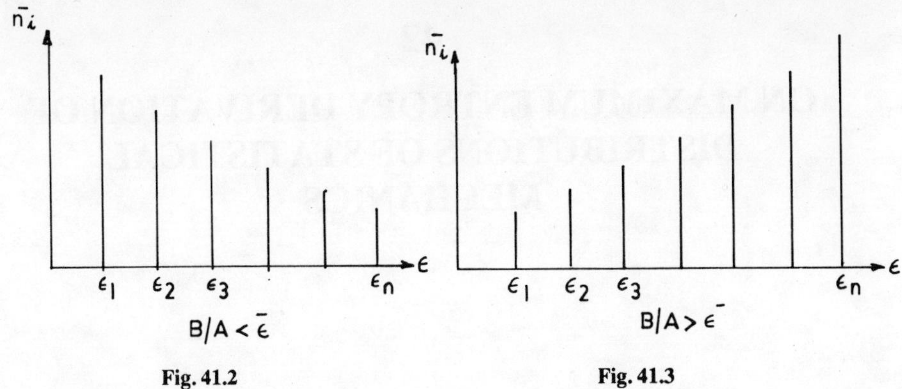

Fig. 41.2 **Fig. 41.3**

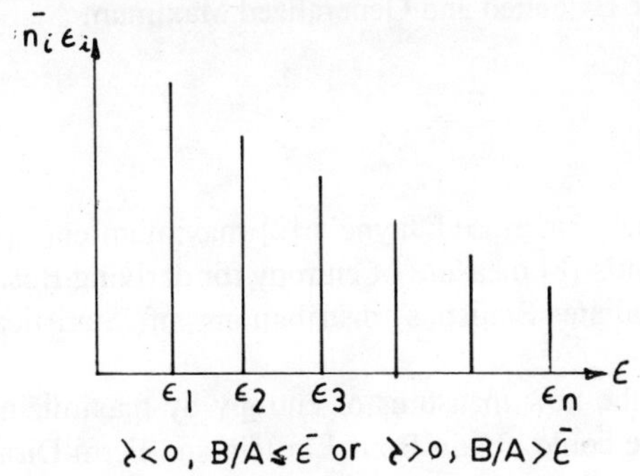

$\bar{n}_i \epsilon_i$ ... ϵ_1 ϵ_2 ϵ_3 ϵ_n ϵ

$\lambda < 0$, $B/A \le \bar{\epsilon}$ or $\lambda > 0$, $B/A > \bar{\epsilon}$

Fig. 41.4

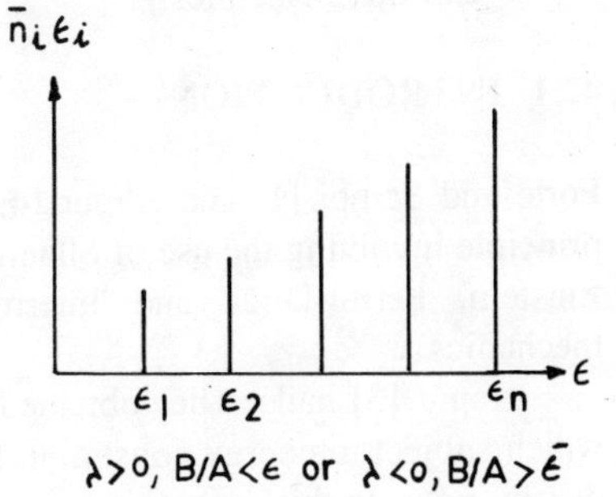

$\bar{n}_i \epsilon_i$ ... ϵ_1 ϵ_2 ϵ_n ϵ

$\lambda > 0$, $B/A < \epsilon$ or $\lambda < 0$, $B/A > \bar{\epsilon}$

Fig. 41.5

doubt useful, but knowledge of variances of these numbers should give us deeper insights. It is shown that it is possible to find these variances.

(ii) It is shown that if the ratio of expected total energy and the expected number of particles is less that $\bar{\varepsilon}$, then the expected number of particles in an energy level decreases, with the rise of energy in that level. The result is reversed when this ratio is greater than $\bar{\varepsilon}$.

(iii) The variation of total expected energy in each level is also studied with the energy level.

REFERENCES

1. B. Forte and C. Sempi (1976), "Maximizing Conditional Entropy : A derivation of quantal statistics", Rendi Conta de Mathematica, Vol. 9, pp 551-566.
2. J.N. Kapur (1972), "Measures of uncertainty, mathematical programming and physics", Jour. Ind. Soc. Agri. Stat. Vol. 24, pp 47-66.
3. J.N. Kapur (1977), " Non-additive measures of entropy and distributions of statistical mechanics", Ind. Jour. Pure and Applied. Math. Vol 14, No 11, pp. 1372-1384.
4. J.N. Kapur (1980), " Maximum-Entropy Models in Science and Engineering", Wiley Eastern, New Delhi & John Wiley, New York.
5. J.N. Kapur and H.K. Kesavan (1989). " Generalised Maximum Entropy Principle with Applications", Sandford Educational Press, Waterloo University, Canada.

42

ON MAXIMUM ENTROPY DERIVATION OF DISTRIBUTIONS OF STATISTICAL MECHANICS

[In the present paper two new Intermediate Statistics distributions have been derived by using the Extended and Generalized Maximum Entropy Principles.]

42.1 INTRODUCTION

Forte and Sempi [1] and Kapur [4,5,6] have used Jayne's [2] maximum entropy principle involving the use of Shannon's [8] measure of entropy for deriving Bose-Einstein, Fermi-Dirac and Intermediate Statistics distributions of Statistical mechanics.

Kapur [3] had earlier obtained the new measures of entropy by maximizing which subject to energy constraint, he could obtain Bose-Einstein and Ferm-Dirac distributions. In this process, he had implicitly used inverse maximum entropy principle [7]

In the present paper we have used the extended maximum entropy principle (where we use proportions in place of probabilities) and the Generalized Maximum Entropy Principle (where we use measures of entropy other than Shannon's) to derive two new Intermediate Statistics Distributions, not obtained earlier from Jaynes maximum entropy principle.

42.2 DERIVATION OF THE FIRST INTERMEDIATE STATISTICS DISTRIBUTION : USE OF GENERALISED MAXIMUM ENTROPY PRINCIPLE

Let n = number of energy levels

N = total number of particles

$N\overline{\in}$ = total energy of the system

M_i = maximum number of particles permissible to the ith level.

N_i = number of particles in the ith energy level

It is given that

$$N_1+N_2+\ldots+N_n=N \tag{1}$$

$$N_1 \in_1 + N_2 \in_2 + \ldots + N_n \in_n = N\overline{\in} \tag{2}$$

$$0 \le N_i \le M_i, \quad i = 1, 2, \ldots, n \tag{3}$$

We are taking N_i's as real numbers rather than as integers. This will be a good approximation, if N is large.

There may be an infinity of sets of values of $N_1, N_2, \ldots, N_n$ consistent with (1), (2) and (3). We have to choose only one of these

For this we need an additional criterion. Let us use the criterion that $N_1, N_2, \ldots, N_n$ should be as equal as possible subject to (1), (2) and (3).

To maximize the equality, we maximize the equality measure or entropy measure

$$-\sum_{i=1}^{n} N_i \ln N_i \tag{4}$$

subject to (1) and (2) by using Lagrangian method, but the maximizing values of $N_1, N_2, \ldots, N_m$ may not satisfy (3).

Lagrangian method is not adequate for maximizing (4) subject to (1), (2) and (3).

We get a complex mathematical programming (optimization) problem, when the constraints (3) are taken into account. This is not easy to solve.

We therefore look for alternative methods.

We also examine the criterion of maximum equality critically.

If $M_1, M_2, \ldots, M_n$ are equal, this criterion will make sense, but when these are unequal, Laplace's principle of insufficient reason does not apply even if (2) is absent.

Thus if only constraint (1) is applied, the inequality of $M_1, M_2, \ldots, M_n$ is sufficient reason to choose $N_1, N_2, \ldots, N_n$ as different. What values shall we give them?

The most sensible set of values has to be proportional to $M_1, M_2, \ldots, M_n$, i.e. in the presence of (1) above, we choose

$$\frac{N_1}{M_1} = \frac{N_2}{M_2} = \ldots = \frac{N_n}{M_n} = \frac{N}{M}, \quad M = \sum_{i=1}^{n} M_i \tag{5}$$

The problem will have a solution only if $M \ge N$ and in that case (5) will give

$$N_1 \le M_1, N_2 \le M_2, \ldots, N_n \le M_n, \tag{6}$$

so that constraint (3) will be satisfied.

We now modify (4) to

$$-\sum_{i=1}^{n} N_i \ln N_i - \sum_{i=1}^{n} (M_i - N_i) \ln (M_i - N_i) \tag{7}$$

This is a concave funtion of $N_1, N_2, \ldots, N_n$.

If we maximize this subject to (1) only, we see

$$\frac{N_i}{M_i - N_i} = \text{const.} \quad \text{or} \quad \frac{N_i}{M_i} = \text{const.}, \tag{8}$$

which meets our modifed criterion.

The function (7) is defined only when $N_i \le M_i$, $i = 1, 2, \ldots, n$.

It may be noticed that (7) is a concave function of $N_1, N_2, ..., N_n$ so that when it is maximized subject to given constraints (1) and (2), the local maximum will be the global maximum.

Maximizing (7) subject to (1) and (2) we get

$$\frac{N_i}{M_i - N_i} = e^{-\lambda - \mu \epsilon_i} \tag{9}$$

or

$$N_i = \frac{M_i}{e^{\lambda + \mu \epsilon_i} + 1} \tag{10}$$

This verifies that $N_i \le M_i$ and the constraints (3) are automatically satisfied. The constants λ and μ are determined by using

$$\sum_{i=1}^{n} \frac{M_i}{e^{\lambda + \mu \epsilon_i} + 1} = N, \quad \sum_{i=1}^{n} \frac{M_i \epsilon_i}{e^{\lambda + \mu \epsilon_i} + 1} = N \overline{\epsilon} \tag{11}$$

If $M_i = 1$ for each i, (10) gives

$$N_i = \frac{1}{e^{\lambda + \mu \epsilon_i} + 1} \tag{12}$$

which is Fermi-Dirac distribution of statistical mechanics in which each energy level can have zero or one particle.

If $M_i > 1$ for each i, we get a new intermediate statistics distribution.

42.3 DERIVATION OF THE SECOND INTERMEDIATE STATISTICS DISTRIBUTION : USE OF GENERALISED MAXIMUM PRINCIPLE

We now modify (7) to

$$-\sum_{i=1}^{n} N_i \ln N_i - \sum_{i=1}^{n} \frac{1}{a_i} (1 - a_i N_i) \ln (1 - a_i N_i) \tag{13}$$

This is also a concave function of $N_1, N_2, ..., N_n$. Maximizing this subject to (1) and (2) we get

$$\frac{N_i}{1 - a_i N_i} = e^{-\lambda - \mu \epsilon_i} \tag{14}$$

It is obvious that each $N_i < \frac{1}{a_i}, i = 1, 2, ..., n,$ (15)

so that there is maximum value for number of particles in each energy level and this number varies from level to level. If each $a_i \to 0$, we get

$$N_i = e^{-\lambda - \mu \epsilon_i}, \tag{16}$$

which is Maxwell-Boltzmann distribution in which each energy level can have any number of particles.

The measure (13) is concave even if a_i is negative. In this case from (14) we find

$$N_i = \frac{1}{e^{\lambda+\mu\epsilon_i} + a_i} \tag{17}$$

If $a_i > 0$ the maximum number of particles in each energy level is finite. If $a_i = 0$ or negative, there is no upper limit on the number of particles in each energy level.

In F.D. distribution $a_i = 1$, in BE distribution $a_i = -1$ and in MB distribution, $a_i = 0$. Thus (17) includes all the three and generalizes all the three. It even allows $a_i > 0$ for some i's and $a_i < 0$ for other i's.

If each $a_i \geq 0$, there is a guarantee that each $N_i > 0$ but if $a_i < 0$ there is no such guarantee. What is guaranted that each $N_i/(1 - a_i N_i) > 0$.

Let $a_i = a$ for each i, then to determine λ and μ we have the equations,

$$\sum_{i=1}^{n} \frac{1}{e^{\lambda+\mu\epsilon_i} + a} = N, \quad \sum_{i=1}^{n} \frac{1}{e^{\lambda+\mu\epsilon_i} + a} = N\bar{\epsilon} \tag{18}$$

or

$$\sum_{i=1}^{n} \frac{1}{e^{\lambda'+\mu\epsilon_i} + 1} = Na, \quad \sum_{i=1}^{n} \frac{1}{e^{\lambda'+\mu\epsilon_i} + 1} = Na\,\bar{\epsilon} \tag{19}$$

Thus if we take the total number of particles as Na and total energy as $Na\,\bar{\epsilon}$ and solve for λ' and μ from (19), the Intermediate statistics distribution is

$$\bar{N}_i = \frac{1}{e^{\lambda+\mu\epsilon_i} + a} = \frac{1/a}{e^{\lambda'+\mu\epsilon_i} + 1} \tag{20}$$

Thus we can deduce the results for the intermediate statistics distribution $[\exp(\lambda + \mu\epsilon_i) + a]^{-1}$ from the results for the ISD $\{\exp[(\lambda + \mu\epsilon_i) + 1]\}^{-1}$.

42.4 CONCLUDIING REMARKS

The two new distribution obtained in this paper are given by (9) and (17)

$$N_i = \frac{M_i}{e^{\lambda+\mu\epsilon_i} + 1}, \quad N_i = \frac{1}{e^{\lambda+\mu\epsilon_i} + a_i} = \frac{1/a_i}{\frac{1}{a_i} e^{\lambda+\mu\epsilon_i} + 1} \tag{21}$$

In the first case $N_i \leq M_i$, in the second case $N_i \leq \frac{1}{a_i}$ but the second distribution cannot be obtained from the first by putting $M_i = 1 / a_i$. The two distributions are distinct and they are also distinct from the ISD obtained by Forte and Sempi [1] and Kapur [4].

REFERENCES

1. B. Forte and C. Sempi (1976),'' Maximising Conditional Entropy : A derivation of quantal statistics,'' Rend Condi de Mathematics, Vol. 9, 551-566.
2. E.T. Jaynes (1957),''Information Theory and Statistical Mechanics'', Physical Reviews, Vol. 106, 620-630.
3. J.N. Kapur (1972), ''Measures of Uncertainty, Mathematical Programmig and Physics'' Journ. Ind. Soc. Ag. Stat. Vol. 24, 47-66.

4. J.N. Kapur (1973), ''Non-additive measures of entropy and distributions of statistical mechanics'', Ind. Jour. Pure App. Maths. Vol. 14 No. 11, 1372-1384.
5. J.N. Kapur (1990),''Maximum Entropy Models in Science and Engineering'', Wiley Eastern, New Delhi, John Wiley, New York.
6. J.N. Kapur and H.K. Kesavan (1987),''Generalised maximum entropy principle (with application), Sandford Educational Press, University of Waterloo, Canada.
7. H.K. Kesavan and J.N. Kapur (1980),''Generalised maximum entropy principle'', IEEE Systems, Mans and Cybernetics 19 (9), 1042-1052.
8. C.E. Shannon (1948), ''Mathematical Theory of Communication'' Bell System Tech. Journ. Vol. 27, 419-423, 623-659.

43

SOME NEW STATISTICAL MECHANICS DISTRIBUTIONS AND THEIR APPLICATIONS

[Some new distributions which are similar to Bose-Einstein, Fermi-Dirac and Intermediate Statistics distributions of statistical mechanics are derived by using the maximum entropy principle. These are compared with Maxwell-Boltzmann distribution and their applications to some areas, specially to those in which Maxwell-Boltzmann distribution is usually applied, are considered].

43.1 INTRODUCTION

The maximum entropy principle of Jaynes [2] has been used frequently to derive the distributions of statistical mechanics by maximizing the entropy of the system subject to some given constraints.

The Maxwell-Boltzmann distribution is obtained when there is only one constraint on the system which prescribes the expected energy per particle of the system (Kapur [6]).

The Bose-Einstein (B.E.), Fermi-Dirac (F, D) and Intermediate Statistics (I.S.) distributions are obtained by maximizing the entropy subject to two constraints (Forte and Sempi [1], Kapur [4, 5, 6] Kapur and Kesavan [7, 8]).

(i) The expected number of particles in the system is prescribed,
(ii) the expected total energy of the system is prescribed.

Though these distributions arose in the first instance in statistical mechanics, they are widely applicable in urban and regional planning, transportation studies, finance, banking and economics (Kapur [6] Kapur and Kesavan [7, 8]).

In the present paper, we obtain the distributions when only one constraint is given. This may prescribe,

(i) the expected number of particles in the system,
or (ii) the total expected energy in the system,
or (iii) the expected energy per particle in the system.

Even in cases (ii) and (iii), where there is only expected energy constraint, as in the case of M.B. distribution, the distributions obtained will differ from the M.B. distribution because of the different structures, of the basic models. We shall therefore compare the new distributions with the M.B. distribution.

The new distributions can have some interesting applications. We shall also study some of these in the present paper.

43.2 THE CLASSICAL STATISTICAL MECHANICS DISTRIBUTIONS

Let p_{ij} be the probability of there being j particles in the ith energy level, and let m_i be the maximum number of particles permissible in the ith energy level, then if the expected number of a particles in the system is prescribed as A and the expected total energy of the system is prescribed as B, we get

$$\sum_{i=1}^{n} \sum_{j=0}^{m_i} j\, p_{ij} = A \tag{1}$$

$$\sum_{i=1}^{n} \varepsilon_i \sum_{j=0}^{m_i} j\, p_{ij} = B \tag{2}$$

Again the ith energy level will have either 0 or 1 or 2 or .... m_i particles, so that we get the constraints

$$\sum_{j=0}^{m_i} p_{ij} = 1, \quad i = 1, 2, \ldots, n \tag{3}$$

By maximizing the entropy measure,

$$-\sum_{i=1}^{n} \sum_{j=0}^{m_i} p_{ij} \ln p_{ij} \tag{4}$$

subject to constraints (1), (2) and (3), by using Lagrange's method, we get

$$p_{ij} = a_i\, e^{-(\lambda + \mu\varepsilon_i) j} = a_i\, x_i^j; \quad x_i = e^{-(\lambda + \mu\varepsilon_i)} \tag{5}$$

Now three cases arise :

(i) In the first case, j can vary from 0 to ∞, so that from (3) and (5),

$$a_i = (1 - x_i),\, p_{ij} = (1 - x_i)\, x_i^j \tag{6}$$

and,

$$\bar{x}_i = \sum_{j=0}^{\infty} j\, p_{ij} = (1 - x_i) \frac{x_i}{(1 - x_i)^2} = \frac{x_i}{1 - x_i} = \frac{1}{\frac{1}{x_i} - 1}$$

$$= \frac{1}{e^{\lambda + \mu\varepsilon_i} - 1}, \tag{7}$$

which is known as the Bose Einstein (B.E.) distribution.

(ii) In the second case, j can take values 0 and 1 only, so that

$$a_i = \frac{1}{1 + x_i}, \quad p_{ij} = \frac{x_i}{1 + x_i} \tag{8}$$

and,

$$\bar{n}_i = \sum_{j=0}^{1} j\, p_{ij} = p_{i1} = \frac{x_i}{1 + x_i} = \frac{1}{\frac{1}{x_i} + 1}$$

$$= \frac{1}{e^{\lambda + \mu\varepsilon_i} + 1}, \tag{9}$$

which is known as Fermi-Dirac distribution.

(iii) In the third case, j can vary from 0 to m_i from the ith level, so that

$$a_i = \frac{1}{1 + x_i + x_i^2 + \ldots + x_i^{m_i}}, \qquad p_{ij} = \frac{x_i^j}{1 + x_i + \ldots + x_i^{m_i}} \tag{10}$$

and,

$$\overline{n}_i = \frac{x_i + 2\,x_i^2 + \ldots + m_i\,x_i^{m_i}}{1 + x_i + x_i^2 + \ldots + x_i^{m_i}} \tag{11}$$

This is called the Intermediate Statistics Distribution.

43.4 THE NEW DISTRIBUTIONS WHEN THERE IS ONE CONSTRAINT ON THE EXPECTED NUMBER OF PARTICLES

Maximizing (4) subject to constraints (1) and (3), we get

$$p_{ij} = b_i\, e^{-\lambda j} = b_i\, y^j, \quad y = e^{-\lambda} \tag{12}$$

(i) When $\quad m_i = \infty, \quad b_i = (1 - y)$

$$\overline{n}_i = \sum_{j=0}^{\infty} j\, p_{ij} = \frac{b_i\, y}{(1-y)^2} = \frac{y}{1-y}$$

and since from (1), $\quad \sum_{i=1}^{n} \overline{n}_i = A,$

we get, $$\overline{n}_i = \frac{A}{n} \tag{13}$$

(ii) When $\quad m_i = 1, \quad b_i = \dfrac{1}{1+y}$

$$\overline{n}_i = \sum_{j=0}^{1} j\, p_{ij} = p_{i1} = b_i\, y = \frac{y}{1+y},$$

so that again $$\overline{x}_i = \frac{A}{n} \tag{14}$$

(iii) When $\quad m_i \neq 1 \;$ or $\; \infty, \quad b_i = \dfrac{1}{1 + y + y^2 + \ldots + y^{m_i}}$

and $$\overline{n}_i = \frac{y + 2\,y^2 + \ldots + m_i\,y^{m_i}}{1 + y + \ldots + y^{m_i}} \tag{15}$$

where, y is obtained by solving

$$\sum_{i=1}^{n} \frac{y + 2\,y^2 + \ldots + m_i\,y^{m_i}}{1 + y + y^2 \ldots + y^{m_i}} = A \tag{16}$$

43.4 THE NEW DISTRIBUTIONS WHEN THERE IS ONLY ONE CONSTRAINT ON THE TOTAL EXPECTED ENERGY OF THE PARTICLES

Maximizing (4) subject to (2) and (13), we get

$$p_{ij} = c_i \quad e^{-\mu j \varepsilon_i} = c_i\, z_i^j; \quad z_i = e^{-\mu \varepsilon_i} \tag{17}$$

(i) When $\quad m_i \to \infty, \quad c_i = (1 - z_i), \quad p_{ij} = (1 - z_i) \quad z_i^j$

$$\bar{n}_i = (1 - z_i) \quad \sum_{j=0}^{\infty} j\, z_i^j = (1 - z_i) \frac{z_i}{(1 - z_i)^2} = \frac{1}{e^{\mu \varepsilon_i} - 1}\,, \tag{18}$$

where μ is determined from

$$\sum_{i=1}^{n} \frac{c_i}{e^{\mu \varepsilon_i} - 1} = B \tag{19}$$

Having determined μ from (19), (18) then gives the expected number of particles in the ith energy level and the expected total number of particles in the system is given by

$$N = \sum_{i=1}^{n} \bar{n}_i \tag{20}$$

(ii) Let $\quad m_i = 1, \quad c_i = \frac{1}{1 + z_i}, \quad p_{ij} = \frac{z_i}{1 + z_i}$

$$\bar{n}_i = p_{i1} = \frac{z_i}{1 + z_i} = \frac{1}{e^{\mu \varepsilon_i} + 1}, \tag{21}$$

where μ is determined from $\quad \sum_{i=1}^{n} \frac{\varepsilon_i}{e^{\mu \varepsilon_i} + 1} = B \quad (22)$

(iii) Let $\quad m_i \neq 1, \infty, \quad c_i = \frac{1}{1 + z_i + z_i^2 + \ldots + z_i^{m_i}}$

$$\bar{n}_i = \frac{z_i + 2\, z_i^2 + \ldots + m_i\, z_i^{m_i}}{1 + z_i + z_i^2 + \ldots + z_i^{m_i}}\,,$$

where z_i is determined by solving

$$\sum_{i=1}^{n} \frac{z_i + 2\, z_i^2 + \ldots + m_i\, z_i^{m_i}}{1 + z_i + z_i^2 + \ldots + z_i^{m_i}}\, \varepsilon_i = B \tag{24}$$

In each of the three cases, we can find the total expected number of particles in the system.

43.5 THE DISTRIBUTIONS WHEN THE EXPECTED ENERGY PER PARTICLE IS PRESCRIBED

In this case constraint in addition to (3) is

$$\frac{\sum_{i=1}^{n} \varepsilon_i \sum_{j=0}^{m_i} j\, p_{ij}}{\sum_{i=1}^{n} \sum_{j=0}^{m_i} j\, p_{ij}} = \hat{\varepsilon}$$

or,
$$\sum_{i=1}^{n} (\varepsilon_i - \hat{\varepsilon}) \sum_{j=0}^{m_i} j\, p_{ij} = 0 \tag{25}$$

Maximizing (4) subject to (3) and (25), we get

$$p_{ij} = k_i\, e^{-j(\varepsilon_i - \hat{\varepsilon})\mu} \tag{26}$$

Case (i) $\quad m_i = \infty, \quad$ so that $\quad k_i \sum_{j=0}^{\infty} e^{-j(\varepsilon_i - \hat{\varepsilon})\mu} = 1 \qquad (27)$

If $\varepsilon_i > \hat{\varepsilon}$, the series has a finite sum if $\mu > 0$

If $\varepsilon_i < \hat{\varepsilon}$, the series has a finite sum if $\mu < 0$,

but μ cannot be both positive and negative . The sum is not finite even when $\mu = 0$. Thus in this case the maximum entropy distribution does not exist.

This is easily seen otherwise since if $\bar{n}_1, \bar{n}_2, \ldots, \bar{n}_n$ could be a *MEPD*, then so could be $k\,\bar{n}_1, k\,\bar{n}_2, \ldots, k\,\bar{n}_n$, when k is any positive real number.

Case (ii) $\quad m_i = 1, \quad k_i \left(1 + e^{-(\varepsilon_i - \hat{\varepsilon})\mu}\right) = 1 \qquad (28)$

$$\bar{n}_i = \sum_{j=0}^{1} j\, p_{ij} = p_{i1} = k_i\, e^{-(\varepsilon_i - \hat{\varepsilon})\mu} = \frac{1}{e^{(\varepsilon_i - \hat{\varepsilon})\mu} + 1}, \tag{29}$$

where μ is determined by using

$$\frac{\sum_{i=1}^{n} \frac{\varepsilon_i}{e^{(\varepsilon_i - \hat{\varepsilon})\mu} + 1}}{\sum_{i=1}^{n} \frac{1}{e^{(\varepsilon_i - \hat{\varepsilon})\mu} + 1}} = \hat{\varepsilon} \tag{30}$$

Case (iii) $\quad m_i \neq 1, \infty,$

$$\bar{n}_i = \frac{\mu_i + 2\mu_i^2 + \ldots + m_i\, \mu_i^{m_i}}{1 + \mu_i + \mu_i^2 + \ldots + \mu_i^{m_i}}, \quad \mu_i = e^{-(\varepsilon_i - \hat{\varepsilon})\mu}, \tag{31}$$

where μ is determined by using

$$\sum_{i=1}^{n} \frac{\mu_i + 2\mu_i^2 + \ldots + m_i\, \mu_i^{m_i}}{1 + \mu_i + \ldots + \mu_i^{m_i}}\, \varepsilon_i \Bigg/ \sum_{i=1}^{n} \frac{\mu_i + 2\mu_i^2 + \ldots + m_i\, \mu_i^{m_i}}{1 + \mu_i + \ldots + \mu_i^{m_i}} = \hat{\varepsilon} \tag{32}$$

43.6 RELATIONSHIP OF NEW DISTRIBUTIONS WITH CLASSICAL DISTRIBUTIONS

We have obtained above eight new distributions – three corresponding to the case when the expected number of particles A alone is prescribed, three corresponding to the case when the expected total energy B alone is prescribed and two corresponding to the case when B/A alone is prescribed.

In the first case, each of three distributions will give a value say $\hat{B}$ for the total expected energy, corresponding to each given value of A. If we had started with A, $\hat{B}$ as prescribed values and obtained the classical distributions, we would have obtained the new distributions obtained above. The three new distributions may thus be regarded as special cases of the classical distributions when A has a prescribed arbitrary value and B has a special value depending on the given value of A.

Similarly, the three new distributions, obtained in the second case can be obtained as special cases of classical distributions when B has a prescribed arbitrary value and A has value $\hat{A}$ dependent on B which is derived from the new distributions.

In the third case, the first distribution does not exist. For the other distribution B/A has a prescribed value and the values of $\hat{A}$ and $\hat{B}$ are derived from it.

For the sake of convenience, we use the following notations. B.E.I., F.D.I., I.S.I. for the special B.E., F.D., and I.S. distributions when A alone is prescribed, and $\hat{B}$ is dependent on A;

B E II, F D II, I S II for the special B.E., F.D. and I.S. distributions when B alone is prescribed and $\hat{A}$ depends on B;

F D III, I S III for the special F.D. and I.S. distributions when B/A alone is prescribed and $\hat{A}, \hat{B}$ depend on it.

43.7 THERMODYNAMICS OF B. E. II DISTRIBUTION

(a) Zeroeth Law of Thermodynamics

This distribution is obtained when the number of particles in each energy level can vary from 0 to ∞ and the total expected energy alone is prescribed. In this case from (19),

$$\frac{dB}{d\mu} = -\sum_{i=1}^{n} \frac{\varepsilon_i^2 e^{\mu\varepsilon_i}}{(e^{\mu\varepsilon_i}-1)^2} < 0 \;\Rightarrow \frac{d\mu}{dB} < 0, \tag{33}$$

so that B is a monotonic decreasing function of μ and at the same time μ is a monotonic decreasing function of B,

When $B = 0$, $\mu = \infty$ and when $B = \infty$, $\mu = 0$. Also,

$$\frac{d^2 B}{d\mu^2} = \sum_{i=1}^{n} \frac{\varepsilon_i^3 e^{\mu\varepsilon_i}}{(e^{\mu\varepsilon_i}-1)^3}\left(1 + e^{\mu\varepsilon_i}\right) > 0, \tag{34}$$

so that B is a convex monotonic decreasing function of μ (Figure 43.1) and there will be a unique positive value of μ for every positive value of B. If we put

$$\mu = \frac{1}{kT} \tag{35}$$

then T is a monotonic increasing function of B. We may call T as the temperature of the system.

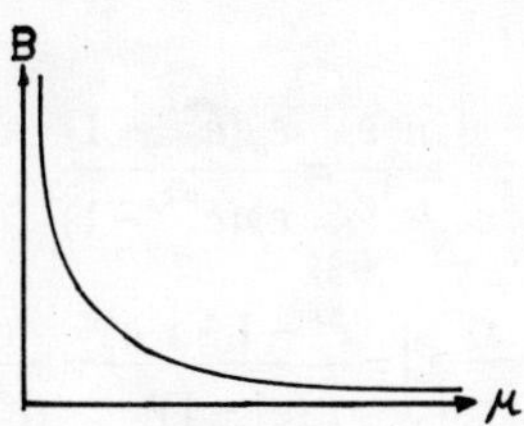

Fig. 43.1

If two systems have the same total expected energy, they would have the same temperature, we may call this as the zeroeth law of our thermodynamics.

We find that greater the total expected energy of the system, the greater is the temperature; the smaller is the value of μ, the larger is the expected number of particles in each energy level and the larger is the total expected number of particles in the system.

(b) Variations of expected number of particles $\overline{n}_i$ and the expected energy $\overline{n}_i\,\varepsilon_i$

(i) Variation with temperature :

$$\overline{n}_i = \frac{1}{e^{\mu\varepsilon_i} - 1} \Rightarrow \frac{d\,\overline{n}_i}{d\,\mu} = -\frac{\varepsilon_i\, e^{\mu\varepsilon_i}}{(e^{\mu\varepsilon_i} - 1)^2} < 0 \Rightarrow \frac{d\,\overline{n}_i}{d\,T} > 0, \tag{36}$$

so that the expected value of number of particles in each energy level increases with T. Now, if the temperature becomes k times, $k > 1$, μ becomes $1/k$ times, so that we get

$$\frac{\overline{n}'_i}{\overline{n}_i} = \frac{e^{\mu\varepsilon_i} - 1}{e^{\mu/k\,\varepsilon_i} - 1} = \frac{x^k - 1}{x - 1}, \quad x = e^{\frac{\mu}{k}\varepsilon_i} \tag{37}$$

Now let $$g\,(x) = x^k - 1 - k\,(x - 1) \tag{38}$$

then $$g\,(1) = 0, \quad g'(x) = k\,(x^{k-1} - 1), \tag{39}$$

so that if $$x > 1, \quad k > 1, \quad g'(x) > 0, \quad g\,(x) > 1 \quad if x > 1 \tag{40}$$

From (37) and (40) $$\frac{\overline{n}'_i}{\overline{n}_i} > k \tag{41}$$

Thus if temperature increases k times , the number of particles in each energy level and in whole system increases by more than k times

(*ii*) Variation with energy level :

From (18), $$\frac{\overline{n}_i}{\overline{n}_j} = \frac{e^{\mu\varepsilon_j} - 1}{e^{\mu\varepsilon_i} - 1} \tag{42}$$

so that, $$\bar{n}_i \gtreqless \bar{n}_j \text{ according as } \varepsilon_j \gtreqless \varepsilon_i \tag{43}$$

Thus the higher the energy level, the smaller is the expected number of particles in it, so that the largest expected number of particles is in the smallest energy level.

Again from (18),

$$\frac{\bar{n}_i \, \varepsilon_i}{\bar{n}_j \, \varepsilon_j} = \frac{\varepsilon_i/(e^{\mu \varepsilon_i} - 1)}{\varepsilon_j/(e^{\mu \varepsilon_j} - 1)} \tag{44}$$

Now, $$\frac{d}{dx}\left(\frac{x}{e^x - 1}\right) = \frac{e^x - 1 - x\, e^x}{(e^x - 1)^2} = \frac{g(x)}{(e^x - 1)^2}, \tag{45}$$

where, $$g(x) = e^x - 1 - x\, e^x, \quad g(0) = 0, \quad g'(x) = -x\, e^x, \tag{46}$$

so that when $x > 0$, $g(x) < 0$, and from (45) $x/(e^x - 1)$ is a decreasing function of x. It follows from (44) that

$$\bar{n}_i \, \varepsilon_i \gtrless \bar{n}_j \, \varepsilon_j \quad \text{according as} \quad \varepsilon_j \gtrless \varepsilon_i \tag{47}$$

Thus the higher the energy level, not only is the expected number of particles in it smaller, the total energy of these particles will also be smaller inspite of the energy of each particle being larger.

(c) Variation of S_{max}

$$S_{max} = -\sum_{i=1}^{n} \sum_{j=0}^{\infty} p_{ij} \ln p_{ij} = -\sum_{i=1}^{n} \sum_{j=0}^{\infty} p_{ij} (\ln c_i - j\, \mu\, \varepsilon_i)$$

$$= -\sum_{i=1}^{n} \ln c_i + \mu \sum_{i=1}^{n} \varepsilon_i \sum_{j=0}^{n} j\, p_{ij} = \sum_{i=1}^{n} \ln\left(1 - e^{-\mu \varepsilon_i}\right) + \mu B \tag{48}$$

$$\frac{d\, S_{max}}{d\, \mu} = -\sum_{i=1}^{n} \frac{\varepsilon_i}{e^{\mu \varepsilon_i} - 1} + B + \mu \frac{d\, B}{d\, \mu} = \mu \frac{dB}{d\mu}$$

or, $$\frac{d\, S_{max}}{d\, B} = \mu, \quad \frac{d^2 S_{max}}{d\, B^2} = \frac{d\, \mu}{d\, B} < 0, \tag{49}$$

so that S_{max} is a monotonic increasing concave function of B. We have next to find S_{max} when $B = 0$ or $B = \infty$. Now,

$$\operatorname*{Lt}_{B \to 0} B\, \mu = \operatorname*{Lt}_{\mu \to \infty} B\, \mu = \operatorname*{Lt}_{\mu \to \infty} \sum_{i=1}^{n} \frac{\mu\, \varepsilon_i}{e^{\mu \varepsilon_i} - 1} = 0 \tag{50}$$

$$\operatorname*{Lt}_{B \to \infty} B\, \mu = \operatorname*{Lt}_{\mu \to 0} B\, \mu = \operatorname*{Lt}_{\mu \to 0} \sum_{i=1}^{n} \frac{\mu\, \varepsilon_i}{e^{\mu \varepsilon_i} - 1} = n \tag{51}$$

From (48), (50) and (51),

$$\operatorname*{Lt}_{B \to 0} S_{max} = 0 \quad \operatorname*{Lt}_{B \to \infty} S_{max} = \infty \tag{52}$$

so that S_{max} increases from 0 to ∞ as B goes from 0 to ∞ in Figure 43.2

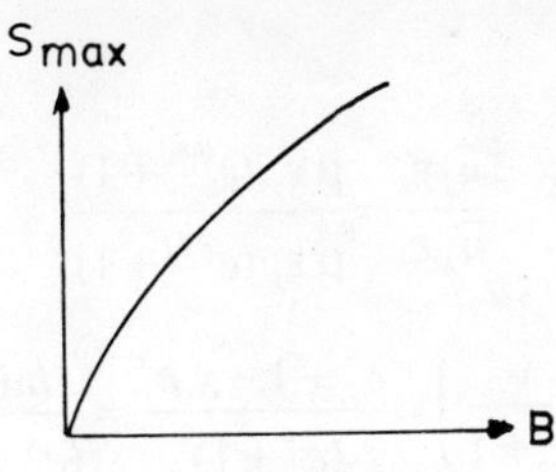

Fig. 43.2

(*d*) *Other Laws of Thermodynamics*

(i) When $$T \to 0, \bar{n}_i \to 0, \sum_{i=1}^{n} \bar{n}_i \to 0, B \to 0 \quad (53)$$

so that when $T \to 0$, *the expected number of particles in each energy level and the total expected energy both tend to zero. This may be called the third law of our thermodynamics.*

(ii) $$B = \sum_{i=1}^{n} \bar{n}_i \varepsilon_i \Rightarrow dB = \sum_{i=1}^{n} d\bar{n}_i \varepsilon_i + \sum_{i=1}^{n} \bar{n}_i d\bar{\varepsilon}_i \quad (54)$$

The first part may be called the heat effect ΔQ and the second part may be called the work effect $(-\Delta W)$ so that

$$-\Delta B = \Delta Q - \Delta W$$

and, $$\oint (\Delta Q - \Delta W) = 0 \quad (55)$$

This is the first law of our thermodynamics.

43.8 THERMODYNAMICS OF FD II DISTRIBUTION

In this case, from (22),

$$\frac{dB}{d\mu} = -\sum_{i=1}^{n} \frac{\varepsilon_i^2 e^{\mu\varepsilon_i}}{(e^{\mu\varepsilon_i}+1)^2} < 0 \Rightarrow \frac{d\mu}{dB} < 0 \quad (56)$$

$$\frac{d^2B}{d\mu^2} = \sum_{i=1}^{n} \frac{\varepsilon_i^3 e^{\mu\varepsilon_i}}{(e^{\mu\varepsilon_i}+1)^3}\left(e^{\mu\varepsilon_i} - 1\right) > 0, \quad (57)$$

so that B is a convex monotonic decreasing function of μ (Figure 43.1). Again we can define temperature by (35) and the zeroeth law of thermodynamics holds. Also,

$$\frac{d\bar{n}_i}{d\mu} = -\frac{\varepsilon_i e^{\mu\varepsilon_i}}{(e^{\mu\varepsilon_i}+1)^2} < 0, \quad (58)$$

so that $\bar{n}_i$ decreases as μ increases or as T decreases, i.e. as the temperature increases, the expected number of particles in each energy level as well as the total expected number of particles increases. Again,

$$\frac{\bar{n}_i}{\bar{n}_j} = \frac{e^{\mu \varepsilon_j} + 1}{e^{\mu \varepsilon_i} + 1}, \quad \text{so that } \bar{n}_i \gtreqless \bar{n}_j \text{ according as } \varepsilon_j \gtreqless \varepsilon_i \tag{59}$$

Also,

$$\frac{\bar{n}_i \varepsilon_i}{\bar{n}_j \varepsilon_j} = \frac{\mu \varepsilon_i/(e^{\mu \varepsilon_i} + 1)}{\mu \varepsilon_j/(e^{\mu \varepsilon_j} + 1)} \tag{60}$$

Now,

$$\frac{d}{dx}\left(\frac{x}{e^x + 1}\right) = \frac{e^x + 1 - x e^x}{(e^x + 1)^2} = \frac{h(x)}{(e^x + 1)^2}, \tag{61}$$

where, $h(x) = e^x + 1 - x e^x, h(0) = 2, \quad h'(x) = -x e^x, h(1.3) = .007$

so that $h(x)$ is always decreasing, but it starts with a positive value at $x = 0$ and remains positive till about $x = 1.3$, so that $d/dx\ (x/e^x + 1)$ increases from 0 to 1.3 and then decreases so that except for small value of μ,

$$\bar{n}_i \varepsilon_i \gtreqless \bar{n}_j \varepsilon_j \quad \text{according as } n_j \gtreqless \bar{n}_i \tag{62}$$

Thus for FD II also as the temperature increases, the number of particles in each energy level decreases as energy of level increases and the total energy in each level also decreases with increases in energy level except when the total prescribed is very large. Again,

$$S_{\max} = B\mu + \sum_{i=1}^{n} \ln\left(1 + e^{-\mu \varepsilon_i}\right) \tag{63}$$

$$\frac{d S_{\max}}{d B} = \mu, \quad \frac{d^2 S_{\max}}{d B^2} = \frac{d\mu}{d B} < 0, \tag{64}$$

so that $S_{\max}$ is here also a concave function of B

$$\underset{B \to \infty}{\text{Lt}}\ S_{\max} = \underset{\mu \to 0}{\text{Lt}}\ S_{\max} = n \ln 2, \quad \underset{B \to 0}{\text{Lt}}\ S_{\max} = \underset{\mu \to \infty}{\text{Lt}}\ S_{\max} = 0 \tag{65}$$

so that $S_{\max}$ increases from 0 to $n \ln 2$ as B goes from 0 to ∞

43.9 THERMODYNAMICS OF IS II DISTRIBUTION

From (26),

$$\frac{dB}{d\mu} = \sum_{i=1}^{n} \varepsilon_i \frac{d}{dz_i}\left(\frac{z_i + 2z_i^2 + \ldots + m_i z_i^{m_i}}{1 + z_i + z_i^2 + \ldots + z_i^{m_i}}\right)\left(-\varepsilon_i e^{-\mu \varepsilon_i}\right) \tag{66}$$

Now,

$$\frac{d}{dz}\left(\frac{z + 2z^2 + \ldots + mz^m}{1 + z + \ldots + z^m}\right) = \frac{X}{(1 + z + z^2 + \ldots + z^m)^2}, \tag{67}$$

Where,

$$Xz = (1 + z + z^2 + \ldots + z^m)(z + 2^2 z^2 + \ldots + m^2 z^m)$$
$$- (z + 2z^2 + \ldots + m z^m)(z + 2z^2 + \ldots + m z^m)$$

$$= (z + 2^2 z^2 + \ldots + m^2 z^m) + \sum_{i=1}^{n} a_i^2 \sum_{i=1}^{n} b_i^2 - \left(\sum_{i=1}^{n} a_i b_i \right)^2 \tag{68}$$

When, $a_i = z^{i/2} \quad b_i = i\, z^{i/2}, \quad i = 1, 2, \ldots n$ (69)

Using Cauchy-Schwarz inequality, we find that $X > 0$ so that

$$\frac{dB}{d\mu} < 0, \quad \frac{d\mu}{dB} < 0 \tag{70}$$

Thus μ is a monotonic decreasing function of B and again we can define a temperature and zeroeth law of thermodynamics holds.

We have also incidentally proved that

$$\frac{d\bar{n}_i}{d\mu} < 0, \frac{d\bar{n}_i}{dT} > 0, \tag{71}$$

so that as T increases, the expected number of particles in each energy level increases. Also,

$$\bar{n}_i \gtreqless \bar{n}_j \quad \text{according as} \quad \varepsilon \lesseqgtr \varepsilon_i \tag{72}$$

Again we find

$$S_{\max} = B\mu + \sum_{i=1}^{n} \ln\left(1 + e^{-\mu \varepsilon_i} + \ldots + e^{-m_i \mu \varepsilon_i}\right) \tag{73}$$

$$\frac{dS_{\max}}{dB} = \mu, \quad \frac{d^2 S_{\max}}{dB^2} = \frac{d\mu}{dB} < 0, \tag{74}$$

so that $S_{\max}$ is a concave function of B

$$\operatorname*{Lt}_{B \to 0} S_{\max} = 0, \quad \operatorname*{Lt}_{B \to \infty} S_{\max} = \sum_{i=1}^{n} \ln(m_i + 1) \tag{75}$$

Also in this case $\bar{n}_i \to 1/2$ as $B \to \infty, \mu \to 0$

From (73) and (74), we can easily deduce that
in FD II case $\vec{n}_i \to 1/2$ and $S_{\max} \to n \ln 2$
and in BE II case $\vec{n}_i \to \infty$ and $S_{\max} \to \infty$.

43.10 COMPARISON OF BE II DISTRIBUTION WITH MAXWELL BOLTZMANN DISTRIBUTION

(a) Maxwell-Boltzmann Distribution

In this case, we do not talk of the number of particles at all. We define p_i as the probability of a particle being in the ith energy level and we prescribe the expected energy per particle as $\hat{\varepsilon}$ so that we get the constraints,

$$\sum_{i=1}^{n} p_i = 1, \quad \sum_{i=1}^{n} p_i \varepsilon_i = \hat{\varepsilon} \tag{76}$$

Maximizing the entropy $-\sum_{i=1}^{n} p_i \ln p_i$ subject to (76), We get

$$p_i = e^{-\mu \varepsilon_i} / \sum_{i=1}^{n} e^{-\mu \varepsilon_i}, \tag{77}$$

where μ is determined by using

$$\sum_{i=1}^{n} \varepsilon_i e^{-\mu \varepsilon_i} / \sum_{i=1}^{n} e^{-\mu \varepsilon_i} = \hat{\varepsilon} \tag{78}$$

It is easily seen that $d\hat{\varepsilon}/d\mu < 0$, so that $\hat{\varepsilon}$ is a monotonic decreasing function of μ and if $\mu = 1/kT$, then T is a monotonic increasing function of $\hat{\varepsilon}$. It is also easily seen that when $\hat{\varepsilon} = \varepsilon_1, \mu = \infty, S_{max} = 0$ and when $\hat{\varepsilon} = \varepsilon_n$, $\mu = -\infty$, $S_{max} = 0$ and when $\hat{\varepsilon} = \bar{\varepsilon}, \mu = 0, S_{max} = \ln n$ so that S_{max} is a concave function of $\hat{\varepsilon}$. (Figure 43.3)

In fact only the first half of the curve is significant because on the other half, T becomes negative.

Let us now compare the MB distribution with the BE II distribution. In both the only constraint on the system is the energy constraint.

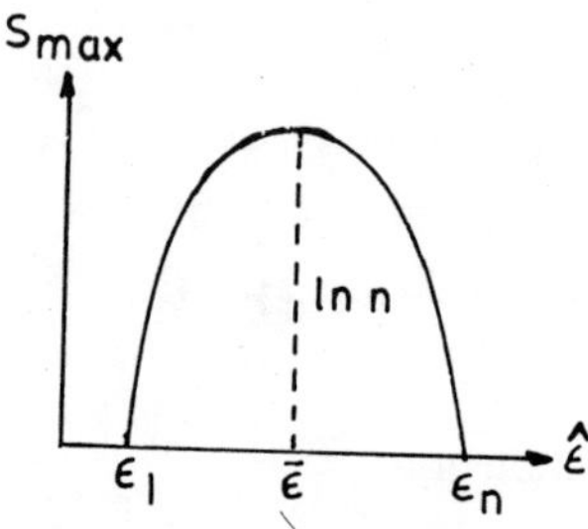

Fig. 43.3

(b) Comparison of the Two Distributions

BE II DISTRIBUTION

(i) Total expected energy is prescribed and there is no other constraint.

(ii) We talk of the expected number of particles in each energy level and we also talk of the total expected number of particles in the system.

(iii) The expected proportion of particles in the ith energy level is given by

$$\frac{\bar{n}_i}{\sum_{i=1}^{n} \bar{n}_i} = \frac{(e^{\mu \varepsilon_i} - 1)^{-1}}{\sum_{i=1}^{n} (e^{\mu \varepsilon_i} - 1)^{-1}}$$

MB DISTRIBUTION

(i) Expected energy per particle is prescribed and there is no other constraint.

(ii) We do not talk of the number of particles. If we are given that the total number of particles is N, we expect Np_i to be in the ith energy level.

(iii) The probability of a particle being in the ith energy level is found to be

$$p_i = e^{-\mu \varepsilon_i} / \sum_{i=1}^{n} e^{-\mu \varepsilon_i} \tag{79}$$

(iv) Here μ is determined by	(iv) Here μ is determined by
$\sum_{i=1}^{n} \frac{\varepsilon_i}{e^{\mu\varepsilon_i}-1} = \sum_{i=1}^{n} \bar{n}_i \varepsilon_i = B = \left(\sum_{i=1}^{n} \bar{n}_i\right)\hat{\varepsilon}$	$\sum_{i=1}^{n} \varepsilon_i e^{-\mu\varepsilon_i} / \sum_{i=1}^{n} e^{-\mu\varepsilon_i} = \hat{\varepsilon}$ (80)
(v) $S_{\max}$ is a concave function of B and increases with B.	(v) $S_{\max}$ is a concave function of $\hat{\varepsilon}$ and $S_{\max}$ increases with $\hat{\varepsilon}$ if $\hat{\varepsilon} \le \bar{\varepsilon}$
(vi) We can define temperature T as a monotonic increasing function of B.	(vi) We can define T as a monotonic increasing function of $\hat{\varepsilon}$.
(vii) When T increases the expected number of particles in each energy level increases and the total expected number of particles also increases.	(vii) When T increases, the particles tend to go to higher energy levels, but the number of particles is independent of T.
(viii) $\bar{n}_i \gtreqless \bar{n}_j$ according as $\varepsilon_j \gtreqless \varepsilon_i$	(viii) $p_i \gtreqless p_j$ according as $\varepsilon_j \gtreqless \varepsilon_i$.
(ix) $\bar{n}_i \varepsilon_i \gtreqless \bar{n}_j \varepsilon_j$ according as $\varepsilon_j \gtreqless \varepsilon_i$	(ix) $N p_i \gtreqless N p_j$ according as $\varepsilon_j \gtreqless \varepsilon_i$
(x) $\oint (\Delta Q - \Delta W) = 0.$	(x) $\oint (\Delta Q - \Delta W) = 0$
(xi) $S_{\max}$ is the maximum of the information-theoretic entropy for a given value of expected total energy,	(xi) $S_{\max}$ is the maximum of the information-theoretic entropy for a given value of expected energy per particle.
(xii) The total number of particles in the system is variable and depends on B and is determined by using the maximum entropy principle.	(xii) The total number of particles in the system is not known and cannot be determined by using Maximum Entropy Principle.
(xiii) As T increases from 0 to ∞, μ decreases from ∞ to 0 and $\bar{n}_i$ increases from 0 to ∞.	(xiii) As T increases from 0 to ∞, μ decreases from ∞ to 0, but $\bar{n}_i$ is not changing.
(xiv) When $T \to 0$, the number of particles in each level tends to zero.	(xiv) When $T \to 0$, all the particles go to the lowest energy level.

We can similarly compare MB distribution with FD II and IS II Distributions. In FD II, the number of particles in each level is 0 or 1. In ISD, the number can go from 0 to m_i, in BE distribution, it can go from 0 to ∞, in MB distribution, the number of particles is either given or is indeterminate.

43.11 VARIANCE OF THE NUMBER OF PARTICLES IN EACH ENERGY LEVEL

We have proved above that for each distribution, the mean number of particles in an energy level decreases as the energy level increases. We now want to investigate the behaviour of the variance of the number of particles in each energy level.

For IS II distribution,

$$p_{ij} = \frac{e^{-\mu\varepsilon_{ij}}}{\sum_{j=0}^{m_i} e^{-\mu\varepsilon_i j}} = \frac{x_i^j}{\sum_{j=0}^{m_i} x_i^j} \quad ; \quad x_i = e^{-\mu\varepsilon_i} \tag{81}$$

$$\bar{n}_i = \frac{x_i + 2x_i^2 + \ldots + m_i x_i^{m_i}}{1 + x_i + \ldots + x_i^{m_i}} \tag{82}$$

$$\sigma_1^2 = \frac{1^2 x_i + 2^2 x_i^2 + \ldots + m_i^2 x_i^{m_i}}{1 + x_i + x_i^2 + \ldots + x_i^{m_i}} - \left(\frac{x_i + 2x_i^2 + \ldots + m_i x_i^{m_i}}{1 + x_i + \ldots + x_i^{m_i}} \right)^2 \tag{83}$$

For BE II distribution : $m_i \to \infty$

$$\sigma_i^2 = \frac{x_i + x_i^2}{(1 - x_i)^3} - \left(\frac{x_i}{1 - x_i} \right)^2 = \frac{x_i}{(1 - x_i)^2} \tag{84}$$

$$\frac{d}{dx_i}(\sigma_i^2) = \frac{1 + x_i}{1 - x_i} > 0, \tag{85}$$

so that σ_i^2 increases with x_i and since $x_i = e^{-\mu\varepsilon_i}$, σ_i^2 decreases with ε_i.

For FD II distribution, $m_i = 1$

$$\sigma_i^2 - \frac{x_i}{1 + x_i} - \left(\frac{x_i}{1 + x_i} \right)^2 = \frac{x_i}{(1 + x_i)^2} \tag{86}$$

$$\frac{d}{dx_i}(\sigma_i^2) = \frac{1 - x_i^2}{(1 + x_i)^4} > 0, \tag{87}$$

so that σ_i^2 again increases with x_i and decreases with ε_i.

When $m_i \neq 1, \infty$ the calculations are more complicated and σ_i^2 decreases with ε_i. Thus when the total expected energy is prescribed, both the mean number of particles and the variance of the number of particles in each energy level decreases as ε_i increases.

43.12 SOME APPLICATIONS OF THE NEW DISTRIBUTIONS

We start with the following example:

A firm sells n items with prices $c_1, c_2, ---, c_n$ per unit, In a certain period the total sales realised amount to S. We have to estimate the numbers of different items sold.

The firm does not know the total number of units sold. If it had this information, it could use the Maxwell-Boltzman's distribution. It could also use an Intermediate Statistics distribution if it knew the stocks of each item it had in the beginning so that it could know the maximum number of units of each type that could be sold. Failing that information it could apply Bose-Einstein distribution as a very good approximation.

Now since the firm does not know the total number of units sold, it can apply either IS II or BE II distribution.

However, if the distribution apply, the following conclusions can be easily drawn when we use BE II distribution.

(i) The number of units of ith type sold is $(e^{\mu c_i} - 1)^{-1}$ where μ is determined by solving

$$\sum_{i=1}^{n} \frac{c_i}{e^{\mu c_i} - 1} = S \tag{88}$$

(ii) If S increases, the expected number of units of each type will increase and the total number of units sold will increase.

(iii) The number of units sold of each type will go on decreasing as the cost increases, so that the smallest number of units are sold of the most costly quality.

(iv) Even the money realised from the sale of each type will be larger for lower cost items than it is for high cost items.

(v) The firm can talk of a market-temperature which increases with S but is different from S. In fact, it is determined from (88) as the reciprocal of μ.

(vi) The change in total sales from one period to another can be due to increase in prices or increase in sales and the other two effects can be discussed separately.

A similar discussion can apply to the following examples :

(i) A university employs person in n grades and the average emoluments in these grades are $c_1, c_2, \ldots c_n$. If the total salary budget is known, estimate the number of employees in each grade.

(ii) A bank accepts deposits in n denominations $c_1, c_2, \ldots, c_n$. If the total deposits are known, estimate the number of deposits in each denomination.

(iii) A company pays overtime to its workers at n different rates $r_1, r_2, \ldots, r_n$ per hour. If the total amount of honorarium is known, estimate the number of hours for each category for which honorarium has been paid.

(iv) Income tax is paid in different slabs. If total income tax received in known, estimate the amounts received in each slab.

(v) Knowing the total population of a country, estimate the population in cities with different population sizes.

(vi) Knowing the different room-rents in a hotel and knowing the total rent income, estimate the number of days for which each type of room has been occupied.

(vii) Knowing the different slabs of fees in a university and knowing the total fee income, estimate the number of students paying in different slabs.

It is obvious that there can be a large number of situations in which distributions discussed in this paper are applicable.

REFERENCES

1. B. Forte and C. Sempi (1976), " Maximizing Conditional Entropy : A derivation of quantal statistics," Rendi Conta de Mathematics, Vol. 9., pp. 551-556.
2. E.T. Jaynes (1957), " Information theory and Statistical Mechanics" Raynal Review vol 106, 620-630.

3. J.N. Kapur (1983), "Comparative assessment of various measures of entropy," Jour. Inf. and opt. Sci., Vol. 4, no 1, pp. 207-232.
4. J.N. Kapur (1983), "Non-additive measures of entropy and distributions of statistical mechanics," Ind. Jour. Pure App. Math Vol. 14. No 11, pp. 1372-1384.
5. J.N. Kapur (1983), "Twenty five years of maximum entropy," Jour. Math. Phy. Sci. Vol. 17, No 2, pp. 103-156.
6. J.N. Kapur (1989), Maximum Entropy Models in Science and Engineering, Wiley Eastern, New Delhi and John Wiley, New York.
7. J.N. Kapur and H.K. Kesavan (1987), "The Generalised Maximum Entropy Principle (with Applications), Sandford Educational Press. Uni of Waterloo, Canada :
8. J.N. Kapur and H.K. Kesavan (1992), "Entropy Optimization Principles and their Application, Academic Press, New York.

44

MAXIMUM ENTROPY PRINCIPLE AND PARAMETER ESTIMATION

[Some new methods, based on the Principle of Maximum Entropy, are given for estimation of parameters of probability density functions.]

44.1 INTRODUCTION

There are many methods for estimation of parameters of a probability distribution, e.g. Pearson's method of moments, Neyman's method of minimum chi-square, Gauss's method of least squares and the most widely used Fisher's [3] method of maximum likelihood. Kapur [7] compared a method of estimation given earlier by Gauss and a method based on the principle of maximum entropy. Recently, Lind and Solana [9] gave an alternative method based on the principle of least information.

The object of the present paper is to examine some of these methods from the vantage point of Jaynes' [5] principle of maximum entropy and Kapur and Kesavan's [8] generalised maximum entropy principle and to propose new methods of estimation based on these principles.

The principles simply state that we should choose the parameters in such a way that the probability distribution is as much spread-out or as much flat or as much uniform as possible, subject to the information available.

In section 2, we discuss the case when no information is available except in the form of the density function. In Section 3, and in subsequent sections, we consider the case when in addition to the knowledge of this function, we have also a random sample $x_1, x_2, \ldots, x_n$ from the population.

44.2 THE CASE WHEN NO RANDOM SAMPLE IS GIVEN

(i) In the case of the normal distribution with density function,

$$f(x\,; m, \sigma^2) = \frac{1}{\sqrt{2\pi}\,\sigma} \exp\left[-\frac{1}{2}\frac{(x-m)^2}{\sigma^2}\right] \tag{1}$$

Entropy,

$$H = -\int_{-\infty}^{\infty} f(x\,; m, \sigma^2) \ln f(x\,; m, \sigma^2)\, dx$$

$$= \frac{1}{2} \ln (2\pi e \sigma^3)$$

and this case can be made as large as we please by making σ sufficiently large. As such, the maximum entropy principle gives m arbitrary and σ equal to infinity. This gives the degenerate uniform distribution over $(-\infty, \infty)$.

(ii) Similarly, in the case of the exponential distribution with density function

$$f(x,a) = \frac{1}{a}\exp\left(-\frac{x}{a}\right), \quad x > 0 \tag{3}$$

$$H = 1 + \ln a \tag{4}$$

and this also can be made as large as we please by making a sufficiently large.

Again the maximum entropy exponential distribution is the degenerate uniform distribution over the interval $[0, \infty)$.

(iii) In the case of the beta distribution with density function

$$f(x\,; m,n) = \frac{1}{B\,(m,n)} x^{m-1}(1-x)^{n-1}; \quad 0 \le x \le 1 \tag{5}$$

$$H = -\ln B\,(m,n) - (m-1)\int_0^1 f(x;m,n)\ln x\, dx$$

$$-(n-1)\int_0^1 f(x;\ m,n)\ln(1-x)\,dx \tag{6}$$

But,

$$\int_0^1 x^{m-1}(1-x)^{n-1}\,dx = B\,(m,n) \tag{7}$$

so that differentiating (7), with respect to m and n respectively under the sign of integration, we get

$$\frac{1}{B\,(m,n)}\int_0^1 x^{m-1}(1-x)^{n-1}\ln x\,dx = \frac{\Gamma'(m)}{\Gamma(m)} - \frac{\Gamma'(m+n)}{\Gamma(m+n)} \tag{8}$$

Similarly,

$$\frac{1}{B\,(m,n)}\int_0^1 x^{m-1}(1-x)^{n-1}\ln(1-x)\,dx = \frac{\Gamma'(n)}{\Gamma(n)} - \frac{\Gamma'(m+n)}{\Gamma(m+n)} \tag{9}$$

From (6), (8) and (9),

$$H = -\ln B\,(m,n) - (m-1)\left[\frac{\Gamma'(m)}{\Gamma(m)} - \frac{\Gamma'(m+n)}{\Gamma(n+n)}\right]$$

$$-(n-1)\left[\frac{\Gamma'(n)}{\Gamma(n)} - \frac{\Gamma'(m+n)}{\Gamma(m+n)}\right] \tag{10}$$

It can be shown that the maximum value of H is zero and it arises when $m = 1, n = 1$.

becomes the uniform distribution over [0, 1] and since no other distribution over [0, 1] can have a larger entropy or spreadout, this must be the distribution with the maximum entropy.

(iv) Consider the geometric distribution with density

$$p_i = \frac{1-\rho}{1-\rho^{m+1}} \quad \rho^i; \quad i = 0, 1, 2, \ldots, m \tag{11}$$

For the entropy to be maximum, the probabilities should be as equal as possible and these would all tend to be equal as $\rho \to 1$ when each probability is $1/(m+1)$. As such the maximum-entropy estimate for ρ is unity and the maximum-entropy distribution is the uniform distribution when the probability is $1/(m+1)$ for each of the values $0, 1, 2, \ldots, m$.

(v) Consider the distribution (Fig. 44.1)

$$f(x, \theta) = \frac{2x}{\theta}, \quad 0 \le x \le \theta \tag{12}$$

$$= \frac{2(1-x)}{1-\theta}, \quad \theta \le x \le 1$$

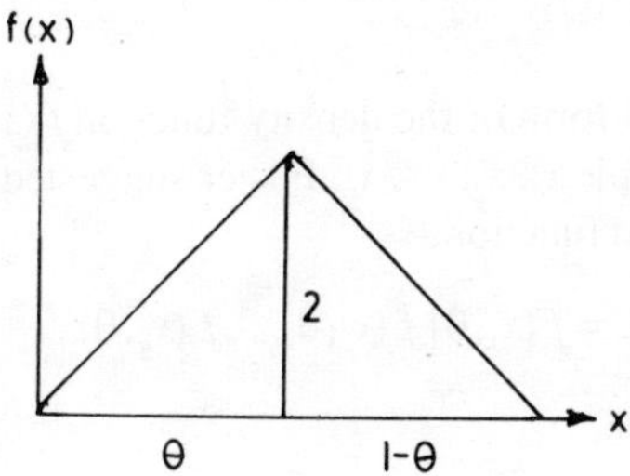

Fig. 44.1

For this distribution,

$$H = -\int_0^\theta \frac{2x}{\theta} \ln \frac{2x}{\theta} \, dx - \int_\theta^1 \frac{2(1-x)}{1-\theta} \ln \frac{2(1-x)}{1-\theta} \, dx$$

$$= \frac{1}{2} - \ln 2 \tag{13}$$

This entropy is independent of θ and thus all positive values of $\theta \le 1$ are equally acceptable.

(vi) For the binomial distribution with n fixed and p a parameter, in order to maximize the entropy, we should choose p so as to make the probabilities

$$nc_0\, p^0 q^{n-0}, nc_1, p^1 q^{n-1}, \ldots\ldots, nc_n\, p^n q^{n-n}, \tag{14}$$

as equal as possible. For this purpose, we can maximize any measure of equality. We choose to maximize Burg's [1] measure, i.e we maximize

$$\sum_{r=0}^{n} \ln\left(n_{c_r} p^r q^{n-r}\right) = \sum_{r=0}^{n} \ln nc_r + \ln p \sum_{r=0}^{n} r + \ln q \sum_{r=0}^{n} (n-r)$$

$$= \sum_{r=0}^{n} \ln nc_r + \frac{n(n+1)}{2} \ln(p\,q) \tag{15}$$

This is maximum when $p = q = 1/2$ in which case the probabilities become

$$nc_0\left(\frac{1}{2}\right)^n, nc_1\left(\frac{1}{2}\right)^n, \ldots\ldots\ldots, nc_n\left(\frac{1}{2}\right)^n \tag{16}$$

These are maximally equal ; though these are not equal.

Thus it appears that if we can find values of parameters for which the probabilities become equal or for which the probability distributions become uniform, degenerate or non-degenerate, we should choose these values of the parameters. If such values of parameters cannot be found, we should find those values which make the probabilities as equal as possible or probability distribution as uniform as possible.

44.3 THE CASE WHEN THERE IS A RANDOM SAMPLE : THE PRINCIPLE OF MAXIMUM LIKELIHOOD

We are given the functional form of the density function $f(x, \theta)$ and to estimate θ, we are given a random sample $x_1, x_2, \ldots, x_n$. Fisher suggested that we should choose θ to maximize the likelihood function,

$$L = f(x_1, \theta) f(x_2, \theta) \ldots f(x_n, \theta) \tag{17}$$

or the log likelihood function,

$$\ln L = \ln f(x_1, \theta) + \ln f(x_2, \theta) + \ldots + \ln f(x_n, \theta) \tag{18}$$

Now we can form a probability distribution

$$p_i = \frac{f(x_i, \theta)}{\sum_{i=1}^{n} f(x_i, \theta)}, \quad i = 1, 2, \ldots, n \tag{19}$$

and to make these as equal as possible, we should choose θ to maximize Burg's measure of entropy of this distribution, viz.

$$\sum_{i=1}^{n} \ln p_i = \sum_{i=1}^{n} \ln f(x_i, \theta) - \ln \sum_{i=1}^{n} f(x_i, \theta) \tag{20}$$

This requires us to solve the equation,

$$\sum_{i=1}^{n} \left(\frac{1}{f}\frac{\partial f}{\partial \theta}\right)_{x=x_i} - \frac{\sum_{i=1}^{n} \left(\frac{\partial f}{\partial \theta}\right)_{x=x_i}}{\sum_{i=1}^{n} f(x_i, \theta)} = 0 \tag{21}$$

As against this the principle of maximum likelihood requires us to solve

$$\sum_{i=1}^{n} \frac{1}{f}\left(\frac{\partial f}{\partial \theta}\right)_{x=x_i} = 0 \tag{22}$$

Equations (21) and (22) are different and would give different results, since $\sum_{i=0}^{n} f(x_i, \theta)$ is not independent of θ.

In fact $f(x_1, \theta), f(x_2, \theta) \ldots f(x_n, \theta)$ are not probabilities, but are only the values of probability density function at $x_1, x_2, \ldots, x_n$.

Even for discrete valued variates, when there are probabilities, their sum is not unity or independent of θ, since $x_1, x_2, \ldots, x_n$ represent only a random sample and not all the values which x can take.

For continuous variates, probabilities are given by the areas under the probability density curve. We use the method of maximum entropy with these probabilities in the next section.

44.4 A METHOD BASED ON GENERALISED PRINCIPLE OF MAXIMUM ENTROPY USING BURG'S ENTROPY

Let the random variate x vary over the interval $[a, b]$ and let the random sample values be arranged so that

$$a = x_0 < x_1 < x_2 < \ldots < x_i < x_{i+1} < \ldots < x_n < x_{n+1} = b \tag{23}$$

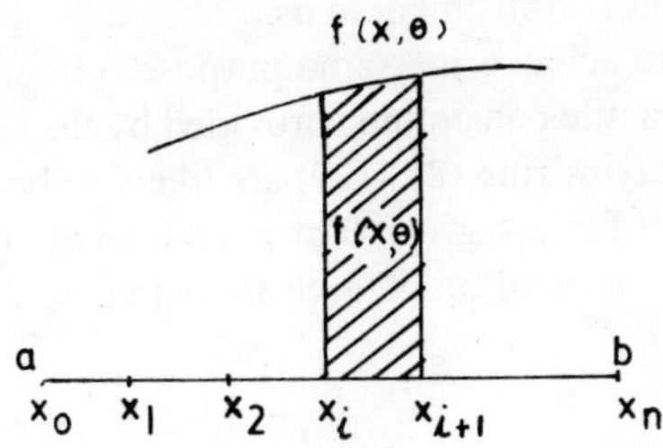

Fig. 44.2

so that the sample points divide the interval $[a, b]$ into $(n+1)$ sub interval (x_i, x_{i+1}), $i = 0, 1, 2, \ldots, n$. We now define

$$P_i = \int_{x_i}^{x_i+1} f(x, \theta)\, dx; \quad i = 0, 1, 2, \ldots, n \tag{24}$$

It is obvious that $\sum_{i=0}^{n} P_i = 1$ and $(P_0, P_1, \ldots, P_n)$ gives us a genuine probability distribution depending upon θ.

Now we choose θ to make these probabilities as equal as possible by using Burg's measure of entropy, i.e by maximizing

$$\sum_{i=0}^{n} \ln P_i = \sum_{i=0}^{n} \ln \int_{x_i}^{x_i+1} f(x,\theta)\, dx \tag{25}$$

This gives the same result as the method of least information principle based on two-stage minimization of cross-entropy under fractile constraints (Lind and Solana [9]) and the principle of Cheng and Amin [2].

44.5 METHOD BASED ON GENERALISED PRINCIPLE OF MAXIMUM ENTROPY USING OTHER MEASURES OF ENTROPY

Now that we have got a probability distribution (24) based on the random sample $x_1, x_2, ..., x_n$ and the given density function $f(x, \theta)$ and our object is to choose θ to make these probabilities as equal as possible, we are free to use any measure of equality. Thus we can choose θ to maximize

$$-\sum_{i=0}^{n} P_i \ln P_i; \quad \text{Shannon} \quad [11] \tag{26}$$

or

$$\frac{1}{\alpha(1-\alpha)}\left(\sum_{i=0}^{n} P_i^{\alpha} - 1\right) \quad \alpha \neq 0, 1, \quad \text{Harvda and Charvat} \quad [4] \tag{27}$$

or

$$\frac{1}{\alpha(1-\alpha)} \ln \sum_{i=0}^{n} P_i^{\alpha}, \quad \alpha \neq 0, 1, \quad \text{Renyi} \quad [10] \tag{28}$$

or

$$-\sum_{i=0}^{n} P_i \ln P_i + \frac{1}{a}\sum_{i=0}^{n} (1 + a\, P_i) \ln (1 + a\, P_i) - \frac{1}{a}(1+a)\ln(1+a), \quad \text{Kapur} \ [6] \tag{29}$$

and we may use whichever is convenient to us.

Since all these aim to achieve the same purpose, viz. making P_i's as equal as possible, subject to the fractile constraints, provided by the random sample, the values of θ that we get by maximizing (25)-(29) are likely to be almost equal. This can be shown by showing that for any given θ or any given $P_0, P_1, ..., P_n$, the measures of entropy (25)-(29), after normalisation, give almost the same results.

REFERENCES

1. J.P. Burg (1972). "The relationship between Maximum Entropy Spectra and Maximum Likelihood Spectra." in Modern Spectral Analysis, ed. D. G. Childers, pp. 130-131.
2. R.C.H. Cheng and N.A.K. Amin (19). "Estimating parameters of continuous univariate distributions with a shifted origin." J.R. Stat. Soc. B 45 no (3), pp. 394-403.
3. R.A. Fisher (1921). "On the Mathematical Foundations of Theoretical Statistics," Phil. Trans. Roy. Soc., vol. 222 (A), pp. 309-368.
4. J.H. Havrda and F. Charvat (1967). "Quantification methods of classification processes : Concept of Structural α Entropy." Kybernetica. Vol. 3, pp 30-35.
5. E.T. Jaynes (1957). "Information Theory and Statistical mechanics." Physical Reviews, Vol. 106, pp. 620-630.
6. J.N. Kapur (1986). "Four families of measures of entropy." Ind. Jour. Pure and App. Maths, Vol. 17, no. 4 pp. 429-449.
7. J.N. Kapur (1988). "On equivalence of Gauss's and maximum entropy principles of estimation of probability distributions, Ganit Sandesh, 2 (1), 4-10.
8. J. N. Kapur and H. K. Kesavan (1987). "Generalised Maximum Entropy Principle with Applications." Sandford Educational Press, Waterloo.

9. N.C. Lind and V. Solana (1988), "Cross-entropy estimation of random variables with fractile constraints," paper No. 11. Institute for Risk Research, University of Waterloo.
10. A. Renyi (1961), " On Measures of Information," Proc. 4th Berkeley Symp. Maths. Stat. Prob, vol. 1, pp 547-561.
11. C.E. Shannon (1948), "A Mathematical Theory of Communication," Bell. System Tech. J., vol. 27, pp. 379-423, 623-659.

45

INVERSE MAXENT AND MINXENT PRINCIPLES AND THEIR APPLICATIONS TO SPECTRAL ANALYSIS

[The object of the present paper is to obtain those measures of entropy (cross-entropy) by maximizing (minimizing) which, subject to knowing values for $2m + 1$ autocovariance functions, we obtain a given or observed spectral density function. In particular these measures are obtained for autoregressive, moving average and mixed autoregressive moving average processes. A new point of view is presented on the controversy about which of the two measures $\int \ln S(f)df$ and $-\int S(f) \ln S(f)df$ is more appropriate in spectral analysis, speech recognition and image processing.]

45.1 INTRODUCTION

There are at least six approaches to the study of maximum entropy and minimum cross entropy spectral analysis (MESA/MINXESA/MCESA).

(*i*) Let S(f) be the spectral density function, then the entropy gain in a stochastic process that is passed through a linear filter that converts white noise to a process with power spectral density $S(f)$ is

$$\int_{-\frac{1}{2}}^{\frac{1}{2}} \ln S\,(f)\,df \tag{1}$$

In the first approach due to Burg [1], we maximize (1) subject to

$$A_k = \int_{-\frac{1}{2}}^{\frac{1}{2}} S(f)\,e^{-2\pi i k f}\,df, k = -m, \ldots, 0, .., m \tag{2}$$

(ii) In the second approach, proposed by Skilling and Gull [9], we maximize

$$-\int_{-\frac{1}{2}}^{\frac{1}{2}} S\,(f) \ln S\,(f)\,df \tag{3}$$

subject to (2). Here f is regarded as a random variable and $S(f)$ is regarded as its probability density function.

(iii) We take any realisation

$$\{y_{-T}, \ldots, y_0, \ldots, y_T\} \tag{4}$$

of a stationary stochastic process and calculate $2m+1$ autocovariance functions A_k ($k = -m, \ldots, m$) from it ($m \ll T$). We then use Jaynes [4] maximum-entropy principle to find the maximum entropy probability distribution of y's. We use this to estimate the remaining autocovariance functions and then use

$$S(f) = \sum_{k=-\infty}^{\infty} A_k e^{2\pi i f k} \tag{5}$$

to estimate $S(f)$. This approach is essentially due to Jaynes [5].

(iv) If we have knowledge of an *apriori* spectral density function $T(f)$, we choose $S(f)$ to minimize Kullback-Leibler's [7] measure

$$\int_{-\frac{1}{2}}^{\frac{1}{2}} S(f) \ln \frac{S(f)}{T(f)}\, df \tag{6}$$

or Itakura-Saito [3] measure

$$\int_{-\frac{1}{2}}^{\frac{1}{2}} \left(\frac{S(f)}{T(f)} - \ln \frac{S(f)}{T(f)} - 1 \right) df \tag{7}$$

subject to (2). This approach is due to Johnson and Shore [6] and Tzannes *et al.* [10].

(v) We either minimize

$$\frac{1}{\alpha(\alpha-1)} \left[\int_{-\frac{1}{2}}^{\frac{1}{2}} S^{\alpha}(f)\, T^{1-\alpha}(f)\, df - 1 \right], \alpha \neq 1, \alpha > 0 \tag{8}$$

or,

$$\frac{1}{\alpha(\alpha-1)} \left[\int_{-\frac{1}{2}}^{\frac{1}{2}} T^{\alpha}(f)\, S^{1-\alpha}(f)\, df - 1 \right], \alpha \neq 1, \alpha > 0 \tag{9}$$

or,

$$\frac{1}{\alpha(\alpha-1)} \left[\int_{-\frac{1}{2}}^{\frac{1}{2}} \left(\frac{S(f)}{T(f)} \right)^{\alpha} - 1 - \left(\frac{S(f)}{T(f)} - 1 \right) \alpha \right] df, \alpha \neq 1, \quad \alpha > 0 \tag{10}$$

or,

$$\frac{1}{\alpha(\alpha-1)} \left[\int_{-\frac{1}{2}}^{\frac{1}{2}} \left(\frac{T(f)}{S(f)} \right)^{\alpha} - 1 - \left(\frac{T(f)}{S(f)} - 1 \right) \alpha \right] df, \alpha \neq 1, \quad \alpha > 0 \tag{11}$$

subject to (2). This approach includes the earlier approaches as limiting cases as $\alpha \to 0$ or $\alpha \to 1$.

(vi) We start with known spectral density function, e.g. we start with

$$S(f) = \frac{1}{\lambda_0 + \sum_{k=-m}^{m} \lambda_k e^{-2\pi ikf}} \text{ (AR Process)} \quad (12)$$

or,

$$S(f) = \lambda_0 + \sum_{k=-m}^{m} \lambda_k e^{-2\pi ikf} \text{ (MA Process)} \quad (13)$$

or,

$$S(f) = \frac{\mu_0 + \sum_{l=-n}^{n} \mu_l e^{-2\pi ilf}}{\lambda_0 + \sum_{k=-m}^{m} \lambda_k e^{-2\pi ikf}} \text{ (ARMA Process)} \quad (14)$$

and find out the measure of entropy or cross-entropy which can give rise to this spectral density function, when it is maximized or minimized subject to (2).

In section 2, we shall discuss the first three approaches, while sections 3 and 4 will be devoted to the fifth and sixth approaches. Section 5 will consider the role of generalised measures of entropy and cross-entropy in spectral analysis. Section 6 obtains a number of possible spectral density functions by using the minimum cross-entropy principle. Finally, Section 7 gives some concluding remarks.

45.2 THE THREE APPROACHES TO MAXIMUM ENTROPY SPECTRAL ANALYSIS

All the first three approaches are based on Shannon's [8] measure of entropy and Jaynes [4] principle of maximum entropy. They differ in their choice of random variable or processes whose entropy is to be maximized.

Skilling and Gull [9] considered $S(f)$ (normalised, if necessary) as a probability density function, found its entropy and sought to maximize it subject to given constraints in a straightforward application of MEP. They did not assume explicitly the process to be Gaussian or stationary. They did assume the knowledge of expected values of exp $(2\pi ikf)$ for $k = -m$ to m, but these can be regarded as special forms of moments.

Burg [1] did not regard $S(f)$ as a probability density function. On the other hand, he considered the entropy gain in a stochastic process that is passed-through a linear filter with white Gaussian noise and maximized this gain subject to (2) and it was not surprising that he got the same spectral density function as arises for an AR process which has a similar underlying process.

Jaynes [5] also used *MEP*, but he applied it to observations in time-domain and not in the frequency domain. Since he assumed that only autocovariances were known, he assumed essentially a stationary Gaussian process and as such he also arrived at the same spectrum density function as Burg, which was again the same as for an autoregressive process.

Thus, Burg and Jaynes got the AR spectral density function (12) and Skilling and Gull got the spectral density function,

$$S(f) = \exp\left(\lambda_0 + \sum_{k=-m}^{m} \lambda_k e^{-2\pi ikf}\right) \quad (15)$$

None of them got the spectral density functions (13) and (14) which were already well-established in the literature.

45.3 THE MINIMUM CROSS-ENTROPY SPECTRAL ANALYSIS APPROACH

The MESA approaches do not presume any knowledge of the prior spectral density function. In MCESA, we assume that we have knowledge of *a prior* spectral density $T(f)$, before the autocovariances were obtaind. Johnson and Shore [6] therefore proposed that out of all those density functions which satisfy (2), we choose that which is in some sense closest to $T(f)$. They proposed that we minimize Kullback-Leibler [7] measure,

$$\int_{-\frac{1}{2}}^{\frac{1}{2}} S(f) \ln \frac{S(f)}{T(f)}\, df \tag{16}$$

subject to (2) to get

$$S(f) = T(f) \exp\left(\lambda_0 + \sum_{k=-m}^{m} \lambda_k e^{-2\pi i f k}\right) \tag{17}$$

This spectrum also however did not give us (13) and (14). They also considered minimization of another measure of cross-entropy

$$\int_{-\frac{1}{2}}^{\frac{1}{2}} \left(\frac{S(f)}{T(f)} - \ln \frac{S(f)}{T(f)} - 1\right) df \tag{18}$$

given by Itakura and Saito [3]. This gives

$$\frac{1}{T(f)} - \frac{1}{S(f)} = \lambda_0 + \sum_{k=-m}^{m} \lambda_k \exp(-2\pi i k f) \tag{19}$$

or,

$$S(f) = \frac{1}{\frac{1}{T(f)} - \lambda_0 - \sum_{k=-m}^{m} \lambda_k \exp(-2\pi i k f)} \tag{20}$$

If $T(f)$ is the spectrum density function for another autoregressive process, this gives

$$S(f) = \frac{1}{\mu_0 + \sum_{l=-n}^{n} \mu \quad \exp(-2\pi i l f) - \lambda_0 - \sum_{k=-m}^{m} \lambda_k \quad \exp(-2\pi i f k)} \tag{21}$$

which is again spectral density function for still another AR process.

45.4 THE INVERSE MAXIMUM ENTROPY AND MINIMUM CROSS-ENTROPY APPROACHES

We naturally ask whether we can bring spectral density functions (13) and (14) within the orbit of maximum entropy and minimum cross-entropy principles, i.e. whether we can find measures of entropy or cross-entropy by maximizing or minimizing which subject to (2) we can get (13) or (14).

Let us consider the entropy measure

$$\int_{-\frac{1}{2}}^{\frac{1}{2}} \phi\,(S\,(f))\; df, \tag{22}$$

where, $\phi(.)$ is a positive concave function. Maximizing it subject to (2) we get

$$\phi'(S\,(f)) = \lambda_0 + \sum_{k=-m}^{m} \lambda_k \exp\quad(-2\pi i f k) \tag{23}$$

If $S(f)$ is given by (12), we get

$$\phi'(S\,(f)) = \frac{K}{S\,(f)}$$

$$\phi'\,(x) = \frac{K}{x} \tag{24}$$

$$\phi\,(x) = k\,\ln x + \text{constant} \tag{25}$$

$$\phi\,(S\,(f)) = k\,\ln S\,(f) + \text{constant} \tag{26}$$

$$\int_{-\frac{1}{2}}^{\frac{1}{2}} \phi\,(S\,(f)\,df) = k \int_{-\frac{1}{2}}^{\frac{1}{2}} \ln S\,(f)\,df + \text{constant} \tag{27}$$

This gives us Burg's measure, so that we have shown that *if we have to get AR spectral density function through entropy maximization, then the only measure of entropy we can use is Burg's measure.* Of course we have to remember that in the maximization process, additive and positive multiplicative constants do not matter.

Again if $S(f)$ is given by (13), we get

$$\phi'\,(S\,(f)) = KS\,(f) \tag{28}$$

or,
$$\phi'\,(x) = Kx$$

or,
$$\phi\,(x) = \frac{1}{2}Kx^2 + \text{constant} \tag{29}$$

or,
$$\phi\,(S\,(f)) = \frac{1}{2}KS^2(f) + \text{constant} \tag{30}$$

or,
$$\int_{-\frac{1}{2}}^{\frac{1}{2}} \phi(S\,(f)\,df) = \frac{1}{2}K \int_{-\frac{1}{2}}^{\frac{1}{2}} S^2(f)\,df + \text{constant} \tag{31}$$

Since we want $\phi(.)$ to be a concave function and since additive and positive multiplicative constants do not matter, we choose the measure

$$\int_{-\frac{1}{2}}^{\frac{1}{2}} (1 - S^2(f))\, df \tag{32}$$

which is Havrda-Charvat [2] entropy of the second order. This entropy of order α is given by

$$\frac{1}{1-\alpha}\left(\int_{-\frac{1}{2}}^{\frac{1}{2}} S^{\alpha}(f)df - 1\right), \alpha \neq 1, \alpha > 0 \tag{33}$$

Thus we have proved that *to get moving average spectral density function, we have to use Havrda-Charvat entropy of second order.*

To get (14), let us write

$$T(f) = \mu_0 + \sum_{l=-n}^{n} \mu_l \exp \quad (-2\pi i l f) \tag{34}$$

so that

$$\frac{T(f)}{S(f)} = \lambda_0 + \sum_{k=-m}^{m} \lambda_k \exp \quad (-2\pi i f k) \tag{35}$$

from (23) and (35)

$$\phi'(S(f)) = k\frac{T(f)}{S(f)} \tag{36}$$

or, $$\phi(S(f)) = k\, T(f) \ln S(f) + \text{constant} \tag{37}$$

This will make the entropy measure dependent on the unknown $T(f)$. This is not desirable. However, in this case we have an alternative approach and that is to use the principle of minimum cross-entropy instead of using the principle of maximum entropy.

Accordingly we manimize the measure

$$\int_{-\frac{1}{2}}^{\frac{1}{2}} T(f)\, \phi\left(\frac{S(f)}{T(f)}\right) df, \tag{38}$$

where $\phi(.)$ is a twice-differentiable convex function for which $\phi(1) = 1$, subject, to (2) to get

$$\phi'\left(\frac{S(f)}{T(f)}\right) = \lambda_0 + \sum_{k=-m}^{m} \lambda_k \exp \quad (-2\pi i f k) \tag{39}$$

From (14), (34) and (39)

$$\phi\left(\frac{S(f)}{T(f)}\right) = k\frac{T(f)}{S(f)} \tag{40}$$

so that,

$$\phi'(x) = \frac{k}{}$$

$$\phi(x) = k \ln x + \text{constant}.$$

However, we want $\phi(x)$ to be a convex function vanishing at $x = 1$, so that

$$\phi\left(\frac{S(f)}{T(f)}\right) = -\ln\frac{S(f)}{T(f)} = \ln\frac{T(f)}{S(f)} \tag{41}$$

$$\int_{-\frac{1}{2}}^{\frac{1}{2}} T(f)\,\phi\left(\frac{S(f)}{T(f)}\right) df = \int_{-\frac{1}{2}}^{\frac{1}{2}} T(f) \ln\frac{T(f)}{S(f)}\, df \tag{42}$$

so that the *desired measure of cross-entropy for getting ARMA spectral density function is the Kullback-Leibler cross-entropy of T(f) from S(f) (and not the other way about).*

Thus we have proved that,

(i) AR model SDF is obtained by maximizing $\int_{-\frac{1}{2}}^{\frac{1}{2}} \ln S(f)\, df$ subject to (2)

(ii) MA model SDF is obtained by maximizing $\int_{-\frac{1}{2}}^{\frac{1}{2}} (1 - S^2(f))\, df$ subject to (2)

(iii) ARMA model SDF is obtained by minimizing $\int_{-\frac{1}{2}}^{\frac{1}{2}} T(f) \ln (T(f)) / (S(f))\, df$ subject to (2)

Thus all the three spectral density functions can be obtained by using MEP and MCEP, provided we choose appropriate measures of entropy and cross-entropy.

In deriving the above results, we have used the following two *inverse principles of maximum entropy and minimum cross-entropy.*

(i) *Given a set of constraints and a spectral density function, choose that measure of entropy such that when this measure is maximized subject to the given constraints, the given spectral density function is obtained.*

(ii) *Given a set of constraints, a priori spectral density function T(f) and a posteriori spectral density function S(f), choose that measure of cross-entropy by minimizing which subject to the given constraints, we get the given posterior spectral density function.*

45.5 THE GENERALISED MEASURES OF ENTROPY AND CROSS-ENTROPY

As noted earlier, Skilling and Gull and Jaynes used Shannon's measure of entropy as the entropy measure. Burg used a measure derived from Shannon's entropy. Johnson and Shore used Kullback-Leibler measure of cross-entropy and Itakura-Saito measure which can be derived again by using Kullback-Leibler measure

However, in our derivation of MA spectral density function, we needed an entirely different measure of entropy (32) and another measure of cross-entropy (42) which is Kullback-Leibler measure of cross-entropy of $T(f)$ from $S(f)$. If we put $T(f) = 1$, (42) gives

$$\int_{-\frac{1}{2}}^{\frac{1}{2}} \ln \frac{1}{S(f)}\, df \tag{43}$$

and minimizing it would mean maximizing entropy (27). The points we wish to emphasise here are the following :

(i) $S(f) \geq 0$ and it can always be normalised by dividing it by $\int_{-\frac{1}{2}}^{\frac{1}{2}} S(f)\, df$ so that the normalised spectral density function can be considered as a probability density function.

(ii) $\int_{-\frac{1}{2}}^{\frac{1}{2}} \ln S(f)\, df$ can be considered as a measure of entropy in its own right irrespective of the fact that it arises as the entropy of a stochastic process. In fact $-\int S(f)\, df$ is the cross-entropy of the flat distribution from the distribution $S(f)$.

If we can consider,

$$-D(S(f):1) = -\int_{-\frac{1}{2}}^{\frac{1}{2}} S(f) \ln \frac{S(f)}{1}\, df = -\int_{-\frac{1}{2}}^{\frac{1}{2}} S(f) \ln S(f)\, df \tag{44}$$

as a measure of entropy, we can also consider

$$-D(1:S(f)) = -\int_{-\frac{1}{2}}^{\frac{1}{2}} \ln \frac{1}{S(f)}\, df = \int_{-\frac{1}{2}}^{\frac{1}{2}} \ln S(f)\, df \tag{45}$$

as a legitimate measure of entropy.

(iii) The measure (7) was originally derived by considering the limit as $N \to \infty$ of $(1/N)\, D(x_n, \bar{x}_n)$ where

$$(x_1, x_2, \ldots, x_n), (\bar{x}_1, \bar{x}_2, \ldots, \bar{x}_n) \tag{46}$$

are two Gaussian stationary processes. However, it is a measure of directed divergence in its own right, since it can be easily shown that

$$\frac{S(f)}{T(f)} - \ln \frac{S(f)}{T(f)} - 1 \geq 0 \tag{47}$$

and it vanishes iff $S(f) = T(f)$ and since (47) is a convex function of both $S(f)$ and $T(f)$.

(iv) If $S(f)$ and $T(f)$ are two normalised spectral density function and $\phi(.)$ is a twice differentiable convex function with $\phi(1) = 0$ then both

$$\int_{-\frac{1}{2}}^{\frac{1}{2}} T(f)\ln\left(\frac{T(f)}{S(f)}\right) df \quad \text{and} \quad \int_{-\frac{1}{2}}^{\frac{1}{2}} S(f)\ln\frac{S(f)}{T(f)}\, df \tag{48}$$

can be considered as cross-entropy measures and by putting $T(f) = 1$,

$$\int_{-\frac{1}{2}}^{\frac{1}{2}} \ln S(f)\, df \text{ and} - \int_{-\frac{1}{2}}^{\frac{1}{2}} S(f) \ln S(f)\, df \tag{49}$$

can be considered as measures of entropy in their own right.

(v) We can and should give up the self-imposed restriction of using only Shannon's and Kullback-Leibler's measures if we want to consider the possibility of getting spectral density functions other than that for AR process.

(vi) If we use (8), we get the spectral density function

$$S(f) = T(f)\left[(\alpha-1)\left(\lambda_0 + \sum_{k=-m}^{m} \lambda_k \exp\ (-2\pi i f k)\right)\right]^{\frac{1}{\alpha-1}} \tag{50}$$

which approaches the exponential form

$$S(f) = T(f)\exp\left(\lambda_o + \sum_{k=-m}^{m} \lambda_k \exp(-2\pi i f k)\right) \tag{51}$$

as $\alpha \to 1$ and which approaches AR spectrum density (12) as $\alpha \to 0$. Thus this gives both spectrum density function obtained by using Shannon's measure or Burg's measure (when $T(f) = 1$), but it also gives an infinity of other spectral density function when $\alpha \neq 0, 1$. In particular it gives MA spectral density function when $\alpha = 2$. Again if we put $\alpha = 0$ and take $T(f)$ to be a moving average spectral density function, we get the ARMA spectral density function. Thus (50) includes all the spectral density functions obtained so far and many more.

(vii) Since (50) gives Burg's spectrum when $\alpha \to 0$ and gives Shannons's measure spectrum when $\alpha \to 1$ and gives a whole range of other spectra when α lies between 0 and 1, it can be used to discuss the continuous transition of Burg's spectrum into Shannon's measure spectrum and vice versa.

(viii) In practice, these spectra will not differ very significantly as α varies from 0 to 1, specially when m is not very small. This result can be shown to be in general true by using the empirical method.

(ix) The three processes studied extensively are AR, MA and ARMA. All these involve the variables $x_t, x_{t-1}, x_{t-2}, \ldots$ and the innovations $\varepsilon_t, \varepsilon_{t-1}, \varepsilon_{t-2}, \ldots$ linearly only. In the spectral density function, the Lagrange's multipliers λ's occur linearly both in the numerator and denominator and these can easily be determined in terms of known autocovariances.

45.6 SOME MINIMUM CROSS-ENTROPY SPECTRAL DENSITY FUNCTIONS

(i) If we minimize

$$\int_{-\frac{1}{2}}^{\frac{1}{2}} S(f) \ln \frac{S(f)}{T(f)}\, df \tag{52}$$

subject to autocovariance functions being prescribed as in (2), we get

$$S((f) = T(f) \exp\left(\lambda_0 + \sum_{k=-m}^{m} \lambda_k \exp \quad (-2\pi i f k)\right), \tag{53}$$

so that we get three possible spectral density functions according as we take $T(f)$ to be given by (12), (13) and (14). If we put $T(f) = 1$, i.e. if we take a flat prior we get the maximum entropy spectral density function given by Skilling and Gull [9].

(ii) If we minimize

$$\int_{-\frac{1}{2}}^{\frac{1}{2}} T(f) \ln \frac{T(f)}{S(f)}\, df \tag{54}$$

subject to (2), we get SDF of the form

$$S(f) = \frac{T(f)}{\lambda_0 + \sum_{k=-m}^{m} \lambda_k \exp(-2\pi i f k)} \tag{55}$$

If we put $T(f) = 1$, we get the SDF for an AR process.
We get three other spectral density functions from (55) according as $T(f)$ is given by (12), (13), (14).

(iii) If we minimize

$$\int_{-\frac{1}{2}}^{\frac{1}{2}} \left(\frac{S(f)}{T(f)} - \ln \frac{S(f)}{T(f)} - 1\right) df, \tag{56}$$

we get

$$S(f) = \frac{T(f)}{1 - T(f)\left[\lambda_0 + \sum_{k=-m}^{m} \lambda_k \exp(-2\pi i f k)\right]} \tag{57}$$

If we put $T(f) = 1$, we get essentially the SDF for an AR process. We get three other special density functions according as we take $T(f)$ to be given by (12), (13) and (14).

(iv) If we minimize

$$\int_{-\frac{1}{2}}^{\frac{1}{2}} \left(\frac{T(f)}{S(f)} - \ln \frac{T(f)}{S(f)} - 1\right) df \tag{58}$$

we get

$$-\frac{T(f)}{S^2(f)} + \frac{1}{S(f)} = \lambda_0 + \sum_{k=-m}^{m} \lambda_k \exp \quad (-2\pi i f k) \tag{59}$$

and we get a rather complicated expression for $S(f)$.

(v) We can similarly get sixteen possible spectral density functions by minimizing (8), (9), (10), (11) and taking $T(f) = 1$ or $T(f)$ given by (12), (13) or (14).

(vi) We can now take each of the 32 spectral density functions obtained so far as priors and generating new possible spectral functions and this process can be continued.

Each of these spectral density functions would satisfy the constraints (1) and out of all those satisfying these constraints, it will be the one closest to the given a priori density functions. We get a large number of measures of closeness and we can have a large number of a prior spectral density functions.

45.7 CONCLUDING REMARKS

In the direct application of maximum entropy principle, we are given the measure of entropy and we find that spectral density function out of all these that are consistent with given autocovariances which maximizes this entropy. We have however also used the inverse maximum entropy principle according to which we are given the spectral density function and we seek to find the measure of entropy by maximizing which we can find the given spectral density function.

In the same way, we use the inverse minimum cross-entropy principle to find the measure of cross-entropy by minimizing which we can get the given *a posteriori* spectral density function when the a priori spectral density function is given.

These principles have been used to obtain the appropriate measure of entropy and cross-entropy for AR, MA and ARMA processes.

We have proposed here a parametric measure of cross-entropy by minimizing which, subject to given autocovariance constraints, we can get AR, MA, ARMA and exponential spectral density functions and which can help us in understanding the transition of spectral density function from Burg and AR spectral density function to Shannon's and Skilling's spectral density function.

Some problems associated with existence, uniqueness and completeness of the solutions obtained by this approach will be discussed in a separate paper.

REFERENCES

1. Burg, J.P. (1972), "The Relationship between Maximum Entropy Spectra and Minimum Likelihood Spectrum" in Modern Spectral Analysis, ed. D.G.Childers, pp. 130-131, M.S.A.
2. Havrda, J.H. and F. Charvat (1967) "Quantification Methods of Classification Processes : Concept of Structural α Entropy," Kybernetica, Vol.3, pp. 30-35,
3. Itakura, F. and S Saito, (1968) "Analysis synthesis telephony based on the maximum likelihood method", in Rep. 6th Int. Cong. Accoust.
4. Jaynes, E.T. (1957) "Information Theory and Statistical Mechanics", Physical Reviews, Vol. 106, pp. 620-630.
5. Jaynes, E.T. (1985), "Where do we go from here '?" in Maximum-Entropy and Bayesian Methods on Inverse Problems, ed.C.R. Smith and W.T. Grandy Jr., pp. 21-58.
6. Johnson, R.W. and J.E. Shore (1984), "Which is the better entropy for speech processors : S In *S* or In *S* ?"IEEE Trans. Acc., Speech and Signal Proc. AASP, Vol. 32, pp. 129-137.
7. Kullback, S. and R.A. Leibler, (1951), "On Information and Sufficiency", Ann. Math. Stat., Vol. 22, pp. 79-86.

8. Shannon, C.E. (1948), " A Mathematical Theory of Communication", Bell System Tech . J., Vol. 27, pp. 379-423, 623-659.
9. Skilling, J. and S.F. Gull (1985), "Algorithms and Application", C. Ray Smith and W.T. Grandy, Jr. (eds.), Maximum-Entropy and Bayesian Methods in Inverse Problems, pp. 83-132,
10. Tzannes, M.A., Dimitris Politis and N.S Tzannes, (1985),"A General Method of Minimum Cross-Entropy Spectral Estimation", IEEE Transactions on Accoustics, Speech and Signal Processing, Vol. ASSP-33,, No. 3, pp. 748-752.

46

MAXIMUM-ENTROPY SOLUTIONS OF SOME ASSIGNMENT PROBLEMS

[A number of optimum assignment problems have been solved by using appropriate versions of the Principle of Maximum Entropy.]

46.1 THE PROBLEMS

Consider the following assignment problems :

(i) There are k colonies and N individuals have to be assigned to these. The lower and upper bounds of the number of individuals that can be assigned to each colony are given. The rent of a house in each colony is known and the total funds that can be spent on rent are also prescribed. Find the assignment which is optimal in the sense that (*a*) the populations in different colonies are maximally equal, or (*b*) the ratios of the deficits from the maximum possible to the excesses over the minimum permissible are maximally equal so that, considering the capacities of the colonies, the colonies are in some sense, maximally equally crowded.

(2) From the k colonies, individuals travel to Central Business District (CBD) and the individual costs are $c_1, c_2, ..., c_k$ and the total travel budget is prescribed. Find the assignment which is optimal in the sense that (*a*) the populations in the k colonies are maximally equal, or (*b*) the ratios of the deficits from the maximum possible population to the excesses over the minimum permissible population are maximally equal, or (*c*) the rents realised from the k colonies are maximally equal.

(3) A number of particles have to be assigned to k energy levels with energies $\varepsilon_1, \varepsilon_2, ..., \varepsilon_k$ and the total energy of the system is prescribed. Find the optimal assignment in which the number of particles in different energy levels are maximally equal when the maximum number of particles in each energy level is known.

46.2 MATHEMATICAL FORMULATIONS OF THE PROBLEMS

Let N be the total number of individuals (particles) and let Nb_i and Na_i be the maximum and minimum number of individuals (particles) that can be assigned to the ith colony (energy level), then we have

$$p_1 \geq 0, p_2 \geq 0, ..., p_k \geq 0 \tag{1}$$

$$\sum_{i=1}^{k} p_i = 1 \tag{2}$$

$$a_i \leq p_i \leq b_i \quad ; \quad i = 1, 2, ..., k \tag{3}$$

$$\sum_{i=1}^{k} p_i c_i = C, \text{ if the travel budget is prescribed} \tag{4}$$

$$\sum_{i=1}^{k} p_i r_i = R, \text{ if the total rent is prescribed} \tag{5}$$

$$\sum_{i=1}^{k} p_i \varepsilon_i = \varepsilon, \text{ if the total energy is prescribed} \tag{6}$$

$p_1, p_2, \ldots\ldots\ldots, p_k$ would be maximally equal if the number of individuals (particles) assigned are maximally equal. (7)

$(b_i - p_i)/(p_i - a_i)$, $i = 1.2, \ldots\ldots, k$ would be maximally equal if the colonies are to be maximally equally crowded. (8)

$p_1 r_1, p_2 r_2, \ldots\ldots, p_k r_k$ would be maximally equal if the rents realised from the k colonies are maximally equal. (9)

In problem 1(a), we have to ensure (7) subject to (1), (2), (3) and (5).
In problem 1(b), we have to ensure (8) subject to (1), (2), (3) and (5).
In problem 2(a), we have to ensure (7) subject to (1), (2), (3), and (4).
In problem 2(b), we have to ensure (8) subject to (1), (2), (3), and (4).
In problem 2(c) we have to ensure (9) subject to (1), (2), (3), and (4).
In problem 3, we have to ensure (7) subject to (1), (2), (3), and (6).

46.3 THE MAXIMUM ENTROPY PRINCIPLE AS A PRINCIPLE OF MAXIMUM EQUALITY

In the classical maximum entropy principle due to Jaynes [2], we maximize the entropy measure of a probability distribution

$$P = (p_1, p_2, \ldots\ldots, p_k) \tag{10}$$

$$H_1(P) = -\sum_{i=1}^{k} p_i \ln p_i \tag{11}$$

due to Shannon [9] subject to the natural constraints (1) and (2) and some additional constraints of the type

$$\sum_{i=1}^{k} p_i g_r(x_i) = a_r, \qquad r = 1, 2, \ldots\ldots, m \tag{12}$$

Now (12) includes (4), (5) and (6) as special cases.

In this principle entropy represents 'uncertainty' and by maximizing entropy, we are getting the most uncertain or most random distribution satisfying the given constraints.

This principle also ensures the maximum equality of p_i's subject to given constraints in the sense that if the constraints (12) are absent and only the natural constraints (1) and (2) are present, then this principle gives

$$p_1 = p_2 = \ldots\ldots = p_k = 1/k, \tag{13}$$

so that in this case $p_1, p_2, \ldots\ldots, p_k$ are all equal. In the presence of (12), these probabilities are maximally equal, *i.e.* are as equal as possible subject to their satisfying the constraints (12).

The Principle of Maximum Entropy can also therefore be interpreted as the Principle of Maximum Equality.

However, in the context of our assignment problems, we do not have any probability distribution and as such we have no uncertainty to be maximized.

In our case,

$$p_i = \frac{\text{number of individual (particles) assigned to the } i^{\text{th}} \text{ colony (energy) level}}{\text{Total number of individuals (particles)}}$$

and, as such P does not represent a probability distribution ; at most it represents a relative frequency distribution. As such $H_I(P)$ is not a measure of uncertainty. We use it as a measure of equality. We maximize it subject to appropriate constraints to get maximally equal proportions $p_1, p_2, ..., p_k$.

46.4 PROPERTIES OF H_1 (P) WHICH MAKE IT A DESIRABLE MEASURE OF EQUALITY

(i) $H_1(P)$ does not change when $p_1, p_2, ..., p_k$ are permuted among themselves.

(ii) $H_1(P)$ is a continuous function of $p_1, p_2, ..., p_k$ if we define $0 \ln 0 = 0$.

(iii) Subject to (1) and (2) $H_1(P)$ is maximum when all the proportions are equal and its maximum value, viz, $\ln k$ increases as k increases.

(iv) $H_1(P)$ has the minimum value 0 when one of the p_i's is unity and all the others are zero, so that $H_1(P) \geq 0$.

(v) When $H_1(P)$ is maximized subject to (2) and (12), the maximizing p_i's are always positive. This is highly desirable since p_i's represent proportions of individuals or particles and cannot be negative. This implies that constraint (1) need not be taken into account explicitly when $H_1(P)$ is used as a measure of equality.

(vi) $H_1(P)$ is a concave function, so that when it is maximized subject to known constraints (2) and (12), its local maximum is its global maximum.

(vii) Maximizing $H_1(P)$ subject to (2) and (12) gives the same result as minimizing

$$D_1(P:U) = \sum_{i=1}^{k} p \ln \frac{p_i}{1/k} = \ln k - H_1(P) \tag{15}$$

subject to the same constraints. Now $D_1(P : U)$ is the Kulback-Leibler [7] measure of directed divergence of P from $U = (1/k, 1/k ...... 1/k)$, i.e. the distribution corresponding to perfect equality so that maximizing $H_1(P)$ subject to given constraints gives a distribution which is as near as possible to perfect equality distribution subject, of course, to the given constraints.

46.5 OTHER VERSIONS OF MAXIMUM ENTROPY PRINCIPLE

We have already given two versions of the Principle of Maximum Entropy, viz.

as Principle of Maximum Uncertainty,
as Principle of Maximum Equality,

It is not difficult to implement these principles if the only inequality constraints on probabilities are given by (1). If however inequality constraints (3) are present, then complicated concave programming methods have to be used. The only case which has been discussed is when only the constraints (1), (2) and (3) are there [1].

However, our criterion (8) does not requires us to get maximally equal proportions; rather it requires us to get maximally equally crowding or maximally equal values of $(b_i - p_i) / (p_i - a_i)$. As such the measure $H_1(P)$ is not appropriate in this case. The following measure (5) is suggested.

$$H_2(P) = -\sum_{i=1}^{k} (p_i - a_i) \ln (p_i - a_i) - \sum_{i=1}^{k} (b_i - p_i) \ln (b_i - p_i) \tag{16}$$

This has the following properties :

(i) It changes if $p_1, p_2, ..., p_k$ are permuted among themselves, but does not change if the triplets (a_i, b_i, p_i), $i = 1.2, ..., k$ are permuted among themselves.

(ii) It is maximum subject to $\sum_{i=1}^{k} p_i = 1$ when

$$\frac{b_i - p_i}{p_i - a_i} = K, \quad p_i = \frac{b_i + Ka_i}{1+K}, \quad K = \frac{B-1}{1-A} \tag{17}$$

where,
$$A = \sum_{i=1}^{k} a_i \quad , \quad B = \sum_{i=1}^{k} b_i \tag{18}$$

provided the conditions

$$1 \leq A, \qquad B \geq 1 \tag{19}$$

are satisfied [1].

(iii) We may call

$$u_i = \frac{b_i + Ka_i}{1+K} \quad , \quad K = \frac{B-1}{1-A} \tag{20}$$

as the modified uniform distribution, modified to take into account constraints (3) on p_i's. We may also call it as a generalised uniform distribution since when $b_i = 1$, $a_i = 0$. It gives

$$A = 0, \quad B = k, \quad K = k - 1, \qquad u_i = 1/k \tag{21}$$

so that in this case, we get the usual uniform distribution

(iv) We may define

$$D(P:Q) = \sum_{i=1}^{k} (p_i - a_i) \ln \frac{p_i - a_i}{q_i - a_i} + \sum_{i=1}^{k} (b_i - p_i) \ln \frac{b_i - p_i}{b_i - q_i} \tag{22}$$

as a new measure of directed divergence provided $a_i \leq q_i \leq b_i$ for all i, since this is a convex function of $p_1, p_2, ...,p_k$ which attains its maximum value when $p_i = q_i$ for all i and as such it is ≥ 0 and vanishes iff $P = Q$, so that

$$\begin{aligned} D(P:U') &= \sum_{i=1}^{k} (p_i - a_i) \ln \frac{p_i - a_i}{u_i - a_i} + \sum_{i=1}^{k} (b_i - p_i) \ln \frac{b_i - p_i}{b_i - u_i} \\ &= \sum_{i=1}^{k} (p_i - a_i) \ln (p_i - a_i) + \sum_{i=1}^{k} (b_i - p_i) \ln (b_i - p_i) \\ &- \sum_{i=1}^{k} (p_i - a_i) \ln (u_i - a_i) - \sum_{i=1}^{k} (b_i - p_i) \ln (b_i - u_i) \end{aligned}$$

$$= -H_2(P) - \sum_{i=1}^{k} (p_i - a_i) \ln \frac{b_i - a_i}{1+K} - \sum_{i=1}^{k} (b_i - p_i) \ln \frac{K(b_i - a_i)}{1+k}$$

$$= -H_2(p) - \sum_{i=1}^{k} (b_i - a_i) \ln \frac{b_i - a_i}{1+K} - \sum_{i=1}^{k} (b_i - p_i) \ln K$$

$$= -H_2(P) - \sum_{i=1}^{k} (b_i - a_i) \ln (b_i - a_i) + (B - A) \ln (1+k) - (B-1) \ln K \quad (23)$$

so that minimizing $D_2(P:U)$ subject to (1), (2), (3) and (4) is equivalent to maximizing $H_2(P)$ subject to the same constraints. Thus maximizing $H_2(P)$ subject to the given constraints would give us P as close to the modified uniform distribution as possible when 'closeness' is defined in terms of the measure $D_2(P:Q)$.
(v) The constraints (3) will automatically be satisfied. This gives us the third version of the principle of maximum entropy, viz.
Principle of Maximum Equality of $(b_i - p_i)/(p_i - a_i)$ *for* $i = 1, 2, ..., k$ *when the constraints (3) are applicable.*

Again to implement criterion (9), we propose the measure

$$H_3(P) = -\sum_{i=1}^{k} (p_i \ln p_i + p_i \ln r_i - p_i) \quad (24)$$

This has the following properties :

(i) it changes when p_i's are permuted among themselves, but it does not when the pairs (p_i, r_i), $(i = 1, 2, ..., k)$ are permuted among themselves;
(ii) it is a concave function of $p_1, p_2, ..., p_k$, which has a global maximum value, subject to $\sum_{i=1}^{k} p_i = 1$,, when

$$p_i r_i = K', \frac{p_i}{1/r_i} = \frac{1}{\sum_{i=1}^{k} 1/r_i} = K', \quad (25)$$

i.e. when the rent realised from each colony is the same;
(iii) When we maximize $H_3(P)$ subject to linear constraints, all the maximizing p_i's are > 0
(iv) *Let*

$$U'' = \left(\frac{1/r_1}{\sum_{i=1}^{k} 1/r_i}, \frac{1/r_2}{\sum_{i=1}^{k} 1/r_i}, ..., \frac{1/r_k}{\sum_{i=1}^{k} 1/r_i} \right) \quad (26)$$

then

$$D_1(P:U'') = \sum_{i=1}^{k} p_i \ln \frac{p_i}{\frac{1}{r_i}} \sum_{i=1}^{k} \frac{1}{r_i}$$

$$= p_i \ln p_i + p_i \ln r_i + \ln \sum_{i=1}^{k} \frac{1}{r_i}$$

$$= -H_3(P) + 1 + \ln \sum_{i=1}^{k} \frac{1}{r_i} \quad (27)$$

Thus maximizing H_3 (P) subject to some constraints is equivalent to minimizing D_1 $(P : U'')$, so that by maximizing H_3 (P) we get a proportion distribution which is as close as possible to the distribution in which the rents realised from the various colonies are the same, subject to whatever constraint (2) and (4) or (5) or (12) that may be given. This gives the fourth version of the principle of maximum entropy, viz.

Maximize $-\sum_{i=1}^{k} (p_i \ln p_i + p_i \ln r_i - p_i)$ subject to the given constraints to get total rents from the colonies as equal as possible. This is the *principle of maximum equality of rents.*

46.5 SOLUTIONS OF PROBLEMS GIVEN IN THE FIRST SECTION

Problem 1 (*a*):

Maximizing

$$-\sum_{i=1}^{k} p_i \ln p_i \quad \text{subject to} \quad \sum_{i=1}^{k} p_i = 1, \qquad \sum_{i=1}^{k} p_i r_i = R \tag{28}$$

we get,

$$p_i = \frac{e^{-\mu r_i}}{\sum_{i=1}^{k} e^{-\mu r_i}}, \quad i = 1, 2, \ldots, k, \tag{29}$$

where μ is determined by using

$$\sum_{i=1}^{k} r_i e^{-\mu r_i} \Big/ \sum_{i=1}^{k} e^{-\mu r_i} = R\,, \tag{30}$$

which is the well-known Maxwell-Boltzmann distribution.

PROBLEM 1 (*b*) : Maximizing H_2 (P) subject to (1), (2), (3) and (5), we get

$$\frac{b_i - p_i}{p_i - a_i} = e^{-\lambda - \mu r_i} \tag{31}$$

or

$$p_i = \frac{b_i + a_i e^{-\lambda - \mu r_i}}{1 + e^{-\lambda - \mu r_i}} = \frac{b_i e^{\lambda + \mu r_i} + a_i}{e^{\lambda + \mu r_i} + 1}, \tag{32}$$

where λ and μ are obtained by solving

$$\sum_{i=1}^{k} \frac{b_i e^{\lambda + \mu r_i} + a_i}{e^{\lambda + \mu r_i} + 1} = 1 \tag{33}$$

$$\sum_{i=1}^{k} r_i \frac{b_i e^{\lambda + \mu r_i} + a_i}{e^{\lambda + \mu r_i} + 1} = R \tag{34}$$

If each $a_i = 0$, $b_i = 1$, this gives

$$p_i = \frac{1}{1 + e^{-\lambda - \mu r_i}}, \tag{35}$$

which is *Fermi-Dirac distribution*. In this sense (32) may be regarded as *a generalised Fermi-Dirac distribution.*

Problem 2 (a) : The solution is the same as that of Problem 1 (*a*) except that r_i and R have to be replaced by c_i and C.

Problem 2 (b) : The solution is the same as that of Problem 1 (b) except that r_i and R have to be replaced by c_i and C.

Problem 2 (c) : Here we maximize $H_3\,(P)$ subject to,

$$\sum_{i=1}^{k} p_i = 1, \quad \sum_{i=1}^{k} p_i\, c_i = C$$

to get

$$p_i\, r_i = e^{-\lambda - \mu c_i} \tag{36}$$

where λ and μ are determined by solving

$$\sum_{i=1}^{k} \frac{1}{r_i}\, e^{-\lambda - u c_i} = 1, \qquad \sum_{i=1}^{k} \frac{c_i}{r_i}\, e^{-\lambda - \mu c_i} = C \tag{37}$$

Problem 3 : The solution of the first part is the same as that of Problems 1 (a) and 2 (a) except that r_i, c_i are replaced by ε_i and R, C are replaced by ε. The solution of the second part is the same as that of 1 (*b*) or 2 (*b*) where c_i, r_i are replaced by ε_i, C, R by ε and a_i is put equal to zero.

46.6 CONCLUDING REMARKS

(i) Some persons want to reserve the use of the word 'entropy' for uncertainty only; some others go even further and want to reserve it only for Shannon's measure. However, the word has already been used in so many senses [6,8] and so many measures of entropy other than Shannon's have already been proposed [3, 4, 6, 10] that it is too late in the day to limit the scope of the word. The major criterion is whether such extensions lead to useful results.

(ii) In the present paper, we have used entropy in the sense of a function of $p_1, p_2, \ldots, p_k$ where $p_1, p_2, \ldots, p_k$ are positive proportions whose sum is unity, whose maximization subject to $\sum_{i=1}^{k} p_i = 1$ leads to desirable proportions and whose maximization subject to additional linear constraints would give us proportions having the desirable property as closely as possible subject to the proportions satisfying the given constraints. We have also required that the function should be concave so that we have not to make a distinction between local and global maxima and we have also limited ourselves to such functions whose maximization always leads to positive proportions.

REFERENCES

1. D. Freund and U. Saxena (1984). "An algorithm for a class of discrete maximum entropy problems", Operations Research Vol 32 (1), 210-215.
2. E.T. Jaynes (1957). "Information Theory and Statistical Physics" Physical Reviews. Vol. 106, 620-630.
3. J.N. Kapur (1983). "A Comparative assessment of various measures of entropy" Jour. Inf. Opt. Sci. Vol. 4 no 1, 203-236.
4. J.N. Kapur (1986). "Four families of measures of entropy", Ind. Jour. Pure App. Math. Vol. 17 (4), 429-449.
5. J.N. Kapur (1990). "Maximum Entropy Probability Distribution when there are inequality constraints on probabilities and equality constraints on moments", Aligarh Joun. Stat. (to appear).
6. J.N. Kapur & H.K. Kesavan (1987). "*Generalised Maximum Entropy Principle* (with Applications), Sandford Educational Press, Waterloo University, Canada.
7. S. Kullback and R.A. Leibler (1951). "On Information and Sufficiency", Ann. Math. Stat. Vol. 22, 79-86.
8. A. Seth (1989). "Prof. J.N. Kapur's Views on Entropy Optimization Principles." Bull. Math. Ass. Ind. Vol 21, 1-38.
9. C.E. Shannon (1948) "A Mathematical Theory of Communication", Bell System Tech. Jour. Vol. 27, 379-423, 623-659.
10. I.J. Taneja (1989). "On Generalised Information Measures and their Applications" in Advances in Electronics and Electronic Physics, Vol. 76, Academic Press, New York.

47

AN IMPORTANT APPLICATION OF GENERALISED MINIMUM CROSS-ENTROPY PRINCIPLE : GENERATING MEASURES OF DIRECTED DIVERGENCE

[The generalised Minimum Cross-Entropy Principle has been used to give a new derivation of Csiszer's measure of directed divergence (cross-entropy) from first principles. It has also been used to derive measures of directed divergence for the situation when both the *a priori* and *a posteriori* probability distributions satisfy certain inequality constraints.]

47.1 INTRODUCTION

The Generalised Minimum Cross-Entropy Principle (GMCEP) [4,5] states that given any three of the following :

(i) a posteriori probability distribution $P\ (p_1, p_2, ..., p_n)$

(ii) Set C of moment constraints of the form

$$\sum_{i=1}^{n} p_i = 1, \quad \sum_{i=1}^{n} p_i\, g_r\,(x_i) = a_r, \quad r = 1,2,\ldots,m \tag{1}$$

(iii) *a priori* probability distribution

$$Q = (q_1, q_2, \ldots, q_n) \tag{2}$$

(iv) a measure of directed divergence or cross-entropy

$$D\ (P{:}Q) = D\ (p_1, p_2, \ldots, p_n; q_1, q_2, \ldots, q_n) \tag{3}$$

satisfying the conditions :

(a) $D\ (P : Q) \geq 0$,

(b) $D\ (P : Q) = 0$ iff $P = Q$ or $p_i = q_i$ for each i,

(c) $D\ (P : Q)$ is a convex function of $p_1, p_2, \ldots, p_n$.

we should find the fourth, so that out of all probability distributions, satisfying C, the probability distribution P has the minimum value for $D\ (P : Q)$.

The object of the present paper is to use this important principle for generating measures of directed divergence satisfying a fourth condition.

(d) when $D\ (P : Q)$ is minimizesd subject to C, the minimizing probabilities are all ≥ 0.

In Section 2, we derive Csiszer's [1] improtant measure of directed divergence from generalised minimum cross-entropy principle and deduce some other important measures due to Kullback-Leibler [6], Havrda-Charvat [2], and Kapur [3] from it.

In Section 3, we show how some of the measures and the Euclidean directed divergence fail to meet the condition (d) given above.

In Section 4, we obtain some directed divergence measures applicable to situations where there are inequality constraints,

$$0 \le a_i \le p_i \le b_i \le 1, \quad 0 \le a_i \le q_i \le b_i \le 1, \quad i = 1, 2, \ldots, n \tag{4}$$

on the probabilities. In Section 5, we find some measures of entropy from these measures of directed divergence. In Section 6, we give particular measures when $a_i = 0, b_i = 1$ in each case. In Section 7, we give concluding remarks.

47.2 DERIVATION OF CSISZER'S MEASURE OF DIRECTED DIVERGENCE

To simplify matters, we confine ourselves to measures of directed divergence of the sum form, i.e. to measure of the type

$$D\,(P:Q) = \sum_{i=1}^{n} \phi\,(p_i, q_i) \tag{5}$$

and for the present, we confine ourselves to only one constraint, i.e, the natural constraint,

$$\sum_{i=1}^{n} p_i = 1 \tag{6}$$

Minimizing (5) subject to (6), we get

$$\frac{\partial \phi}{\partial p_i} = K = \text{constant} \tag{7}$$

Also we want (7) to be satisfied when $p_i = q_i$, i.e by $p_i/q_i = 1$. Thus will be satisfied if

$$\frac{\partial \phi}{\partial p_i} = f'\left(\frac{p_i}{q_i}\right) \tag{8}$$

We want to minimize $D\,(P:Q)$. As such we would like $\partial^2 \phi/\partial p_i^2$ to be positive when $p_i = q_i$, i.e we want $f''(1) > 0$. To be on the safe side, we choose $f(x)$ to be convex function of x. Now integrating (8)

$$\phi\,(p_i, q_i) = q_i\, f\left(\frac{p_i}{q_i}\right) + g\,(q_i), \tag{9}$$

so that,

$$D\,(P:Q) = \sum_{i=1}^{n} \phi\,(p_i, q_i) = \sum_{i=1}^{n} q_i\, f\left(\frac{p_i}{q_i}\right) + \sum_{i=1}^{n} g\,(q_i) \tag{10}$$

Since Q is given to us, $\sum_{i=1}^{n} g\,(q_i)$ is a constant to be determined from the condition that $D\,(P:Q) = 0$ when $p_i = q_i$ for each i, so that

$$0 = \sum_{i=1}^{n} q_i\, f(1) + \sum_{i=1}^{n} g\,(q_i) \tag{11}$$

Equations (10) and (11) give

$$D\,(P{:}Q) = \sum_{i=1}^{n} q_i \left[f\left(\frac{p_i}{q_i}\right) - f(1) \right] = \sum_{i=1}^{n} q_i \bar{f}\left(\frac{p_i}{q_i}\right), \tag{12}$$

where,
$$\bar{f}\left(\frac{p_i}{q_i}\right) = f\left(\frac{p_i}{q_i}\right) - f(1), \quad \bar{f}(1) = 0, \tag{13}$$

so that finally we get the measure,

$$D\,(P{:}Q) = \sum_{i=1}^{n} q_i \bar{f}\left(\frac{p_i}{q_i}\right) \tag{14}$$

where, (*i*) $\bar{f}(x)$ is a twice - differentiable convex function of x, and (*ii*) $\bar{f}(1) = 0$

The minimum value of $D\,(P : Q)$ arises when $p_i = q_i$ for each i and is zero so that $D\,(P : Q) \geq 0$ and vanishes iff $P = Q$.

In our subsequent discussion, we shall omit the bar over f and write,

$$D_1\,(P{:}Q) = \sum_{i=1}^{n} q_i f\left(\frac{p_i}{q_i}\right), \tag{15}$$

where $f(x)$ is a twice - differentiable convex function which vanishes at $x = 1$.

The measure (16) was orginally given by Csiszer [1] from other considerations. However, its derivation from GMCEP is new and shows why this measure is so useful.

This measure satisfies our conditions (*a*), (*b*), (*c*) of Section 1. If it has also to satisfy our condition (*d*), we have to choose for $f(x)$ only such a function that $f'(x)$ is defined for only positive values of x.

Now we consider some special cases of (15)

(*i*) $$f(x) = x \ln x \quad f'(x) = 1 + \ln x, \quad f''(x) = \frac{1}{x} > 0$$

$$D_2\,(P{:}Q) = \sum_{i=1}^{n} p_i \ln \frac{p_i}{q_i} \tag{16}$$

This is the most useful measure of directed divergence and is due to Kullback and Leibler [6]. Here $f'(x)$ is defined for only positive values of x.

(*ii*) $$f(x) - \frac{x^\alpha - x}{\alpha(\alpha - 1)}, \quad \alpha > 0, \alpha \neq 1, \quad f'(x) = \frac{\alpha x^{\alpha-1} - 1}{\alpha(\alpha - 1)}, \quad f''(x) = x^{\alpha - 2} > 0 \tag{17}$$

$$D_3\,(P{:}Q) = \frac{\sum_{i=1}^{n} p_i^\alpha q_i^{1-\alpha} - 1}{\alpha(\alpha - 1)}, \quad \alpha \neq 0, \alpha \neq 1, \tag{18}$$

This is a modified form of the measure due to Havrda and Charvat [2]. If $\alpha \to 1$, this gives in the limit Kullback-Leibler measure (16) and if $\alpha \to 0$, this gives

$$D_4(P:Q) = -\sum_{i=1}^{n} q_i \ln \frac{p_i}{q_i} = \sum_{i=1} q_i \ln \frac{q_i}{p_i} \tag{19}$$

(iii) $f(x) = x \ln x - (1+x) \ln (1+x) + 2 \ln 2, f'(x) = \ln \frac{x}{1+x}, f''(x) = \frac{1}{x(1+x)} > 0$

(20)

$$D_5(P:Q) = \sum_{i=1}^{n} p_i \ln \frac{p_i}{q_i} - \sum_{i=1}^{n} (p_i + q_i) \ln \left(1 + \frac{p_i}{q_i}\right) \tag{21}$$

This is a measure due to Kapur [3].

47.3 WEAKNESS OF EUCLIDEAN MEASURE OF DIRECTED DIVERGENCE

Since in (7), we want $\partial \phi / \partial p_i$ to be constant when $p_i = q_i$ for each i, we can replace (8) by

$$\frac{\partial \phi}{\partial p_i} = f'(p_i - q_i) \tag{22}$$

and since we want $D(P : Q)$ to be minimum, we want $f''(p_i - q_i)$ to be positive so that we want $f(x)$ to be a convex function.

Integrating (22)

$$\phi(p_i, q_i) = f(p_i - q_i) + g(q_i) \tag{23}$$

$$D(P:Q) = \sum_{i=1}^{n} \phi(p_i, q_i) = \sum_{i=1}^{n} f(p_i - q_i) + \sum_{i=1}^{n} g(q_i)$$

Since, $D(P : Q)$ is to vanish when $p_i = q_i$ for each i,

$$0 = \sum_{i=1}^{n} (f(0)) + \sum_{i=1}^{n} g(q_i)$$

$$D_6(P:Q) = \sum_{i=1}^{n} (f(p_i - q_i) - f(0)) \tag{24}$$

In particular choosing $f(x) = x^2$, we get

$$D_7(P:Q) = \sum_{i=1}^{n} (p_i - q_i)^2 \tag{25}$$

This is a convex twice-differentiable function of $p_1, p_2, \ldots, p_n$ which on minimization subject to $\sum_{i=1}^{n} p_i = 1$ gives $p_i = q_i$ for each i. However, in the presence of other constraints in (2) it gives

$$p_i - q_i = \lambda_0 + \lambda_1 g_{1i} + \ldots + \lambda_m g_{mi} \tag{26}$$

and, there is no guarantee that when we solve for $\lambda_0, \lambda_1, \ldots \lambda_m$ by using (1) and substitute is (26), we shall get all p_i's positive, so that this measure violates our condition (d).

The Euclidean distance is square root of $D\ (P : Q)$ in (25), but this does not matter since if $D\ (P : Q)$ can be used as a measure of directed divergence, any one-one monotonic function of it which vanishes when $D\ (P : Q)$ vanishes can also be used as a measure of directed divergence. This fact can enable us to deduce Renyi's [7] measure,

$$D_8\,(P:Q) = \frac{1}{\alpha\,(\alpha-1)} \ln \sum_{i=1}^{n} p_i^{\alpha}\, q_i^{1-\alpha} \tag{27}$$

47.4 MEASURE OF DIRECTED DIVERGENCE WHEN THERE ARE INEQUALITY CONSTRAINTS ON PROBABILITIES

Here we want to develop measures which are applicable when,

$$0 \le a_i \le p_i, \quad q_i \le b_i \le 1 \tag{28}$$

Instead of (8), we now consider

$$\frac{\partial \phi}{\partial p_i} = f'\left(\frac{p_i - a_i}{q_i - a_i}\right) - f'\left(\frac{b_i - p_i}{b_i - q_i}\right) \tag{29}$$

Taking $f(x) = x \ln x$ and integrating,

$$\phi\,(p_i, q_i) = (p_i - a_i) \ln \frac{p_i - a_i}{q_i - a_i} + (b_i - p_i) \ln \frac{b_i - p_i}{b_i - q_i} + g\,(q_i) \tag{30}$$

Using the fact that $D\ (P : Q) = 0$ when $P = Q$, we get

$$D_9\,(P:Q) = \sum_{i=1}^{n} (p_i - a_i) \ln \frac{p_i - a_i}{q_i - a_i} + \sum_{i=1}^{n} (b_i - p_i) \ln \frac{b_i - p_i}{b_i - q_i} \tag{31}$$

Another alternative is to use,

$$\frac{\partial \phi}{\partial p_i} = \sqrt{\frac{p_i - a_i}{q_i - a_i}} - \sqrt{\frac{b_i - p_i}{b_i - q_i}} \tag{32}$$

Integrating and using that $D\ (P : Q) = 0$ when $P = Q$, we get

$$D_{10}\,(P:Q) = \sum_{i=1}^{n} \frac{(p_i - a_i)^{\frac{3}{2}}}{(q_i - a_i)^{\frac{1}{2}}} + \sum_{i=1}^{n} \frac{(b_i - p_i)^{\frac{3}{2}}}{(b_i - q_i)^{\frac{1}{2}}} + \sum_{i=1}^{n} a_i - \sum_{i=1}^{n} b_i \tag{33}$$

47.5 MEASURE OF ENTROPY FOR THE CASE WHEN THERE ARE INEQUALITY CONSTRAINTS ON PROBABILITIES

When there are inequality constraints (28), the generalised uniform distribution is given by

$$q_i = \frac{K\, b_i + a_i}{K+1}, \quad K = \frac{1-A}{B-1}, \quad A = \sum_{i=1}^{n} a_i, \quad B = \sum_{i=1}^{n} b_i \tag{34}$$

$$D\,(P:U) = \sum_{i=1}^{n} (p_i - a_i) \ln \frac{p_i - a_i}{\frac{K}{K+1}(b_i - a_i)} + \sum_{i=1}^{n} (b_i - p_i) \ln \frac{b_i - p_i}{\frac{b_i - a_i}{K+1}} \tag{35}$$

$$= \sum_{i=1}^{n} (p_i - a_i) \ln (p_i - a_i) + \sum_{i=1}^{n} (b_i - p_i) \ln (b_i - p_i)$$

$$- \sum_{i=1}^{n} (b_i - a_i) \ln (b_i - a_i) - (1 - A) \ln \frac{1 - A}{B - A} - (B - 1) \ln \frac{B - 1}{B - A} \tag{36}$$

The corresponding measure of entropy is

$$H_1\,(P) = C - \sum_{i=1}^{n} (p_i - a_i) \ln (p_i - a_i) - \sum_{i=1}^{n} (b_i - p_i) \ln (b_i - p_i) \tag{37}$$

Similarly (35) gives

$$H_2\,(P) = C - \sum_{i=1}^{n} \frac{(p_i - a_i)^{\frac{3}{2}}}{\sqrt{K}\,(b_i - a_i)^{\frac{1}{2}}} - \sum_{i=1}^{n} \frac{(b_i - p_i)^{\frac{3}{2}}}{(b_i - a_i)^{\frac{1}{2}}} \tag{38}$$

47.6 SOME SPECIAL CASES

Putting $a_i = 0$, $b_i = 1$, we get the following measures applicable when the constraints are $0 \leq p_i, q_i \leq 1$.

$$D_{11}\,(P:Q) = \sum_{i=1}^{n} p_i \ln \frac{p_i}{q_i} + \sum_{i=1}^{n} (1 - p_i) \ln \frac{1 - p_i}{1 - q_i} \tag{39}$$

$$D_{12}\,(P:Q) = \sum_{i=1}^{n} \frac{p_i^{3/2}}{q_i^{1/2}} + \sum_{i=1}^{n} \frac{(1 - p_i)^{3/2}}{(1 - q_i)^{1/2}} \tag{40}$$

$$H_3\,(P) = C - \sum_{i=1}^{n} p_i \ln p_i - \sum_{i=1}^{n} (1 - p_i) \ln (1 - p_i) \tag{41}$$

$$H_2\,(P) = C - \sum_{i=1}^{n} p_i^{3/2} - \sum_{i=1}^{n} (1 - p_i)^{3/2} \tag{42}$$

The measures (38) and (40) were given earlier by Kapur.

47.7 CONCLUDING REMARKS

In this present paper, we have given an alternative method of obtaining measures of entropy and directed divergence, both when there are inequality constraints of the type (9) and when the only inequality constraints are $0 \leq p_i, q_i \leq 1$. The method is based on the Generalised Minimum Cross-Entropy Principle and can certainly be used to generate additional measures.

REFERENCES

1. I. Csiszer (1972). "A Class of Measures of Informativity of Observation Channels", Periodic Math. Hungarica, Vol. 2, pp. 191-213.

2. J.H. Havrda and F. Charvat (1967), ''Quantification Methods of Classfication Processes : Concept of Structural α Entropy,'' Kybernatica, Vol. 3, pp. 30-35.
3. J.N. Kapur (1984), ''A Comparative assessment of various measures of directed divergence,'' Advances in Management Studies, Vol 3, pp. 1-16.
4. J.N. Kapur and H.K. Kesavan (1987), ''Generalised Maximum Entropy Principle (with Application), Sandford Educational Press, Waterloo University, Canada.
5. H.K. Kesavan and J.N. Kapur (1989), ''Generalised Maximum Entropy Principle,'' IEEE Trans. Sys Man Crb. Vol. 19 (9), pp. 1042-1052.
6. S. Kullback and R.A. Leibler (1951), ''On Information and Sufficiency,'' Ann. Mth. Stat; Vol. 22, pp. 79-86.
7. A. Renyi (1961), ''On Measures of Entropy and Information,'' Proc. 4th Berkeley Symp. Maths. Stat. Prob, Vol.1, pp. 547-561.

48

SOME NEW APPLICATIONS OF GENERALISED MAXIMUM ENTROPY PRINCIPLE

[A number of new applications of Generalised Maximum Entropy and Generalised Minimum Cross-Entropy Principles are given. New generalised measures of Entropy and Cross-Entropy are derived by using these principles and it is shown how these measures are superior both conceptually and computationally, to the classical measures, when there are inequality constraints on probabilities.]

48.1 INTRODUCTION

According to the Generalised Maximum-Entropy Principle [4], given a probability distribution P and either the measure of entropy $S(P)$ or the set of moment constraints C, we have to find the other, so that P is the maximum-entropy distribution for $S(P)$ and C.

According to the Generalised Minimum Cross-Entropy Principle [4], given a probability distribution P and two of the following,

(i) the set of moment constraints C,
(ii) measure of directed divergence $D(P:Q)$,
(iii) *a priori* probability distribution Q,

we have to find the third so that P is the minimum Cross-Entropy distribution for C. D and Q.

The object of the present paper is to apply these principles to the situations where there are inequality constraints on probabilities, i.e when we are given that

$$0 \le a_i \le p_i \le b_i \le 1; \quad i = 1,2,\ldots,n \tag{1}$$

This immediately leads to

$$\sum_{i=1}^{n} a_i \le \sum_{i=1}^{n} p_i \le \sum_{i=1}^{n} b_i \tag{2}$$

or,

$$A \le 1 \le B, \tag{3}$$

where,

$$\sum_{i=1}^{n} a_i = A, \quad \sum_{i=1}^{n} b_i = B, \tag{4}$$

so that there will be a feasible solution only if

$$A \le 1, \quad B \ge 1, \tag{5}$$

However, if these conditions are satisfied, there may be an infinity of probability distributions satisfying (1). Which one shall we choose?

According to Jayne's maximum-entropy principle [2], we should choose that probability distribution which maximizes Shannon's [6] measure of entropy

$$S(P) = -\sum_{i=1}^{n} p_i \ln p_i \tag{6}$$

subject to (1) and

$$\sum_{i=1}^{n} p_i = 1 \tag{7}$$

Freund and Saxena [1] gave an algorithm for this purpose which is a little cumbersome, but which works. However, if we have, in addition to (1) some moment constraints on the probabilities, no algorithm is available.

If we are given no constraints on the probability distribution,

$$P = (p_1, p_2, \ldots, p_n) \tag{8}$$

except the natural constraint (7), the most unbiased distribution is the uniform distribution

$$U = \left(\frac{1}{n}, \frac{1}{n}, ---, \frac{1}{n}\right), \tag{9}$$

since according to Laplace's principle of insufficient reason, there is no reason in this case to prefer one outcome over the other. This is perfectly consistent with common sense notion of unbiasedness.

Now when we maximize Shannon's measure (6) subject to (7), we get the uniform distribution. In fact in this case, we have to maximize (6) subject to (7) and the additional constraints,

$$p_1 \geq 0, p_2 > 0, \ldots, p_n \geq 0 \tag{10}$$

but we do not consider these explicitly as these are automatically satisfied when we maximize (6).

Laplace did not suggest as to how we should choose the probability distribution when additional information in the form of values of moments is available, i.e when we know

$$\sum_{i=1}^{n} p_i\, g_r(x_i) = a_r; \quad r = 1, 2, \ldots, m \tag{11}$$

Jayne's principle provides the answer that in this case we should choose that probability distribution which maximizes Shannon's measure subject (7), and (11).

Now there are many continuous symmetric concave function of $p_1, p_2, \ldots, p_n$ which when maximized subject to (7) give the uniform distribution (9) which automatically satisfies the non-negativity constraints (10). However, when most of these functions are maximized subject to (7) and (11), by using Lagrange's method, the maximizing probabilities may not satisfy (10).

Shannon's measure has many other useful properties, but the property which makes it specially useful in the application of Jayne's maximum entropy principle is that the maximizing probabilities obtained by Largrange's method, always satisfy (10).

However, when the non-negativity constraints (10) are replaced by inequality constraints (1), the advantage of Shannon's measure is lost completely.

As such we search for a measure of entropy, i.e a suitable continuous concave function of $p_1, p_2, ..., p_n$ whose maximization subject to (7) and (11) will automatically satisfy the constraints (1). We shall also like the measure to be such that when it is maximized subject to (7), it gives us the most 'natural' distribution which we can expect in the presence of constraints (1). Freund and Saxena's [1] distribution obtained by using Shannon's entropy may be unbiased, but it is far from being 'natural'. In the next section, we first find 'a natural' distribution.

48.2 THE NATURAL DISTRIBUTION SUBJECT TO INEQUALITY CONSTRAINTS ON PROBABILITIES, GENERALISATION OF LAPLACE'S PRINCIPLE OF INSUFFICIENT REASON

When the constraints are given by (7) and (10), the most natural distribution is the uniform distribution.

However, when the constraints are given by (1) and (7), the uniform distribution is not the most natural distribution ; in fact, it may not even satisfy the constraints (1).

In the first case, the uniform distribution is such that each probability $1/n$ divides the permissible range (0,1) in the same ratio $1 : (n - 1)$ (Figure 48.1).

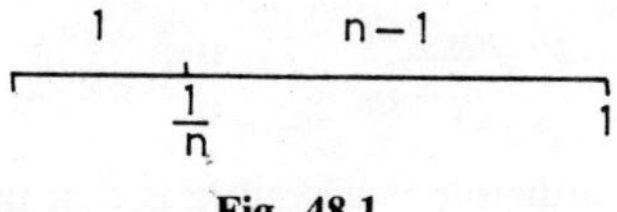

Fig. 48.1

This property was identified with unbiasedness because this ratio is the same for all outcomes.

We apply the same principle of unbiasedness to the second case. We choose p_i to divide the permissible range (a_i, b_i) (Figure 48.2).

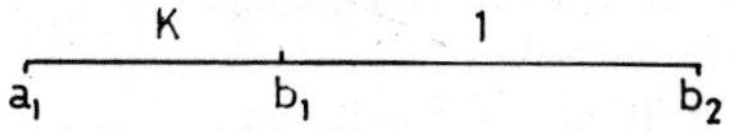

Fig. 48.2

in the same ratio $k : 1$, i.e we choose

$$p_i = \frac{k\, b_i + a_i}{k + 1} \tag{12}$$

Now $$\sum_{i=1}^{n} p_i = 1 \Rightarrow k\,B + A = k + 1 \Rightarrow k = \frac{1 - A}{B - 1},$$

so that,

$$p_i = \frac{(1 - A)\, b_i + (B - 1)\, a_i}{B - A}, i = 1, 2, \ldots, n \tag{13}$$

If $a_i = 0$, $b_i = 1$ for each i, this gives $A = 0$, $B = n$,

$$p_i = \frac{1+(n-1)0}{n} = \frac{1}{n} \tag{14}$$

as expected.

We call the probability distribution defined by (13) as the 'generalised uniform distribution'. It includes the usual uniform distribution as a special case when each $a_i = 0$ and each $b_i = 1$.

This generalised uniform distribution is the most natural distribution when only information in the form of (1) and (7) is available to us.

We can now state the generalised Laplace's Principle of Insufficient Reason as follows:

When the only information available about the probabilities is that $a_i \le p_i \le b_i$, $i = 1, 2, ..., n$) and $\sum_{i=1}^{n} p_i = 1$, then there is no reason why we should not choose each p_i to divide each corresponding interval (a_i, b_i) in the same ratio $k : 1$ whose k is determined by using $\sum_{i=1}^{n} p_i = 1$.

In fact if we choose any p_i, it divides (a_i, b_i) in the ratio

$$k_i = \frac{p_i - a_i}{b_i - p_i}, \quad i = 1, 2, \ldots, n \tag{15}$$

What generalised Laplace principle is asserting is that there is no reason to choose $k_1, k_2, ..., k_n$ different.

Having obtained the most natural distribution, we next want to choose a measure of entropy whose maximization subject to $\sum_{i=1}^{n} p_i = 1$ will lead to this natural distribution and which in the presence of moment constraints leads to maximizing probabilities automatically satisfying the inequality constraints (1).

It is here that we use the generalised maximum entropy principle or the special principle included in it, viz. the following:

Given a probability distribution P and a set of moment constraints C, find the measure of entropy whose maximization subject to C gives us P.

48.3 A GENERALISED ENTROPY MEASURE

We are given the generalised uniform probability distribution

$$p_i = \frac{k\, b_i + a_i}{k+1}, \quad i = 1, 2, \ldots, n \tag{16}$$

where, $$0 \le a_i \le b_i \le 1, \quad \sum_{i=1}^{n} a_i = A \le 1, \quad \sum_{i=1}^{n} b_i = B \ge 1, \quad k = \frac{1-A}{B-1} \tag{17}$$

Our problem is to find the generalised measure of entropy,

$$\sum_{i=1}^{n} \phi\,(p_i) \tag{18}$$

by maximizing which subject to

$$\sum_{i=1}^{n} p_i = 1, \quad i = 1, \ldots, n \tag{19}$$

we get (16) as the maximum entropy probability distribution. Now maximization of (18) subject to (19) gives

$$\phi'(p_i) = \text{constant} \tag{20}$$

Also from (16),

$$\frac{p_i - a_i}{b_i - p_i} = k = \text{constant} \tag{21}$$

so that,

$$\phi'(p_i) = f\left(\frac{p_i - a_i}{b_i - p_i}\right) \tag{22}$$

We now choose this function $f(.)$ subject to the following :

(i) $f(x)$ is defined only for positive values of x so that x is never negative and $a_i \le p_i \le b_i$,

(ii) $\phi(p_i)$ is a concave function, so that local maximum of $\sum_{i=1}^{n} \phi(p_i)$ subject to linear constraints is the global maximum,

(iii) The choice of $f(.)$ leads to maximum entropy distribution which is easy to handle.

One obvious choice is

$$f(x) = -\ln x \tag{23}$$

This gives

$$\phi'(p_i) = -\ln \frac{p_i - a_i}{b_i - p_i} \tag{24}$$

Intergrating,

$$\phi(p_i) = -(p_i - a)\ln(p_i - a_i) - (b_i - p_i)\ln(b_i - p_i) + ci \tag{25}$$

and the measure of entropy is

$$S = -\sum_{i=1}^{n}(p_i - a_i)\ln(p_i - a_i) - \sum_{i=1}^{n}(b_i - p_i)\ln(b_i - p_i) + C \tag{26}$$

where C is a constant.

Its maximum value is

$$S_{\max} = -\sum_{i=1}^{n} \frac{k(b_i - a_i)}{k+1} \ln \frac{k(b_i - a_i)}{k+1} - \sum_{i=1}^{n} \frac{b_i - a_i}{1+k} \ln \frac{b_i - a_i}{k+1} + C$$

$$= -\sum_{i=1}^{n}(b_i - a)\ln\frac{(b_i - a_i)}{k+1} - (B - A)\frac{k}{k+1}\ln\frac{k}{k+1} + C$$

$$= -\sum_{i=1}^{n} (b_i - a_i) \ln (b_i - a_i) + (B - A) \ln (B - A) - (1 - A) \ln (1 - A) - (1 - B) \ln (1 - B) + C \tag{27}$$

It may be noted that it is not a symmetric measure of entropy unless $a_i = a$, $b_i = b$ and using (5)

$$a < \frac{1}{n}, \quad b > \frac{1}{n} \tag{28}$$

in which case the measure becomes

$$S = -\sum_{i=1}^{n} (p_i - a) \ln (p_i - a) - \sum_{i=1}^{n} (b - p_i) \ln (b - p_i) + C, \tag{29}$$

so that the generalised uniform distribution is given by

$$k = \frac{1 - n\,a}{n\,b - 1}, \quad p_i = \frac{(1 - n\,a)\,b + a\,(n\,b - 1)}{n\,(b - a)} = \frac{1}{n}, \tag{30}$$

which is the usual uniform distribution. We also have a special case of generalised Laplace's principle of insufficient reason, viz.

If each probability lies between the same limits $a < (1/n)$ and $b > (1/n)$ then there is no reason to choose any other distribution other than the usual uniform distribution.

If $a = 0$, $b = 1$, we get the measure

$$-\sum_{i=1}^{n} p_i \ln p_i - \sum_{i=1}^{n} (1 - p_i) \ln (1 - p_i) + C \tag{31}$$

first given by Kapur [3]. Thus if $a = 0$, $b = 1$, we do not get Shannon's entropy, but we get what may be called Fermi-Dirac Entropy.

48.4 ANOTHER GENERALISED ENTROPY MEASURE

We now choose $f(x) = -\sqrt{x}$ which also satisfies the three conditions of the last section, so that

$$\phi'(p_i) = -\sqrt{\frac{p_i - a_i}{b_i - p_i}} \tag{32}$$

Integrating,

$$\phi(p_i) = -(b_i - a_i) \tan^{-1} \sqrt{\frac{p_i - a_i}{b_i - p_i}} + \sqrt{(p_i - a_i)(b_i - p_i)}$$

giving,

$$S = \sum_{i=1}^{n} \phi(p_i) = -\sum_{i=1}^{n} (b_i - a_i) \tan^{-1} \sqrt{\frac{p_i - a_i}{b_i - p_i}} + \sum_{i=1}^{n} \sqrt{(p_i - a_i)(b_i - p_i)} \tag{33}$$

This is a trigonometric measure of entropy. As a special case, if we put $a_i = a$, $b_i = b$, we get the symmetric measure

$$S = -(b-a)\sum_{i=1}^{n}\tan^{-1}\sqrt{\frac{p_i-a}{b-p_i}} + \sum_{i=1}^{n}\sqrt{(p_i-a)(b-p_i)} \tag{34}$$

The measures (33) and (34) may not appear to be convenient, but in many entropy maximization problems, we have to deal with (31) which is not so inconvenient to handle.

If we are given only one-way constraints, viz.

$$p_i \ge a_i \ge 0, \quad i = 1,2,\ldots,n \tag{35}$$

then the measures (26), (29), (33), (34) become

$$S = -\sum_{i=1}^{n}(p_i-a_i)\ln(p_i-a_i) + C, \tag{36}$$

$$S = C - \sum_{i=1}^{n}(p_i-a_i)^{3/2} \tag{37}$$

The constraints,

$$0 \le a_i \le p_i \le 1, \quad i = 1,2,\ldots n \tag{38}$$

will give different measures of entropy. The constraints (35) and (38) appear to be the same in view of the constraints $\sum_{i=1}^{n} p_i = 1$. In fact (35) are strictly equivalent to

$$0 \le a_i \le p_i \le 1 + a_i - A, \quad i = 1,2,\ldots,n, \tag{39}$$

so that (26) becomes

$$S = -\sum_{i=1}^{n}(p_i-a_i)\ln(p_i-a_i) - \sum_{i=1}^{n}(1-A+a_i-p_i)\ln(1-A+a_i-p_i) + C \tag{40}$$

If each $a_i = 0$, this gives

$$S = -\sum_{i=1}^{n}p_i\ln p_i - \sum_{i=1}^{n}(1-p_i)\ln(1-p_i) \tag{41}$$

48.5 PHILOSOPHICAL IMPLICATIONS OF THE USE OF TWO MEASURE OF ENTROPY

Consider the following problem :

Given the inequality constraints,

$$0 \le a_i \le p_i \le b_i \le 1, \quad i = 1,2,\ldots,n \tag{42}$$

the natural constraint

$$\sum_{i=1}^{n} p_i = 1 \tag{43}$$

and the moment constraint

$$\sum_{i=1}^{n} i\, p_i = m \tag{44}$$

find the maximum entropy probability distributions.

We have two alternative methods of solving it :

(i) By using Jayne's Principle of Maximum Entropy ; Here we maximize Shannon's measure subject to satisfying (42), (43) and (44);

(ii) By using our generalised Principle of Maximum Entropy. Here we maximize (26) subject to constraints (43) and (44) only.

In the second method, the solution is simple and straight forward. We get

$$\frac{p_i - a_i}{b_i - p_i} = D x^i, \quad i, 2, \ldots, n, \tag{45}$$

where D and x are determined by using

$$\sum_{i=1}^{n} \frac{b_i D x^i + a_i}{D x^i + 1} = 1, \ \sum_{i=1}^{n} i \frac{b_i D x^i + a_i}{D x^i + 1} = m \tag{46}$$

In the first method, the solution is not so simple. A heurstic procedure would be to first maximize $-\sum_{i=1}^{n} p_i \ln p_i$ subject to (43) and (44) and, in case some p_j comes out to be less than corresponding a_j, we put $p_j = a_j$. Similarly, if some p_k comes out to be greater than b_k, we put $p_k = b_k$, then we take remaining outcomes and maximize entropy $-\sum p_i \ln p_i$ for these outcomes for the resulting constraints on these outcomes and see if these satisfy the constraints, otherwise we proceed as before. This heuristic procedure will give almost maximum entropy but there is no guarantee that this will give us the maximum entropy probability distribution.

Even forgetting the difficulty in implementing the first method, let us see what we are trying to achieve by the two methods.

By maximizing $-\sum_{i=1}^{n} p_i \ln p_i$, we try to get a probability distribution in which the probabilities are 'as equal as possible' or in which the probability distribution is 'as near usual uniform distribution as possible' subject to the constraints (42), (43) and (44) being satisfied.

If inequality constrains (42) were absent, this may appear as a natural thing to do. However, in the presence of constraints like

$$.1 \le p_1 \le .3, \quad .8 \le p_2 \le .9 \tag{47}$$

even to talk of the probability distribution (.5, .5) appears unnatural. Suppose there was no mean constraint, even then with (47) at our disposal, we shall not talk of the probability distribution (.5, .5) as natural. However, the generalised uniform distribution (.16, .83) appears as natural and we should try to be as near it as possible subject to the mean constraint being satisfied.

Thus the main difference in the philosophy of the two approaches appear to be as follows :

In Jayne's approach, the inequality constraints and the moment constrains are clubbed together and Shannon's measure which takes into account the natural constraint only, is maximized subject to these two groups of constraints.

In our approach, the inequality constraints are decoupled from moment constraints. The measure of entropy is determined to take into account the inequality constraints i.e when it is maximizing subject to moment constraints, it is such that the maximizing probabilities satisfy the inequality constraints automatically.

In our method, the object is make $(p_i - a_i)/(b_i - p_i)$ $(i = 1, 2, ..., n)$ as equal as possible subject to the moment constraints. We want each p_i to divide (a_i, b_i) as equitably a possible subject to given moment constraints.

In Jayne's method, the inequality constraints were there, viz. we wanted $0 \le p_i \le 1$, but these did not create any problems because fortunately Shannon's measure of entropy has such a form that its maximization subject to linear constraints leads to these constraints being automatically satisfied. In fact, it was because of this reason that Shannon's measure scored over other measures of entropy, so much so that Shannon's measure became the measure of entropy.

However, with inequality constraints of form (1), this advantage is lost completely. In fact in this case the measure (26) scores over that measure both conceptually and computationally.

48.6 A GENERALISED MEASURE OF CROSS-ENTROPY CONSISTENT WITH CONSTRAINTS

In section 48.4 we developed measures of generalised entropy to be used in the principle of maximum entropy when there are inequality constraints (1) on probabilities. We want a similar measure to be used along with the principle of minimum cross-entropy when the constraints (1) are there.

Let $$P = (p_i, p_2, \ldots, p_n), Q = (q_i, q_2, \ldots, q_n) \tag{47a}$$

be two probability distributions satisfying the constraints (1), then a measure of cross-entropy is given by

$$D\,(P:Q) = \sum_{i=1}^{n} (p_i - a_i) \ln \frac{p_i - a_i}{q_i - a_i} + \sum_{i=1}^{n} (b_i - p_i) \ln \frac{b_i - p_i}{b_i - q_i} \tag{48}$$

where whenever $q_i = a_i$ then $p_i = a_i$ and when $q_i = b_i$, $p_i = b_i$ and 0 ln 0/0 is defined as zero. It is easily seen that $D\,(P : Q)$ is a convex function of $p_1, p_2, ..., p_n$ whose minimum values zero occurs when $p_i = q_i$ for each i.

If we minimize it subject to

$$\sum_{i=1}^{n} p_i = 1, \quad \sum_{i=1}^{n} i\, p_i = m \tag{49}$$

we get,

$$\frac{p_i - a_i}{q_i - a_i} \; \frac{b_i - q_i}{b_i - p_i} = E\, x^i \tag{50}$$

where, E and x are obtained by using (49) and the probability distribution P automatically satisfies the constraints (1) :

If we take Q to be the generalised uniform distribution

$$q_i = \frac{k\, b_i + a_i}{k+1}, \quad k = \frac{1-A}{B-1} \tag{51}$$

(48) becomes

$$D\,(P:\overline{U}) = \sum_{i=1}^{n} (p_i - a_i) \ln \frac{p_i - a_i}{\frac{k\,(b_i - a_i)}{k+1}} + \sum_{i=1}^{n} (b_i - p_i) \ln \frac{(b_i - p_i)}{\frac{b_i - a_i}{k+1}}$$

$$= \sum_{i=1}^{n} (p_i - a_i) \ln (p_i - a_i) + \sum_{i=1}^{n} (b_i - p_i) \ln (b_i - p_i)$$

$$- (1 - A) \ln k - (B - A) \sum_{i=1}^{n} \ln \frac{b_i - a_i}{k+1} \tag{52}$$

so that minimizing the generalised cross-entropy of P from the generalised uniform distribution is equivalent to maximizing the generalised measure of entropy.

48.7 SOME FURTHER APPLICATIONS OF GENERALISED MAXIMUM ENTROPY AND GENERALISED MINIMUM CROSS-ENTROPY PRINCIPLES

(i) Let the given probability distribution be

$$p_i = \frac{k\, b_i + a_i}{k+1}, \quad i = 1, 2, \ldots, n;\ k = \frac{1-A}{B-1} \tag{53}$$

then the constraints are given by (1) and the measures of entropy is given by (26)

(ii) Let the given probability distribution be

$$p_i = \frac{k\, b_i + a_i}{k+1} C\, x^i, \quad i = 1, \ldots, n \tag{54}$$

then the constraints are given by (42), (43), (44) and the measure of entropy by (26).

(iii) Alternatively (54) can be regarded as a minimum cross entropy distribution when the *a priori* probability distribution is given by (53), the constraints are given by (43) and (44) and Kullback-Leibler [5] measure of entropy is used.

(iv) Let the given probability distribution be

$$p_i = \frac{a_i + b_i\,(C + D\, x^i)^2}{1 + (C + D\, x^i)^2} \tag{55}$$

the measure is given by (4) and the constraints are (43) and (44).

REFERENCES

1. D. Freund and V. Saxena (1984), " An algorithm for a class of discrete maximum entropy problems", Operations Research, Vol. 3-2, No 1, pp. 210-215.
2. E.T. Jaynes (1957), " Information Theory and Statistical Mechanics", Physical Reviews 106, pp. 620-630.
3. J.N. Kapur (1972), "Measures of Uncertainty, Mathematical Programming and Physics, Jour. Ind. Soc Ag. Stat. 24. pp 47-66.
4. J.N. Kapur and H.K Kesavan (1987) Generalised Maximum Entropy Principle (with Applications), Sandford Educational Press, University of Waterloo, Canada.
5. S. Kullback and R.A. Leibler (1951), " On Information and Sufficiency". Ann. Math. Stat. Vol 22, pp. 79-86.
6. C.E. Shannon (1948), " A Mathematical Theory of Communication", Bell System Tech. Journal. Vol. 27, pp. 379-423, 623-659.

49

APPROXIMATING A GIVEN PROBABILITY DISTRIBUTION BY A MAXIMUM ENTROPY DISTRIBUTION

[Let $g_1(x), g_2(x), ..., g_m(x)$ be the characterizing moment functions of a family of exponential distributions. It is here shown that the member of this family which is closest to a given probability distribution with density function $f(x)$ is that which has the same values for $E(g_1(x)),...,E(g_m(x))$ as their values for the density function $f(x)$. The role of maximum entropy distribution in approximating a given probability distribution is also examined in the light of this result.]

49.1 INTRODUCTION

One of the most important applications of maximum entropy probability distributions is their use in approximating a given probability distribution, which may be

(*i*) exact, but specified by a complicated analytical expression, or

(*ii*) only numerically specified, e.g. a distribution which may be obtained through a simulation process, or

(*iii*) incompletely known, e.g. when only a few of its moments may be known.

In case (*iii*), the true distribution may be any one of the infinity of distributions which have the specified values for the given moments. According to the maximum entropy principle, we choose that distribution, out of all these, which has maximum entropy or maximum uncertainty or which is most objective, most random or most uniform.

Cases (*i*) and (*ii*) have been considered extensively by many authors. Thus Shore [7] approximated many discrete variate queueing theory distributions by maximum entropy distributions having common values for some specified one to four algebraic moments. Similarly Sobezyk and Trebricki [8] approximated many continuous variate probability distributions arising in stochastic dynamics by maximum entropy distributions having 2, 4, or 6 algebraic moments common with the given distributions. Both of them found that in all cases they considered, the 4 or 6 algebraic moment maximum-entropy approximations are quite satisfactory, though in some cases at least, two-moment approximations were unsatisfactory.

None of these authors, however, considered MEPD's based on moments other than algebraic. In fact, if we are prepared to consider non-algebraic moments, we can get much better approximations by using fewer moments. We illustrate this result in Section 2.

When we use Shannon's measure of entropy, the maximum entropy distributions are exponential distributions. In Section 3, we consider the problem of finding the exponential distribution which is, in some sense, closest to a given probability distribution. In section 4, we consider some special cases.

In Section 5, we make some general remarks about approximation of a given distribution by a maximum entropy distributions in the light of our earlier discussion and Jaynes' entropy concentration theorem [3].

49.2 THE ROLE OF CHARACTERISING MOMENTS

We consider the distribution with density function,

$$f(y) = 2y\, e^{-y^2}, \qquad y > 0 \tag{1}$$

approximated by Sobezyk and Tribecki by the MEPD's

$$p_1(y) = \exp(-1.462278 + 3.0499444\, y - 18.51472\, y^2) \tag{2}$$

and $p_2(y) = \exp(-2.200745 + 7.007829\, y - 7.56754\, y^2 + 2.94147\, y^3$

$$-0.4757316\, y^4) \tag{3}$$

They found the value of Kullback-Leibler [5] measure as

$$\int_0^\infty f(y) \ln \frac{f(y)}{p_1(y)}\, dy = 1.533783 \times 10^{-2} \tag{4}$$

and,
$$\int_0^\infty f(y) \ln \frac{f(y)}{p_2(y)}\, dy = 3.624085 \times 10^{-3} \tag{5}$$

and concluded that the four moment approximation $p_2(y)$ gives a very good approximated to $f(y)$.

Now we consider only two moment approximation but use the moments,

$$E(\ln y) \quad \text{and} \quad E(y^2) \tag{6}$$

Now, we know,

$$\int_0^\infty y^c e^{-y^2}\, dy = \int_0^\infty z^{c/2} e^{-z} \frac{1}{2} z^{-1/2}\, dz = \frac{1}{2}\int_0^\infty e^{-z} z^{\frac{c-1}{2}}\, dz$$

$$= \frac{1}{2}\Gamma\frac{(c+1)}{2} \tag{7}$$

Differentiating w.r.t c under the sign of integration, we get

$$\int_0^\infty e^{-y^2} y^c \ln y\, dy = \frac{1}{4}\Gamma'\left(\frac{c+1}{2}\right), \tag{8}$$

so that,

$$\int_0^\infty 2e^{-y^2}\, y \ln y \, dy = \frac{1}{2}\Gamma'(1) \text{ or } E(\ln y) = \frac{1}{2}\Gamma'(1) = -.2886 \tag{9}$$

$$\int_0^\infty 2\, e^{-y^2} y^3 \, dy = \Gamma(2) \text{ or } E(y^2) = 1 \tag{10}$$

Now maximizing,

$$-\int_0^\infty p(y) \ln p(y)\, dy \tag{11}$$

subject to,

$$\int_0^\infty p(y)\, dy = 1, \int_0^\infty p(y) \ln y \, dy = -.2886, \int_0^\infty y^2 p(y)\, dy = 1 \tag{12}$$

we get the distribution

$$p(y) = e^{-\lambda_0 - \lambda_1 \log y - \lambda_2 y^2}, \tag{13}$$

where λ_0, λ_1, λ_2 are determined by using (12). Calculating these and substituting in (13), we get

$$p(y) = 2y\, e^{-y^2} \tag{14}$$

which is exactly the same as $f(y)$ so that

$$\int_0^\infty f(y) \ln \frac{f(y)}{p(y)}\, dy = 0 \tag{15}$$

Thus by using only 2 moments $E(\ln y)$ and $E(y^2)$, we have got a perfect fit which we would not have got by using even a million algebraic moments!

Moments like $E(\ln y)$ and $E(y^2)$ which give a perfect fit will be called the characterising moments of the distribution (1).

In general, characterising moments of a probability distribution $f(y)$ are those moments $E(g_1(y)), E(g_2(y)), \ldots, E(g_m(y))$ which are such that if we maximize,

$$-\int \phi(y) \ln \phi(y)\, dy \tag{16}$$

subject to,

$$\int \phi(y)\, g_r(y)\, dy = \int f(y)\, g_r(y)\, dy, \qquad r = 0, 1, 2, \ldots, m, \tag{17}$$

where $g_0(y) = 1$, we get

$$\phi(y) = f(y) \tag{18}$$

These moments exist (when we use Shannon's measure of entropy) only when $f(y)$ has the form

$$f(y) = \exp(-\lambda_0 - \lambda_1 g_1(y) - \ldots - \lambda_m g_m(y)] \tag{19}$$

$$= \frac{\exp(-\lambda_1 g_1(y) - \ldots - \lambda_m g_m(y))}{\int \exp(-\lambda_1 g_1(y) - \ldots - \lambda_m g_m(y)]\, dy} \tag{20}$$

In this case $g_1(y)$, $g_2(y)$,..., $g_m(y)$ are the characterising moment functions and $E(g_1(y))$, $E(g_2(y))$, ..., $E(g_m(y))$ are the characterising moments.

If we have a probability distribution which is of the exponential form, (19) or (20), to approximate it by another exponential distribution of the form,

$$\exp(-\mu_0 - \mu_1 y - \mu_2 y^2 - \ldots - \mu_k y^k) \tag{21}$$

$$= \frac{\exp(-\mu_1 y - \mu_2 y^2 - \ldots - \mu_k y^k)}{\int \exp(-\mu_1 y - \mu_2 y^2 - \ldots - \mu_k y^k)\, dy} \tag{22}$$

is not very useful since however large k may be, we may never get the exact distribution unless the characterising moments are algebraic.

Similarly when Sobezyk and Tribecki found the 2 ,4 and 6 moment approximations for the probability distribution,

$$f(y) = \frac{1}{\sqrt{2\pi}\, y\sigma} \exp\left[-\frac{1}{2\sigma^2}(\ln y - \ln \beta)^2\right] \tag{23}$$

for $\beta = 1.2$, $\sigma = 0.4$, they found two or four moment approximations very unsatisfactory but the 6 moment approximation was satisfactory. However if they had used the moments $E(\ln y)$ and $E(\ln y)^2$, they would have got a perfect fit with two moments only.

Again in the same way, the best moment approximation for the distribution,

$$f(y) = c \exp\left[-\frac{1}{D^2}\left(y^2 + \frac{1}{2}\mu y^4\right)\right] \tag{24}$$

will be provided by the MEPD based on $E(y^2)$ and $E(y^4)$.

The need for maximum entropy approximation therefore arises only for those distributions which are not of the form (19) or (20), then we can use approximating MEPD's of the form (21) and (22). However, in these cases also there is no guarantee that non-algebraic moments or a combination of algebraic and non-algebraic moments will not give better results.

When the original distribution is not of the exponential family, approximating it by algebraic moments maximum entropy probability distributions of the form (21) or (22) is more a discussion of the comparative degree of approximation with increasing number of algebraic moments, rather than an effort to get a useful approximating distribution.

As such in the next section, we consider approximation of exponential distribution by distributions of the form (19) or (20) or (21) (22). We consider the first type of approximation first in the next section.

49.3 APPROXIMATING A NON-EXPONENTIAL PROBABILITY DISTRIBUTION BY A MEMBER OF EXPONENTIAL FAMILY

Let $f(y)$ be a non-exponential density function. We want to approximate it by a number of the exponential family

$$g(y)=\frac{\exp[-\lambda_1 g_1(y)-\lambda_2 g_2(y)-\ldots-\lambda_m g_m(y)]}{\int_a^b \exp[-\lambda_1 g_1(y)-\ldots-\lambda_m g_m(y)]\,dy} \tag{25}$$

(25) represent an infinity of probability distributions, $g(y)$ for different values of λ_1, ..., λ_m and we want to choose that set of values of $\lambda_1, \lambda_2, \ldots, \lambda_m$ for which $f(y)$ is closest to $g(y)$. For this purpose, we choose $\lambda_1, \lambda_2, \ldots, \lambda_m$ to minimize Kullback-Leibler measure

$$\int_a^b f(y)\ln\frac{f(y)}{g(y)}\,dy \tag{26}$$

Since $f(y)$ does not contain λ's, we have to minimize

$$-\int_a^b f(y)\ln g(y)\,dy \tag{27}$$

or to maximize,

$$\int_a^b f(y)[-\lambda_1 g_1(y)-\ldots-\lambda_m g_m(y)]\,dy$$

$$-\int_a^b f(y)\ln\left[\int_a^b \exp[-\lambda_1 g_1(y)-\ldots-\lambda_m g_m(y)]\,dy\right]dy$$

or to maximize,

$$\phi(\lambda_1,\lambda_2,\ldots,\lambda_m)=-\lambda_1\int_a^b f(y)\,g_1(y)\,dy$$

$$-\lambda_2\int_a^b f(y)\,g_2\,dy-\ldots-\lambda_m\int_a^b f(y)\,g_m(y)\,dy$$

$$-\ln\int_a^b \exp[-\lambda_1 g_1(y)-\ldots-\lambda_m g_m(y)]\,dy \tag{28}$$

Now, $$\frac{\partial\phi}{\partial\lambda_r}=-\int_a^b f(y)\,g_r(y)\,dy+\frac{\int_a^b [g_r(y)\exp(-\lambda_1 g_1(y)-\ldots-\lambda_m g_m(y)]\,dy}{\int_a^b \exp(-\lambda_1 g_1(y)-\ldots-\lambda_m g_m(y))\,dy} \tag{29}$$

$$\frac{\partial^2\phi}{\partial\lambda_r^2}=\frac{\left[\int_a^b g_r(y)Y\,dy\right]^2-\left(\int_a^b Y\,dy\right)\left(\int_a^b g_r^2(y)Y\,dy\right)}{\left[\int_a^b Y\,dy\right]^2}=\text{var}\,(g_r(y)) \tag{30}$$

where,

$$Y=\exp\left[-\lambda_1 g_1(y)-\ldots-\lambda_m g_m(y)\right] \tag{31}$$

$$\frac{\partial^2\phi}{\partial\lambda_r\partial\lambda_r}=\frac{\int_a^b g_r(y)Y\,dy\int_a^b g_s(y)Y\,dy-\int_a^b g_r(y)(g_s(y))Y\,dy}{\left[\int_a^b Y\,dy\right]^2}$$

$$=-\text{cov}\,[g_r(y),g_s(y)] \tag{32}$$

so that the Hessian matrix of second order partial derivatives is negative definite. As such we get the maximum value of $\phi(\lambda_1, \ldots, \lambda_m)$ by putting

$$\frac{\partial\phi}{\partial\lambda_r}=0, \qquad r=1,2,\ldots,m \tag{33}$$

Using (25) and (33) we get

$$\int_a^b f(y)\,g_r(y)\,dy=\int_a^b g(y)\,g_r(y)\,dy, \qquad r=1,2,\ldots,m \tag{34}$$

Equations (34) give m equations and determine $\lambda_1, \lambda_2, \ldots, \lambda_m$. Thus the best choice of $\lambda_1, \lambda_2, \ldots, \lambda_m$ which gives the closest distribution of the family (25) to the density function $f(y)$ is obtained by equating the values of the characterising moments for the distribution $f(y)$ to the corresponding values of the same characterising moments for the distribution $g(y)$.

49.4 SPECIAL CASES OF THE ABOVE RESULT

(i) Let,

$$f(y)=\sum_{r=1}^{m} h_r a_r e^{-a_r x}, \qquad x>0, \qquad \sum_{r=1}^{m} h_r=1 \tag{35}$$

This is not an exponential distribution but is the weighted sum of m exponential distributions with different means. Let

$$g(y)=a\,e^{-ax} \tag{36}$$

then since the characterising moment is the mean,

$$\frac{1}{a}=\sum_{r=1}^{m}\frac{h_r}{a_r} \tag{37}$$

(ii) Let,

$$f(y) = \sum_{r=1}^{m} h_r \frac{1}{\sqrt{2}\pi\sigma_r} \exp\left[-\frac{1}{2}\frac{((x-m_r)^2)}{\sigma_r^2}\right), -\infty < x < \infty,$$

$$\sum_{r=1}^{m} h_r = 1 \tag{38}$$

$$g(y) = \frac{1}{\sqrt{2}\pi\sigma} \exp\left(-\frac{1}{2}\frac{(x-m)^2}{\sigma^2}\right) \tag{39}$$

the best approximation is given by choosing

$$m = \sum_{r=1}^{m} h_r m_r \, \sigma^2 + m^2 = \sum_{r=1}^{m} h_r (\sigma_r^2 + m_r^2) \tag{40}$$

(iii) Let,

$$f(y) = \sum_{r=1}^{m} h_r \frac{a_r^{\gamma_r}}{\Gamma(\gamma_r)} e^{-a_r x} x^{\gamma_r - 1}, \qquad x > 0, \sum_{r=1}^{m} h_r = 1 \tag{41}$$

$$g(y) = \frac{a^{\gamma}}{\Gamma(\gamma)} e^{-ay} (y^{\gamma-1}), \qquad y > 0 \tag{42}$$

then,

$$\frac{\gamma}{a} = \sum_{r=1}^{m} h_r \frac{\gamma_r}{a_r}, \quad \frac{\Gamma'(\gamma)}{\Gamma(\gamma)} \ln a = \sum_{r=1}^{m} h_r \left(\frac{\Gamma'(\gamma_r)}{\Gamma(\gamma_r)} - \ln a_r\right) \tag{43}$$

(iv) Let,

$$f(k) = \sum_{r=1}^{m} h_r \, n_{c_k} p_r^k (1-p_r)^{n-k}, \qquad k = 0, 1, \ldots, n \tag{44}$$

$$g(k) = n_{c_k} p^k (1-p)^{n-k}, \qquad k = 0, 1, \ldots, n \tag{45}$$

then,

$$p = \sum_{r=1}^{m} h_r p_r \tag{46}$$

(v) Let,

$$f(k) = \sum_{r=1}^{m} h_r (1-\sigma_r) \sigma_r^k, \qquad k = 0, 1, 2, \ldots \tag{47}$$

$$g(k) = (1-\sigma)\sigma^k, \qquad k = 0, 1, 2, \ldots \tag{48}$$

then,

$$\frac{\sigma}{1-\sigma} = \sum_{r=1}^{m} h_r \frac{\sigma_r}{1-\sigma_r} \tag{49}$$

(vi) Let,

$$g(Y) = \frac{1}{(2\pi)^{n/2} |\Sigma|^{1/2}} \exp\left\{-\frac{1}{2}(Y-M)' \Sigma^{-1} (Y-M)\right\} \tag{50}$$

where $Y = \{y_1, y_2, \ldots, y_n\}'$ is a multivariate normal vector and $M = (m_1, m_2, \ldots, m_n)'$ is its mean vector.

and
$$\Sigma = \begin{bmatrix} \sigma_1^2 & \rho_{12}\sigma_1\sigma_2 & \ldots & \rho_{1n}\sigma_1\sigma_n \\ \rho_{21}\sigma_1\sigma_2 & \sigma_2^2 & \ldots & \rho_{2n}\sigma_2\sigma_n \\ \ldots & \ldots & \ldots & \ldots \\ \rho_{n_1}\sigma_n\sigma_1 & \rho_{n_2}\sigma_n\sigma_2 & \ldots & \sigma_n^2 \end{bmatrix} \quad (51)$$

is the variance-covariance matrix. Let,

$$f(y) = \sum_{r=1}^{m} h_r \frac{1}{(2\pi)^{n/2} |\Sigma_r|^{1/2}} \exp\left\{-\frac{1}{2}(Y - M_r)' \Sigma_r^{-1} (Y - M_r)\right\} \quad (52)$$

then,
$$M = \sum_{r=1}^{m} h_r M_r \quad (53)$$

$$\Sigma + MM' = \sum_{r=1}^{m} h_r (\Sigma_r + M_r M_r') \quad (54)$$

49.5 USE OF NON-SHANNON MEASURES

The above discussion is based on the use of Shannon measure of entropy, which undoubtedly is most useful since it leads to exponential family of distributions and these are the distributions which arise most frequently. But other distributions can arise and if these have characterising moments in terms of measures of entropy other than Shannon's, it will be better to use these in conjunction with, these measures, than to use algebraic moments in conjunction with Shannon's measure. Thus, considers Burg's measure,

$$B(f(y)) = \int_a^b \ln f(y)\, dy \quad (55)$$

Maximising it subject to,

$$\int_a^b f(y)\, dy = 1, \int_a^b f(y)\, g_r(y)\, dy = a_r, r = 1, 2, \ldots, m, \quad (56)$$

we get the Burg's *MEPD*,

$$f(y) = \frac{1}{\lambda_0 + \lambda_1 g_1(y) + \ldots + \lambda_m g_m(y)}, \quad (57)$$

where $\lambda_0, \lambda_1, \lambda_2, \ldots, \lambda_m$ are determined by using (56). As such for the distribution (57), $g_1(y), \ldots, g_m(y)$ are the characterising moment function when Burg's measure is used.

If a probability distribution is given by (57), it is completely characterised by the moment functions $g_1(y), \ldots, g_m(y)$ and their expected values, since if these are known, the probability distribution can be recovered by maximising. Burgs' entropy subject to given constraints.

The probability distribution will not however be recovered if we use the algebraic moments $E(y), E(y^2), \ldots, E(y^k)$, however, large the value of k may be, whether we use Burg's entropy or Shannon's entropy.

Since in some sense, characterising moments contain all the information about the distribution, it may be used even for approximating (57) by an exponential distribution, these moments may be more useful than algebraic moments.

It may be noted that for (57) all the moments $E(g_1(x)), E\, g_2(x)), ..., E(g_k(x))$ will exist only for finite intervals.

In the same way if we use Kapur's [4] measure of entropy,

$$-\int_0^b f(y)\ln f(y)\,dy+\frac{1}{c}\int_0^b \{1+cf(y)]\ln(1+cf(y))\,dy \tag{58}$$

we get the *MEPD*,

$$f(y)=\frac{1}{\exp[\lambda_0+\lambda_1 g_1(y)+\ldots+\lambda_m g_m(y)]-c}, \tag{59}$$

In the limit as $c\to 0$, this give the exponential family of distributions.

49.6 USE OF MINIMUM CROSS ENTROPY DISTRIBUTIONS TO APPROXIMATE GIVEN DISTRIBUTIONS

Suppose we have to approximate the Poisson distribution with probability distribution,

$$P_r=\frac{e^{-m}m^r}{r!}, \qquad r=0,1,2,\ldots \tag{60}$$

with a maximum entropy distribution with same mean, this would give

$$p_r=\frac{1}{1+\mathrm{m}}\left(\frac{\mathrm{m}}{1+\mathrm{m}}\right)^r, \qquad r=0,1,2,\ldots \tag{61}$$

Clearly (61) will not give a good approximation to (60), Even if we use a maximum entropy distribution with a large number of moments same as those of Poisson distribution, the approximation will not be good.

However, if we take an a priori probability distribution Q

$$q_r=\frac{1}{e}\frac{1}{r!}, \qquad r=0,1,2,\ldots \tag{62}$$

and then out of all those distribution which have the same mean as (60), we choose that probability distribution which gives minimum cross-entropy relative to Q, we shall get exactly the Poisson distribution.

In this case approximating in terms of minimum cross-entropy probability, distribution could be more appropriate.

Similarly, for approximating a binomial or a negative binomial distribution, minimum cross-entropy approximation will give the exact distribution in one step.

49.7 INTERPRETATION IN TERMS OF FITTING OF DISTRIBUTIONS TO GIVEN DATA

If a probability distribution has been obtained from some frequency distribution, then approximating it by a *MEPD* based on the mean is equivalent to fitting a geometric or exponential distribution; approximating the distribution based on the first two moments is equivalent to fitting a discrete or a continuous normal distribution; approximating by a *MEPD* based on arithmetic and geometric means is equivalent to fitting the data using a gamma distribution and so on.

Thus when we fit specialized distributions to given data, we should preferably use the characterising moments of the distribution.

49.8 GENERAL REMARKS

(i) Let,

$$f(y) = \frac{1}{a + by^2 + cy^4}, \qquad y > 0 \tag{63}$$

This is not of the form, $\exp(-\lambda_0 - \lambda_1 g_1(y) - \lambda_2 g_2(y))$ of course, we can write $f(y) = e^{-\ln(a + by^2 + cy^4)}$, but we want $g_1(y)$, $g_2(y)$,... to be functions of y only and not to contain parameters. In this cases, we can consider approximating (63) by maximum entropy distribution based on algebraic moments. Of course, if x goes from 0 to ∞, we can use only the first three moments, since higher moments do not exist.

Thus, the M.E. approximation based on algebraic moments will be useful only if sufficient algebraic moments exist.

(ii) Let $f(y)$ be a given density function and let $h_r(y)$ be the maximum entropy probability distribution based on r moments. Out of all distributions having specific values for the r moments $h_r(y)$ is maximum entropy distribution and $f(y)$ is any other distribution with the same moments, so that

$$S(h_r(y)) \geq S(f(y)) \tag{64}$$

i.e., entropy in the approximating distribution $h_r(y) \geq$ entropy of the original distribution $f(y)$ for all value of r.

Again, let $s < r$, then out of all distribution having same s moments, $h_s(y)$ has the maximum entropy and $h_r(y)$ is any other distribution with these s moments, so that

$$S(h_s(x)) \geq S(h_r(x)), \tag{65}$$

so that we get,

$$S(h_1(y)) \geq S(h_2(y) \geq \ldots \geq S(h_r(y)) \geq \ldots \geq S(f(y)) \tag{66}$$

$\{S(h_r(y)\}$ is a monotonic decreasing sequence which is bounded below by $S(f(y))$ and approaches a limit $\geq S(f(y))$.

$$S(f(y)) \ldots \leq S(h_r(y)) \leq S\{h_{-1}(y)\} \ldots \leq S(h_3(y)) \leq S(h_2(y)) \leq S\{h_1(y)\}$$

In fact, it can be shown that it approaches $S(f(y)$ under very general conditions. It was shown earlier that

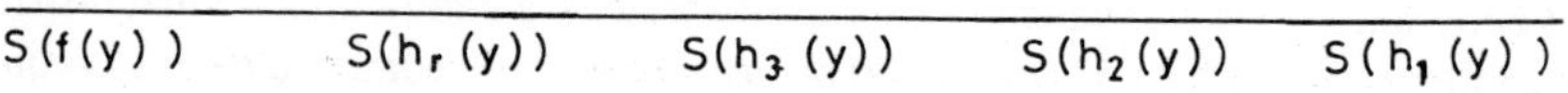

S(f(y)) S(h_r(y)) S(h_3(y)) S(h_2(y)) S(h_1(y))

Fig. 49.1

$$S(h_r(y)) - S(f(y)) = D((f(y)):h_r(y)) = \int f(y) \ln \frac{f(y)}{h_r(y)} dy \tag{67}$$

and,

$$S(h_s(y)) - S(h_r(y)) = D((h_r(y)):h_s(y)) = \int h_r(y) \ln \frac{h_r(y)}{h_s(y)} dy, \qquad s < r \tag{68}$$

so that distances in Figure 49.1 do not represent Euclidean distances, but these do represent directed divergences.

(iii) For an exponential probability distribution, all information is contained in its characterising moments in the sense that if we know the probability distribution, we can find all characterising moments, and if we know the characterizing moments, we can recover the probability distribution by using the maximum-entropy principle. If, however, distribution is not of the form (19) or (20) then there are no characterising moments containing all the information, but its moments, algebraic or not, do contain information about the distribution and the more independent moments we have, the more information, we will get. In general, we choose the algebraic moments as the simplest moments and go on increasing their number to get more and more information and reach nearer and nearer the given distribution.

REFERENCES

1. J.P. Burg (1972), "The Relationship between Maximum Entropy Spectra and Maximum Likelihood Spectra" in Modern Spectral Analysis ed. D.G. Childers, pp.130-131.
2. G.T. Jaynes (1957), "Information Theory and Statistical Mechanics" Physical Reviews Vol. 106, 620-630.
3. J.N. Kapur (1986), "Four Families of measure of entropy", Ind. Jour. Pure and Appl Maths. Vol. 17, no. 4, pp. 429-449.
4. J.N. Kapur, "Maximum Entropy Models in Science and Engineering." Wiley Eastern, New Delhi and John Wiley, New York."
5. S. Kullback and R.A Leibler. (1951), "on Information and Sufficiency", Am. Math. Stat. Vol. 22, pp. 79-86.
6. C.E. Shannon (1948), "A Mathematical Theory of Communication", Bell System Tech. Jour. Vol. 27, 327-426, 623-659.
7. J.E. Shore (1982), "Information Theoretic Approximation for M|G|1 and G|G|1 Queueing System, Acta Informatica Vol. 17, 41-61.
8. K. Sobezyk and Tribecki (1991), "Maximum Entropy Principle in Stochastic Dynamics" Probability Engineering Mechanics, Vol. 5 No. 9, pp. 101-110.

50

CLOSEST APPROXIMATION TO A MIXTURE OF DISTRIBUTIONS

[It is shown that the parameters of maximum-entropy probability distribution which is closest to a weighted mean of a number of probability distributions of the same family, are obtained by equating the characterising moments of the distribution to the corresponding characterising moments of the mixture of distributions.]

50.1 INTRODUCTION

The hyperexponential distribution density function is given by

$$\sum_{i=1}^{k} h_i a_i e^{-a_i x}, \qquad h_i > 0, \quad \sum_{i=1}^{k} h_i = 1, \qquad x \geq 0 \tag{1}$$

In the same way can define hypernormal or hypergamma distributions by

$$\sum_{i=1}^{k} h_i \frac{1}{\sqrt{2\pi}\,\sigma_i} \exp\left[-\frac{1}{2\sigma_i^2}(x - m_i)^2\right], \qquad -\infty < x < \infty \tag{2}$$

and,

$$\sum_{i=1}^{k} h_i \frac{a_i^{\gamma_i}}{\Gamma(\gamma_i)} e^{-a_i x} x^{\gamma_i - 1}, \qquad x \geq 0 \tag{3}$$

These essentially give weighted means of a number of distributions or a mixture of distributions of the same type, i.e. distributions differing in the values of their parameters only.

Our object is to find a single distribution of the same type which will be as close as possible to the weighted mean of the distribution.

For this purpose, we need a measure of closeness of probability distribution and we use Kullback-Leibler [5] measure of directed divergence of a density function $f(x)$ from the density function $g(x)$ defined by

$$\int_a^b f(x) \ln \frac{f(x)}{g(x)}\, dx \tag{4}$$

If $f(x)$ is the density function for the mixture of distributions and $g(x)$ is the density for the single approximating distribution, we shall choose the parameters of $g(x)$ so as to minimize (4).

In Section 2, we shall discuss some particular cases and then in Section 3, we shell generalise the special cases to get a general theorem.

50.2 SOME SPECIAL CASES

(i) Hyper-exponential distribution

Let, $$f(x) = \sum_{i=1}^{k} h_i\, a_i\, e^{-a_i x}, \qquad x > 0 \tag{5}$$

and, $$g(x) = c\, e^{-cx}, \qquad x > 0 \tag{6}$$

then to minimize (4), we have to minimize

$$\int_0^\infty f(x) \ln f(x)\, dx - \int_0^\infty f(x) \ln g(x)\, dx \tag{7}$$

Since the first integral does not involve c, we have to choose c so the minimize

$$-\int_0^\infty f(x) \ln g(x)\, dx \tag{8}$$

or maximize, $$\int_0^\infty f(x) \ln g(x)\, dx \tag{9}$$

Let $$I = \int_0^\infty f(x)\, [\ln c - cx]\, dx$$

$$= \ln c - c \int_0^\infty x \sum_{i=1}^{k} h_i\, a_i\, e^{-a_i x}\, dx$$

$$= \ln c - c \sum_{i=1}^{k} \frac{h_i}{a_i} \tag{10}$$

$$\frac{dI}{dc} = \frac{1}{c} - \sum_{i=1}^{k} h_i/a_i, \quad \frac{d^2 I}{dc^2} = -1/c^2 < 0, \tag{11}$$

so that I is maximum when

$$1/c = \sum_{i=1}^{n} h_i/a_i \tag{12}$$

Thus the mean of the exponential distribution (6) which is closest to the hyper exponential distribution (5), is equal to the weighted mean of the component exponential distributions.

(II) Hypernormal Distribution

Let, $$f(x) = \sum_{i=1}^{k} h_i \frac{1}{\sqrt{2\pi}\sigma_i} \exp\left[-\frac{1}{2}\frac{(x-m_i)^2}{\sigma_i^2}\right], -\infty < x < \infty \tag{12a}$$

$$g(x) = \frac{1}{\sqrt{2\pi}\sigma} \exp\left[-\frac{1}{2}\frac{(x-m^2)}{\sigma^2}\right], \tag{13}$$

then using (9) we have to choose m and σ^2 by maximizing

$$\int_{-\infty}^{\infty} f(x)\left[-\ln\sqrt{2}\pi\sigma - \frac{1}{2}\frac{(x-m)^2}{\sigma^2}\right]dx$$

$$= -\ln\sqrt{2\pi}\sigma - \frac{1}{2\sigma^2}\sum_{i=1}^{k}\frac{h_i}{\sqrt{2}\pi\sigma_i}\int_{-\infty}^{\infty}(x^2-2mx+m^2)\exp\left[-\frac{1}{2}\frac{(x-m_i)^2}{\sigma_i^2}\right]dx$$

$$= -\ln\sqrt{2}\pi\sigma - \frac{1}{2\sigma^2}\left[\sum_{i=1}^{k}h_i(m_i^2+\sigma_i^2) - 2m\sum_{i=1}^{k}h_i m_i + m^2\right] \tag{14}$$

Differentiating w.r.t. m and σ, we get

$$\sum_{i=1}^{k} h_i m_i = m \tag{15}$$

$$-\frac{1}{\sigma}+\frac{1}{\sigma^3}\left[\sum_{i=1}^{k}h_i(m_i^2+\sigma_i^2) - 2m\sum_{i=1}^{k}h_i m_i + m^2\right] = 0 \tag{16}$$

From (15), the mean of the single approximating normal distribution is equal to the weighted mean of the means of the component normal distributions.

From (16) and (15)

$$\sigma^2 = \sum_{i=1}^{k} h_i\,(m_i^2+\sigma_i^2) - m^2$$

or
$$\sigma^2 + m^2 = \sum_{i=1}^{k} h_i\,(m_i^2+\sigma_i^2), \tag{17}$$

so that the second moment about the origin of the approximating normal distribution = weighted mean of the second moments about the origin of the component distributions.

Again from (17),

$$\sigma^2 = \sum_{i=1}^{k} h_i\,\sigma_i^2 + \left[\sum_{i=1}^{k} h_i\,m_i^2 - \left(\sum_{i=1}^{k} h_i\,m_i\right)^2\right] \tag{18}$$

$$= \sum_{i=1}^{k} h_i\,\sigma_i^2 + \sum_{i=1}^{k} h_i(m_i - m)^2, \tag{19}$$

so that the variance of the approximating distribution ≥ weighted mean of the variance of the component distributions and the equality sign holds iff all the means of the component distributions are equal.

(iii) Hyper Gamma Distribution

Let,
$$f(x) = \sum_{i=1}^{k} h_i \frac{a_i^{\gamma_i}}{\Gamma(\gamma_i)} e^{-a_i x} x^{\gamma_i - 1}, \qquad x > 0 \tag{20}$$

$$g(x) = \frac{a^{\gamma}}{\Gamma(\gamma)} e^{-ax} x^{\gamma-1}, \qquad x > 0 \tag{21}$$

We choose a and γ by maximizing

$$I = \int_0^\infty f(x) \ln g(x)\, dx$$

$$= \int_0^\infty f(x) [\gamma \ln a - \ln \Gamma(\gamma) - ax + (\gamma - 1) \ln x]\, dx$$

$$= \gamma \ln a - \ln \Gamma(\gamma) - a \sum_{i=1}^{k} h_i \frac{\gamma_i}{a_i} + (\gamma - 1) \left[\sum_{i=1}^{k} h_i \left(\frac{\Gamma'(\gamma_i)}{\Gamma(\gamma_i)} - \ln a_i \right) \right] \tag{22}$$

Differentiating w.r.t. to a and γ, we get

$$\ln a - \frac{\Gamma'(\gamma)}{\Gamma(\gamma)} + \sum_{i=1}^{k} h_i \left[\frac{\Gamma'(\gamma_i)}{\Gamma(\gamma_i)} - \ln a_i \right] = 0 \tag{23}$$

$$\frac{\gamma}{a} - \sum_{i=1}^{k} h_i \frac{\gamma_i}{a_i} = 0 \tag{24}$$

(23) shows that the geometric mean of the approximating distribution = weighted mean of the geometric means of component gamma distributions (24) shows that the arithmetic mean of approximating distribution = weighted mean of the arithmetic means of the component distributions.

(iv) Hyper Geometric Distribution

$$f(r) = \sum_{i=1}^{k} h_i (1 - \sigma_i) \sigma_i^r, \qquad r = 0, 1, 2, 3, \ldots.. \tag{25}$$

$$g(r) = (1 - \sigma) \sigma^r, \qquad r = 0, 1, 2, 3, \ldots..$$

To determine σ, we maximize

$$\sum_{r=0}^{\infty} f(r) \ln g(r) = \sum_{r=0}^{\infty} f(r) [\ln (1 - \sigma) + r \ln \sigma]$$

$$= \ln (1 - \sigma) + \ln \sigma \sum_{i=1}^{k} h_i \frac{\sigma_i}{1 - \sigma_i} \tag{26}$$

Differentiating w.r.t. σ, we get

$$-\frac{1}{1 - \sigma} + \frac{1}{\sigma} \sum_{i=1}^{k} h_i \frac{\sigma_i}{1 - \sigma_i} = 0$$

or

$$\frac{\sigma}{1 - \sigma} = \sum_{i=1}^{k} h_i \frac{\sigma_i}{1 - \sigma_i} \tag{27}$$

Thus the mean of the approximating hypergeometric distribution = weighted mean of the means of the component geometric distributions.

50.3 THE GENERAL RESULT

$$\int_0^\infty x\,f(x)\,dx = \int_0^\infty x\,g(x)\,dx \text{ in the first example} \tag{28}$$

$$\int_{-\infty}^{\infty} x\,f(x)\,dx = \int_{-\infty}^{\infty} x\,g(x)\,dx;\ \int_{-\infty}^{\infty} x^2 f(x)\,dx = \int_{-\infty}^{\infty} x^2\,g(x)\,dx,$$

in the second example.

(29)

$$\int_0^\infty x\,f(x)\,dx = \int_0^\infty x\,g(x)\,dx,\ \int_0^\infty \ln x\,f(x)\,dx = \int_0^\infty \ln x\,g(x)\,dx$$

in the third example

(30)

$$\sum_{r=0}^{\infty} r f(r) = \sum_{r=0}^{\infty} r\,g(r) \text{in the fourth example} \tag{31}$$

i.e., we have equated the value of $E(x)$, $E(x^2)$, $E(\ln x)$, $E(r)$ for the mixture of distributions to the same expected values for the approximating distribution. This suggests the general result which we proceed to prove.

Theorem : Let $f_i(x)$ be a maximum-entropy probability distribution obtained by maximizing Shannon's [6] entropy

$$-\int_a^b f_i(x) \ln f_i(x)\,dx, \qquad i = 1,2,\ldots,k \tag{32}$$

subject to,

$$\int_a^b f_i(x)\,dx = 1,\ \int_a^b f_i(x)\,g_r(x)\,dx = a_{ir}, \qquad r = 1,2,\ldots,m$$

$$i = 1,2,\ldots,k \tag{33}$$

and, let $g(x)$ be the maximum entropy probability distribution obtained by maximizing,

$$-\int_a^b g(x) \ln g(x)\,dx \tag{34}$$

subject to,

$$\int_a^b g(x)\,dx = 1,\ \int_a^b g(x)\,g_r(x)\,dx = a_r, \qquad r = 1,2,\ldots,m \tag{35}$$

If the characterising functions $g_r(x)$ are the same in both cases, then the parameters of MEPDF $f(x)$ which is closest to the hyper distribution or mixture of distributions

$$\sum_{i=1}^{k} h_i f_i(x) \tag{36}$$

are given by

$$a_r = \sum_{i=1}^{k} h_i a_{ir}, \qquad r = 1, 2, \ldots, m \tag{37}$$

Proof : The MEPDF $f_1(x)$ is given by,

$$f_i(x) = \frac{\exp[-\mu_{1i} g_1(x) - \mu_{2i} g_2(x) - \ldots - \mu_{mi} g_m(x)]}{\int_a^b \exp[-\mu_{1i} g_1(x) - \mu_{2i} g_2(x) - \ldots - \mu_{mi} g_m(x)]\, dx}$$

$$i = 1, 2, \ldots, k, \tag{38}$$

where the Langrange's multipliers $\mu_{1i}, \mu_{2i}, \ldots, \mu_{mi}$ are obtained from the m equations

$$\frac{\int_a^b g_r(x) \exp[-\mu_{1i} g_1(x) - \ldots - \mu_{mi} g_m(x)]\, dx}{\int_a^b \exp[-\mu_{1i} g_1(x) - \ldots - \mu_{mi} g_m(x)]\, dx} = a_{ir} \tag{39}$$

Similarly, $$g(x) = \frac{\exp[-\lambda_1 g_1(x) - \ldots - \lambda_m g_m(x)]}{\int_a^b \exp[-\lambda_1 g_1(x) - \ldots - \lambda_m g_m(x)]\, dx} \tag{40}$$

where, $$\frac{\int_a^b g_r(x) \exp[-\lambda_1 g_1(x) - \ldots - \lambda_m g_m(x)]\, dx}{\int_a^b \exp[-\lambda_1 g_1(x) - \ldots - \lambda_m g_m(x)]\, dx} = a_r \tag{41}$$

To obtain the parameters $\lambda_1, \lambda_2, \ldots, \lambda_m$ of the distribution $g(x)$, we have to maximize

$$\int_a^b f(x) \ln g(x)\, dx$$

$$\int_a^b f(x)[-\lambda_1 g_1(x) - \ldots - \lambda_m g_m(x)]\, dx$$

$$-\int_a^b f(x)\left[\ln \int_a^b \exp(-\lambda_1 g_1(x) - \ldots - \lambda_m g_m(x))\, dx\right] dx$$

$$= -\ln \int_a^b \exp(-\lambda_1 g_1(x) - \ldots - \lambda_m g_m(x))dx$$

$$- \sum_{r=1}^{m} \lambda_r \left[\int_a^b g_r(x) \sum_{i=1}^{k} h_i f_i(x) \right]$$

$$= -\ln \int_a^b \exp[-\lambda_1 g_1(x) - \ldots - \lambda_m g_m(x)]dx - \sum_{r=1}^{m} \lambda_r \sum_{i=1}^{k} h_i a_{ir} \tag{42}$$

Differentiating w.r.t. to $\lambda_1, \lambda_2, \ldots, \lambda_m$, we get

$$\frac{\int_a^b g_r(x) \exp[-\lambda_1 g_1(x) - \ldots - \lambda_m g_m(x)]dx}{\int_a^b \exp[-\lambda_1 g_1(x) - \ldots - \lambda_m g_m(x)]dx} - \sum_{i=1}^{k} h_i a_{ir} = 0 \tag{43}$$

or, $$a_r = \sum_{i=1}^{k} h_i \; a_{ir}, \qquad r = 1, 2, \ldots, m \tag{44}$$

This shows that for the approximating maximum entropy distribution, the value of the rth characterising moment = weighted mean of the rth charactering moments of the component maximum entropy distributions characterised by the same moment functions.

50.4 CHARACTERISING MOMENTS OF A PROBABILITY DISTRIBUTION

We restrict ourselves to the probability distribution of the exponential family. We recently [1,2,3] defined characterising moments of such a distribution in two ways :

(i) Characterising moments of a distribution are those moments of the distribution which lead to the given distribution as MEPD when these moments have same prescribed values, and

(ii) Characterising moments of a distribution are those moments of the distribution such that if we take a random sample from the population represented by the distribution and equate the sample values of the moments to the population values of these moments, we get the maximum likelihood estimates for the parameters of the distribution.

We now have a third charactersation of characterising moments.

(iii) The characterising moments are those moments of the probability distribution such that if we take a weighted mean of the distributions with different parameters and want the parameter of the distribution which is closest to the weighted mean, then these moments of the approximating closest distribution are equal to the weighted mean of the values of the corresponding moments of the component distributions.

50.5 MINIMUM CROSS-ENTROPY DISTRIBUTIONS

Let,
$$f(r) = \sum_{i=1}^{k} h_i\, n_{c_r}\, p_i^r\, q_i^{n-r}, \qquad r = 0, \ldots, n \tag{45}$$

$$g(r) = n_{c_r}\, p^r\, q^{n-r}, \qquad r = 0, \ldots, n \tag{46}$$

If we try to choose p by maximizing $\sum_{i=1}^{n} f(r) \ln g(r)$, we have to maximize,

$$\sum_{i=}^{n} f(r)\,[\ln n_{c_r} + r \ln p + (n - r) \ln q] \tag{47}$$

and the first term cannot be evaluated in a closed form, but this does not matter since it does not depend on p. As such we choose p to maximize,

$$\ln p \sum_{i=1}^{k} h_i\, np_i + \ln q \sum_{i=1}^{k} h_i\, nq_i \tag{48}$$

which give,
$$\frac{\sum_{i=1}^{k} h_i\, p_i}{p} = \frac{\sum_{i=1}^{k} h_i\,(1 - p_i)}{1 - p} = \frac{1}{1} \tag{49}$$

or,
$$p = \sum_{i=1}^{k} h_i\, p_i \quad \text{or} \quad np = \sum_{i=1}^{k} h_i\, np_i,$$

so that the binomial distribution which is closest to the weighted mean of binomial distributions with same n has its mean = weighted mean of the means of independent binomial distributions.

The binomial distribution is not a maximum entropy distribution in the sense that it cannot be obtained by maximizing entropy measure subject to some moment constraints. It is really a minimum cross-entropy distribution which can be obtained by minimising the cross-entropy relative to the standard binomial distribution with $p = q = 1/2$, subject to its mean being prescribed.

Thus, we find our above theorem will hold not only for maximum entropy distributions, but also for minimum cross-entropy distributions provided the prior probability distribution is the same for every component distribution as well as for the approximating distribution.

Thus, the familiar Poisson distribution is also a minimum cross-entropy distribution. where the a priore probability distributions is given by

$$q_i = \frac{1}{e}\frac{1}{r!}; \qquad r = 0, 1, 2, \ldots, \tag{50}$$

and the mean of the Poisson distribution which is closest to the weighted mean of the component distributions is equal to the weighted mean of the means of the individual Poisson distributions.

50.6 SOME EXAMPLES

(i) Queueing Theory

We consider an $M|H_2|1$ system for which the arrival and service-time distributions are given by

$$a(t) = \lambda e^{-\lambda t}, s(t) = \frac{1}{4}\lambda e^{-\lambda t} + \frac{3}{4}2\lambda e^{-2\lambda t} \tag{51}$$

so that the service-time distribution is hyper-exponential and is the weighted mean of the two exponential distributions.

We have two alternatives here :

(i) We can first find the exact queue-size distribution based as (51) and then approximate the mixture of distributions so obtained by a single distribution, or

(ii) We can first approximate the service-time distribution by a single exponential distribution and then find the queue-size distribution.

In the present case, fortunately an exact solution by the problem for (i) is available and is given by [4,7]

$$q(k) = \frac{3}{32}\left(\frac{2}{5}\right)^k + \frac{9}{32}\left(\frac{2}{3}\right)^k, \qquad k = 1, 2, \ldots, \tag{52}$$

where $q(k)$ is the probability of there being k persons in the queue. This is hypergeometric distribution. Since

$$q(k) = \frac{5}{32}\left(\frac{3}{5}\left(\frac{2}{5}\right)^k\right) + \frac{27}{32}\left(\frac{1}{3}\left(\frac{2}{3}\right)^k\right) \tag{53}$$

we have,

$$h_1 = \frac{5}{32}, h_2 = \frac{27}{32}, \sigma_1 = \frac{2}{5}, \sigma_2 = \frac{2}{3} \tag{54}$$

Using (27),

$$\frac{\sigma}{1-\sigma} = \frac{5}{32}\frac{2/5}{(1-2/5)} + \frac{27}{32}\frac{2/3}{(1-2/3)} = \frac{5}{48} + \frac{27}{16} = \frac{43}{24}$$

or, $$\sigma = \frac{43}{67} \tag{55}$$

so that according to (ii) we replace $s(t)$ by $\mu e^{-\mu t}$ where,

$$\frac{1}{\mu} = \frac{1/4}{\lambda} + \frac{3/4}{2\lambda} = \frac{1}{4\lambda}(1 + 3/2) = 5/8\lambda, \tag{56}$$

so that, $\sigma = \lambda/\mu = 5/8$

and,

$$q_2(k) = \frac{3}{8}(5/8)^k, \qquad k = 1, 2, \ldots \tag{57}$$

We get the table :

k	0	1	2	3	4	5
$q_1(k)$	.3982	.2299	.1475	.0947	.0608	.0390
$q_2(k)$	.3750	.2344	.1465	.0916	.0572	.0385

The two distributions are very close to each other, but are not identical.

Thus the two operations of obtaining exact queueing result and approximating by a single distribution are not commutative and these are not expected to be commutative.

However, in the present case, we have the exact theory for the hyper exponential distribution. In general, very few exact results can be obtained for hyper distributions, but we can obtain good approximate results by approximating hyper-distributions by a single distribution.

(ii) Reliability Theory

Let mean time to failure of k similar looking components from different factories be $T_1, T_2, T_3, \ldots, T_k$ and let these be mixed in proportion $h_1, h_2, \ldots, h_k$, so that we get

$$f(t) = \sum_{i=1}^{k} h_i \frac{1}{T_i} e^{-t/T_i} \tag{58}$$

This can be approximated by a single exponential distribution with mean time to failure T so that

$$g(t) = \frac{1}{T} e^{-t/T} \tag{59}$$

where,
$$\frac{1}{T} = \frac{h_1}{T_1} + \frac{h_2}{T_2} + \ldots + \frac{h_k}{T_k} \tag{60}$$

The proportion which fail till time t_0 is according to (58) is

$$\sum_{i=1}^{k} h_i\left(1 - e^{-t_o/T_i}\right) = 1 - \sum_{i=1}^{k} h_i e^{-t_o/T_i} \tag{61}$$

while the corresponding probability for $g(t)$ is

$$1 - e^{-t_o/T} = 1 - e^{-t_o \Sigma \frac{h_i}{T_i}} \tag{62}$$

The results are different except for small values of t_o. Now suppose we are given $T_1, T_2, \ldots, T_k$ but we are not given $h_1, h_2, \ldots, h_k$, but we can experimentally find T, than we can estimate $h_1, h_2, \ldots, h_k$ by maximizing,

$$-\sum_{i=1}^{k} h_i \ln h_i \tag{63}$$

subject to,
$$\sum_{i=1}^{k} h_i = 1, \sum_{i=1}^{k} \frac{h_i}{T_i} = \frac{1}{T} \tag{64}$$

to get,
$$h_i = \frac{e^{-\mu/T_i}}{\sum_{i=1}^{k} e^{-\mu/T_i}}, \tag{65}$$

where, μ is defined by using

$$\frac{1}{T}=\frac{\sum_{i=1}^{k}\frac{1}{T_i}e^{-\mu/T_i}}{\sum_{i=1}^{k}e^{-\mu/T_i}} \tag{66}$$

50.7 THE INVERSE PROBLEM

(i) We have solved above the direct problem of finding the parameters of the single distribution which is closest to a mixture of distribution with known parameter values and known proportions,

The inverse problem arises when the value of the parameters of the component and the composite distributions are known and we have to estimate the proportions, $h_1, h_2, \ldots, h_k$.

We solved one problem of this type in the last section where the mean times to failure $T_1, T_2, \ldots, T_k$ and T were known and we got the most unbiased estimates,

$$h_i=\frac{e^{-\mu/T_i}}{\sum_{i=1}^{k}e^{-\mu/T_i}},\ \frac{1}{T}=\frac{1/T_i e^{-\mu/T_i}}{\sum_{i=1}^{k}e^{-\mu/T_i}}, i=1,2,\ldots,k$$

This gives the well-known Maxwell-Boltzmann distribution.

Now,

$$-\frac{1}{T^2}\frac{dT}{d\mu}=\frac{-\sum_{i=1}^{k}e^{-\mu/T_i}\sum_{i=1}^{k}\frac{1}{T_i^2}e^{-\mu/T_i}+\left(\sum_{i=1}^{k}\frac{1}{T_i}e^{-\mu/T_i}\right)^2}{\left(\sum_{i=1}^{k}e^{-\mu/T_i}\right)^2} \tag{67}$$

By using Cauchy-Schwartz inequality, it is easily seen that μ is an increasing function of T.

Let us number the components so that

$$T_i<T_2<\ldots<T_k \tag{68}$$

When $\mu=0$

$$h_i=h_2=\ldots=h_k=\frac{1}{k} \tag{69}$$

and so all type of components are mixed in equal proportions and in this case

$$\frac{1}{T}=\frac{\sum_{i=1}^{k}1/T_i}{k} \tag{70}$$

so that if T is the Harmonic mean of $T_1, T_2, \ldots T_k$ then the proportions are all equal. Now let $j>i$ so that $T_j>T_i$, and

$$\frac{h_i}{h_j}=\frac{e^{-\mu/T_i}}{e^{-\mu/T_j}}=e^{-\mu\left[\frac{1}{T_i}-\frac{1}{T_j}\right]}\ \substack{\le\\=\\>}\ 1 \text{ according as } \mu\ \substack{\ge\\=\\<}\ 0 \tag{71}$$

If $\mu > 0$ then $T > H.M.$ of $T_1, T_2, ..., T_k$ then $T_i < T_j$ and $h_i < h_j$

Thus, if $T > H.M.$ of $T_1, T_2, ..., T_k$ then $T_i < T_j ==> h_i < h_j$, i.e the proportions of components with smaller time to failure are smaller and the proportions of components with larger time to failure are larger.

If $T < H.M.$ of $T_1, T_2, ..., T_k$ then $\mu < 0$, $T_i < T_j ==> h_i > h_j$ and the components with smaller times to failure are in larger proportion.

(ii) As a second example consider heterogeneous class of k groups with known mean intelligence quotients $m_1, m_2, ..., m_k$ and with the standard deviations as $\sigma_1, \sigma_2, ..., \sigma_k$ and with known group mean intelligence quotient as m and with known standard derivation σ. Also, we known that intelligence quotients are distributed normally, then we have the relations (15) and (17) so that we now have to estimate $h_1, h_2, ..., h_k$ subject to the constraints,

$$\sum_{i=1}^{k} h_i = 1, \sum_{i=1}^{k} h_i m_i = m, \sum_{i=1}^{k} h_i(\sigma_i^2 + m_i^2) = \sigma^2 + m^2 \tag{72}$$

so that we get the maximum entropy estimate

$$h_i = e^{-\lambda_0 - \lambda_1 m_i - \lambda_2\left(\sigma_i^2 + m_i^2\right)} \tag{73}$$

where, $\lambda_0, \lambda_1, \lambda_2$ are obtained by using (72).

(iii) As a third example, consider income distribution for a heterogeneous group containing k relatively homogeneous groups. The distribution are

$$f(x) = \sum_{i=1}^{k} h_i \frac{a_i}{x^{a_i + 1}}, \qquad x \geq 1$$

$$g(x) = \frac{a}{x^{a+1}}, \qquad x \geq 1 \tag{74}$$

In the direct problem, we have to choose a to maximize,

$$\int_1^\infty f(x) \ \ln \frac{a}{x^{a+1}} dx = \ln a - \sum_{i=1}^{k} \int_1^\infty (a+1) h_i a_i x^{-(a_i+1)} \ln x \ dx \tag{75}$$

Also

$$\int_1^\infty x^{-(a+1)} \ dx = \frac{1}{a} \Rightarrow \int_1^\infty x^{-(a+1)} \ln x = 1/a^2 \tag{76}$$

As such we have to choose a to maximize

$$I = \ln a - (a+1) \sum_{i=1}^{k} h_i/a_i \tag{77}$$

$$\frac{dI}{da} = \frac{1}{a} - \sum_{i=1}^{k} h_i/a_i \qquad \frac{d^2I}{da^2} = -\frac{1}{a^2},$$

so that,

$$1/a = \sum_{i=1}^{k} h_i/a_i. \tag{78}$$

The inverse problem is now that we are given $a_1, a_2, ..., a_k$ and we have to estimate $h_1, h_2, ..., h_k$ by maximising,

$$-\sum_{i=1}^{k} h_i \ln h_i \quad \text{subject to} \quad \sum_{i=1}^{k} h_i = 1, \sum_{i=1}^{k} h_i/a_i = 1/a \tag{79}$$

This is the same problem as solved with (63) and (64) and gives the same Maxwell-Boltzmann distribution which will become the uniform distribution, if

$$\frac{1}{a} = \frac{1/a_1 + 1/a_2 + \ldots + 1/a_k}{k} \text{ or } \ln G = \frac{\ln G_1 + \ln G_2 + \ldots + \ln G_k}{k}$$

or
$$G = k\sqrt{G_1 G_2 \ldots G_k} \tag{80}$$

where G, G_1, G_2, ... G_k are the geometric means of composite and component groups so that for the uniform distribution the geometric mean income of composite group = geometric mean of geometric means of incomes of component sub-groups.

50.8 CONCLUDING REMARKS

(i) From (10) it appears that the variance of the single approximating normal distribution is not equal to the weighted mean of the variance of the individual component normal distributions, though variance is also taken as a characterising moment of the normal distribution, along with the mean of the distribution.

In fact, $E(x)$, $E(x)^2$, $E(x-m)^2$ are all characterising moments (though not independent) of the normal distribution. What we proved in our general theorem is,

$$\int_{-\infty}^{\infty} xg(x)\,dx = \int_{-\infty}^{\infty} xf(x)\,dx \Rightarrow \int_{-\infty}^{\infty} x \sum_{i=1}^{k} h_i f_i(x)\,dx = \sum_{i=1}^{k} h_i \int_{-\infty}^{\infty} x\,f(x)\,dx \tag{81}$$

$$\int_{-\infty}^{\infty} x^2 g(x)\,dx = \int_{-\infty}^{\infty} x^2 f(x)\,dx \Rightarrow \int_{-\infty}^{\infty} x^2 \sum_{i=1}^{k} h_i f_i(x)\,dx = \sum_{i=1}^{k} h_i \int_{-\infty}^{\infty} x^2 f_i(x)\,dx \tag{82}$$

$$\int_{-\infty}^{\infty} (x-m)^2 g(x)\,dx = \int_{-\infty}^{\infty} (x-m)^2 f(x)\,dx \Rightarrow \int_{-\infty}^{\infty} (x-m)^2 \sum_{i=1}^{k} h_i f_i(x)\,dx =$$

$$\sum_{i=1}^{k} h_i \int_{-\infty}^{\infty} (x-m)^2 f_i(x)\,dx \tag{83}$$

but since,
$$m = \sum_{i=1}^{k} h_i m_i \tag{84}$$

$$\sum_{i=1}^{k} \int_{-\infty}^{\infty} (x-m)^2 h_i f_i(x)\,dx \neq \sum_{i=1}^{k} h_i \int_{-\infty}^{\infty} (x-m_i)^2 f_i(x)dx \tag{85}$$

unless the means are all equal.

It will be recalled that in the proof of our theorem, we took $E(g_1(x))$, $E(g_2(x))$, ...,$E(g_m(x))$ as prescribed moments. Though when $E(g_1(x))$, $E(g_2(x))$ are both prescribed, $E(g_2(x)) - [E(g_1(x))]^2$ is also prescribed, but $E(g_2(x)) - [E(g_1(x))]^2$ cannot be written as $E(h(x))$. In our case $E(x)^2 - [E(x))]^2$ could be written as $E(x - E(x))^2$ but $(x - E(x)^2$ is just not a function of x.

Thus, in order to get the parameters of the approximating distribution, we should consider characterising functions of x only.
(ii) All the results can be extended to a multivariate distribution with density function $g(x)$. Thus, to get the parameter of the multivariate normal distribution which is closest to the mixture of distribution with density function,

$$f(X) = \sum_{i=1}^{k} h_i \frac{1}{(2n)^{n/2} |\Sigma_i|^{1/2}} \exp\left(-\frac{1}{2}(X - M_i)' \Sigma_i^{-1}(X - M_i)\right) \qquad (86)$$

We have to use the equations,

$$\int x_j \, g(X) \quad dX = \int x_j \, f(X) \quad dX, \qquad j = 1, 2, \ldots, n \qquad (87)$$

$$\int x_j^2 \, g(X) \quad dX = \int x_j^2 \, f(X) \quad dX, \qquad j = 1, 2, \ldots, n \qquad (88)$$

$$\int x_j x_k \, g(X) \, dX = \int x_j \, x_k \, f(X) \quad dX \qquad j, k = 1, 2, \ldots, n, j \neq k \qquad (89)$$

where $X = (x_i, x_2, \ldots, x_n)'$ (90)

Thus if the approximating normal distribution is $N(M, \Sigma)$, then

$$M = \sum_{i=1}^{k} h_i M_i \qquad (91)$$

$$\Sigma + M\,M' = \sum_{i=1}^{k} h_i (\Sigma_i + M_i\, M_i') \qquad (92)$$

so that,

$$\Sigma = \sum_{i=1}^{k} h_i \, \Sigma_i + \sum_{i=1}^{k} h_i (M_i M_i' - MM') \qquad (93)$$

$$= \sum_{i=1}^{k} h_i \, \Sigma_i + \sum_{i=1}^{k} h_i (M_i - M)(M_i - M)' \qquad (94)$$

since, $\sum_{i=1}^{k} h_i (M_i - M)(M_i - M)' = \sum_{i=1}^{k} h_i (M_i M_i' - M_i M' - MM_i' + MM')$

$$= \sum_{i=1}^{k} h_i \, M_i \, M_i' - MM' - MM' + MM'$$

$$= \sum_{i=1}^{k} h_i \quad M_i \, M_i' - MM'$$

$$= \sum_{i=1}^{k} h_i (M_i \, M_i' - MM') \qquad (95)$$

Thus,
(i) Mean vector of the single multivariate normal distribution = weighted mean of the mean vectors of the component distributions.
(ii) $\Sigma + MM'$ for the single multivariate normal distribution = weighted mean of the $\Sigma_i + M_i M_i'$ for component multivariate normal distributions = weighted mean of variance – covariance matrices of the component distributions + weighted mean of

the matrices arising due to difference of the component mean vectors from the mean vector of the mixture distribution.

(iii) The approximating distribution $g(X)$ is not identical with $f(X)$. In fact, only the characterising moments are the same and other moments are different. Thus, for hyperexponential distribution.

$$f(x) = \sum_{i=1}^{k} h_i\, a_i\, e^{-a_i x},\; g(x) = ce^{-cx} \tag{96}$$

we get,

$$\int_0^\infty x\, f(x)\, dx = \int_0^\infty x\, g(x)\, dx \text{ giving } 1/c = \sum_{i=1}^{k} h_i/a_i \tag{97}$$

but,

$$\int_0^\infty x^2 f(x)\, dx = 2\sum_{i=1}^{k} h_i/a_i^2,\; \int_0^\infty x^2 g(x)dx = \frac{2}{c^2} = \left(\sum_{i=1}^{k} \frac{h_i}{a_i}\right)^2 \tag{98}$$

so that,

$$\int_0^\infty x^2 f(x)\, dx - \int_0^\infty x^2 g(x)\, dx = 2\left[\sum_{i=1}^{k} \frac{h_i}{a_i^2} - \left(\sum_{i=1}^{k} \frac{h_i}{a_i}\right)^2\right] \geq 0 \tag{99}$$

and the equality sign holds iff $a_1 = a_2 = \ldots = a_k$. (97) and (99) give

$$\text{variance of } g(x) \leq \text{variance of } f(x), \tag{100}$$

so that the single distribution is less spread-out than the mixture of distribution.

On the other hand, in the normal distribution case, we found that

$$\sigma^2 \geq \sum_{i=1}^{k} h_i \sigma_i^2, \tag{101}$$

so that the single distribution is less compact than the mixture of distributions.

(iv) Suppose a person has invested proportions $h_1, h_2, \ldots h_k$ of his wealth in k securities which gives returns which are normally distributed with means m_i and variance σ_i^2. He wants to sell all the securities and invest in a single security whose distribution of return is $N(m, \sigma^2)$. It appears from (101) that while he can ensure that his mean return does not change, the variance of his return will exceed the weighted mean of the variance of his earlier returns.

REFERENCE

1. J. N. Kapur & H.K. Kesavan, (1987), "Generalised Maximum Entropy Principle with Applications," Sandfort Educational Press, University of Waterloo, Canada.
2. J.N Kapur (1989),"Maximum Entropy Models in Science and Engineering, Wiley Eastern," New Delhi & John Wiley, New York.
3. J.N Kapur & H.K Kesavan (1992), "Entropy Optimization Principles & their Applications" Academic Press, New York.
4. Kleinvrck (1975), "Queueing Systems Theory" Vol. 1. John Wiley, New York.
5. S. Kullback & R.A Leibler (1951), "On Information and Sufficiency, Ann. Math. Stat., Vol. 22, PP. 79-86.
6. C.E. Shannon (1948), "A Mathematical Theory of Communication," Bell System Tech. Joun. Vol. 27, 327-423, 623-659.
7. J.E.Shore (1982), "Information-Theoretic Approximation for M | G | 1 and G | G | 1 Queueing Systems, Acta Inform. Vol. 17, 41-61.

51

SOME ENTROPY-BASED LOGISTIC-TYPE GROWTH MODELS

[Some new logistic-type growth models are proposed. These may be found useful as population growth models or as innovation diffusion models. These are based on some new measures of entropy developed recently.]

51.1 INTRODUCTION

In a recent paper [5], some properties of the function $\phi(f)$ in the innovation diffusion model (IDM),

$$\frac{1}{c}\frac{df}{dt} = \phi(f) \tag{1}$$

were made use of to generate some possible new IDM's. The properties used were :

$$(i)\quad \phi(0) = 0,\quad (ii)\quad \phi(1) = 0,\quad (iii)\quad \phi'(f) > 0 \quad \text{when} \quad f_0 < f < f^*$$

$$(iv)\quad \phi'(f) < 0 \quad \text{when} \quad f^* < f < 1 \quad (v)\quad \phi'(f) = 0 \quad \text{when} \quad f = f^*, \tag{2}$$

where, f_0 is the value of f at $t = 0$. These properties require that $\phi(f)$ increases from 0 to a maximum value $\phi(f^*)$ and then again decreases to 0.

These properties ensure that the equation (1) will give an *S*-shaped growth model of the logistic type with a point of inflexion at f^*.

These properties do not imply that $\phi(f)$ has to be necessarily a concave function of f, but if $\phi(f)$ is a differentiable concave function, it can satisfy all these properties.

Now a measure of entropy of a probability distribution $p_1, p_2, \ldots, p_n$ is in general of the form

$$\sum_{i=1}^{n} \phi(p_i) \tag{3}$$

and the corresponding measure of entropy for a continuous-variate distribution with density function $f(x)$ is of the form

$$\int_a^b \phi(f(x))\, dx \tag{4}$$

where $\phi(x)$ is a concave function.

As such, the concave functions found useful for generating measures of entropy may also be found useful for generating logistic-type growth models.

In an earlier paper [5], the following concave functions were used :

$$\phi(f) = -f \ln f + \frac{1}{a}(1 + a f) \ln (1 + a f) - \frac{1}{a}(1 + a) \ln (1 + a) \tag{5}$$

$$\phi(f) = -f^{\alpha} \ln f^{\alpha} + \frac{1}{a}(1 + a f^{\alpha}) \ln (1 + a f^{\alpha}) - \frac{1}{a}(1 + a) \ln (1 + a) \tag{6}$$

$$\phi(f) = -f^{\beta} \ln f^{\beta} - (1 - f^{\beta}) \ln (1 - f^{\beta}), \quad 0 < \beta < 1 \tag{7}$$

In the present paper, we draw our inspiration from other measures of entropy to get new logistic-type growth models (LTGM's).

51.2 USE OF KAPUR-BURG MEASURE OF ENTROPY

Burg [1] developed his measure of entropy

$$B(S(f)) = \int_{-1/2}^{1/2} \ln S(f)\, df, \tag{8}$$

where $S(f)$ is the spectral density function for use in time-series analysis. This measure has been found extremely useful in geophysics, economics, hydraulics, medicine etc. Its discrete-variate version is

$$B(P) = \sum_{i=1}^{n} \ln p_i \tag{9}$$

It can be derived from the consideration that if U is the uniform distribution and

$$D(P:Q) = \sum_{i=1}^{n} p_i \ln \frac{p_i}{q_i} \tag{10}$$

is the well-known Kullback-Leibler [8] measure of directed divergence, then

$$D(U:P) = \sum_{i=1}^{n} \frac{1}{n} \ln \frac{1/n}{p_i} = -\ln n - \frac{1}{n} \sum_{i=1}^{n} \ln p_i \tag{11}$$

so that minimizing $D(U : P)$ => maximizing B(P).

Now, minimizing $D(U : P)$ means finding a probability distribution closest to the uniform distribution which happens to be the most uncertain or most random distribution so that by maximizing $B(P)$ subject to given constraints, we can get the most uncertain or most uniform or most random or most unbiased distribution subject to given constraints.

This is in accordance with the generalised maximum entropy principle of Kapur and Kesavan [7].

Now $B(P)$ is quite a useful measure since when it is maximized subject to linear constraints by using Lagrange's method, it always gives positive probabilities [4].

However, $B(P)$ is always negative and it may be difficult to interpret it as a measure of uncertainty, though this does not matter when we want to maximize uncertainty in using the principle of maximum entropy.

The maximum value of this measure is $-n \ln n$ which decreases (rather than increases) with n, which is not a desirable property.

Accordingly, Kapur [6] modified Burg's entropy to give the measure

$$K(P) = \sum_{i=1}^{n} (\ln(1 + a\, p_i) - \ln(1+a)\, p_i) \tag{12}$$

He showed that it is always ≥ 0, its maximum value increases with n and it vanishes for all degenerate distributions.

For the continuous-variate case, we get the entropy measure

$$\int_b^d (\ln(1 + a\, f(x) - \ln(1+a)\, f(x))\, dx \tag{13}$$

The corresponding LTGM is

$$\frac{1}{c}\frac{df}{dt} = \ln(1 + a\, f) - f \ln(1+a) \tag{14}$$

so that,

$$\phi(f) = \ln(1 + a\, f) - f \ln(1+a)$$

$$\phi(0) = 0, \quad \phi(1) = 0$$

$$\phi'(f) = \frac{a}{1 + a\, f} - \ln(1+a) \tag{15}$$

$$\phi''(f) = -\frac{a^2}{(1 + a\, f)^2} < 0 \tag{16}$$

Thus, $\phi(f)$ is a concave function. From (15) the point of inflexion of this model arises when $f = f^*$, where

$$1 + a\, f^* = \frac{a}{\ln(1+a)}$$

or,

$$f^* = \frac{1}{a}\left(\frac{a}{\ln(1+a)} - 1\right) = \frac{1}{\ln(1+a)} - \frac{1}{a} \tag{17}$$

Now,

$$\underset{a \to -1}{\mathrm{Lt}}\; f^* = 1 \tag{18}$$

$$\underset{a \to 0}{\mathrm{Lt}}\; f^* = \underset{a \to 0}{\mathrm{Lt}}\; \frac{a - \ln(1+a)}{a \ln(1+a)} = \frac{1}{2} \tag{19}$$

$$\underset{a \to \infty}{\mathrm{Lt}}\; f^* = \underset{a \to \infty}{\mathrm{Lt}}\; \left(\frac{1}{\ln(1+a)} - \frac{1}{a}\right) = 0 \tag{20}$$

so that f^* varies from 1 to 1/2 as a varies from -1 to 0 and f^* varies from 1/2 to 0 and a varies from 0 to ∞

Thus for any given value f^* between 0 and 1, there is a unique value of a and there is a LTGM.

This model is not suitable when $a = -1$, but is all right when $a > -1$

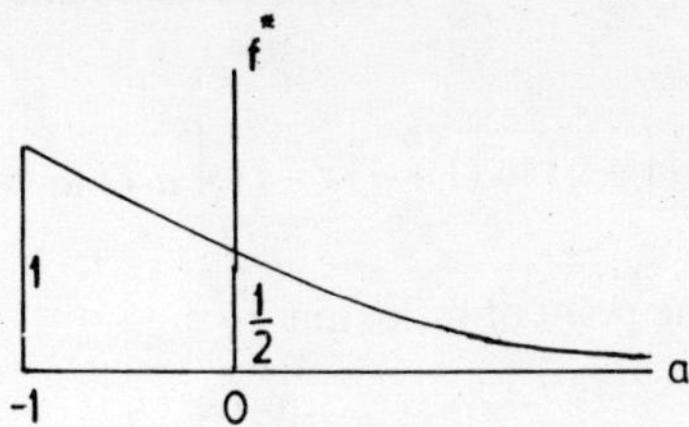

Fig 51.1

In fact we can modify the model (14) to

$$\frac{1}{c}\frac{df}{dt}=\frac{\ln(1+af)-f\ln(1+a)}{a^2} \tag{21}$$

For this model

$$\underset{a\to 0}{\mathrm{Lt}}\ \frac{\ln(1+af)-f\ln(1+a)}{a^2}=\frac{f(1-f)}{2} \tag{22}$$

so that when $a \to 0$, (21) gives the model

$$\frac{1}{c}\frac{df}{dt}=\frac{1}{2}f(1-f) \tag{23}$$

which gives the Fisher-Pry [2] model of innovation of diffusion or Mckendric-Pai [9] logistic model.

However, while Fisher-Pry model is not flexible, since it can represent only those LTGM's for which $f^* = 1/2$, the generalised model (21) can have a point of inflexion for any specified value of f^* for some value of a.

Of course from (17) for a point of inflexion to occur, we should require that

$$\frac{1}{\ln(1+a)}-\frac{1}{a}>f_0 \tag{24}$$

51.3 USE OF MODIFIED SHANNON ENTROPY

We can use,

$$-\sum_{i=1}^{n}(1+a\,p_i)\ln(1+a\,p_i)+(1+a)\ln(1+a) \tag{25}$$

as a measure of entropy which is a modified version of Shannon's [10] measure of entropy. We consider the corresponding function

$$\phi(f)=(1+a)\ln(1+a)f-(1+af)\ln(1+af) \tag{26}$$

so that

$$\phi(0)=0,\quad \phi(1)=0 \tag{27}$$

$$\phi'(f)=(1+a)\ln(1+a)-a\,(1+\ln(1+af)) \tag{28}$$

$$\phi''(f)=-\frac{a^2}{1+af}<0 \tag{29}$$

Thus, $\phi(f)$ as a concave function of f which vanishes at $f = 0$ and $f = 1$. The corresponding LTGM is

$$\frac{1}{c}\frac{df}{dt} = (1+a)\ln(1+a)f - (1+af)\ln(1+af) \tag{30}$$

Putting $\phi'(f) = 0$, we get the point of inflexion

$$f^* = \frac{e^{-1}(1+a)^{(1+a)/a} - 1}{a} \tag{31}$$

Let us find the limiting models as $a \to -1, a \to 0, a \to \infty$.

(i) When a $\to$ 1, the model is

$$\frac{1}{c}\frac{df}{dt} = -(1-f)\ln(1-f) \tag{32}$$

giving,

$$t = -\int_{f_0}^{f} \frac{df}{(1-f)\ln(1-f)}$$

$$= \ln \frac{\ln(1-f)}{\ln(1-f_0)} \tag{33}$$

The point of inflexion is given by

$$\underset{a\to -1}{\text{Lt}}\ f^* = 1 - e^{-1} \underset{a\to -1}{\text{Lt}}\ (1+a)^{(1+a)/a}$$

If $\quad x = (1+a)^{(1+a)/a}$, then, $\ln x = \dfrac{1+a}{a}\ln(1+a)$

$$\underset{a\to -1}{\text{Lt}}\ \ln x = 0 \Rightarrow \underset{a\to -1}{\text{Lt}}\ x = 1$$

$$\underset{a\to -1}{\text{Lt}}\ f^* = 1 - \frac{1}{e} = .62 \tag{34}$$

Also the time of reaching the point of inflexion is given by

$$c\,t^* = \ln \frac{\ln 1/e}{\ln(1-f_0)} = \ln\left[-\frac{1}{\ln(1-f_0)}\right] \tag{35}$$

The values of $c\,t^*$ are given in Table 50.1.

Table 50.1

f_0	.025	.050	.075	.100	.125	.150
$c\,t^*$	3.671	2.971	2.552	2.250	2.013	1.817

(ii) when $a \to 0$, we consider the limit of the model

$$\frac{1}{c}\frac{df}{dt} = \underset{a \to 0}{\text{Lt}} \frac{(1+a)\ln(1+a)f - (1+af)\ln(1+af)}{a^2} \tag{36}$$

$$= \frac{1}{2} f(1-f) \tag{37}$$

which gives $f^* = 1/2$. Alternatively, it can be shown that

$$\underset{a \to 0}{\text{Lt}} \frac{e^{-1}(1+a)^{(1+a)/a} - 1}{a} = \frac{1}{2} \tag{38}$$

(iii) when $a \to \infty$, we consider the limit of the model

$$\frac{1}{c}\frac{df}{dt} = \underset{a \to \infty}{\text{Lt}} \frac{(1+a)\ln(1+a)f - (1+af)\ln(1+af)}{a} \tag{39}$$

$$= f \ln\frac{1}{f} \tag{40}$$

which is Gompertz [3] model. The time to point of inflexion is given by

$$c\,t^* = \int_{f_0}^{f^*} \frac{df}{-f\ln f} = \ln\left(-\frac{1}{\ln f^*}\right) - \ln\left(-\frac{1}{\ln f_0}\right) \tag{41}$$

but, $$f^* = \underset{a \to \infty}{\text{Lt}} \frac{e^{-1}(1+a)^{(1+a)/a} - 1}{a} = \frac{1}{e} \tag{42}$$

so that $$c\,t^* = -\ln\left(-\frac{1}{\ln f_0}\right) = \ln(-\ln f_0) \tag{43}$$

which gives table 50.2

Table 50.2

f_o	.025	.050	.075	.100	.125	.150
ct^*	1.8053	1.097	0.952	.834	.732	.640

Thus we find that for the LTGM based on modified Shannon Entropy Measure f^* varies from $1 - 1/e$ to $1/e$ or from .632 to .3678 as a varies from -1 to ∞.

This model also gives another generalised logistic-model.

51.4 CONJUGATE MODELS

For every model,

$$\frac{df}{dt} = \phi(f), \tag{44}$$

we have the conjugate model

$$\frac{df}{dt} = \phi(1-f) \tag{45}$$

If f^* gives the point of inflexion of the first model, then $1-f^*$ gives the point of inflexion of the conjugate model. Thus corresponding to (21), we have the conjugate model,

$$\frac{1}{c}\frac{df}{dt} = \frac{\ln(1+a-af)+f\ln(1+a)-\ln(1+a)}{a^2} \tag{46}$$

for which
$$f^* = 1 - \frac{1}{\ln(1+a)} + \frac{1}{a} \tag{47}$$

and varies for 0 to 1 as a goes from -1 to ∞.

Models (32) and (40) are conjugate of each and corresponding to (26) we have the model,

$$\frac{1}{c}\frac{df}{dt} = (1+a)\ln(1+a)(1-f) - (1+a-af)\ln(1+a-af) \tag{48}$$

for which f^* varies from $\frac{1}{e}$ to $1-\frac{1}{e}$.

51.5 CONCLUDING REMARKS

In the present paper, we have given essentially two new logistic type growth models. These models can be used as innovation diffusion models when the rate of growth of the number of adopters of a new innovation depends on some average of number of adopters n and the number of non-adopters $N-n$, i.e on

$$\begin{aligned} A(f) &= \lambda n + (1-\lambda)(N-n) = (1-\lambda)N + (2\lambda-1)n \\ &= (1-\lambda)N\left[1+\frac{2\lambda-1}{1-\lambda}\frac{n}{N}\right] \\ &= (1-\lambda)N[1+af], \quad 0 \le \lambda \le 1 \end{aligned} \tag{49}$$

where,
$$a = \frac{2\lambda-1}{1-\lambda}, \quad f = \frac{n}{N} \tag{50}$$

and when λ varies from 0 to 1, a varies from -1 to ∞.

However, instead of the rate depending on $A(f)$, it is depending on $\ln A(f)$. Also some adjustment has to be made to see that the rate vanishes as $f \to 1$.

An alternative hypothsis can be that instead of df/dt depending on $\ln A(f)$, it is $d/dt(\ln(A(f))$ which is a linear function of $\ln A(f)$, so that

$$\frac{d}{dt}(\ln(1+af)) = A\ln(1+af)+B$$

or,
$$\frac{a}{1+af}\frac{df}{dt} = A\ln(1+af)+B \tag{51}$$

or, $$a\frac{df}{dt}=A\,(1+af)\ln(1+af)+B\,(1+af)$$

but since. $df/dt=0$ when $f=1$, this gives

$$0=A\,(1+a)\ln(1+a)+B\,(1+a),$$

so that $$B=-A\,\ln(1+a)$$

and $$a\frac{df}{dt}=A\,(1+af)\ln(1+af)-A\,\ln(1+a)\,(1+af)$$

or, $$\frac{1}{c}\frac{df}{dt}=-(1+af)\ln(1+af)+(1+af)\ln(1+a) \qquad (52)$$

This is only slightly different from the model given by modified Shannon entropy.

REFERENCES

1. J.P. Burg (1972), "The Relationship between Maximum Entropy Spectra and Maximum Likelihood Spectra," in Modern Spectra Analysis ed by D. G Childers, M.S.A.
2. J.C. Fisher and R.C. Pry (1971), "A simple substitution model for technological change". Technological Forecasting and Social Change 3, 75-88.
3. B. Gompertz (1935), "On the Science Connected with Human Mortality'. Trans. Roy. Soc. 115, 5-11.
4. J.N. Kapur (1993), "Maximum entropy probability distribution when mean of a random variate is prescribed and Burg's entropy is used". Chapter 27 of the present book.
5. J.N. Kapur, Uma Kumar and Vinod Kaumar (1992), "Some Possible Models for Technological Innovation Diffusion : Exploiting Analogous Characteristics of Entropy Measures", CSIR Journal of Scientific and Industrial Research Vol 51, 202-208.
6. J.N. Kapur (1993), "A New Parametric Measures of Entropy. Chapter 2 of the present book.
7. H.K. Kesavan and J.N. Kapur (1989), "Generalised Maximum Entropy Principle, IEEE Trans Systs Man Cybernetics 19, 1042-1050.
8. S. Kullback and R.A. Leibler (1951), "Information and Sufficiency". Ann. Math Stat. 22, 79-86.
9. A. Mekendric and M.K. Pai (1911), "The Rate of Multiplication of Micro Organisms : A Mathematical Study", Proc. Roy. Soc. Edinburgh 31, 649-653.
10 C.A. Shannon (1948), "A Mathematical Theory of Communication". Bell Syst. Tech. Journal 27, 379-423, 623-656.

52

INFORMATION–THEORETIC PROOFS OF SOME ALGEBRAIC GEOMETRIC AND TRIGONOMETRIC INEQUALITIES

[Proofs of some inequalities involving sines and cosines of angles of a convex polygon of n sides are given by making use of some concepts from information theory. Information-theoretic proofs are also given of some otherwise well-known inequalities in algebra and geometry.]

52.1 INTRODUCTION

There seems to be no apparent relation between algebra, geometry and trigonometry on the one hand and information theory on the other. Nevertheless one relationship arises because measures of entropy used in information theory are concave functions and some algebraic and trigonometric functions used in geometry and trigonometry are also concave functions. We intend to exploit this relationship and establish some inequalities between sines and cosines of angles of a convex polygon and between areas and perimeters of some polygons. We also give an information-theoretic proof of the arithmetic-geometric mean inequality.

Let,
$$P = (p_1, p_2, \ldots, p_n) \tag{1}$$

be a probability distribution so that

$$p_1 \geq 0, p_2 \geq 0, \ldots.., p_n \geq 0\,;\; \sum_{i=1}^{n} p_i = 1, \tag{2}$$

then measure of equality or uniformity of the probabilities is given by Shannon's [7] measure of entropy,

$$S_1(P) = -\sum_{i=1}^{n} p_i \ln p_i \tag{3}$$

This is a concave function which is maximum when

$$p_1 = p_2 = \ldots\ldots\ldots = p_n = 1/n \tag{4}$$

and has its minimum value when one of the probabilities is unity and all others are zero.

A measure of entropy having these properties and involving trigonometric functions has been recently given by Behara and Chornykov, Kapur [3] and Kapur and Tripathi [6]. This measure is given by :

$$S_2(P) = \sum_{i=1}^{n} \sin(\pi p_i) \tag{5}$$

Another measure of entropy having same properties and involving logarithmic functions is given by

$$S_3(P) = \sum_{i=1}^{n} \ln p_i \tag{6}$$

This is applicable for probability distributions, none of whose components are zero. Its properties have been discussed and compared with those of $S_1(P)$ by Kapur [4].

These measures are useful even when $p_1, p_2, ..., p_n$ are not probabilities but are just proportions, i.e. if $p_1, p_2, ..., p_n$ are all non-negative numbers whose sum is unity. These still measure the degree of equality of various proportions.

Now let $A_1, A_2, ..., A_n$ be the angles measured in radians of a convex polygon of n sides, so that

$$A_1 + A_2 + ... + A_n = (n-2)\pi \tag{7}$$

Let,

$$p_i = \frac{A_i}{(n-2)\pi}; \quad i = 1, 2, ..., n \tag{8}$$

so that $p_1, p_2, ..., p_n$ are proportions whose sum is unity. In the next section, we use a generalised form of $S_2(P)$ for the distribution (8) to obtain some trigonometric inequalities.

52.2 BASIC TRIGONOMETRIC INEQUALITIES

Consider the measure of entropy

$$S_4(P) = \sum_{i=1}^{n} \sin[k(n-2)\pi p_i + \alpha] = \sum_{i=1}^{n} \phi(p_i) \text{ (say)}, \tag{9}$$

where k and α are two parameters satisfying,

$$0 < \alpha < \pi \quad , \quad 0 < k \le 1 - (\alpha/\pi) \tag{10}$$

Since in a convex polygon of n sides each $A_i < \pi$, from (8) and (10) we get

$$(n-2)\pi p_i < \pi \text{ or } p_i < \frac{1}{n-2}$$

$$(n-2)k\pi p_i + \alpha < k\pi + \alpha = \pi(k + \alpha/\pi) \le \pi, \tag{11}$$

and each $\phi(p_i) \ge 0$. Also,

$$\frac{\partial \phi}{\partial p_i} = k(n-2)\pi \cos[k(n-2)\pi p_i + \alpha]$$

$$\frac{\partial^2 \phi}{\partial p_i^2} = -k^2 (n-2)^2 \pi^2 \sin[k(n-2)\pi p_i + \alpha] \tag{12}$$

Since from (11), $0 < (n-2)k\pi p + \alpha \le \pi$ and since $0 \le \sin\theta \le 1$ when $0 \le \theta \le \pi$, $\frac{\partial^2 \phi}{\partial p_i^2} < 0$ and $\phi(p_i)$ is a concave function of p_i. As such from equation (9), $S_4(P)$ is a concave function of $p_1, p_2, ..., p_n$. Its maximum value arises for the uniform distribution U given by (4) so that

$$S_4(P) \leq S_4(U) \tag{13}$$

or,

$$S_4(p_1, p_2, ..., p_n) \leq S_4\ (1/n, 1/n, ..., 1/n) \tag{14}$$

or

$$\sum_{i=1}^{n} \sin\,[k\,(n-2)\,\pi\,p_i + \alpha] \leq \sum_{i=1}^{n} \sin\,[k\,(n-2)\,\pi/n + \alpha] \tag{15}$$

or

$$\sum_{i=1}^{n} \sin(kA_i + \alpha) \leq n \sin\,[k\,(n-2)\,\pi/n + \alpha] \tag{16}$$

Moreover the equality sign in (13) holds only when $p_i = 1/n$ for each i so that the equality sign in (16) holds also only when

$$A_1 = A_2 = ... = A_n = (n-2)\,\pi/n \tag{17}$$

Inequality (16) is our basic trigonometric inequality involving trigonometric functions of the angles $A_1, A_2, ..., A_n$ of a convex polygon of n sides. The equality sign will hold in (16) when all the angles are equal.

52.3 SOME SPECIAL CASES

The inequality (16) represents a triple infinity of inequalities since it involves three parameters n, α, and k. Firstly, we can give any integral value, ≥ 3 to n. Secondly, we can give any real value to α lying between 0 and π. Thirdly, corresponding to any value of α we can give any positive value to k less than $1 - (\alpha/\pi)$

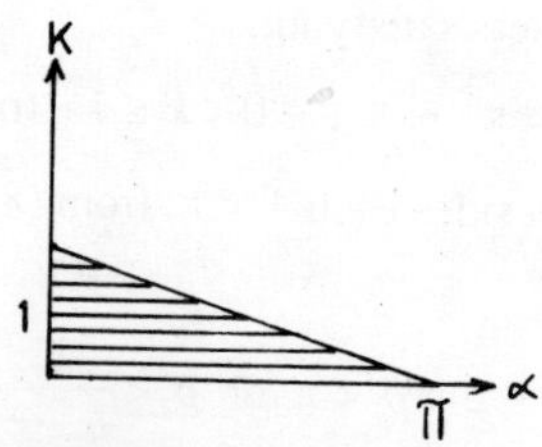

Fig. 52.1 Permissible values for α and k.

We can get special inequalities by giving particular values to α, n and k.

First Special Case : $n = 3$

For a triangle with angles A_1, A_2, A_3 (16) gives

$$\sin\,(kA_1 + \alpha) + \sin\,(kA_2 + \alpha) + \sin\,(kA_3 + \alpha) \leq 3\sin\,(k\pi/3 + \alpha); \quad 0 < k \leq 1 - (\alpha/\pi) \tag{18}$$

This gives the inequalities,

$$\sin A_1 + \sin A_2 + \sin A_3 \leq \frac{3(3)^{1/2}}{2} \tag{19}$$

$$\sin(A_1/2)+\sin(A_2/2)+\sin(A_3/2)\le 3/2 \tag{20}$$

$$\cos(A_1/2)+\cos(A_2/2)+\cos(A_3/2)\le \frac{3(3)^{1/2}}{2} \tag{21}$$

Second Special Case : $n = 4$

For a convex quadrilateral with angles A_1, A_2, A_3, A_4,

$$\sin(kA_1,+\alpha)+\sin(kA_2+\alpha)+\sin(kA_3+\alpha)+\sin(kA_4+\alpha)$$
$$\le 4\sin(k\pi/2+\alpha),\ 0<k<1-\alpha/\pi \tag{22}$$

This gives the inequalities

$$\sin A_1+\sin A_2+\sin A_3+\sin A_4\le 4 \tag{23}$$

$$\cos(A_1/2)+\cos(A_2/2)+\cos(A_3/2)\ \le\ 2(2)^{1/2} \tag{24}$$

$$\sin(A_1/2)+\sin(A_2/2)+\sin(A_3/2)+\sin(A_4/2)\ \le\ 2(2)^{1/2} \tag{25}$$

$$\sin(A_1/3)+\sin(A_2/3)+\sin(A_3/3)+\sin(A_4/3)\ \le\ 2 \tag{26}$$

Some General Results :

(i) The sum of the sines of angles of a convex polygon of any number of sides is maximum when all the angles are equal.

(ii) The sum of sines of k times the angles $(0 < k \le 1)$ of a convex polygons of any number of sides is maximum when all the angles are equal.

(iii) The sum of cosines of k times the angles $(0 < k \le 1/2)$ of a convex polygon of any number of sides is maximum when all the angles are equal.

The more equal the angles, the nearer will be these sums to the corresponding maximum values.

It may be noted that all the angles of a convex polygon being equal does not necessarily imply that the polygon is regular (except when n = 3).

52.4 ANOTHER BASIC INEQUALITY

The inequality (16) has been used to obtain inequalities involving sines of angles (or proper fractions of angles) or cosines of half angles (or fractions of half angles) of a convex polygon.

We can get similar inequalities for tangents of proper fractions of half angles of a convex polygon.

We use the entropy function :

$$S_5(P)=C-\sum_{i=1}^{n}\tan[(n-2)k\pi p_i+\alpha]=C+\sum_{i=1}^{n}f(p_i) \tag{27}$$

where C is a constant, and $f(p_i)=-\tan[(n-2)k\pi p_i+\alpha]$

$$f'(p_i)=-(n-2)k\pi\sec^2[(n-2)k\pi p_i+\alpha] \tag{28}$$

$$f''(p_i) = -2(n-2)^2 k^2 \pi^2 \sec^2 [(n-2) k\pi p_i + \alpha] \tan [(n-2) k\pi p_i + \alpha] \quad (29)$$

Since we shall take,

$$p_i = \frac{A_i}{(n-2)\pi} \quad \text{and} \quad A_i < \pi, \quad p_i < \frac{1}{n-2}, \quad (30)$$

we have $(n-2)\, k\, \pi\, p_i + \alpha < k\,\pi + \alpha < \pi/2$, if

$$k < 1/2 - \alpha/\pi \quad (31)$$

In this case $f''(p_i) < 0$, and, $S_5(P)$ is a concave function of $p_1, p_2, \ldots, p_n$ so that,

$$S_5(P) \le S_5(U), \quad (32)$$

so that,

$$\sum_{i=1}^{n} [\tan(n-2)\, k\pi p_i + \alpha] \ge n \tan [(n-2)\, k\pi/n + \alpha]$$

or,

$$\sum_{i=1}^{n} \tan (k\, A_i + \alpha) \ge n \tan [(n-2)\, k\pi/n + \alpha] \quad (33)$$

This is our basic inequality involving tangents of angles of a convex polygon. As special cases we have

$$\sum_{i=1}^{n} \tan k\, A_i \ge n \tan(n-2)\, k\pi/n; \quad k < 1/2 \quad (34)$$

$$\tan(A_1/3) + \tan(A_2/3) + \tan(A_3/3) \ge 3 \tan \pi/9; \quad A_1 + A_2 + A_3 = \pi \quad (35)$$

$$\tan(A_1/4) + \tan(A_2/4) + \tan(A_3/4) + \tan(A_4/4) \ge 4 \tan \pi/8\,; \quad A_1 + A_2 + A_3 + A_4 = 2\pi \quad (36)$$

Again (33) gives a triple infinity of inequalities since it involves three parameters n, α, k. The parameter α can take all integral values ≥ 3, α can take all real values less than π and k can take all real values $< (1/2 - \alpha / \pi)$.

It may be noted that unlike the case of sums of sines and cosines which were maximum when all the angles were equal, the sum of tangents is minimum when all the angles are equal. Also while we could take sines of angles of the form kA_i ($k \le 1$), we can take cosines of angles of the form kA_i ($k \le 1/2$) and tangents of angles of the form kA_i ($k < 1/2$).

52.5 MINIMUM AREA OF A TRIANGLE WITH GIVEN PERIMETER

It is well known that the area A of a triangle with given semi-perimeter s is maximum when the triangle is equilateral. Its dual result is also true, i.e. the minimum perimeter of a triangle with given area arises when the triangle is equilateral. Many proofs of these results are known. We give below information-theoretic proofs of the results. From Hero's formula

$$A = [s\,(s-a)\,(s-b)\,(s-c)]^{1/2} \quad (37)$$

where a, b, c are the lengths of the sides of the triangle, so that

$$\begin{aligned} \ln A &= 1/2 \ln s + 1/2 \ln (s-a) + 1/2 \ln (s-b) + 1/2 \ln (s-c) \\ &= 2 \ln s + 1/2 \ln (s-a)/s + 1/2 \ln (s-b)/s + 1/2 \ln (s-c)/s \end{aligned} \quad (38)$$

$$= 2\ln s + 1/2\ln p_1 + 1/2\ln p_2 + 1/2\ln p_3$$

$$= 2\ln s + 1/2\sum_{i=1}^{3}\ln p_i = 2\ln s + 1/2\, S_3(P) \tag{39}$$

where

$$p_1 = (s-a)/s,\ p_2 = (s-b)/s,\ p_3 = (s-c)/s$$

and, $$\sum_{i=1}^{3} p_i = \frac{3s-(a+b+c)}{s} = 1 \tag{40}$$

Now if s is given, ln Δ is maximum when 1/2 S_3 (P) is maximum. Also S_3 (P) is maximum when $p_1 = p_2 = p_3$, i.e. when $a = b = c$. Again if Δ is given, *ln s* is minimum when S_3 (P) is maximum i.e. again when the triangle is equilateral.

Thus we have proved both the results stated above. For a general quadrilateral, a formula like (27) is not available. However, for cyclic quadrilaterals, we have Brahmgupta's formula,

$$\Delta = [(s-a)(s-b)(s-c)(s-d)]^{1/2},\ s = \frac{a+b+c+d}{2}$$

and the above information-theoretic method can be used to show that *out of all cyclic quadrilaterals with a given perimeter, the square has the maximum area and out of all cyclic quadrilaterals with a given area, the square has the minimum perimeter.*

52.6 ARITHMETIC – GEOMETRIC MEAN INEQUALITY

Since rectangle is a cyclic quadrilateral, we have the result that *out of all rectangles with a given perimeter, the square has the maximum area and that out of all rectangles with a given area, the square has the minimum perimeter.*

We now want to extend the result to rectangular hyperparallelopipeds in three and higher dimensional spaces.

If $x_1, x_2, \ldots, x_n$ are the lengths of the edges of a rectangular hyperparallelopiped, then its perimeter L and content V are given by

$$\overline{L} = \frac{L}{2^{n-1}} = x_1 + x_2 + \ldots\ldots + x_n \tag{42}$$

$$V = x_1 x_2 \ldots\ldots x_n \tag{43}$$

so that,

$$\ln V = \ln x_1 + \ln x_2 + \ldots\ldots\ldots + \ln x_n$$

$$= \ln \overline{L} p_1 + \ln \overline{L} p_2 \ldots\ldots\ldots \ln \overline{L} p_n$$

$$= n\ln \overline{L} + \sum_{i=1}^{n}\ln p_i = n\ln\overline{L} + S_3(P), \tag{44}$$

where,

$$p_i = 2^{n-1}x_i/L = \frac{x_i}{\overline{L}}, \qquad i = 1, 2, \ldots, n \tag{45}$$

If L is given, V is maximum when $S_3(P)$ is maximum, i.e. when

$$p_1 = p_2 \ldots = p_n \quad \text{or} \quad \text{when} \quad x_1 = x_2 = \ldots\ldots = x_n \tag{46}$$

If V is given, L is minimum when $S_3(P)$ is maximum i.e. when (36) is satisfied. *Thus, out of all rectangular hyperparallelopipeds, with a given perimeter that one has maximum content for which the edges are all equal. Also out of all rectangular* hyperparallelopipeds with a given content, that one has minimum *perimeter for which the edges are all equal.*

Now, $S_3(P)$ is maximum when P is the uniform distribution so that

$$S_3(P) \le n \ln 1/n = -n \ln n \tag{47}$$

From (44) and (47),

$$\ln V - n \ln \overline{L} \le -n \ln n \tag{48}$$

Now arithmetic mean of $x_1, x_2, \ldots, x_n$ is

$$A = \frac{x_1 + x_2 + \ldots\ldots + x_n}{n} = \frac{\overline{L}}{n} \tag{49}$$

and geometric mean of $x_1, x_2, \ldots, x_n$ is

$$G = (x_1 x_2 \ldots\ldots x_n)^{1/n} \tag{50}$$

From (42), (43), (48), (49), (50)

$$\ln G^n - n \ln n A = -n \ln n$$

or,

$$n \ln G - n \ln A \le 0 \tag{51}$$

so that,

$$\ln G \le \ln A \text{ or } G \le A \tag{52}$$

Also the equality sign will hold only when $p_1 = p_2 = \ldots = p_n$, i.e. when

$$x_1 = x_2 = \ldots = x_n$$

This gives an information-theoretic proof of the result that *arithmetic mean of n positive numbers ≥ the geometric mean and the equality sign holds only when all the numbers are equal.*

This information-theoretic proof is different from the proof of the same inequality given by Kapur [5] and Shier [8] because their proofs were based on the use of Shannon's measure of entropy $S_1(P)$, while the present proof is based on the use of logarithmic measure of entropy $S_3(P)$.

52.7 A MORE GENERAL INEQUALITY

A more general entropy is given by Havrda and Charvat [2],

$$S_5(P) = \frac{1}{\alpha(1-\alpha)} \left(\sum_{i=1}^{n} p_i^{\alpha} - 1 \right), \quad \alpha \ne 0, 1 \tag{53}$$

This approaches $S_1(P)$ as $\alpha \to 1$ and $S_3(P)$ as $\alpha \to 0$. This is also maximum when $P = U$, so that

$$\frac{1}{\alpha(1-\alpha)}\left(\sum_{i=1}^{n} p_i^{\alpha}-1\right)\leq\frac{1}{\alpha(1-\alpha)}(n^{1-\alpha}-1) \tag{54}$$

Putting,

$$p_i=\frac{x_i}{x_1+x_2+\ldots\ldots+x_n};\quad i=1,2,\ldots,n \tag{55}$$

we get,

$$\frac{1}{\alpha(1-\alpha)}\left[\frac{x_1^{\alpha}+x_2^{\alpha}+\ldots\ldots+x_n^{\alpha}}{n}-\frac{(x_1+x_2+\ldots\ldots x_n)^{\alpha}}{n^{\alpha}}\right]\leq 0 \tag{56}$$

This gives the inequality,

$$\frac{x_1^{\alpha}+x_2^{\alpha}+\ldots..+x_n^{\alpha}}{n}\lesseqgtr\left(\frac{x_1+x_2+\ldots.+x_n}{n}\right)^{\alpha} \tag{57}$$

according as $\alpha \gtreqless 1$

As $\alpha\to 0$, (56) gives,

$$\frac{\ln x_1+\ln x_2+\ldots.+\ln x_n}{n}-\ln\left(\frac{x_1+x_2+\ldots..+x_n}{n}\right)\leq 0$$

or

$$\ln G-\ln A\leq 0 \tag{59}$$

or

$$G\leq A \tag{60}$$

so that $AM-GM$ inequality is a special case of (54).

REFERENCES

1. M.Behara and I.Z.Chornykov (1988), ''On Some Trigonometric Entropies'' Jour Orissa Math Society.
2. J.H. Havrda and F. Charvat (1967), ''Quantification Methods of Classification Processes : Concept of Structural α Entropy'', Kybernatica Vol. 3, pp 30-35.
3. J.N. Kapur (1988), ''On Generating Measures of Entropy and Directed Divergence''MSTS Res Rep # 455.
4. J.N. Kapur (1988),''Which is better Entropy $\sum_{i=1}^{n}\ln p_i$ or $-\sum_{i=1}^{n}p_i\ln p_i$?'' MSTS Res Rep # 445.
5. J.N. Kapur,(1983), ''On maximum entropy estimation of missing values'', Nat Acad Sci Letters, Vol. 6, No 2, pp 59-65.
6. J.N.Kapur and G.P. Tripathi (1990), ''On Trigonometric Measures of Information'' Journ. Math. Phy. Sci. 24, 1-10
7. C.E. Shannon (1948), ''A Mathematical Theory of Communication'', Bell Syst Tech J Vol. 27, 378-423, 623-653.
8. D.R. Shier, ''The Monotonicity of Power Means using Entropy'', American Statistician Vol. 42 (3), pp 203-205.

53

ON USE OF INFORMATION THEORETIC CONCEPTS IN PROVING INEQUALITIES

[A dozen methods of obtaining inequalities by using propeties of measures of entropy and directed divergence are given. Some of these are illustrated by deriving special inequalities with their help.]

53.1 INTRODUCTION

(i) Let $P = (p_1, p_2, \ldots p_n)$ be a probability distribution and let $H_n(p_1, p_2, \ldots, p_n)$ be any measure of entropy, then since entropy is always maximum for the uniform distribution, we get the inequality,

$$H_n(p_1, p_2, \ldots, p_n) \le H_n\left(\frac{1}{n}, \frac{1}{n}, \ldots, \frac{1}{n}\right) \tag{1}$$

In particular by using measures of entropy due to Shannon [7], Havrda and Charvat [2], Renyi [6], Kapur [3,4] Sharma and Mittal [8], Sharma and Taneja [9], we get the inequalities

$$-\sum_{i=1}^{n} p_i \ln p_i \le \ln n \tag{2}$$

$$\left(\sum_{i=1}^{n} p_i^{\alpha} - 1\right) / (1-\alpha) \le (n^{1-\alpha} - 1)/(1-\alpha), \quad \alpha \ne 1, \quad \alpha > 0 \tag{3}$$

$$\left(\ln \sum_{i=1}^{n} p_i^{\alpha}\right) / (1-\alpha) \le \ln n \quad \alpha \ne 1, \quad \alpha > 0 \tag{4}$$

$$\frac{1}{\beta - \alpha} \ln \frac{\sum_{i=1}^{n} p_i^{\alpha}}{\sum_{i=1}^{n} p_i^{\beta}} \le \ln n, \quad \alpha < 1, \beta > 1 \quad \text{or}$$

$$\alpha > 1, \beta < 1 \tag{5}$$

$$-\sum_{i=1}^{n} p_i \ln p_i + \frac{1}{a} \sum_{i=1}^{n} (1 + a\, p_i) \ln (1 + a\, p_i) \le \ln n + \frac{1}{a}(n + a) \ln (1 + a/n), \quad a \ge -1$$

$$\tag{6}$$

$$\left(\sum_{i=1}^{n} p_i^{\alpha} - \sum_{i=1}^{n} p_i^{\beta}\right)/(\beta-\alpha) \le (n^{(1-\alpha)} - n^{1-\beta})/(\beta-\alpha) \quad \alpha > 1, \beta < 1$$

$$\text{or} \quad \alpha < 1, \beta > 1 \qquad (7)$$

$$\frac{\left(\sum_{i=1}^{n} p_i^{\alpha}\right)^{\beta} - 1}{(1-\alpha)\,\beta} \le \frac{n^{(1-\alpha)\beta} - 1}{(1-\alpha)\,\beta}; \alpha \ne 1, \beta \ne 0 \qquad (8)$$

Here $(p_1, p_2, \ldots, p_n)$ may be any positive numbers whose sum is unity. Alternatively, we may take any n arbitrary positive numbers $a_1, a_2, \ldots a_n$ and replace in each of the inequalities (2) – (8) p_i by $a_i / \sum_{i=1}^{n} a_i$ to get inequalities holding between any n postive numbers.

(ii) Similarly, let $f(x)$ be the density function for a continuous random variable defined over the interval [a, b] and let a measure of entropy for the distribution be given by

$$H(f) = \int_a^b \phi(f(x))\, dx \qquad (9)$$

then since the maximum entropy occurs when the distribution is uniform, we get the inequality

$$\int_a^b \phi(f(x))\, dx \le (b-a)\, \phi\left(\frac{1}{b-a}\right) \qquad (10)$$

In particular by using the continuous variate versions of the entropy measures used in (1), we get the inequalities

$$-\int_a^b f(x) \ln f(x)\, dx \le \ln(b-a) \qquad (11)$$

$$\int_a^b f^{\alpha}(x)\, dx/(1-\alpha) \le (b-a)^{1-\alpha}/(1-\alpha), \quad \alpha > 0, \alpha \ne 1 \qquad (12)$$

$$\ln \int_a^b f^{\alpha}(x)\, dx/(1-\alpha) \le \ln(b-a), \quad \alpha > 0 \quad \alpha \ne 1 \qquad (13)$$

$$\frac{1}{\beta-\alpha} \ln \frac{\int_a^b f^{\alpha}(x)\, dx}{\int_a^b f^{\beta}(x)\, dx} \le \ln(b-a), \ \alpha, \beta > 0, \quad \alpha \ne \beta, \qquad (14)$$

$$\alpha > 1, \beta < 1 \quad \text{or} \quad \alpha < 1, \beta > 1$$

$$-\int_a^b f(x)\ln f(x)\,dx+\frac{1}{c}\int_a^b (1+c\,f(x)\ln(1+c\,f(x))\,dx \le \ln(b-a)$$

$$+\frac{1}{c}((b-a)+c)\ln\frac{(b-a)+c}{b-a} \quad (15)$$

$$\left[\int_a^b f^{\alpha}(x)dx-\int_a^b f^{\beta}(x)\,dx\right]/(\beta-\alpha)\le[(b-a)^{1-\alpha}-(b-a)^{1-\beta}]/(\beta-\alpha)$$

$$\alpha>1,\beta<1 \quad \text{or} \quad \alpha<1,\beta>1 \quad (16)$$

$$\left[\int_a^b (f^{\alpha}(x))^{\beta}\,dx-1\right]/(1-\alpha)\,\beta\le[(b-a)^{(1-\alpha)\beta}-1]/(1-\alpha)\,\beta,\quad \alpha\ne 1,\beta\ne 0 \quad (17)$$

Since we can have an infinity of density functions $f(x)$, each of these gives us an infinity of inequalities.

(iii) Let $P=(p_1,p_2,\ldots,p_n)$ and $Q=(q_1,q_2,\ldots,q_n)$ be two probability distributions and let $D(P:Q)$ be any measure of directed divergence of P from Q, then since $D(P:Q)\ge 0$, we get the inequalities [3, 4, 5, 6, 7, 8]

$$\sum_{i=1}^{n} p_i \ln p_i/q_i \ge 0 \quad (18)$$

$$\left(\sum_{i=1}^{n} p_i^{\alpha} q_i^{1-\alpha}-1\right)/(\alpha-1)\ge 0,\quad \alpha\ne 1 \quad (19)$$

$$\left(\ln\sum_{i=12}^{n} p_i^{\alpha} q_i^{1-\alpha}\right)/(\alpha-1)\ge 0,\quad \alpha\ne 1 \quad (20)$$

$$\sum_{i=1}^{n} p_i \ln\frac{p_i}{q_i}-\frac{1}{c}\sum_{i=1}^{n}(1+c\,p_i)\ln\frac{(1+c\,p_i)}{1+c\,q_i}\ge 0,\quad c\ge -1 \quad (21)$$

$$\frac{1}{\alpha-\beta}\ln\frac{\sum_{i=1}^{n} p_i^{\alpha} q_i^{1-\alpha}}{\sum_{i=1}^{n} p_i^{\beta} q_i^{1-\beta}}\ge 0,\quad \alpha>1,\beta<1$$

$$\text{or} \quad \alpha<1,\beta>1 \quad (22)$$

$$\frac{\sum_{i=1}^{n} p_i^{\alpha} q_i^{1-\alpha}-\sum_{i=1}^{n} p_i^{\beta} q_i^{1-\beta}}{\alpha-\beta}\ge 0 \quad \alpha\ne 0 \quad \beta>0 \quad \alpha\ne\beta \quad (23)$$

$$\frac{\left(\sum_{i=1}^{n} p_i^{\alpha} q_i^{1-\alpha}\right)^{\beta}-1}{(\alpha-1)\,\beta}\ge 0 \quad \alpha\ne 1,\quad \beta\ne 0 \quad (24)$$

These will hold even if $n \to \infty$. In all of these we can replace p_i by $a_i / \sum_{i=1}^{n} a_i$, q_i by $b_i / \sum_{i=1}^{n} b_i$ where $a_1, a_2, ..., a_n, b_1, b_2, ..., b_n$ are arbitrary positive numbers.

Since, there is an infinity of probability distributions P and an infinity of probability distributions Q, we get a double infinity of inequalities from each of above inequalities.

(iv) Similarly, for continuous variate distributions with density, functions $f(x)$ and $g(x)$, we get the inqualities,

$$\int_a^b f(x) \ln \frac{f(x)}{g(x)} dx \geq 0 \tag{25}$$

$$\left(\int_a^b f^{\alpha}(x)\, g^{1-\alpha}(x)\, dx - 1 \right) / (\alpha - 1) \geq 0, \quad \alpha > 0, \quad \alpha \neq 1 \tag{26}$$

$$\int_a^b f(x) \ln \frac{f(x)}{g(x)} dx - \frac{1}{c} \int_a^b (1 + c\, f(x)) \ln \frac{1 + c\, f(x)}{1 + c\, g(x)} dx \geq 0, \quad c \geq -1 \tag{28}$$

$$\frac{1}{\alpha - \beta} \ln \frac{\int_a^b f^{\alpha}(x)\, g^{1-\alpha}(x)\, dx}{\int_a^b f^{\beta}(x)\, g^{1-\beta}(x)\, dx} \geq 0, \quad \alpha > 0, \quad \beta > 0, \quad \alpha \neq \beta$$

$$\alpha > 1, \beta < 1, \quad \text{or} \quad \alpha < 1, \beta > 1 \tag{29}$$

$$\frac{\int_a^b f^{\alpha}(x)\, g^{1-\alpha}(x)\, dx - \int_a^b f^{\beta}(x)\, g^{1-\beta}(x)\, dx}{\alpha - \beta} \geq 0, \quad \alpha > 0, \quad \beta > 0, \alpha \neq \beta \tag{30}$$

$$\frac{\left(\int_a^b f^{\alpha}(x)\, g^{1-\alpha}(x)\, dx \right)^{\beta} - 1}{(\alpha - 1)\, \beta} > 0, \quad \alpha \neq 1, \beta \neq 0 \tag{31}$$

Again, since we have an infinity of density function $f(x)$ and $g(x)$, we get an infinity of inequalities from each of these.

(v) Let $S_{\max}$ be the maximum value of Shannon's entropy when Shannon entropy is maximized subject to linear constraints,

$$\sum_{i=1}^{n} p_i = 1 \quad \sum_{i=1}^{n} p_i\, g_{ri} = a_r, \quad r = 1, \ldots, m \tag{32}$$

$$\text{or} \qquad \int_a^b f(x)\,dx = 1 \quad \int_a^b g_r(x) f(x)\,dx = a_r, \quad r = 1,\ldots m, \tag{33}$$

then $S_{\max}$ is a concave function of $a_1, a_2, \ldots a_m$ which gives the inequalities,

$$\frac{\partial^2 S_{\max}}{\partial a_r^2} \le 0 \quad \frac{\partial S_{\max}}{\partial a_r^2} \cdot \frac{\partial^2 S_{\max}}{\partial a_s^2} - \left(\frac{\partial^2 S_{\max}}{\partial a_r \partial a_s}\right)^2 \ge 0 \tag{34}$$

(vi) Similarly, if $D_{\min}$ is the minimum value of Kullback-Leibler's directed divergence of probability distribution with density function $f(x)$ from given probability distribution $g(x)$ subject to constraints (32), then $D_{\min}$ is a convex function of $a_1, a_2, \ldots a_m$, so that we get inequalities,

$$\frac{\partial^2 D_{\min}}{D\, a_r^2} \ge 0 \quad \frac{\partial^2 D_{\min}}{\partial a_s^2} \frac{\partial^2 D_{\min}}{\partial^2 a_r^2} - \left(\frac{\partial^2 D_{\min}}{\partial a_r \partial a_s}\right)^2 \ge 0 \tag{35}$$

(vii) Let $f(x)$ be the probability diensity function of a probability density function with parameter α and let the density function become the uniform distribution when $\alpha = \alpha_0$, then

$$H(f(x,\alpha)) \le H(f(x,\alpha_0)) \tag{36}$$

(viii) Even if $f(x, \alpha)$ does not give the uniform distribution for any value of α, let α_1 be the value of α for which the entropy of the distribution is maximum then,

$$H(f(x,\alpha)) \le H(f(x,\alpha_1)) \tag{37}$$

(ix) It has been proved that Renyi measure of entropy, for a given probability distribution P or for a given density function, is a monotonic decreasing function of α, so that we get the inequality,

$$\frac{d}{d\alpha}\left[\frac{1}{1-\alpha} \ln \sum_{i=1}^{n} p_i^{\alpha}\right] \le 0 \tag{38}$$

$$\frac{d}{d\alpha}\left[\frac{1}{1-\alpha} \ln \int_a^b f(x)\,dx\right] \le 0 \tag{39}$$

(x) Similar results hold for Havrda-Charvat measures of entropy so that

$$\frac{d}{d\alpha}\left[\frac{1}{1-\alpha}\left(\sum_{i=1}^{n} p_i^{\alpha} - 1\right)\right] \le 0 \tag{40}$$

$$\frac{d}{d\alpha}\left[\frac{1}{1-\alpha}\left[\int_a^b f^{\alpha}(x)\,dx - 1\right]\right] \le 0 \tag{41}$$

(xi) It is similarly known that Renyi's directed divergence is a monotonic increasing function of α, so that we have the inequalities,

$$\frac{d}{d\alpha}\left(\frac{1}{\alpha-1} \ln \sum_{i=1}^{n} p_i^{\alpha} q_i^{1-\alpha}\right) \ge 0 \tag{42}$$

$$\frac{d}{d\alpha}\left(\frac{1}{\alpha-1} \ln \int_a^b f^{\alpha}(x)\, g^{1-\alpha}(x)\,dx\right) \ge 0 \tag{43}$$

(xii) A similar result hold for Havrda-Charvat measure of directed divergence, so that we get the inequalities,

$$\frac{d}{d\alpha}\left[\frac{1}{\alpha-1}\left[\sum_{i=1}^{n} p_i^{\alpha} q_i^{1-\alpha} - 1\right]\right] \geq 0 \qquad (44)$$

$$\left[\frac{d}{d\alpha}\left[\frac{1}{\alpha-1}\int_a^b f^{\alpha}(x)\, g^{1-\alpha}(x)\, dx - 1\right]\right] \geq 0 \qquad (45)$$

53.2 INTERDEPENDENCE OF THE INEQUALITIES

Suppose (3) is satisfied, then taking the limit of both side as $\alpha \to 1$, we get (2). Again,

$$(3) \Rightarrow \sum_{i=1}^{n} p_i^{\alpha} \lesseqgtr n^{1-\alpha} \text{ according as } \alpha \lessgtr 1 \qquad (46)$$

$$\Rightarrow \ln \sum_{i=1}^{n} p_i^{\alpha} \lesseqgtr (1-\alpha)\ln n \text{ according as } \alpha \lessgtr 1$$

$$\Rightarrow \frac{1}{1-\alpha} \ln \sum_{i=1}^{n} p_i^{\alpha} \leq \ln n \Rightarrow (4) \qquad (47)$$

$$\Rightarrow \frac{1}{1-\beta} \ln \sum_{i=1}^{n} p_i^{\beta} \leq \ln n \qquad (48)$$

Now for (47) and (48) if $\alpha < 1, \beta > 1$ then

$$(3) \Rightarrow \ln \sum_{i=1}^{n} p_i^{\alpha} \leq (1-\alpha)\ln n, \quad -\ln \sum_{i=1}^{n} p_i^{\beta} \leq (\beta-1)\ln n$$

$$\Rightarrow \ln \frac{\sum_{i=1}^{n} p_i^{\alpha}}{\sum_{i=1}^{n} p_i^{\beta}} \leq (\beta-\alpha)\ln n$$

$$\Rightarrow \frac{1}{\beta-\alpha} \ln \frac{\sum_{i=1}^{n} p_i^{\alpha}}{\sum_{i=1}^{n} p_i^{\beta}} \leq \ln n \qquad (49)$$

Similarly when $\alpha > 1, \beta < 1$ (3) = (5)

In the same way (7) and (8) can be deduced from (3).

Thus, out of the seven inequalities (2)- (8) only two are independent, viz (3) & (6). In the same way of the seven inequalities in each of the sets (ii) (iii) (iv), only two are independent.

Again let $$R_{\alpha}(P) = \frac{1}{1-\alpha} \ln \sum_{i=1}^{n} p_i^{\alpha}, \quad H_{\alpha}(P) = \frac{1}{1-\alpha}\left(\sum_{i=1}^{n} p_i^{\alpha} - 1\right) \qquad (50)$$

so that, $$R_{\alpha}(P) = \frac{1}{1-\alpha} \ln[(1-\alpha) H_{\alpha}(P) + 1] \qquad (51)$$

and, $$(1-\alpha)^2 \frac{d}{d\alpha}[R_\alpha(P)] = [(1-\alpha)H_\alpha(P)+1]^{-1}\{(1-\alpha)^2 H'_\alpha(P)$$

$$-(1-\alpha)H_\alpha(P) + \ln[(1-\alpha)H_\alpha(P)+1]$$

$$= \frac{1}{x}(1-\alpha)^2\frac{d}{d\alpha}H_\alpha(P) + \frac{1}{x}(x\ln x - x + 1) \tag{52}$$

where $$x = (1-\alpha)H_\alpha(P) + 1 = \sum_{i=1}^{n} p_i^\alpha > 0 \tag{53}$$

since $x \ln x - x + 1 \geq 0$ and vanishes iff $x = 1$, (52) gives

$$R'_\alpha(P) < 0 \Rightarrow H'_\alpha(P) < 0 \tag{54}$$

or $$\frac{d}{d\alpha}\left(\frac{1}{1-\alpha}\ln\sum_{i=1}^{n} p_i^\alpha\right) < 0 \Rightarrow \frac{d}{d\alpha}\left(\frac{1}{1-\alpha}\left(\sum_{i=1}^{n} p_i^\alpha - 1\right)\right) < 0, \tag{55}$$

so that (40) can be deduced from (38), and (41) can be deduced from (39). In fact, we get from (52) another inequality,

$$(1-\alpha)^2\frac{d}{d\alpha}\left(\frac{1}{1-\alpha}\left(\sum_{i=1}^{n} p_i^\alpha - 1\right)\right) \leq -\left(\sum_{i=1}^{n} p_i^\alpha \ln\sum_{i=1}^{n} p_i^\alpha - \sum_{i=1}^{n} p_i^\alpha + 1\right) \tag{56}$$

which is stronger than (40).

since, $$H_0(P) = n-1 \quad H_1(P) = -\sum_{i=1}^{n} p_i \ln p_i, \quad H_\infty(P) = 0 \tag{57}$$

$$R_0(P) = \ln n \quad R_1(P) = -\sum_{i=1}^{n} p_i \ln p_i, \quad R_\infty(P) = -\ln p_{\max} \tag{58}$$

we have the graphs

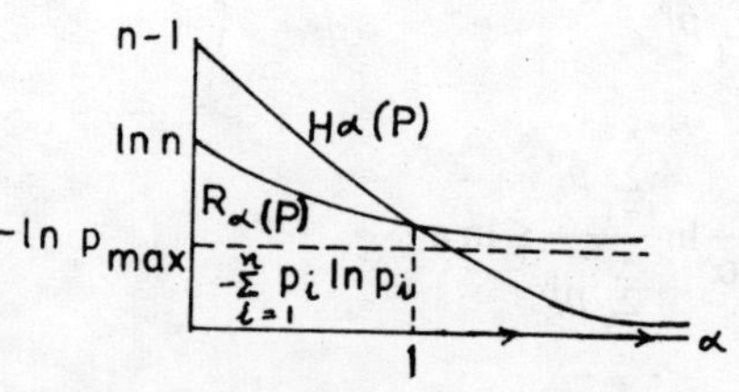

Fig. 53.1

$$H'_0(P) = \sum_{i=1}^{n}\ln p_i + (n-1) \quad H'_1(P) = -\frac{1}{2}\sum_{i=1}^{n} p_i(\ln p_i)^2 \tag{59}$$

$$R'_0(P) = \sum_{i=1}^{n}\frac{\ln p_i}{n} + \ln n \quad R'_1(P) = -\frac{1}{c}\left[\sum_{i=1}^{n} p_i(\ln p_i)^2 - \left(\sum_{i=1}^{n} p_i \ln p_i\right)^2\right]$$

$$R'_1(P) > H'_1(P) \tag{60}$$

Since, $\sum_{i=1}^{n} \ln p_i$ is a concave function, its maximum value subject to $\sum_{i=1}^{n} p_i = 1$ is $-n \ln n$. As such the maximum value of $\sum_{i=1}^{n} \ln p_i + n - 1$ is $-n \ln n + n - 1$

and the maximum value of $\sum_{i=1}^{n} \frac{\ln p_i}{n} + \ln n$ is $-\ln n + \ln n = 0$ (62)

We thus get two more inequlities

$$R_\alpha(P) < H_\alpha(P) \quad \text{when} \quad \alpha < 1$$

$$R_\alpha(P) > H_\alpha(P) \quad \text{when} \quad \alpha > 1 \tag{63}$$

These give,

$$\frac{1}{1-\alpha}\left(\ln \sum_{i=1}^{n} p_i^\alpha - \sum_{i=1}^{n} p_i^\alpha + 1\right) < 0 \quad \text{when} \quad \alpha < 1 \tag{64}$$

$$\frac{1}{1-\alpha}\left(\ln \sum_{i=1}^{n} p_i^\alpha - \sum_{i=1}^{n} p_i^\alpha + 1\right) > 0 \quad \text{when} \quad \alpha > 1 \tag{65}$$

which are equivalent to the inequality,

$$\ln x - x + 1 < 0 \quad \text{when} \quad x > 0 \tag{66}$$

Thus we find :

(a) In each of the four sets of inequalities in (i), (ii), (iii) (iv), only two are independent

(b) In each of the pairs (38), (39) ; (40), (41) ; (42), (43) ; (44), (45) only one is independent.

(c) (viii) includes (vii).

These show that the measures of entropy and directed divergence given above, though given by different authors from different considerations, are dependent among themselves.

53.3 SOME INEQUALITIES

(*a*) Consider the beta distribution with density function,

$$f(x) = \frac{1}{B(m,n)} x^{m-1}(1-x)^{n-1}, \quad 0 \le x \le 1 \tag{67}$$

It becomes uniform when $m = 1$, $n = 1$ and then it has maximum entropy. Its entropy for the values m, n of parameters is given by

$$S = -\int_0^1 f(x)\,[-\ln B(m,n) + (m-1)\ln x + (n-1)\ln(1-x)]\,dx \tag{68}$$

Now,

$$\int_0^1 x^{m-1}(1-x)^{n-1}\,dx = B(m,n) \tag{69}$$

Differentiating under the sign of integration with respect to m,

$$\int_0^1 x^{m-1}(1-x)^{n-1}\ln x\,dx = \frac{\partial}{\partial m}B(m,n) \tag{70}$$

or,
$$\frac{1}{B(m,n)}\int_0^1 x^{m-1}(1-x)^{n-1}\ln x\,dx = \frac{\partial}{\partial m}\ln B(m,n) \tag{71}$$

$$\int_0^1 f(x)\ln x\,dx = \frac{\partial}{\partial m}\left(\ln\frac{\Gamma(m)\Gamma(n)}{\Gamma(m+n)}\right) = \frac{\Gamma'(m)}{\Gamma(m)} - \frac{\Gamma'(m+n)}{\Gamma(m+n)} \tag{72}$$

Similarly,
$$\int_0^1 f(x)\ln(1-x)\,dx = \frac{\partial}{\partial n}\left(\ln\frac{\Gamma(m)\Gamma(n)}{\Gamma(m+n)}\right) = \frac{\Gamma'(n)}{\Gamma(n)} - \frac{\Gamma'(m+n)}{\Gamma(m+n)} \tag{73}$$

so that we get the inequality

$$\boxed{\ln B(m,n) - (m-1)\frac{\partial}{\partial m}\ln B(m,n) - (n-1)\frac{\partial}{\partial n}\ln B(m,n) \le 0}$$

$$\boxed{\left[1-(m-1)\frac{\partial}{\partial m} - (n-1)\frac{\partial}{\partial n}\right]\ln B(m,n) \le 0} \tag{74}$$

and the equality holds only when $m = 1$, $n = 1$.

$$\ln\Gamma(m) + \ln\Gamma(n) - \ln\Gamma(m+n) - (m-1)\left(\frac{\Gamma'(m)}{\Gamma(m)} - \frac{\Gamma'(m+n)}{\Gamma(m+n)}\right) - (n-1)\left[\frac{\Gamma'(n)}{\Gamma(n)} - \frac{\Gamma'(m+n)'}{\Gamma(m+n)}\right] \le 0$$

Now $\Gamma(m+n) = (m+n-1)\ldots m\,\Gamma(m)$ (75)

so that,
$$\frac{\Gamma'(m+n)}{\Gamma(m+n)} = \frac{1}{\Gamma(m+n-1)} + \frac{1}{\Gamma(m+n-2)} + \ldots + \frac{1}{m} + \frac{\Gamma'(m)}{\Gamma(m)} \tag{76}$$

$$\frac{\Gamma'(m)}{\Gamma(m)} - \frac{\Gamma'(m+n)}{\Gamma(m+n)} = -\left(\frac{1}{m} + \ldots + \frac{1}{m+n-1}\right) \tag{77}$$

This gives

$$\boxed{\ln B(m,n) \le -(m-1)\left[\frac{1}{m} + \ldots + \frac{1}{m+n-1}\right) - (n-1)\left(\frac{1}{n} + \frac{1}{n+1} + \ldots + \frac{1}{m+n-1}\right]\Bigg]} \tag{78}$$

(*b*) Now Burg's entropy is given by

$$\overline{S} = \int_0^1 [-\ln B(m,n) + (m-1)\ln x + (n-1)\ln(1-x)]\,dx$$

$$= -\ln B(m,n) + (m-1)(-1) + (n-1)(-1) \le 0 \tag{79}$$

$$\ln B(m,n) + (m+n-2) \ge 0$$

or,
$$\boxed{B(m,n) \ge e^{-(m+n-2)}} \tag{80}$$

(*c*) Harvda-Charvat entropy is

$$\frac{1}{1-\alpha}\left[\int_0^1 \frac{1}{B^\alpha(m,n)} x^{(m-1)\alpha}(1-x)^{(n-1)\alpha}\,dx-1\right]$$

$$=\frac{1}{1-\alpha}\left[\frac{1}{B^\alpha(m,n)}\beta(m\,\alpha-\alpha+1,n\,\alpha-\alpha+1)-1\right]$$

$$\leq\frac{1}{1-\alpha}\left[\int_0^1 1^\alpha\,dx-1\right]=0 \tag{81}$$

$$\begin{array}{l} \text{if } \alpha<1 \quad B(m\,\alpha-\alpha+1,\quad n\,\alpha-\alpha+1)<B^\alpha(m,n) \\ \text{if } \alpha>1, \quad B(m\,\alpha-\alpha+1,\quad n\,\alpha-\alpha+1)>B^\alpha(m,n) \end{array} \tag{82}$$

(*d*) For the Binomial distribution,

$$P_r=n_{c_r}\,p^r\,q^{n-r}, \quad r=0,1,2,\ldots,n \tag{83}$$

Burg's entropy $=\sum_{r=0}^{n}\ln n_{c_r}\,p^r\,q^{n-r}$

$$=\sum_{r=0}^{n}\ln n_{c_r}+\ln p\sum_{r=0}^{n} r+\ln q\sum_{r=0}^{n}(n-r)$$

$$=\sum_{r=0}^{n}\ln n_{c_r}+\frac{n\,(n+1)}{2}(\ln p+\ln q) \tag{84}$$

This is maximum when p = q = 1/2, so that

$$\sum_{r=0}^{n}\ln p^r\,q^{n-r}\leq n\,(n+1)\ln 1/2 \tag{85}$$

which is otherwise obvious.

Its Shannon's entropy

$$=-\sum_{r=0}^{n}P_r\,(\ln n_{c_r}+r\,\ln p+(n-r)\ln q)$$

$$=-\sum_{r=0}^{n}n_{c_r}\,p^r\,q^{n-r}\ln n_{c_r}-n\,p\,\ln p-n\,q\,\ln q)$$

is maximum when p = q = 1/2

$$-\sum_{r=0}^{n}n_{c_r}\,p^r q^{n-r}\ln n_{c_r}-n\,p\,\ln p-n\,q\,\ln q\leq -1\sum_{r=0}^{n}n_{c_r}\left(\frac{1}{2}\right)^n\ln n_{c_r}+n\,\ln 2 \tag{86}$$

(*e*) Again,

$$\frac{1}{1-\alpha}\left(\sum_{r=0}^{n}(n_{c_r})^\alpha\,p^{r\alpha}\,q^{(n-r)\alpha}-1\right)\leq\frac{1}{1-\alpha}\left(\sum_{r=0}^{n}(n_{c_r})^\alpha\,(1/2)^{n\alpha}-1\right)$$

or if $\alpha < 1$ $$\sum_{r=0}^{n} (n_{c_r})^\alpha p^{r\alpha} q^{(n-r)\alpha} \le \sum_{r=0}^{n} (n_{c_r})^\alpha (1/2)^{n\alpha}$$

$$\alpha < 1 \Rightarrow \sum_{r=0}^{n} (n_{c_r})^\alpha \ge \sum_{r=0}^{n} (n_{c_r})^\alpha (2p)^{r\alpha} (2q)^{(n-r)\alpha} \tag{87}$$

$$\alpha > 1 \Rightarrow \sum_{r=0}^{n} (n_{c_r})^\alpha \le \sum_{r=0}^{n} (n_{c_r})^\alpha (2p)^{r\alpha} (2q)^{(n-r)\alpha}$$

In particular, when $p = 1, q = 0$

$$\sum_{r=0}^{n} (n_{c_r})^\alpha \gtreqless 2^{n\alpha} \quad \text{according as} \quad \alpha \lesseqgtr 1 \tag{88}$$

Again thte maximum entropy of binomial distribution for Burg's measure is $\le (n+1)\ln(n+1)$,

$$\sum_{r=0}^{n} \ln n_{c_r} + \frac{n(n+1)}{2} \ln pq \le -(n+1)\ln(n+1)$$

so that, $$\sum_{r=0}^{n} \ln n_{c_r} \le -(n+1)\ln(n+1) + \frac{n(n+1)}{2} \ln 4$$

or, $$\sum_{r=0}^{n} \ln n_{c_r} \le (n+1)(n \ln 2 - \ln(n+1)) \tag{89}$$

Now, $$-\sum_{r=0}^{n} \ln n_{c_r} \left(\frac{1}{2}\right)^n \left(\ln n_{c_r} + \ln (1/2)^n\right) \le \ln(n+1)$$

or $$-\frac{1}{2^n}\left[\sum_{r=0}^{n} n_{c_r} \ln n_{c_r} - \sum_{r=0}^{n} n_{c_r} \ln 2^n\right] \le \ln(n+1)$$

or $$-\sum_{r=0}^{n} n_{c_r} \ln n_{c_r} \le 2^n \ln(n+1) - \ln 2^n$$

$$\sum_{r=0}^{n} n_{c_r} \ln n_{c_r} \ge n \ln 2 - 2^n \ln(n+1) \tag{90}$$

(f) In Geometric distribution,

$$p_r = \frac{(1-\rho)\rho^r}{1-\rho^{N+1}}, \quad r = 0, 1, 2, \ldots, N \tag{91}$$

using Burg's measure we get

$$(N+1)\ln(1-\rho) + \sum_{r=0}^{N} r \ln \rho - (N+1)\ln(1-\rho^{N+1}) \le -(N+1)\ln(N+1)$$

$$\ln \frac{(1-\rho)\rho^{N/2}}{1-\rho^{N+1}} \le -\ln(N+1) = \ln \frac{1}{N+1}$$

$$\frac{(1-\rho)\,\rho^{N/2}}{(1-\rho^{N+1})} \le \frac{1}{N+1} \quad \text{or} \quad \boxed{\frac{\rho^{N/2}}{1+\rho+\ldots+\rho^{N}} \le \frac{1}{N+1}} \tag{92}$$

The maximum value of L.H.S occur when $\rho = 1$

(*g*) *In the Maxwell-Boltzmann distribution,*

$$p_i = \frac{e^{-\mu\varepsilon_i}}{\sum_{i=1}^{n} e^{-\mu\varepsilon_i}} \quad \text{where,} \quad \hat{\varepsilon} = \frac{\sum_{i=1}^{n} \varepsilon_i\, e^{-\mu\varepsilon_i}}{\sum_{i=1}^{n} e^{-\mu\varepsilon_i}} \tag{93}$$

$$S_{\max} = -\sum_{i=1}^{n} p_i \left(-\mu\varepsilon_i - \ln \sum_{i=1}^{n} e^{-\mu\varepsilon_i}\right)$$

$$= \mu\hat{\varepsilon} + \ln \sum_{i=1}^{n} e^{-\mu\varepsilon_i} \tag{94}$$

$$\frac{d\,S_{\max}}{d\,\hat{\varepsilon}} = \mu + \hat{\varepsilon}\frac{d\,\mu}{d\,\hat{\varepsilon}} - \sum_{i=1}^{n} \varepsilon_i \frac{e^{-\mu\varepsilon_i}}{\sum_{r=1}^{n} e^{-\mu\varepsilon_i}} \frac{d\,\mu}{d\,\hat{\varepsilon}} = \mu \tag{95}$$

$$\frac{d^2 S_{\max}}{d\,\hat{\varepsilon}^2} = \frac{d\,\mu}{d\,\hat{\varepsilon}} = \frac{1}{\frac{d\,\hat{\varepsilon}}{d\,\mu}} = \left(\frac{\left(\sum_{i=1}^{n} e^{-\mu\varepsilon_i}\right)^2}{-\sum_{i=1}^{n} e^{-\mu\varepsilon_i\,\varepsilon_i} \sum_{i=1}^{n} \varepsilon_i^2\, e^{-\mu\varepsilon_i} + \left(\sum_{i=1}^{n} \varepsilon_i\, e^{-\mu\varepsilon_i}\right)^2}\right) < 0 \tag{96}$$

$S_{\max}$ is maximum when $\mu = 0$, $\hat{\varepsilon} = \bar{\varepsilon}$ and the maximum value of, $S_{\max} = \ln n$

$$\boxed{\mu \frac{\sum_{i=1}^{n} \varepsilon_i\, e^{-\mu\varepsilon_i}}{\sum_{i=1}^{n} e^{-\mu\varepsilon_i}} + \ln \sum_{i=1}^{n} e^{-\mu\varepsilon_i} \le \ln n} \quad \text{for all value of } \varepsilon_1, \varepsilon_2, \ldots, \varepsilon_n. \tag{97}$$

(*h*) Consider Bose-Einstein distribution,

$$p_i = \frac{1}{e^{\lambda+\mu\varepsilon_i} + a} \qquad \sum_{i=1}^{n} p_i = 1, \qquad \sum_{i=1}^{n} p_i\, \varepsilon_i = \hat{\varepsilon}, -1 \le a \le 1 \tag{98}$$

then,

$$\boxed{\begin{gathered} \sum_{i=1}^{n} \ln\left(e^{\lambda+\mu\varepsilon_i} + a\right) \ge n \ln n \\ \sum_{i=1}^{n} \frac{\ln(e^{\lambda+\mu\varepsilon_i} + a)}{e^{\lambda+\mu\varepsilon_i} + a} \le \ln n \end{gathered}} \tag{99}$$

for all value of λ, μ satisfying $\sum_{i=1}^{n} \frac{1}{e^{\lambda+\mu\varepsilon_i}+a} = 1, \quad \sum_{i=1}^{n} \frac{\varepsilon_i}{e^{\lambda+\mu\varepsilon_i}+a} = \hat{\varepsilon}$ (100)

(ii) Now consider the exponential distribution

$$f(x) = \frac{a\,e^{-ax}}{1-e^{-ab}}, \quad 0 \le x \le b$$

$$\int_0^b [\ln a - a\,x - \ln(1-e^{-ab})]\,dx \le b \ln 1/b$$

or,
$$b \ln a - a\,b^2/2 - b \ln(1-e^{-ab}) \le -b \ln b$$

or
$$\ln a - \frac{a\,b}{2} - \ln(1-e^{-ab}) \le -\ln b$$

or
$$\ln \frac{a\,b}{1-e^{-ab}} \le \frac{a\,b}{2} \quad \text{or} \quad \frac{a\,b}{1-e^{-ab}} \le e^{ab/2}$$

or
$$\frac{x}{1-e^{-x}} \le e^{x/2}$$

or
$$x \le e^{x/2} - e^{-x/2}$$

or
$$x \le 2 \sinh x/2 \quad \text{when} \quad x > 0 \tag{102}$$

This can be easily proved otherwise since

if
$$f(x) = x - 2 \sin h\,x/2, \quad f(0) = 0$$

$$f'(x) = 1 - \cos h\,x/2, \quad f'(0) = 0, \quad f'(x) < 0, \quad \text{when} \quad x < 0 \quad \text{or} \quad x > 0$$

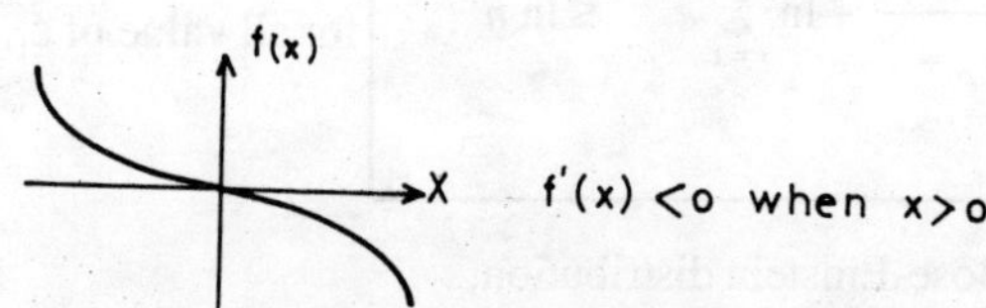

Fig. 53.2

(g) Again
$$\frac{1}{1-\alpha}\left[\int_0^b \frac{a^\alpha e^{-a\alpha x}}{(1-e^{-ab})^\alpha}\,dx - 1\right] \le \frac{1}{1-\alpha}\left[\int_a^b \left(\frac{1}{b}\right)^\alpha dx - 1\right) \tag{103}$$

or
$$\frac{1}{1-\alpha}\,\frac{a^\alpha}{(1-e^{-ab})^\alpha}\,\frac{1-e^{-a\alpha b}}{a\,\alpha} \le \frac{1}{1-\alpha}\,\frac{b}{b^\alpha}$$

or
$$\frac{a^{\alpha-1}}{\alpha}\,\frac{(1-e^{-a\alpha b})}{(1-e^{-ab})^\alpha} \underset{>}{<} b^{1-\alpha} \quad \text{according as} \quad \alpha \underset{>}{<} 1$$

or, $$c^{\alpha-1}\frac{(1-e^{-\alpha c})}{(1-e^{-c})^{\alpha}} \lesseqgtr \alpha \quad \text{according as} \quad \alpha \lesseqgtr 1, \quad a\,b = c > 0$$

or, $$\left(\frac{c}{1-e^{-c}}\right)^{\alpha} \lesseqgtr \alpha \frac{c}{1-e^{-\alpha c}} \qquad \text{according as} \quad \alpha \lesseqgtr 1; c > 0 \tag{104}$$

(*k*) For Shannon measure it gives

$$-\int_0^b \frac{a\,e^{-a\,x}}{1-e^{-a\,b}} \ln \frac{a\,e^{-a\,x}}{1-e^{-a\,b}}\,dx \le -\int_0^b \frac{1}{b}\ln\frac{1}{b}\,dx \tag{105}$$

or $$-\int_0^b \frac{a\,e^{-a\,x}}{1-e^{-a\,b}}((\ln a - a\,x)\,dx - \ln(1-e^{-a\,b}))\,dx \le \ln b$$

or $$-\ln a \frac{1}{1-e^{-a\,b}}[a\,e^{-a\,x}\,x + e^{-a\,x}]_0^b + \ln(1-e^{-a\,b}) \le \ln b$$

or $$\frac{-1}{1-e^{-a\,b}}[e^{-a\,b}\,(a\,b+1)-1] \le \ln\frac{a\,b}{1-e^{-a\,b}}$$

or $$\frac{1}{1-e^{-c}}[1-e^{-c}\,(c+1)] \le \ln\frac{c}{1-e^{-c}}$$

or $$\frac{1}{1-e^{-c}}[1-e^{-c} - c\,e^{-c}] \le \ln\frac{c}{1-e^{-c}}$$

$$1 - \frac{c\,e^{-c}}{1-e^{-c}} \le \ln\frac{c}{1-e^{-c}}, \quad c > 0 \tag{106}$$

(*l*) Consider truncated normal distribution,

$$f(x) = \frac{e^{-\frac{1}{2}x^2}}{\int_0^a e^{-\frac{1}{2}x^2\,dx}} \quad 0 \le x \le a \tag{107}$$

Using Burg's entropy,

$$\int_0^a \left(-\frac{1}{2}x^2 - \ln\phi(a)\right)dx \le \int_0^a \ln 1/a\,dx, \quad \phi(a) = \int_0^a e^{-1/2\,x^2}\,dx$$

or $$-\frac{1}{2}\frac{a^3}{3} - a\ln\phi(a) \le -a\ln a$$

or $$\frac{a^3}{6} + a\ln\phi(a) \le a\ln a \quad \text{or} \quad \phi(a) \ge a\,e^{-a^3/6} \tag{108}$$

$$\text{or} \quad \int_0^a e^{-1/2 x^2}\, dx \geq a\, e^{-a^2/6} \tag{109}$$

(*m*) $$\frac{1}{1-\alpha}\left[\int_0^a \frac{e^{-1/2\,\alpha x^2}}{\phi^\alpha(a)}\, dx - 1\right] \leq \frac{1}{1-\alpha}\left[\int_0^a \left(\frac{1}{a}\right)^\alpha dx - 1\right] \tag{110}$$

or

$$\frac{\phi(a\sqrt{\alpha})}{\phi^\alpha(a)} \;\substack{\leq \\ >}\; \sqrt{\alpha}\, a^{1-\alpha} \quad \text{according as} \quad \alpha \;\substack{\leq \\ >}\; 1 \tag{111}$$

(*n*) Using Shnnon's entropy, we get

$$-\int_0^a f(x)\left[-\frac{1}{2}x^2 - \ln\phi(a)\right] dx \leq -\int_0^a \frac{1}{a}\ln 1/a\, dx \tag{112}$$

or

$$\ln\phi(a) + \frac{1}{\phi(a)}\int_0^a \frac{1}{2}x^2 e^{-1/2 x^2}\, dx \leq \ln a$$

$$\ln\frac{\phi(a)}{a} + 1/2 - 1/2\,\frac{a\, e^{-a^2/2}}{\phi(a)} \leq 0 \tag{113}$$

(*o*) Now Consider truncated binomial distribution,

$$P_r = \frac{n_{c_r}\, p^r q^{n-r}}{1-p^n-q^n}, \quad r = 1, \ldots, n-1 \qquad \ldots(114)$$

$$\text{Burg's entropy} = \sum_{r=1}^{n-1} \ln\frac{n_{c_r}\, p^r q^{n-r}}{1-p^n-q^n}$$

$$= \sum_{r=1}^{n-1} \ln n_{c_r} + \ln p \sum_{i=1}^{n-1} r + \ln q \sum_{r=1}^{n-1}(n-r) - (n-1)\ln(1-p^n-q^n)$$

$$= \text{const} + \frac{(n-1)\,n}{2}(\ln p + \ln q) - (n-1)\ln(1-p^n-q^n) \tag{115}$$

let $$F(p) = \text{Const} + (n-1)\left[\frac{n}{2}\ln p(1-p) - \ln(1-p^n-(1-p)^n\right]$$

(116)

then $$F'(p) = (n-1)\left[\frac{n}{2}\frac{1}{p} - \frac{n}{2}\frac{1}{1-p} - \frac{-n\,p^{n-1} + n(1-p)^{n-1}}{1-p^n-(1-p)^n}\right] = 0$$

when $$p = \frac{1}{2} \tag{117}$$